Robotic Radiosurgery

Harold C. Urschel, Jr. (Editor-in-Chief)
John J. Kresl · James D. Luketich
Lech Papiez ·Robert D. Timmerman (Co-Editors)
Raymond A. Schulz (Contributing Editor)

Treating Tumors that Move with Respiration

With Contributions by Numerous Experts

Foreword by

E. Thomson

With 116 Figures in 168 Separate Illustrations, 120 in Color and 31 Tables

Springer

Editor-in-Chief:

Harold C. Urschel Jr., MD
Chair of Cardiovascular and Thoracic Surgical Research, Education and Clinical Excellence
Baylor University Medical Center
1201 Barnett Tower
3600 Gaston Avenue
Dallas, TX 75246
USA

Co-Editors:

John J. Kresl, MD, PhD
Arizona Oncology Services at
St. Joseph's Hospital & Medical Center
Department of Radiation Oncology
CyberKnife Center
Barrow Neurological Institute
Gamma Knife Center
350 West Thomas Road
Phoenix, Arizona 85013
USA

James D. Luketich, MD
Sampson Family Endowed Professor of Surgery
Chief, The Heart, Lung and
Esophageal Surgery Institute
University of Pittsburgh Medical Center, PUH, C-800
200 Lothrop Street
Pittsburgh, PA 15213
USA

Lech Papiez, PhD
Associate Professor
Department of Radiation Oncology
University of Texas Southwestern Medical Center
5801 Forest Park Road
Dallas, TX 75390
USA

Robert D. Timmerman, MD
Professor and Vice-Chairman
Effie Marie Cain Distinguished Chair in
Cancer Therapy Research
Department of Radiation Oncology
University of Texas Southwestern Medical Center
5801 Forest Park Road
Dallas, TX 75390
USA

ISBN 978-3-662-50105-4 Springer Berlin Heidelberg New York
DOI 10.1007/978-3-540-69886-9

Springer is part of Springer Science+Business Media

http//www.springer.com

Softcover re-print of the Hardcover 1st edition 2007

Medical Editor: Dr. Ute Heilmann, Heidelberg
Desk Editor: Ursula N. Davis, Heidelberg
Production Editor: Kurt Teichmann, Mauer
Cover-Design and Typesetting: Verlagsservice Teichmann, Mauer

Printed on acid-free paper – 21/3151xq – 5 4 3 2 1 0

Foreword

Tracking the Moving Target: The Heart of the CyberKnife Robotic Radiosurgery System

From its inception, when the CyberKnife® Robotic Radiosurgery System (Accuray Incorporated, Sunnyvale, CA) was used to treat only intracranial lesions, the fundamental problem that had to be solved to make frameless stereotactic radiosurgery a reality was how to hit a moving target. If the painful, cumbersome stereotactic head frame was to be eliminated, a new method for targeting brain tumors would have to be devised. Use of a stereotactic frame for radiosurgery treatments reflects the recognition that radiosurgery must be supremely accurate. Unlike radiation therapy, little or no fractionation is used for radiosurgery treatments. This means that any tissue enclosed in the high-dose region will be ablated. High-dose radiation must be delivered precisely to the targeted tissue and only the targeted tissue; generally a total clinical accuracy less than 1 mm is accepted for intracranial applications. The stereotactic frame has become the accepted route to achieving such accuracy. Use of a stereotactic frame is based on two assumptions. The first is that the target tissue will remain in a fixed position relative to the skull. If this is not true, the coordinates of the frame system are irrelevant. The second assumption is that the frame will not move relative to the skull for the duration of treatment planning and delivery. Although questionable, this assumption is generally accepted among practitioners of frame-based radiosurgery.

When an image-guided approach to radiosurgery was developed it was widely assumed that doing away with stereotactic frames would compromise accuracy. In reality, even greater accuracy is possible. CyberKnife image guidance allows direct targeting based on the skull itself, without an intermediary frame. As a result, only the first assumption above is made; target tissue will remain in a constant spatial relationship to the skull. Recognition of this aspect of targeting is the key to understanding how the CyberKnife can potentially achieve greater "end-to-end" clinical accuracy than frame-based systems. It is also the key to understanding the potential for CyberKnife radiosurgery at other body sites, including sites where tumors move with respiration.

The adaptation of other technologies for application to extracranial radiosurgery was based on the assumption that lesions could be targeted based on an external coordinate system, such as a "body frame", in the same way as intracranial tissues could be targeted via a stereotactic frame. This is an error; *neither* of the key assumptions for intracranial radiosurgery using stereotactic frames applies to extracranial targets. First, extracranial targets do not, in most cases, remain in a fixed relation to skeletal or other body landmarks. CyberKnife treatments have demonstrated that even spine tumors move during treatment. Certainly targets in the lung, liver, prostate and pancreas cannot be fixed relative to an external reference frame. Second, a frame system cannot be fixed to the body in the same way that a cranial stereotactic frame can be fixed to the skull.

These are not problems for the CyberKnife System. The beam alignment strategy for the CyberKnife System is based on direct, image-guided tracking of the targeted tissue or radiographically visible landmarks that are in a fixed relationship to the targeted tissue (such as implanted fiducials). This concept builds on the initial approach to intracranial radiosurgery by eliminating external frames of reference and rigid immobilization of the patient. It is *the enabling approach for radiosurgery at almost any body site,* and it is unique to the CyberKnife System. CyberKnife treatments never require clinicians to assume that a patient or targeted tissue inside a patient will remain static throughout the treatment process. The CyberKnife assumption is, instead, that patients are living, breathing, moving beings, and this movement must be taken into account if ultimate treatment accuracy is to be achieved. Constant imaging and a robotic system that automatically responds to motion of the target have evolved as the core fundamentals of the CyberKnife; hence our consistent message that the CyberKnife System "tracks, detects and corrects" for patient and tumor motion.

At the heart of all of the innovations to the CyberKnife System is the concept that patients should be unrestrained, comfortable, and free to breathe naturally. Tumor motion, whether it results from patient motion or natural motion of the tumor within the patient is accounted for not by *fixing* the tumor in space, but by continuously *finding* the tumor in space. Other less sophisticated systems will attempt to mold the patient to fixed coordinates. Our approach is to adjust our system to the living, breathing patient.

As we present the second volume in the Robotic Radiosurgery series, a compelling medical story is unfolding. In this story the substantial risks of surgical resection of tumors in the lungs, liver, and pancreas, and the fact that a tragically large percentage of patients are not surgical candidates because their disease is too far advanced or their health is poor, inspire the search for an effective and safe alternative to surgery [1–3]. A flurry of dose escalation studies show the benefits of treatment with higher radiation doses, but the risk of complications to tissue outside the tumor volume is significant and targeting accuracy is of paramount importance [4–7]. True radiosurgery requires delivery of a high dose of radiation that conforms precisely to the tumor volume in just a few fractions with the objective of tumor ablation. CyberKnife researchers are active in this movement [8–12], bringing radiosurgical accuracy to the treatment of moving lesions throughout the body. In this volume, subtitled "Treating Tumors that Move with Respiration", is described the Synchrony® Respiratory Tracking System (Accuray Incorporated, Sunnyvale, CA), a continuous tumor tracking technology that allows highly conformal delivery of radiation to planning target volumes (PTVs) with the smallest possible margins. Patients lie comfortably and breathe freely as the treatment beam is moved in "synchrony" with respiration. This clearly represents the state of the art in extracranial tumor treatment.

A second key feature of the CyberKnife treatment system is the delivery of hundreds of radiation beams to the tumor from almost any angle, rather than attempting to pick up the shape of a target structure from a small number of coplanar beams. Treatments delivered by the CyberKnife System are demonstrably more conformal and able to minimize radiation outside the PTV [13], thus sparing critical structures near the tumor, than those generated by other systems. This is another highly desirable feature of a radiosurgery system since, as stated above, the objective is ablation of tissue enclosed by the high dose region. As an example, Figure 1 shows dose plans for a lesion in the left mid-lung constructed for image-guided radiotherapy (IGRT), a tomographic approach (Tomo), and the CyberKnife. For each plan the same image set was used to construct treatment plans that covered 100% of the gross tumor volume to the 80% isodose line. Beyond these requirements the plans were not extensively optimized. The figure in the lower right corner shows the ratio of lung volumes covered by IGRT (blue line) and a Tomographic delivery

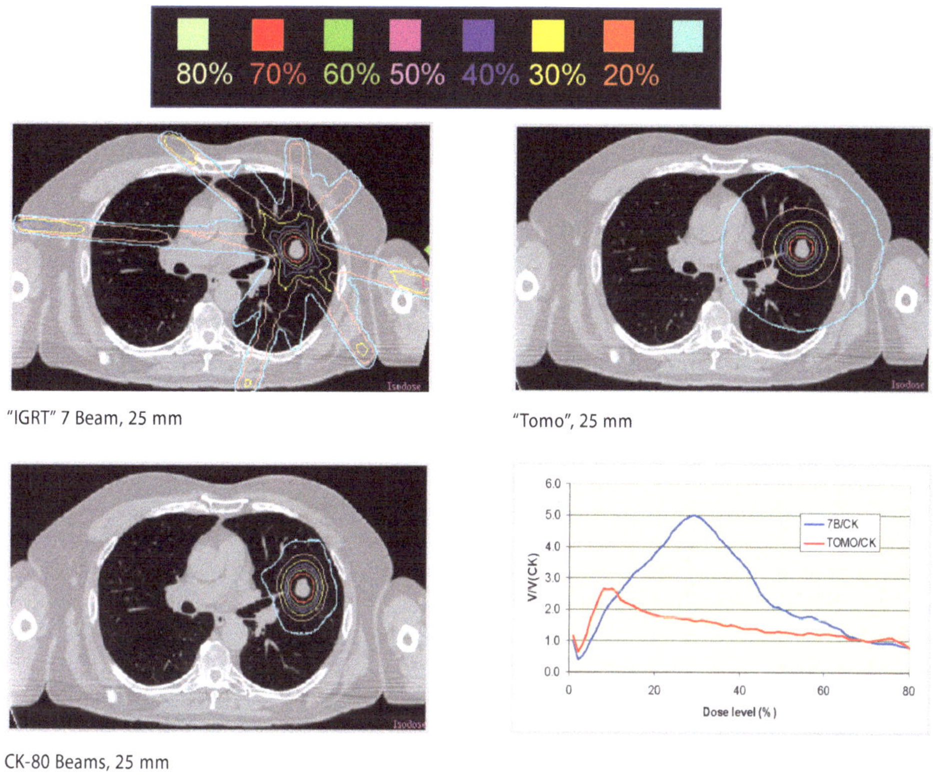

Fig. 1 Dose plans for a lesion in the left mid-lung constructed for IGRT, a tomographic approach, and CyberKnife System. The same image set was used to construct each plan. Plans covered 100% of the gross tumor volume to the 80% isodose line. The lower right panel shows the ratio of lung volumes covered by IGRT (blue line) and tomography (red line) to volumes covered by CyberKnife (CK). (Figure courtesy of John Kresl, MD, PhD, Co-Director of the Stereotactic Radiosurgery Center, Barrow Neurological Institute and St. Joseph's Hospital & Medical Center, Phoenix, AZ).

system (red line) to volumes covered by CyberKnife (CK). Values above 1.0 indicate that, along a wide range of isodoses, IGRT and the Tomographic approach irradiate larger portions of lung (as much as 5 times larger at the 30% isodose with IGRT) than CyberKnife System.

It is apparent that even the static plan generated for the CyberKnife System conforms more closely to the treated volume than the plans generated for the other platforms. What is not apparent is that, because this lung lesion will move with respiration, only the CyberKnife can actually deliver the plan that is pictured. With respiratory gating or abdominal compression, processes used by radiation therapy devices, tumor motion cannot be compensated for to the extent that is possible when the treatment beam moves with the tumor, as it does when the CyberKnife is used. The combined advantage of the CyberKnife System is therefore extreme conformality *and* the ability to track tumors that move with respiration. Since the key to successful radiosurgery is to treat the tumor and only the tumor to a high dose of radiation in a single fraction, or a small number of fractions, both of these characteristics combine to make the CyberKnife System highly appropriate for extracranial radiosurgery applications and using this approach, it is increasingly clear that many solid tumors can be ablated.

The field of CyberKnife radiosurgery is evolving rapidly and the clinical and technical studies presented in this volume are representative of a huge worldwide effort. However, inherent in this clinical paradigm shift are both opportunity and risk. The opportunity is to change the standard of care for solid tumors throughout the body. As described, the accuracy and conformality of the radiation dose delivered by the CyberKnife are unprecedented and cannot be matched by less sophisticated systems; the clinical successes recorded in this volume and elsewhere simply may not apply to any other technology. Use of systems based on body frames or reliance on set-up images and patient immobilization will not produce the same accuracy as is found with the continuous "track, detect and correct" approach of the CyberKnife. Nowhere is this more apparent than with treatment of tumors that move with respiration. To observe the CyberKnife tracking a tumor in the lung, for example, is to observe a fusion of science and art. The result of that treatment may well be unique.

I would like to take this opportunity to congratulate all the contributors to *Robotic Radiosurgery: Treating Tumors that Move with Respiration*, and to thank them for their belief in and support of this unique technology.

Sunnyvale, California

Euan Thomson, PhD
President and Chief Executive Officer
Accuray Incorporated

References

1. Cha C, DeMatteo RP, Blumgart LH. Surgery and ablative therapy for hepatocellular carcinoma. J Clin Gastroenterol 2002; 35:S130–137.
2. Handy JR, Jr., Asaph JW, Skokan L, et al. What happens to patients undergoing lung cancer surgery? Outcomes and quality of life before and after surgery. Chest 2002; 122:21–30.
3. Kelsen DP, Portenoy R, Thaler H, et al. Pain as a predictor of outcome in patients with operable pancreatic carcinoma. Surgery 1997; 122:53–59.
4. Treatment of locally unresectable carcinoma of the pancreas: comparison of combined-modality therapy (chemotherapy plus radiotherapy) to chemotherapy alone. Gastrointestinal Tumor Study Group. J Natl Cancer Inst 1988; 80:751–755.
5. Fuss M, Thomas CR, Jr. Stereotactic body radiation therapy: an ablative treatment option for primary and secondary liver tumors. Ann Surg Oncol 2004; 11:130–138.
6. Onimaru R, Shirato H, Shimizu S, et al. Tolerance of organs at risk in small-volume, hypofractionated, image-guided radiotherapy for primary and metastatic lung cancers. Int J Radiat Oncol Biol Phys 2003; 56:126–135.
7. Park HC, Seong J, Han KH, et al. Dose-response relationship in local radiotherapy for hepatocellular carcinoma. Int J Radiat Oncol Biol Phys 2002; 54:150–155.
8. Koong AC, Christofferson E, Le QT, et al. Phase II study to assess the efficacy of conventionally fractionated radiotherapy followed by a stereotactic radiosurgery boost in patients with locally advanced pancreatic cancer. Int J Radiat Oncol Biol Phys 2005; 63:320–323.
9. Koong AC, Le QT, Ho A, et al. Phase I study of stereotactic radiosurgery in patients with locally advanced pancreatic cancer. Int J Radiat Oncol Biol Phys 2004; 58:1017–1021.
10. Le QT, Loo BW, Ho A, et al. Results of a phase I dose-escalation study using single-fraction stereotactic radiotherapy for lung tumors. Journal of Thoracic Oncology 2006; 1:802–809.
11. Nuyttens JJ, Prevost JB, Praag J, et al. Lung tumor tracking during stereotactic radiotherapy treatment with the CyberKnife: Marker placement and early results. Acta Oncol 2006; 45:961–965.
12. Whyte RI, Crownover R, Murphy MJ, et al. Stereotactic radiosurgery for lung tumors: preliminary report of a phase I trial. Ann Thorac Surg 2003; 75:1097–1101.
13. Casamassima F, Cavedon C, Francescon P, et al. Use of motion tracking in stereotactic body radiotherapy: Evaluation of uncertainty in off-target dose distribution and optimization strategies. Acta Oncol 2006; 45:943–947.

Introduction

Treating Tumors that Move with Respiration: Thoughts of a Surgeon

The constant challenge facing those who develop and apply new medical technology is well articulated by Hippocrates in his aphorism, which is as pertinent today as it was when he wrote it.

"Life is short,
The art long,
The occasion instant,
The experiment perilous and
The decision difficult."

This is certainly true for the CyberKnife® (Accuray Incorporated, Sunnyvale, CA), as this volume and a growing number of articles in peer-reviewed journals make clear. In building on 30 years of experience using the Gamma Knife to the present day, when CyberKnife radiosurgery for extracranial targets that move with respiration is a reality, not only is a creative approach essential, but acceptance of the principle implied by Hippocrates' aphorism is critical. Hippocrates intended it to promote open-mindedness, with the second line emphasizing continuity and the first and the last lines signaling change. Those of us pushing the boundaries of stereotactic radiosurgery (SRS) appreciate its continuity with the technology and principles of both radiotherapy and surgery, but to maximize the potential of SRS to treat patients, change is necessary. Still, "the experiment is perilous and the decision difficult."

The CyberKnife is cutting-edge radiosurgery technology which targets a restricted, localized area for ablation while causing minimal damage to adjacent normal tissue. Because it is a frameless radiosurgical system it is able to destroy target lesions anywhere in the body, and with the recent introduction of the Synchrony® Respiratory Tracking System (Accuray Incorporated, Sunnyvale, CA), an increasing number of treated lesions are extracranial lesions that move as the patient breathes.

It is certainly the right time to ask, "Where do we go from here?" Asking the right questions is crucial for success. The "right" questions asked at the beginning of the 20th century by mathematicians led Einstein to develop the theory of relativity, leading to the atomic and hydrogen bombs, and all the changes, beneficial and disastrous, that have resulted. Currently a committee assembled by Bill Gates and led by Nobel Prize winner Harold Varmus is trying to design the "right" questions to move medicine toward a comparable level of success over the next several decades. This "Grand Challenges in Global Health" initiative is intended to devise strategies for removing roadblocks to solving major public health problems [1]. The committee evaluates solutions based not only on their immediate health impact, but on the short- and long-term consequences of the solutions for societies and the environment. This is a perspective from which the right questions may be asked.

The Right Questions

Perhaps we have begun with the most obvious questions. Having shown that the CyberKnife can do what the Gamma Knife can do in the brain [2], the frameless nature of the CyberKnife naturally suggested its use in the spine, where lesions are relatively stationary [3–6] (but not entirely [7]). Its safety and efficacy in these applications led to the use of the CyberKnife for other extracranial lesions, such as pancreas [8, 9] and lung [10–12]. In Section I of this volume the nature and extent of the problem of targeting, tracking, and treating moving tumors is described, as are methods for dealing with the problem, including Synchrony respiratory tracking.

The chapters in this book attest to the feasibility of CyberKnife treatment throughout the body, and the extension of this technology to other extracranial sites is underway. In this effort CyberKnife investigators have benefited greatly from the findings reported in the broader field commonly known as stereotactic body radiotherapy (SBRT). To place work with the CyberKnife in historical context, we have asked leading researchers in SBRT to review clinical findings and technical concerns in delivering high-dose, hypofractionated radiation to lung, liver, and pancreas (see Sections IV and V). These reviews are accompanied by commentaries on multidisciplinary treatment of cancer; multidisciplinary approaches are critical to patient care and, I would argue, to asking the right questions (see my commentary below). Sections IV and V also present the latest clinical outcomes data from CyberKnife users. These results remain promising and begin to help us establish optimal treatment parameters that will maximize clinical efficacy while minimizing risk of toxicity. Sections II and III highlight science, methodology, and technology that help us understand how high-dose hypofractionated radiation works and how to improve the treatment of patients. Section VI describes how the CyberKnife is evolving and what further investigations are needed to establish the relative value of this technology. Altogether, the question, "Can the CyberKnife be used to treat extracranial lesions, even ones that move with respiration," has been answered with a resounding yes.

As a cardiothoracic surgeon there are several other, less obvious questions that I might ask. Currently, the CyberKnife supplements or replaces standard therapies for carcinoma, such as radiation, chemotherapy, and surgery, all of which have the disadvantage of reducing host resistance while they attack the cancer. Far greater potential may lie in forging a synergistic union between the CyberKnife and current investigative techniques to eradicate cancer which do not reduce host resistance. These new therapies include vaccines, gene therapy and angiogenesis blockade. Combining vaccines with CyberKnife SRS may lead to a truly non-surgical approach to cancer treatment. Cancer could be cured by techniques that minimize the damage to the host's resistance. Angiogenesis blockade prevents the tumor from growing by interrupting development of the tumor's vasculature, thus depriving the tumor of blood. This could eliminate the tumor without damaging the immediate surrounding normal tissue. By combining the CyberKnife with gene therapy, the same synergistic result could possibly be achieved.

Employing the CyberKnife for preoperative radiation therapy, an approach that has been widely successful with conventional radiation techniques, may be explored more fully. That has been beneficial for carcinomas of the superior pulmonary sulcus. One of my associates, Dr Robert Shaw, while treating a Pancoast Tumor patient with conventional irradiation, was confronted by the patient halfway through his treatment who felt that his pain was beyond his ability to bear [13]. He talked Dr Shaw into surgically removing the cancer after only one half of the irradiation treatment (3000 cGy). This tumor was resected along with the lower trunk of the brachial plexus, chest wall, and part of the lung. The patient subsequently lived longer than Dr. Shaw (well over 40 years). This set the stage for a large series of patients (the largest in the world) who had preoperative radiation followed by an interval phase to allow the tumor to shrink, the lymphatics to be blocked and tumor cells that might fall into the wound at surgery weakened [14, 15]. Previously, radiation therapy alone was the treatment of choice (6-8 weeks). The addition of surgery shortened this period to two weeks of irradiation preoperatively plus the operative and post-operative time of one week. Pre-operative use of the CyberKnife will reduce the irradiation time even further by at least one week or more and may provide a similar or possibly even better results.

Similarly, conventional pre-operative irradiation has been effective in tumors requiring bronchoplastic resection to preserve lung tissue and avoid pneumonectomy [16]. The "lower dose" conventional preoperative irradiation reduces the size of the tumor, blocks lymphatics, and weakens any cells that might be left in the wound at surgery without jeopardizing healing. This allows a bronchoplastic procedure to be performed, removing only the proximal lobe with a bronchial resection followed by a re-anastomosis, thus preserving the distal lobe of the same lung. Lung tissue is conserved and the deleterious effects of total pneumonectomy eliminated. The success of preoperative irradiation is similar to that used in a superior pulmonary sulcus area, and should be even better with the CyberKnife. Pneumonectomy has been relegated to a very small role in the surgical armamentarium; in fact, pneumonectomy is considered by some to be more like a "disease" than a therapeutic procedure.

Postoperative CyberKnife SRS for lung, liver and pancreas tumors is also feasible. With further study it may be that, for certain tumors, CyberKnife monotherapy will be shown to result in clinical outcomes comparable to that obtained with surgery. Totally replacing interventional surgery with a non-invasive procedure – this is an audacious dream. Time and good research will determine if it will become a reality.

Thoughts on the Evolution of our Discipline in the Age of CyberKnife

Optimal extracranial SRS requires a multi-disciplinary team of radiation therapists, oncologists, surgeons and other organ-specific specialists. The team approach improves the communication of these professionals so necessary in the advancement of tumor management. The combination of CyberKnife SRS with other conventional therapies, such as radiation therapy, surgery and chemotherapy, should improve current results. In particular, specialists in the use of cancer vaccines, gene therapy, and angiogenesis blockade, included in the multi-disciplinary team, may offer further synergistic benefit as the technology and its applications evolve.

Conclusions

I have been asked by colleagues what it is about the CyberKnife that I find exciting, speaking as a cardiothoracic surgeon. The CyberKnife is one of the most extraordinary adjuncts to surgery that has come along in many years. It markedly shortens the time for radiation therapy from six weeks to one week or less. It is much more accurate than any other kind of radiation, and thereby minimizes damage to the surrounding sensitive tissue. Thus, it has the potential to be far more synergistic with surgery, markedly increasing its efficiency. I have considered possible analogies – to what is the technological achievement of the CyberKnife comparable? The CyberKnife is analogous to a Lugar pistol compared to other handguns of the late 1800s. The best previous handgun had six shots, which were individually and awkwardly loaded and were fired separately, requiring the hammer to be cocked prior to each shot. In contrast, the

Lugar has eight shots in a clip, which could be loaded expeditiously and fired in a few seconds – clearly a tremendous advance over previous handguns. My colleagues have suggested that comparing a medical instrument to a weapon may not be entirely appropriate, but they also see the point of the analogy – and after all, do we not commonly refer to medical tools as part of an "armamentarium"? From a surgeon's perspective the technological advance of the CyberKnife may be more like the marked difference between closed heart operations performed with fingers and instruments versus open-heart procedures where everything can be visualized and managed much more expeditiously. Better for the surgeon, better for the patient.

Some see things as they are and say "why"?
CyberKnife clinicians dream things that never were and say "why not"?

The chapters in this book document some of the areas where the CyberKnife is being utilized currently. With "CyberKnife Vision" we will contemplate possibilities for the future in the treatment of cancers that move with respiration.

Dallas, Texas

Harold C. Urschel Jr. MD
Chair of Cardiovascular & Thoracic Surgical Research
Education & Clinical Excellence
Baylor University Medical Center

References

1. Varmus H, Klausner R, Zerhouni E, et al. Public health. Grand Challenges in Global Health. Science 2003; 302:398–399.
2. Chang SD. The CyberKnife: potential in patients with cranial and spinal tumors. American Journal of Cancer 2005; 4:383–393.
3. Degen JW, Gagnon GJ, Voyadzis JM, et al. CyberKnife stereotactic radiosurgical treatment of spinal tumors for pain control and quality of life. J Neurosurg Spine 2005; 2:540–549.
4. Gerszten PC, Burton SA, Ozhasoglu C, et al. Stereotactic radiosurgery for spinal metastases from renal cell carcinoma. J Neurosurg Spine 2005; 3:288–295.
5. Gerszten PC, Burton SA, Welch WC, et al. Single-fraction radiosurgery for the treatment of spinal breast metastases. Cancer 2005; 104:2244–2254.
6. Ryu SI, Chang SD, Kim DH, et al. Image-guided hypo-fractionated stereotactic radiosurgery to spinal lesions. Neurosurgery 2001; 49:838–846.
7. Murphy MJ, Chang SD, Gibbs IC, et al. Patterns of patient movement during frameless image-guided radiosurgery. Int J Radiat Oncol Biol Phys 2003; 55:1400–1408.
8. Koong AC, Christofferson E, Le QT, et al. Phase II study to assess the efficacy of conventionally fractionated radiotherapy followed by a stereotactic radiosurgery boost in patients with locally advanced pancreatic cancer. Int J Radiat Oncol Biol Phys 2005; 63:320–323.
9. Koong AC, Le QT, Ho A, et al. Phase I study of stereotactic radiosurgery in patients with locally advanced pancreatic cancer. Int J Radiat Oncol Biol Phys 2004; 58:1017–1021.
10. Le QT, Loo BW, Ho A, et al. Results of a phase I dose-escalation study using single-fraction stereotactic radiotherapy for lung tumors. Journal of Thoracic Oncology 2006; 1:802–809.
11. Nuyttens JJ, Prevost JB, Praag J, et al. Lung tumor tracking during stereotactic radiotherapy treatment with the CyberKnife: Marker placement and early results. Acta Oncol 2006; 45:961–965.
12. Whyte RI, Crownover R, Murphy MJ, et al. Stereotactic radiosurgery for lung tumors: preliminary report of a phase I trial. Ann Thorac Surg 2003; 75:1097–1101.
13. Shaw RR. Pancoast's tumor. Ann Thorac Surg 1984; 37:343–345.
14. Urschel HC, Jr. Superior pulmonary sulcus carcinoma. Surg Clin North Am 1988; 68:497–509.
15. Urschel HC, Jr. New approaches to Pancoast and chest wall tumors. Chest 1993; 103:360S–361.
16. Paulson DL, Urschel HC, Jr., McNamara JJ, et al. Bronchoplastic procedures for bronchogenic carcinoma. J Thorac Cardiovasc Surg 1970; 59:38–48.

Acknowledgement
The Editors and I extend our deep appreciation to everyone who contributed to this book, including authors, reviewers, and the staff at Springer Publishing. Thanks especially to Robert Morton and Dr. Chad Lee for reviewing chapters, and to Dr. Pam Thuman-Commike for editing and final formatting. I personally wish to thank Mrs. Rachel Montano and Mrs. Brenda Knee, whose contributions were of immeasurable value in the pursuit of excellence during the editing of this book.

Harold C. Urschel Jr.

Contents

Author Index

Section 1

Respiration Motion and Tumor Tracking Techniques

Tumor Motion Ranges Due to Respiration and Respiratory Motion Characteristics

1

SONJA DIETERICH and YELIN SUH

CONTENTS

1.1 Abstract

Many soft-tissue tumors targeted with extracranial SRS move during respiration. New imaging technologies, motion compensation strategies, and treatment planning algorithms are being developed which enable tracking and treatment of moving tumors in real-time. For this chapter we reviewed the literature to determine known tumor motion amplitudes for lung, liver, and pancreas. Then we analyzed predicted tumor motion for 36 patients and 117 treatment fractions that were previously saved in CyberKnife® (Accuray Incorporated, Sunnyvale, CA) treatment logfiles. These represent 27 tumors in the lung (16 upper lung, 4 middle lung, 7 lower lung) and 9 pancreas patients. For each treatment, the location of the target at end inspiration and end expiration was determined in the patient coordinate system. The origin of the patient coordinate system is at the center of mass of the fiducials as marked on the simulation CT, +x is patient inferior, +y patient left, and +z anterior in a right-handed coordinate system. The mean and variance of respiratory cycle extrema positions were calculated using a program written in MatLab code. Observed motion ranges for all sites except pancreas agree very well with the literature. The largest motion amplitudes of up to 38.7 mm were observed in the lower lung. Twenty-five percent of tumors in the upper lung could have been treated without Synchrony® (Accuray Incorporated, Sunnyvale, CA) with a PTV margin of 2 mm, because the uncertainty is in the range of the technical tracking accuracy of Synchrony of 1.5 mm. Possible causes of large fluctuations around the mean motion could be fiducial tracking errors or irregular breathing. We concluded that a subset of all patients could have been treated using skeletal structure tracking, rather than implanted fiducials, and a PTV margin in the range of the stated tracking accuracy for Synchrony. Defining meaningful parameters to characterize the effects of free breathing is part of ongoing research, since published data from non-dynamic SBRT is limited to short fluoroscopic studies or Cine-CT. The results can be transferred to other treatment modalities to determine PTV margins in standard external beam treatments as well as defining the PTV in the third dimension for 2D motion compensation [1].

1.2 Introduction

1.2.1 Extra-Cranial SRS and the Problem of Tumor Motion

Soon after stereotactic radiosurgery (SRS) applications expanded from brain to extra-cranial targets such as the spine, there was an interest in treating soft-tissue tumors within the lung and pancreas. The challenge was that SRS technologies were initially designed to deliver very precise treatments for non-moving targets, but soft tissue tumors can move considerably during respiration. Therefore, methods to compensate for respiratory motion needed to be developed. But before a tumor motion compensation method can be employed, the extent of respiratory motion for tumors in various organs needs to be observed. Technologies such as fluoroscopy, surrogate markers (spirometry, fiducials), 4D-CT and dynamic MRI are used to achieve that goal.

1.2.2 Methods to Observe Directly Respiratory Motion, Advantages and Disadvantages

The most straightforward imaging tool to study respiratory motion is fluoroscopy. Fluoroscopy acquisition, however, is time-limited to avoid excess exposure to ionizing radiation. In addition, many tumors cannot be directly observed by fluoroscopy, but require surrogate markers such as implanted fiducials. The accuracy of fiducial markers as tumor surrogates is still being investigated [2, 3].

In recent years, 4D-CT imaging has moved into clinical applications, and several 4D-CT models are commercially available. Contrary to breath-hold CT acquisition, the image acquisition occurs during free breathing and data are sorted into bins according to position within the respiratory cycle based on anatomical information (e.g., diaphragm position) or surrogate data (e.g., spirometry or optical skin markers). In this way a series of 3D-CT datasets are obtained, each corresponding to a different part of the respiratory cycle. Also, 4D-CT imaging is very useful to observe the motion amplitude of lung tumors and to determine margins based on motion amplitude, target deformation, and motion compensation method chosen for the radiation treatment [4, 5]. The disadvantage of 4D-CT lies in the fact that it will show tumor motion averaged over multiple respiratory cycles. For patients with irregular breathing patterns, or patients for which correlation of tumor position and surrogate marker may vary [6], averaging respiratory cycles introduces a potential source of error.

A more direct observation of respiratory motion is 4D-MR [7]. The advantage is the use of non-ionizing radiation for multiple image acquisition compared to 4D-CT, and better image contrast for soft tissue. However, the technology is still under investigation. Current pulse sequences are not yet fast enough to collect data in real-time. A conventional treatment planning CT still needs to be obtained. In addition, radiation therapy departments usually do not have MR scanners in their department, which causes a significant logistical overhead and also makes the use of this technology expensive.

An interesting combination of fiducial-based and non-ionizing radiation tracking of tumors is a technology which employs implantable, wireless electromagnetic markers for tumor tracking [8]. These markers were recently FDA-cleared for use in the prostate. Approval for use in other organs is expected in the future.

1.2.3 Respiratory Motion Compensation

Methods to compensate for respiratory motion fall into two categories: real-time adaptive motion compensation and non-adaptive methods. The latter include techniques such as using large PTV margins, abdominal compression, treating during breath-hold, or respiratory gating. Large PTV margins are the method of choice if more advanced technology is not available in a clinic.

The advent of modern feedback devices for breath-hold aids the patient to hold his breath more reproducibly at the predefined respiratory phase. Breath-hold techniques have the advantage that the treatment respiratory motion phase matches the breath-hold treatment planning CT very well. PTV margins can be reduced by 7–12 mm compared to treatments with free breathing [9]. A summary of intrafraction reproducibility studies is provided in a study by Sarrut et al. [10], where residual uncertainties of up to 6 mm were reported. Other than this uncertainty in the breath-hold position, the breath-hold method has two additional disadvantages. First, elderly patients or patients with advanced disease may not be able to tolerate the breath-hold procedure or may not be able to comply with instructions. Therefore, careful patient selection is required for this treatment method. Second, breath-hold may prolong treatment times depending on the treatment delivery technique, monitor units per treatment, and dose rate of the treatment machine. A German group recently reported a frameless, stereotactic treatment delivery to a liver tumor in about 30 minutes using breath-hold [11].

In gating, an external marker or other secondary respiratory indicator is used to determine the phase of respiration, based on the assumption that the respiratory indicator has the same correlation to the tumor position as during simulation. In some centers, X-ray or on-board imagers can be used to verify this assumption before the start of the treatment, assuming that the tumor or a suitable surrogate can be visualized [12]. The user defines the gating window starting point, typically shortly before the maximum exhale, and the percentage of the respiratory cycle, with typical ranges from 10% to 40%. Depending on the range of the gating window, the treatment time is prolonged and residual motion is always present. The advantage of gating over breath-hold is the ability of the patient to breathe freely. Further improvements can be made using visual feedback to the patient to improve regular breathing [13].

Real-time methods include approaches using dynamic multileaf collimators [14], a moving couch [15], or real-time respiratory motion tracking using a robot-mounted linac [16]. Of these three methods, only the latter with adaptive SRS using the CyberKnife Synchrony system has been widely established in clinical use since 2004.

1.2.4 Adaptive SRS as Solution

With Synchrony, tumors are tracked based on the correlation model between surrogate respiratory markers and the tumor motion. This correlation model is updated throughout the treatment using orthogonal X-rays to verify the actual tumor position to the predicted position, and measure the correlation error. Because the tumor cannot be directly and continuously visualized, the clinical accuracy of the delivered treatment is hard to assess. Respiratory motion phantoms [17] can replay patient data and simulate the treatment, but these rely on known internal motion input to reflect true anatomical motion. Such continuous motion data does not exist for time intervals long enough to deliver an adaptive SRS treatment. It is conceivable that in the near future, tumors could be directly and continuously tracked using non-ionizing radiation to follow electromagnetic fiducial markers [8]. Shorter observed intervals of tumor motion could provide an approximation, but do not reflect changes in patient status during the treatment duration, e.g., relaxation. A first-order approach to estimate the accuracy and safety of the treatment delivery is to study the tumor motion range as predicted by the adaptive SRS delivery system and compare the motion range to published literature to see if those motions are within similar ranges. The results are indicative of the quality of the correlation model, if it predicts clinically reasonable motion amplitudes compared to what has been observed during shorter time intervals using continuous direct visualization techniques such as fluoroscopy.

1.3 Methods and Materials

1.3.1 CyberKnife – Synchrony Technology

Synchrony [16] is, at this time, the only FDA-cleared dynamic SRS device in the US. Its technology is based on the CyberKnife SRS technology [18–20]. The CyberKnife SRS system consists of a 6 MV X-band linear accelerator mounted on a robotic arm (KUKA, Augsburg, Germany). The patient is placed on a treatment couch. Images from two orthogonal X-ray cameras with amorphous silicon detectors are compared to digitally reconstructed radiographs (DRRs) from the treatment planning CT. Internal gold fiducial markers or bony landmarks can be used for co-registration.

During the treatment, images can be taken before every treatment beam. The robot will correct for patient motions of up to 10 mm in each translational direction, 1 degree of roll and pitch rotation, and 3 degrees of yaw. The relevant clinical accuracy combines the robot pointing accuracy, camera image tracking system and target localization accuracy, and DRR generation accuracy. For fiducial tracking and CT slice thickness between 0.625 mm and 1.25 mm, the system accuracy has been shown to be 0.7 +/- 0.3 mm [21].

To perform respiratory tracking the Synchrony system builds a correlation model between external optical marker locations, visualized continuously using a camera array, and internal fiducial marker locations placed within the tumor, visualized intermittently using the X-ray cameras (Fig. 1.1). The optical markers and the model are then used to calculate the tumor location continuously throughout treatment. X-ray images acquired between treatment beams during treatment allow the model to adapt to intra-fraction changes in external-internal marker correlation. This system is described in greater detail in the following chapter.

1.3.2 Treatment Procedure

All patients were placed on the treatment table in supine position. The predicted tumor location was analyzed for 36 patients utilizing the data saved in the Synchrony file "Modeler.log" throughout treatment. They represent 27 tumors in the lung (16 upper lung, 4 middle lung, 7 lower lung) and 9 pancreas patients. Each lesion contained 2–4 gold fiducials to mark the tumor location. The skin motion was tracked with visible light-emitting beacons placed on the patient's abdomen in the area of largest motion amplitude. During the treatment, X-rays of the fiducial positions were taken to update the correlation model between skin and tumor motion.

The predicted internal tumor motion from the skin-tumor correlation model was analyzed. These log files, however, also contain data that might be from sources other than tumor motion, and include data calculated during periods when the treatment delivery is suspended. Examples of data due to sources other than tumor motion are time intervals when the treatment may have been interrupted due to patient intervention, a misidentified fiducial, or skin marker-camera line of sight interruptions. Therefore, the predicted internal tumor motion data set was extracted manually for each fraction, removing time intervals with obvious outside interference according to a set of predefined rules. This ensured that the motion analysis was based only on predictions made when a valid model was constructed, and treatment was enabled.

Data points excluded from the study data set were those (1) showing no motion (noise only), during which the treatment could have been stopped or the correlation model reset, (2) at the beginning of the treatment showing significantly larger motions, which correspond to the predictions of an incomplete correlation model made during its construction, (3) showing large time intervals (more than 5 sec) between model predictions, during which treatment could have been stopped, typically because there could have been problems tracking the optical markers (note that usually the prediction

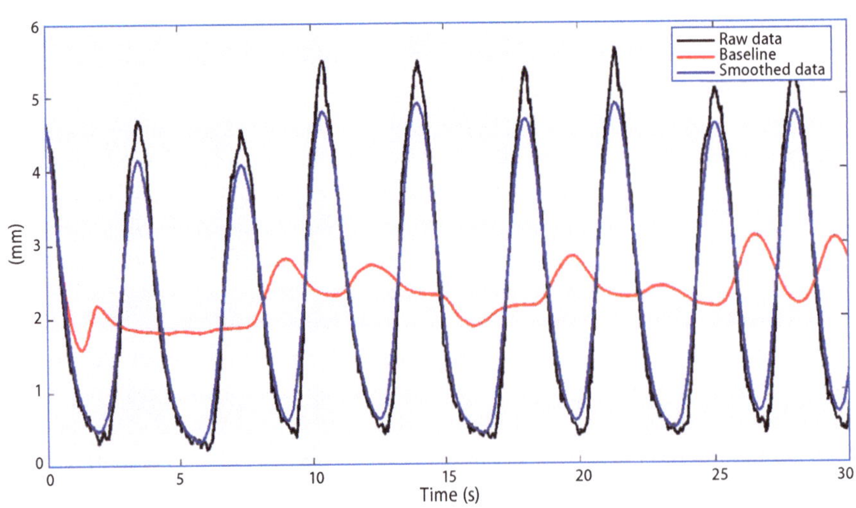

Fig. 1.1 To calculate the peak-trough ranges of patient respiratory tumor motion (*black curve*), a moving average of 25 points (*blue curve*) was used for filtering high-frequency noise and a moving average of 100 points was used for establishing the baseline (*red curve*).

time intervals were around 0.04 sec), and (4) showing a large motion change in only one direction out of three, potentially corresponding to incorrect identification of internal fiducial or large changes in the patient's respiratory pattern, such as coughing fits during which time the treatment was paused (note that data points showing the large motion changes in two directions simultaneously while keeping the motion in the third direction were included).

As shown in Figure 1.1, in order to calculate the peak-trough ranges of patient respiratory tumor motion (black curve), a moving average of 25 points (blue curve) was used for filtering high-frequency noise and a moving average of 100 points was used for establishing the baseline (red curve). This baseline was used to divide each breathing cycle into the peak part (mid-inspiration to mid-expiration) and trough part (mid-expiration to mid-inspiration). The peak and trough parts were identified by finding the intersections of the filtered motion curve (blue) and baseline (red), and then maxima and minima in the peak and trough part, respectively, were calculated as described below. The peak-trough ranges were determined by the differences between the adjacent maxima and minima.

Figure 1.2 shows the predicted internal tumor motion for a different patient. At 1.3 minutes an X-ray was taken in which one fiducial was misidentified by the software. The error was not caught by the treatment team. This misidentified fiducial caused a sharp spike in the predicted tumor motion paths. With our method to determine motion amplitudes, these spikes have been removed in the analysis.

1.4 Results

1.4.1 Lung

Many studies using a variety of imaging modalities to quantify motion amplitudes for lung tumors have been published in the last decade (e.g., see Chen et al. [22], Mageras et al. [23], and Shirato et al. [24]). In addition, tumor motion has been observed in various patient treatment modes ranging from free breathing to breath-hold studies. The number of patients observed spans the range from fewer than ten to several dozen. No comprehensive conclusions have been reached, but general patterns have been observed:

- Respiratory motion patterns and amplitudes cannot be predicted by subject age, gender, weight, height, and other conditions [24].
- Motion ranges tend to increase from upper lung (smallest) to lower lung (largest), although tumor location alone is not a reliable indicator of mobility [25].
- Motion patterns are less linear closer to the bronchi [24].

The data for lower lung (Table 1.1), middle lung (Table 1.2), and upper lung (Table 1.3) follow these general trends. We listed the mean motion for each direction, fraction, and patient. Unlike in other studies, we think that calculating the mean of the

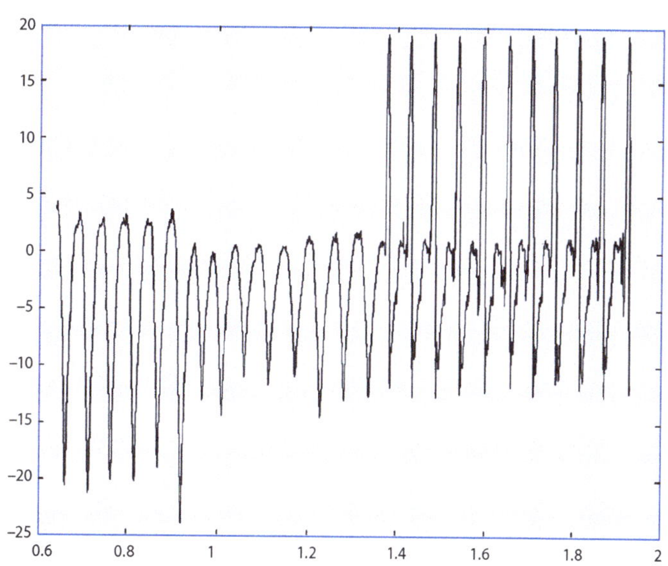

Fig. 1.2 Problematic tumor motion pattern for patient #2 caused by fiducial mistracking in the X-ray image taken 1.4 minutes into the treatment.

Table 1.1 Extent of motion for lower lung.

		Superior-Inferior [mm]				Left-Right [mm]				Anterior-Posterior [mm]			
Pt #	Fx #	mean	max	min	stdev	mean	max	min	stdev	mean	max	min	stdev
5	2	**10.1**	21.6	2.0	2.4	**2.4**	6.2	0.3	0.8	**3.2**	6.2	0.0	1.3
5	4	**8.4**	11.8	0.6	1.5	**2.0**	2.8	0.1	0.4	**3.4**	4.7	0.2	0.7
5	5	**4.3**	7.4	0.5	1.4	**0.4**	0.9	0.0	0.2	**0.7**	1.2	0.1	0.2
5	7	**9.0**	15.6	0.8	2.1	**2.2**	4.1	0.1	0.6	**2.0**	4.1	0.0	0.8
15	1	**0.9**	1.4	0.0	0.3	**12.6**	19.6	0.9	2.2	**3.7**	5.8	0.0	0.9
15	2	**9.8**	14.2	0.4	1.6	**0.6**	1.3	0.0	0.3	**3.3**	6.9	0.8	1.2
15	3	**14.0**	18.8	2.5	2.1	**1.3**	3.0	0.0	0.7	**5.2**	7.6	1.0	1.0
20	1	**12.4**	38.7	5.6	3.0	**0.5**	1.2	0.0	0.3	**2.0**	7.4	0.6	0.7
20	2	**11.5**	37.3	1.9	2.2	**1.0**	2.7	0.5	0.3	**1.1**	2.6	0.5	0.3
20	3	**10.6**	35.4	4.7	2.1	**1.3**	4.7	0.6	0.4	**1.2**	3.1	0.4	0.4
27	1	**6.2**	13.3	1.1	2.0	**0.2**	0.9	0.0	0.1	**0.5**	1.8	0.0	0.3
27	2	**9.8**	16.1	0.4	1.6	**0.7**	1.6	0.0	0.2	**0.7**	1.3	0.0	0.2
27	3	**10.1**	18.6	3.2	1.4	**0.1**	0.6	0.0	0.1	**0.6**	0.9	0.0	0.2
32	1	**1.6**	4.7	0.0	0.6	**3.5**	7.5	0.0	1.2	**1.1**	3.5	0.0	0.6
32	2	**1.5**	4.8	0.0	0.9	**2.1**	5.9	0.0	1.1	**0.6**	1.6	0.0	0.3
32	3	**1.6**	13.5	0.0	1.2	**2.5**	10.7	0.0	1.4	**1.2**	5.5	0.0	0.7
33	1	**5.1**	16.2	0.5	1.7	**7.1**	21.6	0.0	1.9	**6.1**	18.3	0.2	1.8
33	2	**5.7**	13.5	2.2	1.5	**1.1**	2.3	0.0	0.4	**4.7**	11.0	0.3	1.3
33	3	**3.1**	6.3	0.1	1.3	**7.6**	25.3	0.6	2.6	**4.3**	8.3	0.0	1.2
33	4	**8.8**	32.1	1.6	2.8	**3.2**	10.4	0.0	1.1	**4.8**	21.7	0.0	1.4
33	5	**5.4**	12.2	0.6	1.4	**6.4**	18.3	0.1	2.1	**5.8**	17.0	1.8	1.6
43	1	**6.1**	18.6	0.3	1.5	**1.4**	4.0	0.0	0.5	**2.8**	6.0	0.1	1.0
43	2	**6.2**	20.1	0.9	2.1	**1.1**	4.8	0.1	0.5	**7.0**	17.8	1.2	2.6
43	3	**6.0**	13.6	1.1	1.9	**1.0**	3.2	0.0	0.7	**3.6**	7.7	0.4	1.8
43	4	**4.9**	19.3	2.4	1.6	**1.1**	5.3	0.1	0.5	**3.4**	11.3	1.5	1.0
43	5	**6.1**	34.1	0.0	2.9	**0.5**	5.8	0.0	0.5	**4.3**	21.2	0.2	1.8
Range of means		1.5 mm – 14.0 mm				0.1 mm – 12.6 mm				0.6 mm – 7.0 mm			

Table 1.2 Extent of motion for middle lung.

		Superior-Inferior [mm]				Left-Right [mm]				Anterior-Posterior [mm]			
Pt #	Fx #	mean	max	min	stdev	mean	max	min	stdev	mean	max	min	stdev
3	1	**3.1**	8.9	1.0	0.5	**1.0**	2.5	0.1	0.4	**1.4**	2.5	0.0	0.3
3	2	**1.8**	4.3	0.1	0.5	**1.1**	2.8	0.0	0.7	**0.7**	2.0	0.0	0.6
3	3	**3.9**	11.4	1.8	0.7	**1.1**	2.8	0.2	0.3	**0.8**	3.5	0.4	0.2
3	4	**3.3**	8.9	1.4	1.3	**2.5**	10.5	0.0	2.0	**2.0**	8.4	0.0	1.3
11	1	**3.5**	26.1	0.0	2.0	**0.5**	4.8	0.0	0.4	**1.7**	13.0	0.0	1.0
11	2	**7.4**	25.9	0.0	2.5	**1.3**	4.9	0.0	0.7	**3.4**	16.4	0.0	1.3
11	3	**5.7**	19.9	0.0	2.2	**0.9**	4.1	0.0	0.4	**2.1**	6.4	0.0	0.7
12	1	**0.6**	1.6	0.0	0.3	**1.0**	3.5	0.1	0.5	**1.9**	8.8	0.0	1.2
12	2	**0.7**	1.6	0.0	0.4	**0.8**	1.9	0.0	0.4	**2.2**	5.6	0.0	1.1
12	3	**0.8**	2.8	0.0	0.4	**0.9**	2.3	0.0	0.3	**2.3**	6.6	0.2	1.0
13	1	**3.2**	10.1	0.5	1.0	**0.7**	3.8	0.0	0.5	**0.7**	3.3	0.0	0.4
13	2	**3.2**	14.3	0.2	1.2	**0.9**	3.4	0.0	0.4	**0.6**	3.0	0.0	0.3
13	3	**4.0**	15.8	0.4	1.3	**0.9**	5.5	0.0	0.5	**1.2**	5.6	0.0	0.6
Range of means		0.7 mm – 7.4 mm				0.5 mm – 2.5 mm				0.6 mm – 3.4 mm			

Table 1.3 Extent of motion for upper lung.

		Superior-Inferior [mm]				Left-Right [mm]				Anterior-Posterior [mm]			
Pt #	Fx #	mean	max	min	stdev	mean	max	min	stdev	mean	max	min	stdev
5	1	2.9	5.7	0.6	1.4	2.2	4.2	0.4	0.7	4.4	8.4	1.6	1.0
5	3	3.2	5.5	0.4	0.8	1.9	3.1	0.2	0.5	3.0	4.9	0.0	0.8
5	6	2.9	5.5	0.8	1.1	1.1	2.4	0.2	0.5	1.9	3.0	0.0	0.7
6	1	3.8	8.2	0.4	1.1	0.3	1.1	0.0	0.2	4.2	9.8	0.5	2.0
6	2	4.7	17.8	0.1	1.9	0.6	1.3	0.0	0.3	2.7	11.9	0.0	1.2
8	1	1.8	4.2	0.0	0.5	1.5	3.4	0.0	0.9	2.0	5.1	0.3	0.6
8	2	0.8	2.7	0.0	0.4	0.4	1.1	0.0	0.2	1.3	5.2	0.0	0.6
8	3	0.7	2.6	0.0	0.3	0.5	1.4	0.0	0.2	1.3	4.3	0.0	0.6
9	1	4.1	18.9	0.1	1.5	1.1	2.4	0.0	0.5	1.7	6.3	0.0	0.8
9	2	3.9	8.8	0.0	2.0	1.6	3.4	0.0	0.5	1.3	3.0	0.0	0.6
9	3	3.9	7.4	0.0	1.6	0.7	2.2	0.0	0.5	1.4	2.7	0.0	0.6
10	1	0.5	3.0	0.1	0.4	3.8	11.2	0.0	1.1	2.3	5.0	0.0	0.7
10	2	3.6	7.3	0.0	0.9	2.6	4.7	0.0	0.7	2.7	5.1	0.1	0.9
10	3	2.1	3.7	0.3	0.7	6.3	12.5	0.7	1.8	6.2	11.8	0.9	1.8
14	1	3.1	7.6	1.2	0.6	0.9	2.2	0.0	0.5	1.2	2.5	0.3	0.2
14	2	2.9	6.2	0.1	0.7	1.0	2.5	0.0	0.5	1.6	3.0	0.0	0.3
14	3	3.1	4.2	0.5	0.5	0.7	1.4	0.2	0.3	1.2	1.8	0.1	0.3
14	4	3.7	5.6	2.8	0.3	1.5	3.0	0.2	1.0	2.3	4.1	0.2	1.2
14	5	3.0	5.6	1.2	0.5	0.5	1.4	0.1	0.3	0.6	1.1	0.2	0.1
21	1	0.4	1.3	0.0	0.3	0.6	1.3	0.0	0.2	0.5	1.6	0.0	0.3
21	2	0.3	1.0	0.0	0.2	0.6	1.7	0.0	0.3	0.7	2.2	0.0	0.4
21	3	0.2	1.3	0.0	0.2	0.4	1.3	0.1	0.2	0.9	3.5	0.0	0.4
22	1	8.1	41.5	0.1	4.8	1.4	8.5	0.0	1.1	1.7	5.6	0.1	1.0
22	2	7.4	48.2	0.4	4.3	1.8	14.5	0.1	1.5	1.3	8.5	0.1	1.0
22	3	9.1	26.2	0.3	4.6	4.8	12.0	0.0	2.7	3.1	8.9	0.1	1.7
25	1	2.5	7.6	0.8	0.6	2.5	4.8	0.0	0.8	4.4	6.8	0.0	1.2
25	2	2.9	6.4	0.1	0.6	2.5	7.2	0.1	0.6	3.0	6.7	0.0	1.3
25	3	2.9	6.7	1.1	0.6	2.5	4.8	0.0	0.9	4.0	9.8	0.0	2.3
28	1	5.6	16.1	0.1	2.0	2.9	6.1	0.0	1.1	1.9	3.3	0.0	1.0
28	2	4.2	9.8	0.0	2.0	2.3	4.7	0.0	1.1	1.7	6.9	0.0	1.3
28	3	6.6	12.6	0.1	2.2	2.9	7.6	0.1	1.4	1.6	4.4	0.0	1.0
31	1	0.3	0.8	0.0	0.2	0.3	1.0	0.0	0.2	0.4	0.9	0.0	0.2
31	2	0.1	0.4	0.0	0.1	0.1	0.9	0.0	0.1	0.3	0.7	0.0	0.1
31	3	0.4	0.7	0.1	0.1	0.3	0.7	0.0	0.1	0.2	0.7	0.0	0.2
34	1	1.3	3.1	0.4	0.5	0.9	2.5	0.0	0.3	2.6	8.0	0.7	1.0
34	2	1.3	3.6	0.3	0.5	1.5	5.2	0.2	0.5	2.9	14.2	0.4	1.3
34	3	1.7	3.5	0.0	0.6	0.7	2.2	0.1	0.4	2.3	7.5	0.3	0.7
35	1	2.9	9.1	0.6	0.6	0.6	1.4	0.1	0.3	1.3	3.6	0.2	0.2
35	2	0.9	2.5	0.3	0.4	3.3	8.3	1.9	0.8	1.0	3.8	0.0	0.5
35	3	3.8	5.5	1.7	0.7	0.4	0.7	0.0	0.2	0.8	1.4	0.0	0.2
37	1	2.6	18.7	0.0	2.5	1.1	8.3	0.0	0.8	0.5	3.6	0.0	0.6
37	2	4.2	9.3	1.7	0.8	0.5	1.1	0.0	0.3	0.6	1.3	0.2	0.2
37	3	5.6	10.3	1.1	1.5	1.2	1.8	0.0	0.3	1.5	2.5	0.3	0.4
40	1	2.1	7.1	0.0	1.4	1.1	3.3	0.0	0.7	4.7	15.0	0.1	3.2
40	2	0.9	3.1	0.0	0.7	0.8	3.7	0.0	0.5	1.1	4.4	0.0	1.0
40	3	1.2	3.3	0.0	0.7	1.8	7.7	0.0	0.9	1.5	3.5	0.0	0.8
42	1	7.0	31.6	0.2	6.1	1.0	5.0	0.1	0.8	5.1	20.5	0.1	3.5
42	2	3.6	30.3	0.0	2.6	0.5	5.4	0.0	0.6	4.2	37.0	0.0	3.6
42	3	1.3	8.0	0.0	0.8	1.3	6.0	0.0	0.8	2.5	17.4	0.0	1.4
42	4	3.7	12.5	0.0	1.4	0.8	3.0	0.0	0.7	2.6	8.2	0.0	1.9
Range of means		0.2 mm – 8.1 mm				0.1 mm – 6.3 mm				0.2 mm – 6.2 mm			

mean motion amplitudes does not provide any valuable information. The variance in the mean motion range between patients is quite large, and therefore we want to emphasize that even though tumor locations may be similar between the patients, mean tumor motion ranges may be quite different.

1.4.2 Pancreas

The literature on respiratory motion of the pancreas is sparse. Two older publications by Suramo et al., who studied 50 patients [26] and Bryan et al. in 36 patients [27], reported average motions of 20 mm and 18 mm for free breathing, respectively. A newer study by Gierga et al. on 7 patients found much smaller motion amplitudes with averages in cranio-caudal direction ranging from 4.4–9.6 mm [28].

An analysis of 9 pancreas patients treated at Georgetown University Hospital with a total of 28 fractions resulted in predicted tumor motions that agreed with the study by Gierga et al. (Table 1.4). The mean motion ranged from 0.2 to 9.4 mm in superior/inferior (SI), 0.3–4.8 mm in left/right (LR), and 0.9–4.8 mm in anterior/posterior (AP). Again, we did not list the average of the mean motion,

Table 1.4 Data from 9 pancreas patients treated with Synchrony at GUH.

		Superior-Inferior [mm]				Left-Right [mm]				Anterior-Posterior [mm]			
Pt #	Fx #	mean	max	min	stdev	mean	max	min	stdev	mean	max	min	stdev
2	1	8.6	13.3	2.2	1.4	2.2	5.0	0.1	1.2	1.9	4.1	0.4	0.5
2	2	5.1	8.3	3.3	0.6	1.0	1.4	0.1	0.2	1.6	3.6	0.1	0.6
2	3	6.8	8.9	4.7	0.6	0.7	0.9	0.4	0.1	2.0	2.5	1.3	0.2
4	1	0.2	0.7	0.0	0.1	3.2	7.8	0.3	0.8	0.9	3.0	0.1	0.3
4	2	1.4	4.0	0.0	0.6	4.8	17.5	0.2	2.2	1.4	5.8	0.0	1.0
4	3	3.7	10.1	0.0	1.7	1.0	2.4	0.0	0.5	1.3	3.5	0.3	0.5
4	4	4.3	17.7	0.5	1.2	0.9	4.9	0.0	0.4	1.1	6.4	0.0	0.4
4	5	2.2	7.0	0.3	0.6	0.6	1.6	0.1	0.3	1.3	4.1	0.6	0.4
4	6	4.0	10.3	0.2	1.1	1.0	3.0	0.0	0.3	1.5	3.5	0.1	0.5
17	1	6.3	15.2	0.0	3.3	2.9	6.4	0.0	1.5	4.8	9.8	0.0	2.2
17	2	7.8	19.4	0.1	4.2	2.5	9.1	0.1	1.4	4.2	11.3	0.0	2.5
17	3	7.4	28.8	0.1	4.3	1.0	3.1	0.0	0.5	2.6	9.8	0.0	1.4
18	1	5.7	16.5	0.1	2.6	1.4	5.5	0.0	1.0	2.2	7.8	0.0	1.0
26	1	3.3	19.4	0.2	2.1	1.4	10.4	0.0	1.0	1.7	13.1	0.0	1.5
26	2	3.9	11.3	0.8	1.2	0.8	5.1	0.0	0.4	3.4	9.6	1.0	0.8
26	3	3.3	10.7	1.1	1.2	0.8	2.6	0.1	0.6	3.2	8.9	0.6	1.2
39	2	5.2	14.4	0.1	2.1	0.7	2.4	0.0	0.5	1.4	4.5	0.1	0.8
39	3	4.7	35.7	0.0	5.0	0.8	9.3	0.0	1.2	1.7	17.3	0.0	2.0
39	4	0.9	2.1	0.0	0.3	2.1	3.9	1.0	0.7	1.9	3.7	0.5	0.6
41	1	2.1	12.4	0.5	0.9	0.9	1.7	0.0	0.3	1.1	6.8	0.1	0.6
41	2	2.4	10.4	0.1	0.6	0.3	1.0	0.0	0.2	1.4	6.3	0.1	0.5
41	3	3.6	17.6	0.9	1.1	0.7	4.3	0.1	0.3	1.9	6.6	0.3	0.5
45	1	9.3	23.4	0.7	2.3	1.6	4.6	0.1	0.5	2.7	8.0	0.1	0.8
45	2	9.4	18.2	2.9	1.9	1.2	2.3	0.3	0.3	3.2	5.4	0.9	0.7
45	3	7.9	13.1	1.9	1.7	1.3	3.1	0.1	0.6	2.4	4.9	0.9	0.7
46	1	3.5	13.9	0.0	2.9	0.4	2.0	0.0	0.4	1.4	5.0	0.0	1.1
46	2	2.4	7.7	0.0	1.9	1.5	5.1	0.0	1.3	1.2	4.1	0.0	1.1
46	3	4.6	8.6	0.0	2.3	1.0	2.6	0.0	0.7	1.2	3.0	0.0	0.8
Range of means		0.2 mm – 9.4 mm				0.3 mm – 4.8 mm				0.9 mm – 4.8 mm			

because the variance between individual patients is quite large. One should notice that the maximum observed motion range was quite large, particularly in the superior/inferior direction, which can most likely be attributed to occasional deep breaths.

1.5 Discussion

We studied the extent of tumor motion as predicted by the correlation model in CyberKnife Synchrony treatments. For lung patients, the predicted tumor motions compare very well with motion ranges cited in previously published data. Even though the motion patterns varied significantly from patient to patient in the lung, the variability within a patient was usually small.

For lower lung (Table 1.1), the mean motion was largest in the SI direction. The maximum motion amplitude in SI can be as much as eight times larger than the mean amplitude for an individual patient. As a next analytical step, we plan to analyze the distribution of motion amplitudes to determine if these large SI amplitudes can be attributed to occasional deep breaths or if they are an indicator of large-amplitude fluctuations within the patients' actual breathing pattern. If the former is true, care should be taken to exclude these unusual breathing patterns for an individual patient when a 4D imaging study is done for treatment planning purposes. If the latter is the case, these patterns need to be included in the study. The LR motion in the lower lung is generally smaller than 2 mm. The mean AP motion amplitudes fluctuate between 0.5 mm and 5 mm between patients. While 6 of the 7 patients in the study showed the expected pattern of a large range of SI motion, smaller AP motion range, and almost negligible LR motion, Patient 33 was the notable exception. The tumor motion range in this patient was of the same order of magnitude in all directions, and the maximum range was three times as large as the mean. So far, we have found no indicators in tumor size, location, medical history or fiducial placement for this patient that could explain the unusual motion pattern.

Our data for middle lung patients is limited to 3 patients. The extent of motion and the fluctuation in each direction for patients 11 and 13 most resemble the patterns seen in the lower lung patients, although with a slightly smaller motion range. Patient 12, on the other hand, could be grouped with the upper lung patients.

As expected, motion ranges in the upper lung were much smaller than in the lower lung. Patients 9, 22, 28, 37, and 43, however, showed large fluctuations. In 2 of our 16 upper lung patients, the motion ranges in all directions were smaller than the stated accuracy for Synchrony motion tracking. For those patients, a non-Synchrony, non-fiducial stereotactic body radiosurgery treatment would have resulted in a similar quality treatment with no risk from fiducial placement. Therefore, 4D imaging studies such as 4D-CT or fluoroscopy should be performed on upper lung patients to determine if fiducial placement is necessary.

For pancreas patients, our data presents one of the largest samples of pancreas motion studied under free breathing. The observed tumor motion was significantly smaller than in the cases presented by Suramo et al. [26] and Bryan et al. [27], while they agree with measurements of Gierga et al. [28]. The reason for the discrepancies between the older and newer studies could not be determined. The motion was most significant in the superior-inferior direction, with the mean motion range being below 10 mm for all cases. Nevertheless, the motion fluctuations in this direction were quite large, in one case reaching 35 mm. The mean motion amplitude in the LR direction exceeded the stated 1.5 mm accuracy of Synchrony in only 7 of the 28 fractions analyzed. In the AP direction, 6 of the 9 patients had a mean motion range exceeding the 1.5 mm tracking accuracy.

There are artifacts in the log file data that had to be manually removed. One of the factors causing these artifacts is misidentification of fiducials in the X-ray images. If a misidentified fiducial cannot be properly localized by changing the imaging parameters and re-analyzing the X-ray, the skin-tumor correlation model has to be reset. Development of direct, continuous, non-ionizing electromagnetic tracking techniques or fiducial-less tumor tracking could potentially alleviate this problem.

1.6 Conclusion

In Radiation Oncology, parameters that characterize patterns of free breathing have not yet been well defined. Our first attempt to define "motion amplitude" for a quasi-periodic signal rests on very simple mathematical principles and will be refined. As we learn more about complex, quasi-periodical patterns, signal-processing methods developed in other fields such as electrical engineering will be integrated in our mathematical toolset.

Our data analysis on the first FDA-cleared technology for real-time respiratory motion tracking shows that the predicted tumor motion, as verified by X-ray images throughout the treatment, agrees well with independently published data in peer-reviewed publications. In addition, we have reported the first large set of observed pancreas respiratory tumor motions for a treatment-equivalent time interval. We have confirmed that the influence of diaphragmatic motion on lower lung and pancreatic tumors creates a dominant SI motion component for these tumors.

In the absence of imaging modalities that can study respiratory tumor motion under comparable patient settings for an extended time period, the method presented is an independent crosscheck of the technology. We therefore conclude that the Synchrony real-time adaptive motion compensation technology is an efficient way to treat moving tumors.

References

1. Nill S, Unkelbach J, Dietrich L, *et al.* Online correction for respiratory motion: evaluation of two different imaging geometries. *Phys Med Biol* 2005; 50:4087–4096.
2. Shirato H, Harada T, Harabayashi T, *et al.* Feasibility of insertion/implantation of 2.0-mm-diameter gold internal fiducial markers for precise setup and real-time tumor tracking in radiotherapy. *Int J Radiat Oncol Biol Phys* 2003; 56:240–247.
3. Poggi mm, Gant DA, Sewchand W, *et al.* Marker seed migration in prostate localization. *Int J Radiat Oncol Biol Phys* 2003; 56:1248–1251.
4. Leter EM, Cademartiri F, Levendag PC, *et al.* Four-dimensional multislice computed tomography for determination of respiratory lung tumor motion in conformal radiotherapy. *Int J Radiat Oncol Biol Phys* 2005; 62:888–892.
5. Underberg RW, Lagerwaard FJ, Cuijpers JP, *et al.* Four-dimensional CT scans for treatment planning in stereotactic radiotherapy for stage I lung cancer. *Int J Radiat Oncol Biol Phys* 2004; 60:1283–1290.
6. Ahn S, Yi B, Suh Y, *et al.* A feasibility study on the prediction of tumour location in the lung from skin motion. *Br J Radiol* 2004; 77:588–596.
7. Plathow C, Ley S, Fink C, *et al.* Analysis of intrathoracic tumor mobility during whole breathing cycle by dynamic MRI. *Int J Radiat Oncol Biol Phys* 2004; 59:952–959.
8. Balter JM, Wright JN, Newell LJ, *et al.* Accuracy of a wireless localization system for radiotherapy. *Int J Radiat Oncol Biol Phys* 2005; 61:933–937.
9. Engelsman M, Sharp GC, Bortfeld T, *et al.* How much margin reduction is possible through gating or breath hold? *Phys Med Biol* 2005; 50:477–490.
10. Sarrut D, Boldea V, Ayadi M, *et al.* Nonrigid registration method to assess reproducibility of breath-holding with ABC in lung cancer. *Int J Radiat Oncol Biol Phys* 2005; 61:594–607.
11. Boda-Heggemann J, Walter C, Mai S, *et al.* Frameless stereotactic radiosurgery of a solitary liver metastasis using active breathing control and stereotactic ultrasound. *Strahlenther Onkol* 2006; 182:216–221.
12. Willoughby TR, Forbes AR, Buchholz D, *et al.* Evaluation of an infrared camera and X-ray system using implanted fiducials in patients with lung tumors for gated radiation therapy. *Int J Radiat Oncol Biol Phys* 2006; 66:568–575.
13. Kini VR, Vedam SS, Keall PJ, *et al.* Patient training in respiratory-gated radiotherapy. *Med Dosim* 2003; 28:7–11.
14. Keall PJ, Cattell H, Pokhrel D, *et al.* Geometric accuracy of a real-time target tracking system with dynamic multileaf collimator tracking system. *Int J Radiat Oncol Biol Phys* 2006; 65:1579–1584.
15. D'Souza WD, Naqvi SA, Yu CX. Real-time intra-fraction-motion tracking using the treatment couch: a feasibility study. *Phys Med Biol* 2005; 50:4021–4033.
16. Schweikard A, Shiomi H, Adler J. Respiration tracking in radiosurgery. *Med Phys* 2004; 31:2738–2741.
17. Zhou TTJ, Dieterich S, Cleary K. A robotic 3-D motion simulator for enhanced accuracy in CyberKnife stereotactic radiosurgery. In: Computer Aided Radiology and Surgery. London, UK: Elsevier; 2004. pp. 323–328.
18. Adler JR, Jr., Chang SD, Murphy MJ, *et al.* The Cyberknife: a frameless robotic system for radiosurgery. *Stereotact Funct Neurosurg* 1997; 69:124–128.
19. Kuo JS, Yu C, Petrovich Z, *et al.* The CyberKnife stereotactic radiosurgery system: description, installation, and an initial evaluation of use and functionality. *Neurosurgery* 2003; 53:1235–1239; discussion 1239.
20. Quinn AM. CyberKnife: a robotic radiosurgery system. *Clin J Oncol Nurs* 2002; 6:149, 156.
21. Yu C, Main W, Taylor D, *et al.* An anthropomorphic phantom study of the accuracy of Cyberknife spinal radiosurgery. *Neurosurgery* 2004; 55:1138–1149.
22. Chen QS, Weinhous MS, Deibel FC, *et al.* Fluoroscopic study of tumor motion due to breathing: facilitating precise radiation therapy for lung cancer patients. *Med Phys* 2001; 28:1850–1856.

23. Mageras GS, Pevsner A, Yorke ED, *et al.* Measurement of lung tumor motion using respiration-correlated CT. *Int J Radiat Oncol Biol Phys* 2004; 60:933–941.
24. Shirato H, Seppenwoolde Y, Kitamura K, *et al.* Intrafractional tumor motion: lung and liver. *Semin Radiat Oncol* 2004; 14:10–18.
25. de Koste JR, Lagerwaard FJ, de Boer HC, *et al.* Are multiple CT scans required for planning curative radiotherapy in lung tumors of the lower lobe? *Int J Radiat Oncol Biol Phys* 2003; 55:1394–1399.
26. Suramo I, Paivansalo M, Myllyla V. Cranio-caudal movements of the liver, pancreas and kidneys in respiration. *Acta Radiol Diagn (Stockh)* 1984; 25:129–131.
27. Bryan PJ, Custar S, Haaga JR, *et al.* Respiratory movement of the pancreas: an ultrasonic study. *J Ultrasound Med* 1984; 3:317–320.
28. Gierga DP, Chen GT, Kung JH, *et al.* Quantification of respiration-induced abdominal tumor motion and its impact on IMRT dose distributions. *Int J Radiat Oncol Biol Phys* 2004; 58:1584–1595.

Respiratory Motion Tracking for Robotic Radiosurgery

2

Sohail Sayeh, James Wang, William T. Main, Warren Kilby, and Calvin R. Maurer, Jr.

CONTENTS

2.1 Abstract

Tumors in the thorax and abdomen move during respiration. One way to manage respiratory motion is to move or shape the radiation beam to dynamically follow the tumor's changing position, an approach that is often referred to as real-time tracking. The Synchrony® Respiratory Tracking System, which is an integrated subsystem of the CyberKnife® Robotic Radiosurgery System (Accuray, Incorporated, Sunnyvale, CA) is a realization of real-time tracking for tumors that move with respiration. Alignment of each treatment beam with the moving target is maintained in real time by moving the beam dynamically with the target. An advantage of the Synchrony system is that patients can breathe normally during treatment while the robotic manipulator moves the linear accelerator dynamically. The primary concept in the Synchrony system is a correlation model between internal tumor position and external marker position. The position of external optical markers, which are attached with Velcro to a snugly fitting vest that the patient wears during treatment, are measured continuously with a stereo camera system. At the start of treatment, the internal tumor position is measured at multiple discrete time points by acquiring orthogonal X-ray images. A linear or quadratic correlation model is generated by fitting the 3D internal tumor positions at different phases of the breathing cycle to the simultaneous external marker positions. An important feature of this method is its ability to fit different models to the inhalation and exhalation breathing phases, which enables accurate tracking even when the tumor or external marker motions exhibit hysteresis. During treatment, the internal tumor position is estimated from the external marker positions using the correlation model, and this information is used to move the linear accelerator dynamically with the target. The model is checked and updated regularly during treatment by acquiring additional X-ray images. This chapter presents the concepts and methods of the Synchrony Respiratory Tracking System. Experimental measurements and retrospective analysis of clinical data show that the accuracy of the Synchrony System is approximately 1.5 mm.

2.2 Introduction

Tumors in the thorax and abdomen, including in particular the lung, liver, pancreas, and kidneys, move up to several centimeters during respiration [1–4]. This intrafraction motion impacts all forms of external beam radiation therapy and is an issue that is becoming increasingly important in the era of image-guided radiotherapy [1]. For example, the European Organization for the Research and Treatment of Cancer (EORTC) guidelines recommend that "an assessment of 3D tumor mobility is essential for treatment planning and delivery in lung cancer" [5]. The recent report of the American Association of Physicists in Medicine (AAPM) Task Group 76 on the management of respiratory motion in radiation oncology recommends that "respiratory management techniques be considered if either of the following conditions occur: a greater than 5 mm range of motion is observed in any direction, or significant normal tissue sparing (as determined by your clinic) can be gained through the use of a respiration management technique" [1]. The report further notes that the 5 mm motion-limit criterion value may be reduced for special procedures such as stereotactic body radiotherapy.

Several approaches have been developed to manage the effect of respiratory motion in radiation oncology [1, 4, 6]. One common approach is to enlarge the clinical target volume (CTV) by adding an internal margin to account for the range of tumor positions during respiration. In addition to the set-up margin which accounts for static external beam alignment uncertainty, this method gives a planning target volume (PTV) for which a treatment plan can be designed [7]. The variation in target position associated with breathing can be determined by examining the range of target motion with fluoroscopy. A related motion-encompassing approach is the slow-scanning method, in which the CT scanner is operated very slowly, or multiple CT scans are averaged such that multiple respiration phases are recorded per slice [8]. The image of the tumor should show the full extent of the respiratory motion that occurred during the scanning process, provided the acquisition time at each couch position is longer than the breathing cycle. The disadvantage of slow-scan methods is the loss of resolution due to motion blurring, which potentially leads to larger observer errors in tumor and normal organ delineation, as well as estimated dose delivered to the patient. Another disadvantage is the increased dose from slow CT scanning compared with conventional CT scanning. Another solution to obtaining a tumor-encompassing volume is to acquire CT images during the end-exhale and end-inhale phases of a breathing cycle [9]. The resulting pair of CT images provides information about the lower and upper limits of target position, which is used to produce an anisotropic enlargement of the target volume. Finally, the tumor range of motion and a motion-encompassing PTV can be determined using a 4D-CT image study [10]. Although compensating for respiratory motion by increasing target volume margins can ensure that tumor motion during breathing will not affect the dose delivered to the target, it will also lead to increased dose delivered to normal tissues, which can be a particular problem when the lesion is located close to organs at risk.

Another set of approaches attempt to minimize the margin by delivering radiation when the tumor is at a relatively fixed and reproducible position. Breath-hold methods minimize the effects of breathing motion on radiation treatment by means of tumor immobilization [1]. Breath holding has long been used in diagnostic radiology to reduce the blurring of images. For radiation therapy, the goal is to attain the same breath-hold position between beams delivered during a single treatment fraction, and also between treatment fractions. Treatment is delivered during the breath-hold period. Because various ventilation muscles may or may not be involved in normal breathing, and because the tidal volumes between breaths are variable, it is difficult for the patient to achieve a reproducible breath-hold position during normal ventilation. Thus breath-hold methods are typically applied at normal end-exhalation [11, 12] or deep inhalation [13, 14]. Deep inspiration actively recruits all ventilation muscles to expand the lungs, whereas the lung volume is at its most neutral state at the end of normal exhalation [15]. More sophisticated methods such as active breathing control facilitate a reproducible breath-hold position at a phase of the normal breathing cycle [16, 17]. These methods provide higher accuracy of dose delivery than motion-encompassing methods, but breath holding is physically demanding and uncomfortable, and breath-hold repeatability and patient compliance are challenges, especially for elderly patients or patients with compromised pulmonary capacity, which is often the case for patients with lung cancer or other pulmonary disease [1]. Thus breath-hold methods may not be applicable to a significant population of patients (one study reported that a pre-selection process was required even with active breathing control, resulting in a 33% rejection rate [11]). For respiratory gating approaches, the patient continues breathing normally. The radiation beam is turned on only within a specified portion of the patient's breathing cycle, which is commonly referred to as the "gate." The position and width of the gate are determined by monitoring the patient's respiratory motion using either an external respiration signal or internal fiducial markers. Both displacement-based and phased-based approaches are used [18]. Typically the gate is chosen over a range of the breathing cycle during which the tumor motion is minimal (such as at exhale) or the lung volume is maximal (such as at inhale). The delivery of radiation during a limited portion of the breathing cycle can substantially reduce the duty cycle (the ratio of the gate width to the respiratory cycle period) and thus increase the treatment time. The duty cycle is typically about 25%. Lesion motion and gating model stability, which can adversely impact the planned dose distribution, are also challenges for gating methods [19].

2.3 Respiratory Motion Tracking

Another way to manage respiratory motion is to move or shape the radiation beam dynamically to follow the tumor's changing position, an approach that is often referred to as real-time tracking [1]. Continuous real-time tracking can potentially eliminate the need for or substantially reduce the size of the tumor-motion margin added to the CTV while maintaining a 100% duty cycle for efficient dose delivery.

Real-time tracking requires a method to move or shape the radiation beam relative to the moving target. For photon beams there are three main ways to achieve this: move the patient using the treatment couch, change the aperture of the collimator, and move the beam by physically repositioning the radiation source (e.g., linear accelerator (LINAC)) [1, 6]. For charged-particle beams, the beam can also be redirected electromagnetically. Treatment couch-based motion compensation has been studied and shown to be feasible for making intermittent adjustments to the patient's position [20]. More recently, robotic couch-based motion tracking has been shown to be technically feasible for real-time compensation of intra-fraction respiratory motion [21]. However, continuous couch motion associated with real-time respiratory motion tracking has the practical issues of patient comfort and treatment tolerance. Alternatively, the beam can be effectively moved by changing the aperture of the collimator. The technical feasibility of this approach has been demonstrated for a multileaf collimator (MLC) [22–24]. However, there are several potential limitations to this approach. The constant motion for real-time tumor tracking causes additional wear-and-tear on the MLC hardware. For intensity-modulated radiation therapy (IMRT), the MLC motion required for target tracking superimposes on that required for intensity modulation, increasing the chances of exceeding the physical speed limitations of the MLC. If the MLC is part of a conventional gantry-based LINAC system, beam alignment with the moving target can be maintained only in the plane of the treatment field. Finally, MLC-based tracking is limited in resolution in one direction by the leaf width, and there may be a trade-off between setting the MLC orientation to obtain the maximum conformality to the target volume and setting the orientation to align the direction of greatest MLC resolution along the primary axis of motion.

The other approach is to physically reposition the radiation source to follow the tumor's changing position. The remainder of this chapter discusses a realization of this approach by describing the concepts and methods of the Synchrony Respiratory Tracking System.

2.4 Synchrony Respiratory Tracking System

2.4.1 Overview

The CyberKnife Robotic Radiosurgery System moves the radiation beam by physically repositioning the radiation source [25]. A miniature lightweight 6 MV X-band LINAC is mounted to an industrial multi-jointed robotic manipulator that can move freely and accurately aim the radiation beam with six degrees of freedom. Two digital X-ray imaging systems are orthogonally configured. The X-ray generators and amorphous silicon X-ray detectors are rigidly fixed so that their projection camera geometry is calibrated and known with high accuracy. Computer algorithms automatically compare the projection images of the target region with the patient's treatment planning CT image.

Using image registration methods, the target position is computed and sent to the robotic manipulator, which maintains the alignment of each treatment beam with the moving target by moving the beam dynamically with the target. The Synchrony Respiratory Tracking System is an integrated subsystem of the CyberKnife System that allows irradiation of extracranial tumors that move due to respiration. An advantage of the Synchrony system is that patients can breathe normally during treatment and there is no reduction in the duty cycle as there is for gated approaches.

Detecting the tumor position is one of the most important and challenging tasks in real-time respiratory motion tracking. The relatively high speed of breathing-induced target motion requires near-continuous target position information. Ideally, the tumor itself can be tracked directly and continuously. This can be accomplished by implanting gold fiducial markers in or around the tumor and tracking the fiducials using multiple fluoroscopes [19, 26]. This approach has provided detailed information about the motion trajectories of lung tumors [2, 3]. However, although the high radio-opacity of gold fiducials makes them detectable at relatively low exposures, continuous imaging delivers a skin dose of up to 2 cGy/min of tracking [26]. For lengthier treatments such as hypofractionated stereotactic body radiotherapy and radiosurgery, the additional radiation dose from direct and continuous tracking with fluoroscopy is an issue that needs to be considered.

In order to reduce radiographic imaging exposure, hybrid techniques have been proposed that combine episodic radiographic imaging with continuous measurement of an external breathing signal [6, 27–30]. The Synchrony Respiratory Tracking System uses such an approach. Respiratory motion tracking is based on a correlation model between internal tumor position and external marker position. At the start of treatment, the internal tumor position is measured at multiple discrete time points by acquiring orthogonal X-ray images. A linear or quadratic correlation model is generated by fitting the 3D internal tumor positions at different phases of the breathing cycle to the simultaneous external marker positions. During treatment, the internal tumor position is estimated from the external marker positions using the correlation model, and this information is used to move the linear accelerator dynamically with the target. This concept is illustrated in Fig. 2.1. The model is checked and updated regularly during treatment by acquiring additional X-ray images.

A schematic block diagram of respiratory motion tracking in the Synchrony system is shown in Fig. 2.2. There is a separate correlation model for each external marker. The external marker positions are measured continuously and input to the corresponding correlation models. Each model provides an estimate of the target position from the external marker variable. The individual estimates are averaged to get the final estimate of the target position. This value represents the position of the target at the present time. Ideally, this value can be sent to the robotic manipulator as a position command without any delay. However, communication latencies and robotic manipulator and LINAC inertia cause delays; if the present time estimate of target position is sent to the robot, there will be a lag in the robot manipulator's motion. A predictor is used that will, using the history of the target movement, compensate for the delays in the system. The predictor is adaptive and is designed to have a quick response to changes in the breathing pattern and target movement. Finally, the output of the motion predictor is passed through a smoothing filter before it is sent to

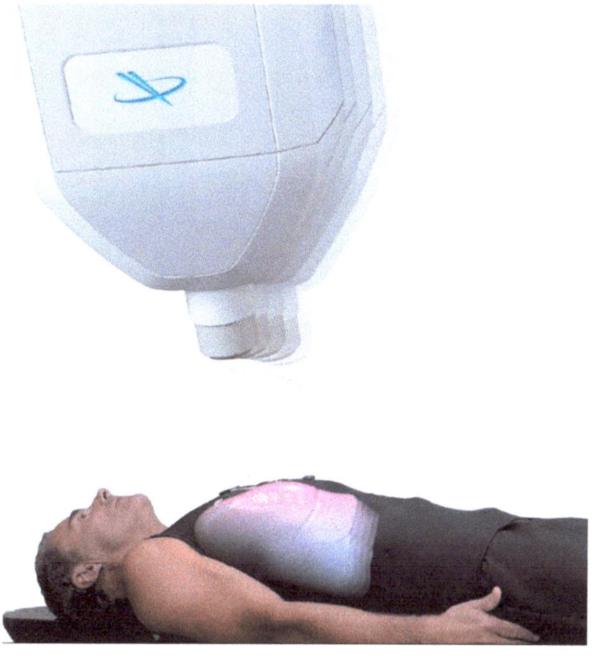

Fig. 2.1 Illustration of the Synchrony Respiratory Tracking System. Respiratory motion tracking is based on a correlation model between internal tumor position and external marker position. During treatment, the internal tumor position is estimated from the external marker positions using the correlation model. This information is used to move the linear accelerator dynamically with the target.

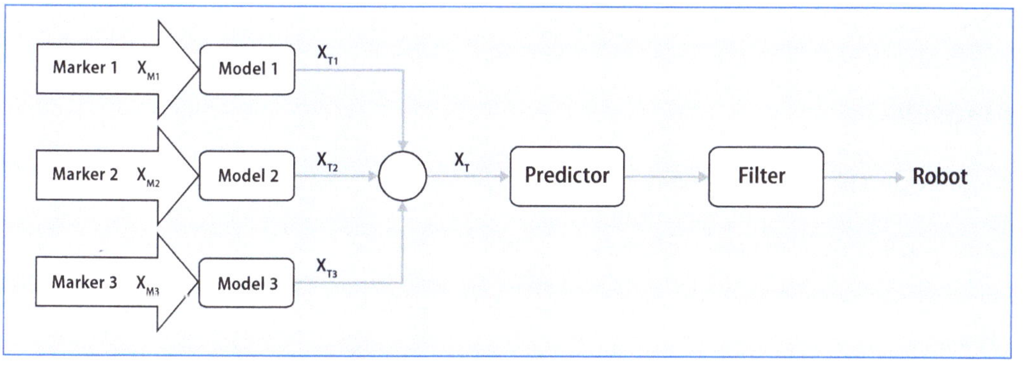

Fig. 2.2 Schematic block diagram of the Synchrony Respiratory Tracking System. For each external marker, there is a correlation model between the position of the internal target and the position of the external marker. The outputs of the individual models are averaged to obtain the present time estimate of the target position. A predictor is used to compensate for communication latencies and robotic manipulator inertia. Finally, the predicted position is filtered and sent to the robotic manipulator as a position command.

the robot as a position command. Each one of the blocks depicted in Fig. 2.2 will be explained in more detail in the following sections.

2.4.2 External Markers

The Synchrony system uses external optical markers to provide a breathing signal. Three markers are attached to a snugly fitting vest that the patient wears during treatment (Fig. 2.3). The marker positions reflect the chest wall position. Light-emitting diodes (LEDs) transmit light through optical fibers that terminate at the cylindrical optical marker. This approach was chosen over directly attaching LEDs to the vest to avoid the presence of copper wire in the tracking X-ray images. The markers are sequentially strobed and a stereo camera system, which consists of three linear charge-coupled device (CCD) detector arrays, measures the 3D marker positions continuously at a frequency of approximately 30 Hz.

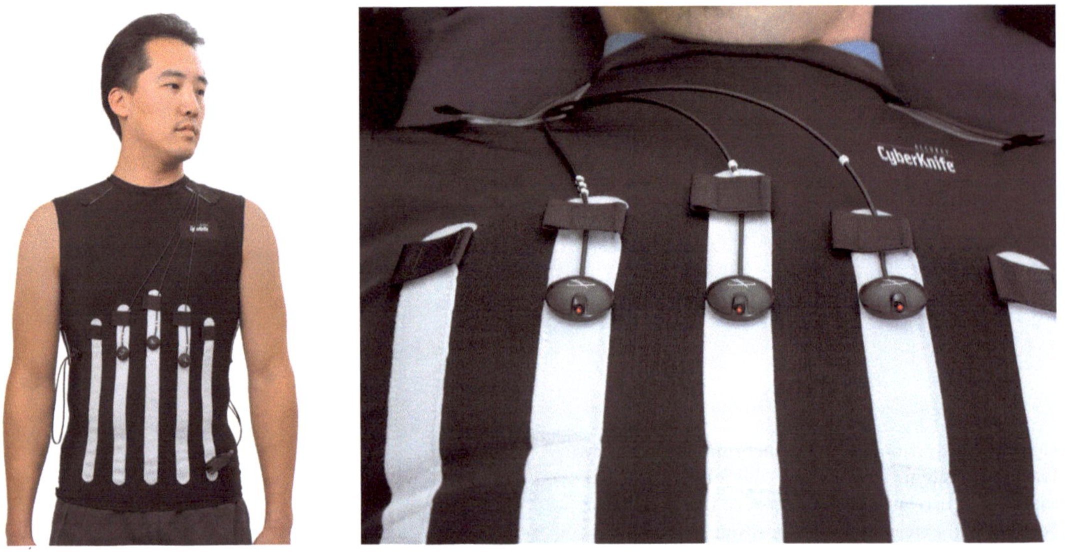

Fig. 2.3 External optical markers used by the Synchrony system to provide a breathing signal. Three markers whose positions reflect the chest wall position are attached with Velcro to a snugly fitting vest that the patient wears during treatment.

2.5 Correlation Model

There is a separate correlation model for each external marker. Each model provides an estimate of the internal target position from the external marker position:

$$\mathbf{X}_{\mathrm{T}i} = f_i\,(\mathbf{X}_{\mathrm{M}i}) \qquad (1)$$

where $\mathbf{X}_{\mathrm{M}i}$ is the position vector of the ith marker and $\mathbf{X}_{\mathrm{T}i}$ is the position vector of the target estimated from the ith marker. Each marker has an independent function f_i that maps the marker's position to the target position. When more than one marker is used, which is almost always the case, the Synchrony system averages the individual target position estimates from all active and visible markers to obtain the final estimate of the target position.

Each point on the chest wall moves on an approximately linear trajectory (Fig. 2.4). Because of this, the 3D vector position of each external marker $\mathbf{X}_{\mathrm{M}i}$ can be replaced by the distance along the marker's principal axis of motion. This distance is the projection of the position vector on the principal axis of motion. This simplification reduces the complexity of the correlation function f_i and reduces correlation

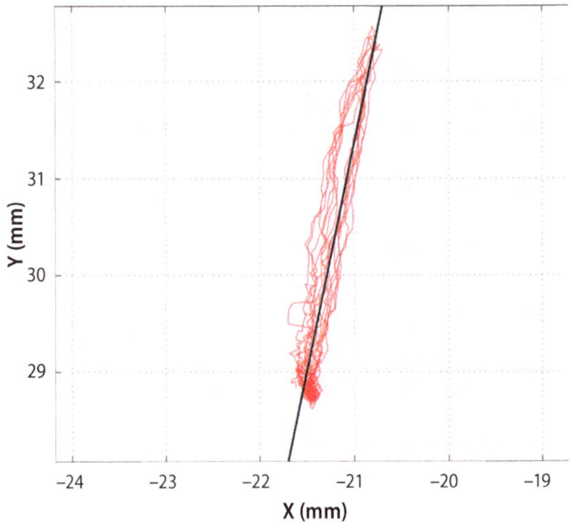

Fig. 2.4 Typical external marker motion for a Synchrony patient. The red curves in this plot show the positions of an external optical marker placed on the chest of a patient during eight breathing cycles. The 3D vector positions have been projected onto a plane. The vertical graph direction (y-axis) corresponds to the anterior-posterior direction of the patient. The horizontal direction (x-axis) corresponds to the left-right direction. The black line is the principal axis of motion. Because the external markers move on approximately linear trajectories, the 3D vector position is replaced by a scalar value that is the projection of the position vector on the principal axis.

model noise. Using this simplification, the correlation model can be written as:

$$\mathbf{X}_{Ti} = f_i(r_i) \tag{2}$$

where r_i represents the distance traveled along each marker's principal axis of motion.

The initial release of the Synchrony system used only a linear function for f_i:

$$\mathbf{X}_{Ti} = \mathbf{A}r_i + \mathbf{B} \tag{3}$$

The linear coeficients **A** and **B** are vector-valued quantities. The correlation function f_i is effectively three correlation models, one for each component of motion. The linear correlation model is easy to build and robust. Though it is a relatively simple model, it provides accurate tracking results for many patients.

During normal breathing, the deflating lung volume is larger than the inflating volume at the same transpulmonary pressure [15]. This is called hysteresis and is attributable to the complex respiratory pressure-volume relationship of the lung and chest wall. Hysteresis in 3D tumor trajectories was observed in half of the patients in one study [2]. The hysteresis was a 1–5 mm separation of the trajectories during inhalation and exhalation. Hysteresis is one source of nonlinearity in the correlation between external markers and an internal target. Even if the motions of the markers and target follow linear trajectories, the correlation between the marker and target positions will be nonlinear if there is a phase difference (time delay) between their motions (Fig. 2.5). Phase differences up to 30 degrees have been observed in several studies [3, 31–33]. To address the cases where the linear correlation model is not adequate, two nonlinear forms of the model were introduced in 2005. One is a simple curvilinear form that introduces higher order terms in Eq. (3). This form of the model allows for nonlinear correlation of the external marker and tumor that is the same for inhalation and exhalation. The most general correlation model available in the Synchrony system is a dual-curvilinear form where two polynomials are used to separately model the inhalation and exhalation phases of the breathing cycle:

$$\mathrm{X}_{Ti} = \begin{cases} \sum_{j=0}^{N} \mathrm{A}_j^+ r_i^j & \dot{r} \geq -\dot{r}_{min} \\ \sum_{j=0}^{N} \mathrm{A}_j^- r_i^j & \dot{r} \leq \dot{r}_{min} \end{cases} \tag{4}$$

The phase of the breathing cycle can be determined from the external marker velocity $\dot{r}$. The current version of the Synchrony System allows $N = 1$, which is the linear model in Eq. (3), and $N = 2$, which is

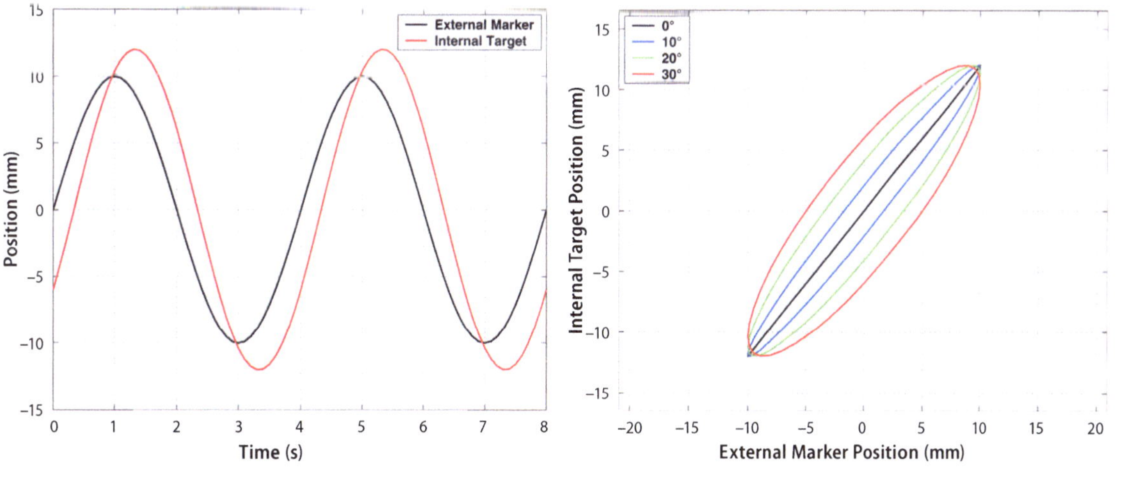

Fig. 2.5 Illustration of phase difference and its effect on correlation between external marker and target positions. The left graph shows sinusoidal motions of an external marker (*black*) and internal target (*red*). There is a 30 degree phase difference (time delay) between their motions. The right graph shows the nonlinear correlation between the external marker and target positions for multiple phase differences between their motions. The marker and target each move on a linear trajectory. If the phase difference is zero, the correlation between the marker and target positions is linear (*black*). As the phase difference between their motions increases, the correlation between their positions becomes increasingly nonlinear (*blue, green, red*).

a quadratic model. The nonlinear models provide flexibility and potentially more accurate representation of the internal movement. These models can model hysteresis as well as nonlinear correlation due to phase differences between external marker and target motion. However, relative to the linear correlation model, these models require additional data points to be defined, and due to their nonlinear form they are more sensitive to the distribution of the data points.

2.5.1 Building the Correlation Model

At the start of treatment, the internal tumor position is measured at multiple discrete time points by acquiring orthogonal X-ray images. The correlation model is generated by fitting the 3D internal tumor position at different phases of the breathing cycle to the simultaneous external marker position (the distance along the marker's principal axis of motion). The target position is determined by automatically detecting fiducial markers, which are implanted in or near the tumor, in orthogonal X-ray images. Alternatively, for some lung tumors that are visible in the X-ray images, the target position can be determined by direct tumor tracking using the Xsight™ Lung Tracking System, which recently became available for the CyberKnife system.

Each successful image acquisition yields a set of spatial coordinates for the target. The target position plus the time at which the image pair was acquired is sent to the Synchrony system computer. The continuously measured external marker positions are stored in a buffer. The image acquisition time is used to find the corresponding position and speed of all the markers. For each marker, a data point consisting of the marker position, marker velocity, and target position is added to its data set.

As each new data point is added, the parameters for each correlation model type are computed using a least-squares fit of all available data points. The three models considered are linear, single quadratic, and dual quadratic. Since the linear model requires the fewest data points, is less sensitive to the distribution of data points, and is generally more robust than the other model types, it is favored over the other types if the standard error of the model is small enough. If the standard error for the linear model is less than 1 mm, the other model types are not evaluated. If the error is larger than 1 mm, parameters for all other model types are determined and the model errors computed. The modified standard error of the model can be written as:

$$e_{comp} = \frac{1}{n-m}\sqrt{\sum_{k=1}^{n} e_k^2} \qquad (5)$$

where e_k is the distance between the model and data point k, n is the total number of data points, and m is the minimum number of data points needed to define the particular model mathematically. This error value is sometimes referred to as the degrees-of-freedom adjusted standard error because m is the degrees of freedom of the model. For the linear model, $m = 2$. For the dual quadratic model, $m = 2 \times 3 = 6$. The Synchrony system requires more data points than the theoretical minimum: a model type is a candidate only if $n > m$. Thus the minimum number of data points required to be a candidate model is 3 for a linear model and $2 \times (3+1) = 8$ for a dual quadratic model. In practice, 5–6 data points are typically used to build a linear model and 10–12 points to build a dual quadratic model. The model with the smallest modified standard error is generally chosen. However, sometimes the linear model is chosen over a nonlinear model even though the linear model has a slightly larger error. This is due to the fact that the errors are weighted before they are compared, which is done to slightly favor the linear model in the selection process.

Each component of motion has its own correlation model. The motion component models are determined independently. It is not uncommon for one component (e.g., anterior-posterior or left-right) to have a linear model while another component (e.g., superior-inferior) has a dual quadratic model.

The parameters for the linear model are always saved, even if a different model was chosen. The linear model is used for blending the two curves for the dual quadratic model as well as estimating target positions for values of r that are slightly outside the range of the data points. Blending of the curves and handling of out-of-range values of r will be discussed in the next section.

2.5.2 Using the Correlation Model

The primary use of the correlation model is to estimate the real-time target position from the continuously measured external marker positions. This information is used to move the linear accelerator dynamically with the target (Fig. 2.2). The details of the process are slightly different depending on the type of correlation model used. For each of the markers:

- If the marker is visible, its 3D vector position $\mathbf{X}_{Mi}$ is acquired.
- If the marker is enabled, its scalar position r_i along its principal axis of motion is computed by projecting the vector position on the principal axis. The velocity $\dot{r}_i$ of the marker is also calculated.
- If r_i is within the range of the data points used to build the correlation model, the target position $\mathbf{X}_{Ti}$ is computed from the equation of the model. If the model is linear, the rest of this section is skipped and the calculated position $\mathbf{X}_{Ti}$ is the final estimate of the target position for this marker.
- If r_i is outside the range of the data points used to build the correlation model, the linear model is used to compute $\mathbf{X}_{Ti}$ and the rest of the section is skipped.
- If the model is dual quadratic and the external marker velocity $\dot{r}_i$ has a very small value, this indicates that the target position is near end-inspiration or end-exhalation, which corresponds to the boundary between the two branches of the model function. At these points in the respiratory cycle the two branches of the function are evaluated and averaged. This provides a blending mechanism for the ends of the two branches that may not necessarily meet at the same point in space.
- Additional blending is performed if the value of r_i is very close to the boundaries of model. This is accomplished by blending the value obtained from the previous step with the output of a linear model. This step ensures a smooth transition from a linear to a dual quadratic model. The handling of out-of-range values of r and the blending of curves for the dual quadratic model is illustrated in Fig. 2.6.

The above steps are repeated continuously for all active external markers. A marker is considered active if it is visible to the Synchrony system camera and enabled. Markers are enabled by default and must be explicitly disabled by the user. Markers should be disabled for a variety of reasons, including small marker motion compared to the internal target motion, high noise level of the marker position signal (small signal-to-noise ratio), and incomplete visibility during the respiration cycle. The individual marker estimates are averaged to provide the final estimate for the target position $\mathbf{X}_T$.

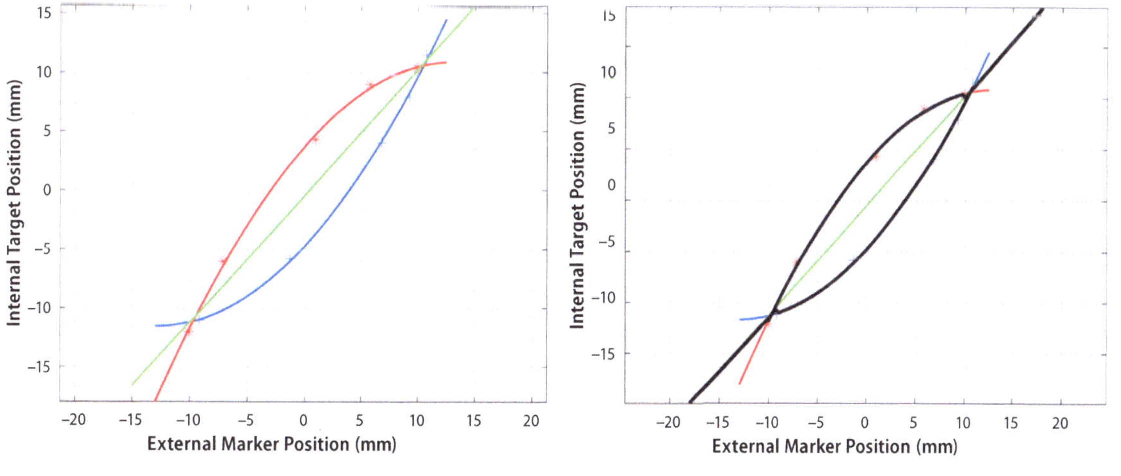

Fig. 2.6 Estimation of target position using the dual quadratic correlation model. The left graph shows the individual quadratic models for the inhalation and exhalation phases (*blue* and *red*) as well as the linear model (*green*). If the external marker position is outside of the range captured during the building of the correlation model, the target position is estimated using the linear model to prevent large extrapolation error that is possible with a quadratic fit. If the marker position is near the boundary between the two branches of the correlation model, the two branches are evaluated and averaged. The right graph shows the overall correlation model (*black*).

2.5.3 Checking and Updating the Correlation Model

Inter- and intra-fraction changes in position and motion are common and well known [1–3]. A correlation model is generated at the beginning of every treatment, which addresses the issue of inter-fraction variability. However, the target position and motion typically changes during the treatment. This could be caused by gradual patient relaxation throughout the treatment period. In the lung, this could be attributed to gravity action on compliant lung tissue. Thus it is important to regularly check and update the correlation model during treatment. This is accomplished in the Synchrony system by acquiring additional X-ray images. In practice, additional X-ray images are typically acquired every 1–5 min.

When a new X-ray image pair is acquired, the time of the image acquisition is used to find the corresponding position of the external markers. The marker positions are used to compute the predicted target position from the correlation model. This information is first used by the target localization software to provide a better initial estimate for the automatic detection of the target position in the X-ray images. This step reduces the number of unsuccessful target detections. If the target position in the X-ray images is successfully determined, the model-based estimated target position is compared with the image-based actual position. The correlation model error, which is the distance between the estimated and actual positions, is computed and displayed in a graph including the previous prediction errors. If the error is larger than a predefined value, the treatment is paused and the user is informed about the discrepancy; the model can then be checked with additional X-ray image acquisitions or completely regenerated. If the correlation model accuracy is adequate, the newly acquired data point is used to update the model as described in the previous section. Thus the correlation model adapts to gradual changes in target position and motion during the treatment. The maximum number of data points for a model is 15. If there are already 15 data points when a model is updated, the most recently acquired data point is added and the oldest data point is deleted (a first-in, first-out strategy).

2.6 Target Motion Prediction

The correlation model provides a spatial model between external marker and internal target positions. It estimates the current position of the target. Positioning the radiation source at this position cannot happen instantly. The time between acquisitions of the external markers is 35 ms. Error in the surrogate breathing signal (external marker positions) introduces error in the estimated target position. Thus the correlation model output needs to be passed through a filter to smooth out uncertainty in the estimated target position. Communicating the position data to the robot controller introduces a delay. The robot controller performs its own filtering of the position commands. Robotic manipulator and LINAC inertia introduce further delays. The total time delay of the current version of the Synchrony system is approximately 115 ms. For a target that is moving at 2 cm/s, this time delay will cause a lag in the robotic manipulator's position of up to 2 mm behind the actual target position.

Target motion prediction is used to compensate for this time delay and thereby minimize lag in the robot manipulator's motion. Although the breathing cycle is nominally periodic, the motion prediction problem is complicated by the fact that a typical breathing cycle varies in amplitude and frequency from one cycle to the next [34]. An example of this cycle-to-cycle variation is illustrated in Fig. 2.7. However, the variations are not purely random, which means that in principle it should be possible to predict future respiratory motion from past motion. Both model-based [2, 23] and adaptive filter [35, 36] approaches have been used. Adaptive filters, which are commonly used in control processes where the input signals are complex and variable, predict the next value of a signal based on a running sample of past values.

The Synchrony system compensates for time delays in the system by using an adaptive predictor that provides an estimate of future target position given the past patterns of target motion. The predictor is designed to respond quickly to changes in the breathing pattern and target motion. It relies on heuristic rules and tuned parameters to accomplish its task. The current implementation uses a window

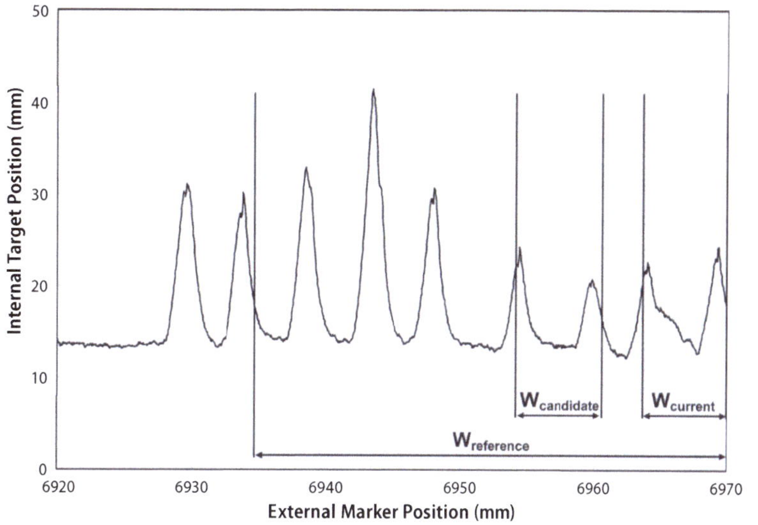

Fig. 2.7 Illustration of the adaptive predictor used by the Synchrony system to compensate for time delays in the system. The predictor provides an estimate of future target position given the past patterns of target motion. An explanation of the window pattern-matching algorithm is given in the text.

pattern-matching algorithm. The approach is illustrated in Fig. 2.7. A large window that stores a running sample of past target positions over many breathing cycles is taken as the reference data ($W_{reference}$). A small window that contains part of the last breathing cycle is considered as the current data ($W_{current}$). For every sample, a small window with identical size to $W_{current}$ called $W_{candidate}$ is moved across the larger window (from the newest data to the oldest). Error metrics are used to assess how well the contents of $W_{candidate}$ match with $W_{current}$. When the best match is found, the pattern following that window is used to predict the future target position after the current window $W_{current}$. This matching process is repeated at every sampling time.

2.7 Treatment Delivery Accuracy

The first reported study of the Synchrony Respiratory Tracking System was performed in 2004 by physicists at three CyberKnife centers [37]. They conducted an "end-to-end" test procedure in which a CT scan was acquired of a test object, a treatment plan was designed to deliver a spherical dose distribution to a target within the object, the treatment was delivered, and the dose distribution centroid was measured on radiochromic dosimetry film placed inside the object. The total system error was computed as the distance between the centroids of the planned and delivered dose distributions and thus represents all possible errors in the treatment planning and delivery process including error in the tracking system, the CT scanner, the treatment planning software, the robot, and the LINAC. Programmable motion tables were used to simulate respiratory motion of the object and the external optical markers. The motion patterns reproduced extreme examples of the motion measured by real-time fluoroscopic examination of lung tumors with implanted fiducial markers [2, 3]. Specifically, the amplitude of motion was 25, 8, and 3 mm for superior-inferior, anterior-posterior, and left-right directions, respectively; the motion pattern was a $\sin^4(\omega T/2)$ waveform; the period was 3.6 s; and the phase difference between the object and marker motions was 0, 15, and 30 degrees for different experiments. Relative to a static treatment case, the mean error observed during treatment with the Synchrony system across all motion patterns was 0.7 ± 0.3 mm.

System improvements have been made since this study was conducted, including the introduction of nonlinear correlation models which improve the tracking accuracy for motions involving non-zero phase differences. A more recent study performed using nearly identical motion amplitudes, waveforms, and phase differences but with an improved version of the Synchrony system reported total system errors less than 1 mm for all measurements made using respiratory motion tracking [38]. Accuray recently released the Synchrony motion table, which was developed for performing quality assurance of Synchrony respiratory motion tracking. The waveform is approximately $|\sin^3(\omega T/2)|$ with a single linear axis target platform excursion of 25 mm and a platform for the external optical markers that moves orthogonal to the target motion through a distance of about 10 mm. The phase difference between the target and marker motions may be adjusted zero to 180 degrees. Preliminary results from three of six centers participating in a multi-institutional study of Synchrony respiratory motion tracking accuracy using this quality assurance motion table show a total system error (distance between the centroids of the planned and delivered dose distributions) of less than 0.5 mm for phase differences of 0, 10, and 20 degrees and less than 1 mm for 30 degrees [39].

In addition to causing a shift in the centroid of the dose distribution, respiratory motion may blur the dose distribution (i.e., reduce the steepness of the dose gradient around the target). This effect was studied using the technique described above. The distance between the 20% and 80% isodose lines (normalized to maximum dose) was measured in the superior-inferior direction (the axis of greatest motion) at the edges of the target [38]. Motion-induced blurring was quantified by the change in the 20–80% distance for treatments with and without motion. When using the Synchrony system to track a target and correct a target with linear motion, no additional blurring resulted; tracking and correcting a target with extreme nonlinear correlation (30 degree phase difference between object and external optical marker motions) resulted in 1 mm blurring (Fig. 2.8). This compares to dose blurring of more than 8 mm when no respiratory

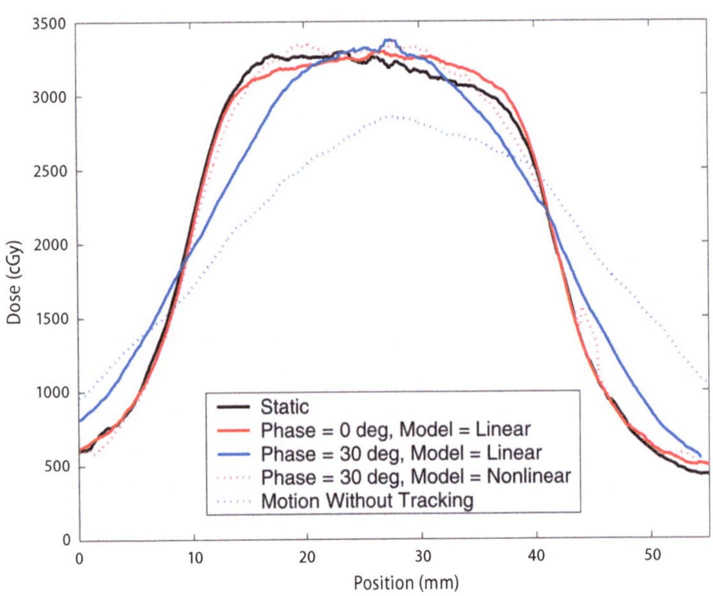

Fig. 2.8 Dose profiles for simulated respiratory motion with and without using the Synchrony System [38]. A programmable motion table was used to simulate respiratory motion of a test object and external optical markers. The motion patterns reproduced extreme examples of the motion measured by real-time fluoroscopic examination of lung tumors with implanted fiducial markers [2, 3]. The amplitude of motion was 25, 8, and 3 mm for superior-inferior, anterior-posterior, and left-right directions, respectively; the motion pattern was a $\sin^4(\omega T/2)$ waveform; the period was 3.6 s; and the phase difference between the object and marker motions was 0 (linear correlation between object and marker motion) or 30 degrees (nonlinear correlation) for different experiments. The total system error (distance between the centroids of the planned and delivered dose distributions) was less than 1 mm for all measurements made using respiratory motion tracking. Dose profiles were measured in the superior-inferior direction (the axis of greatest motion). Relative to the static treatment case (*black line*), motion without using tracking causes very substantial blurring (*blue dotted line*, the 20–80% width is not measurable). The dose profiles show that the linear correlation model works well for linear motion without a phase difference between object and marker motions (*red line*, 20–80% width is same as for static case) and the nonlinear (dual quadratic) correlation model works well with a phase difference of 30 degrees (*red dotted line*, 20-80% width is 1 mm greater than static case). The dose profile is relatively more blurred using linear correlation model with a phase difference of 30 degrees (*blue line*, 20–80% width is 6 mm larger than static case).

tracking was used. The CyberKnife team at Erasmus Medical Center–Daniel den Hoed Center, Rotterdam, The Netherlands, conducted a similar experiment using a treatment plan with far greater isodose line complexity than typically encountered clinically. Treatment was delivered using simulated respiratory motion with and without Synchrony respiratory tracking; the amplitude of motion was 20 mm, the motion pattern was a $\sin^2(\omega T/2)$ waveform, and the period was 7 s. The results show isodose line agreement to be generally better than 1 mm, with a maximum displacement of 2 mm (Fig. 2.9).

Most of the reported testing of the Synchrony system has been performed using motion phantoms. There has also been some retrospective analysis of clinical data. As mentioned previously, during a CyberKnife radiosurgery treatment with the Synchrony system, the correlation model is checked and updated regularly by acquiring additional X-ray images. The correlation model error, which is the distance between the model-based estimated and image-based actual positions, is computed, displayed, and stored in a log file. This error is a measure of the accuracy of Synchrony tracking in actual clinical application. The log files from 14 treatments delivered at three CyberKnife centers using the initial release of Synchrony, in which only the linear correlation model was available, were collected and analyzed [40]. The average of 510 correlation model error values contained in these log files was 1.4 ± 1.0 mm (mean ± standard deviation). More recently, Seppenwoolde et al. [41] examined the correlation model error for eight lung cancer patients treated with respiratory gating [19]. All of these patients had simultaneous and continuous recordings of internal tumor and external marker positions. This data was used to simulate a CyberKnife treatment with Synchrony tracking. The continuous internal tumor position data was used to compute the continuous correlation model error. Although the published abstract does not include quantitative results, the authors concluded that the Synchrony system reduced errors due to breathing motion "largely and consistently over treatment time for all studied patients."

2.8 Conclusion

The CyberKnife Robotic Radiosurgery System with the integrated Synchrony Respiratory Tracking System is the only currently available radiation delivery

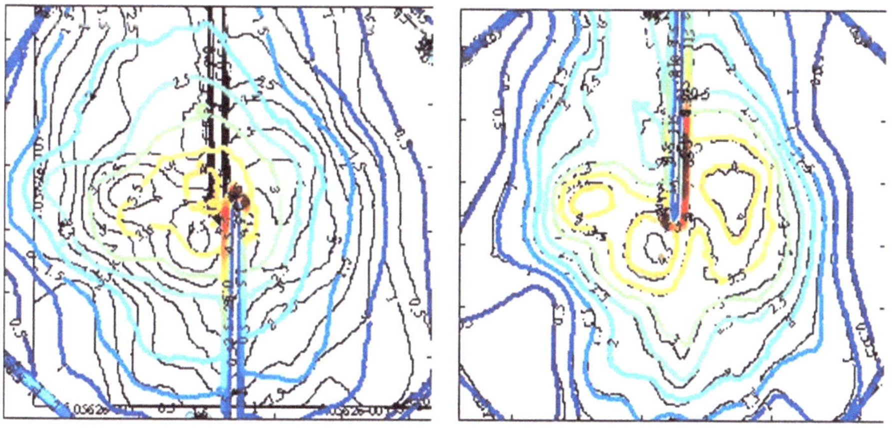

Fig. 2.9 Dose distributions for simulated respiratory motion with (*right*) and without (*left*) using the Synchrony System. The underlying black curves are isodose lines generated by the treatment planning system. The colored curves are the same isodose lines measured after treatment delivery with respiratory motion. The isodose lines are displayed in 0.5 Gy increments, with the highest being 4.0 Gy. The scale is defined by the length of a side of the square dosimetry film, which is 6.3 cm. The amplitude of motion was 20 mm, the motion pattern was a $\sin^2(\omega T/2)$ waveform, and the period was 7 s. The results show isodose line distance to agreement to be generally better than 1 mm, with a maximum displacement of 2 mm. (Images courtesy of Drs. J. P. A. Marijnissen and Y. Seppenwoolde, Erasmus Medical Center–Daniel den Hoed Center, Rotterdam, The Netherlands.)

platform that offers real-time tracking for tumors that move with respiration. Alignment of each treatment beam with the moving target is maintained in real time by physically repositioning the radiation source, which is a LINAC mounted on a multi-jointed robotic manipulator, to follow the target. Respiratory motion tracking combines episodic radiographic imaging with continuous measurement of external optical markers; it is based on a correlation model between internal tumor position and external marker position that is generated by fitting the internal tumor positions at different phases of the breathing cycle to the simultaneous external marker positions. Inter-fraction variability in tumor position and motion is addressed by building the model at the start of every treatment. The model is checked and updated regularly during treatment by acquiring additional X-ray images, thereby adapting to intra-fraction changes in respiratory motion. Reported experimental measurements and retrospective analysis of clinical data demonstrate that the accuracy of Synchrony tracking is approximately 1.5 mm. The Synchrony system is a convenient and practical alternative to other methods for managing respiratory motion. Continuous real-time tracking and treatment maintains a 100% duty cycle for efficient dose delivery. The patient breathes normally throughout the treatment without the need for breath holding, which can be physically demanding and uncomfortable for elderly patients or patients with compromised pulmonary capacity as is often the case for patients with lung cancer. As of April 2007, the Synchrony system is installed on more than 80 CyberKnife systems worldwide and has been used to treat more than 2000 patients with tumors in the lung, liver, pancreas, and kidney. Several other chapters in this volume discuss the clinical experience with this technology.

References

1. Keall PJ, Mageras GS, Balter JM, *et al. Report of AAPM Task Group 76: The Management of Respiratory Motion in Radiation Oncology.* American Association of Physicists in Medicine, College Park, MD, 2006.
2. Seppenwoolde Y, Shirato H, Kitamura K, *et al.* Precise and real-time measurement of 3D tumor motion in lung due to breathing and heartbeat, measured during radiotherapy. *Int J Radiat Oncol Biol Phys* 2002; 53:822–834.
3. Shirato H, Seppenwoolde Y, Kitamura K, *et al.* Intrafractional tumor motion: Lung and liver. *Semin Radiat Oncol* 2004; 14:10–18.
4. Wong J. Methods to manage respiratory motion in radiation treatment. In: Palta JR, Mackie TR, editors, *Intensity Modulated Radiation Therapy Medical Physics*, Madison, WI, 2003; pp. 663–702.
5. Senan S, De Ruysscher D, Giraud P, *et al.* Radiotherapy Group of European Organization for Research and Treatment of Cancer: Literature-based recommendations for treatment planning and execution in high-dose radiotherapy for lung cancer. *Radiother Oncol* 2004; 71:139–146.
6. Murphy MJ. Tracking moving organs in real time. *Semin Radiat Oncol* 2004; 14:91–100.
7. George R, Keall PJ, Kini VR, *et al.* Quantifying the effect of intrafraction motion during breast IMRT planning and dose delivery. *Med Phys* 2003; 30:552–562.
8. Lagerwaard FJ, Van Sornsen de Koste JR, NijssenVisser MR, *et al.* Multiple 'slow' CT scans for incorporating lung tumor mobility in radiotherapy planning. *Int J Radiat Oncol Biol Phys* 2001; 51:932–937.
9. Aruga T, Itami J, Aruga M, *et al.* Target volume definition for upper abdominal irradiation using CT scans obtained during inhale and exhale phases. *Int J Radiat Oncol Biol Phys* 2000; 48:465–469.
10. Underberg RW, Lagerwaard FJ, Slotman BJ, *et al.* Use of maximum intensity projection (MIP) for target volume generating in 4DCT scans for lung cancer. *Int J Radiat Oncol Biol Phys* 2005; 63:253–260.
11. Dawson LA, Brock KK, Kazanjian S, *et al.* The reproducibility of organ position using active breathing control (ABC) during liver radiotherapy. *Int J Radiat Oncol Biol Phys* 2001; 51:1410–1421.
12. Wagman R, Yorke E, Ford E, *et al.* Respiratory gating for liver tumors: Use in dose escalation. *Int J Radiat Oncol Biol Phys* 2003; 55:659–668.
13. Hanley J, Debois mm, Mah D, *et al.* Deep inspiration breath-hold technique for lung tumors: The potential value of target immobilization and reduced lung density in dose escalation. *Int J Radiat Oncol Biol Phys* 1999; 45:603–611.
14. Mah D, Hanley J, Rosenzweig KE, *et al.* Technical aspects of the deep inspiration breath-hold technique in the treatment of thoracic cancer. *Int J Radiat Oncol Biol Phys* 2000; 48:1175–1185.
15. West JB. *Respiratory Physiology: The Essentials.* Waverly Press, Baltimore, MD, 1974.
16. Dawson LA, Eccles C, Bissonnette JP, *et al.* Accuracy of daily image guidance for hypofractionated liver radiotherapy with active breathing control. *Int J Radiat Oncol Biol Phys* 2005; 62:1247–1252.
17. Wong JW, Sharpe MB, Jaffray DA, *et al.* The use of active breathing control (ABC) to reduce margin for breathing motion. *Int J Radiat Oncol Biol Phys* 1999; 44:911–919.
18. Vedam SS, Keall PJ, Kini VR, *et al.* Determining parameters for respiration-gated radiotherapy. *Med Phys* 2001; 28:2139–2146.
19. Berbeco RI, Nishioka S, Shirato H, *et al.* Residual motion of lung tumours in gated radiotherapy with external respiratory surrogates. *Phys Med Biol* 2005; 50:3655–3667.
20. Bel A, Petrascu O, Van de Vondel I. A computerized remote table control for fast online patient repositioning: Implementation and clinical feasibility. *Med Phys* 2000; 27:354–358.

21. D'Souza WD, Naqvi SA, Yu CX. Real-time intra-fraction-motion tracking using the treatment couch: A feasibility study. *Phys Med Biol* 2005; 50:4021–4033.
22. Keall PJ, Kini V, Vedam SS, *et al.* Motion adaptive x-ray therapy: A feasibility study. *Phys Med Biol* 2001; 46:1–10.
23. Neicu T, Shirato H, Seppenwoolde Y, *et al.* Synchronized moving aperture radiation therapy (SMART): Average tumor trajectory for lung patients. *Phys Med Biol* 2003; 48:587–598.
24. Papiez L. DMLC leaf-pair optimal control of IMRT delivery for a moving rigid target. *Med Phys* 2004; 31:2742–2754.
25. Adler Jr JR, Murphy MJ, Chang SD. Image-guided robotic radiosurgery. *Neurosurgery* 1999; 44:1299–1307.
26. Shirato H, Shimizu S, Kunieda T, *et al.* Physical aspects of a real-time tumor-tracking system for gated radiotherapy. *Int J Radiat Oncol Biol Phys* 2000; 48:1187–1195.
27. Ozhasoglu C, Murphy MJ. Issues in respiratory motion compensation during external-beam radiotherapy. *Int J Radiat Oncol Biol Phys* 2002; 52:1389–1399.
28. Schweikard A, Glosser G, Bodduluri M, *et al.* Robotic motion compensation for respiratory movement during radiosurgery. *Comput Aided Surg* 2000; 5:263–277.
29. Schweikard A, Shiomi H, Adler Jr JR. Respiration tracking in radiosurgery. *Med Phys* 2004; 31:2738–2741.
30. Sharp GC, Jiang SB, Shimizu S, *et al.* Prediction of respiratory tumour motion for real-time image-guided radiotherapy. *Phys Med Biol* 2004; 49:425–440.
31. Ahn S, Yi B, Suh Y, *et al.* A feasibility study on the prediction of tumor location in the lung from skin motion. *Br J Radiol* 2004; 77:588–596.
32. Hoisak JD, Sixel KE, Tirona R, *et al.* Correlation of lung tumor motion with external surrogate indicators of respiration. *Int J Radiat Oncol Biol Phys* 2004; 60:1298–1306.
33. Tsunashima Y, Sakae T, Shioyama Y, *et al.* Correlation between the respiratory waveform measured using a respiratory sensor and 3D tumor motion in gated radiotherapy. *Int J Radiat Oncol Biol Phys* 2004; 60:951–958.
34. Benchetrit G. Breathing pattern in humans: Diversity and individuality. *Respir Physiol* 2000; 122:123–129.
35. Isaksson M, Jalden J, Murphy MJ. On using an adaptive neural network to predict lung tumor motion during respiration for radiotherapy applications. *Med Phys* 2005; 32:3801–3809.
36. Murphy MJ, Jalden J, Isaksson M. Adaptive filtering to predict lung tumor breathing motion during image-guided radiation therapy. In: *Computer Assisted Radiology and Surgery (CARS 2002).* Springer-Verlag, Berlin, 2002; pp. 539–544.
37. Dieterich S, Taylor D, Chuang C, *et al. The CyberKnife Synchrony Respiratory Tracking System: Evaluation of systematic targeting uncertainty (white paper presented at ASTRO 2004).* Accuray, Inc., Sunnyvale, CA, 2004.
38. Main W, Taylor D, Hatipoglu S, *et al. The CyberKnife Synchrony Respiratory Tracking System: Initial evaluation of nonlinear correlation model (white paper presented at ASTRO 2005).* Accuray, Inc., Sunnyvale, CA, 2005.
39. Cavedon C, Chuang C, Dieterich S, *et al.* A multi-institutional validation of Synchrony motion tracking accuracy using the Accuray Synchrony quality assurance motion generation tool – A work in progress. In: *Proc. 6th Annual CyberKnife Users' Meeting.* CyberKnife Society, Sunnyvale, CA, 2007.
40. Kilby W, Main WM, Dieterich S, *et al. In-vivo assessment of the Synchrony Respiratory Tracking System accuracy (white paper presented at 5th Annual CyberKnife Users' Meeting).* Accuray, Inc., Sunnyvale, CA, 2005.
41. Seppenwoolde Y, Berbeco RI, Nishioka S, *et al.* Accuracy of tumour tracking to compensate respiratory motion, assessed with continuously recorded internal and external markers. *Radiother Oncol* 2006; 81(Suppl 1):S42–S43.

Patient Alignment and Target Tracking in Radiosurgery of Soft-Tissue Tumors Using Combined Fiducial and Skeletal Structures Tracking Techniques

3

Xiaodong Wu, Dongshan Fu, Alberto de la Zerda, Elizabeth Bossart, Hua Shao, Joseph Both, Walter Nikesch, Zhicong Huang, Arnold M. Markoe, and James G. Schwade

CONTENTS

3.1 Introduction

When using the CyberKnife® (Accuray Incorporated, Sunnyvale, CA) Image-Guided Stereotactic Radiosurgery (SRS) System to treat soft-tissue tumors in anatomic sites other than intracranial or spinal locations – such as in the lung, liver, kidney, prostate, and pancreas – fiducial placement in or close to the tumors is necessary to assist patient alignment and target tracking for precise treatment delivery. Under the assumption of rigid transformation, at least three fiducial markers are required to obtain six-degrees-of-freedom transformation parameters, i.e., three translations and three rotations. However, in most cases, soft tissue is highly deformable and non-rigid. This results in three possible scenarios: 1) the rigid body criteria fail and the rotational transformation cannot be obtained, 2) the tumor deformation results in unreliable computed rotational information, and 3) even when the fiducial array meets the rigid body criteria, the orientation of the tumor often has poor correlation with the global body orientation and thus results in dosimetric deviation. In these cases, the rigid rotations computed by fiducial tracking have no clear meaning and it is questionable if the rotation corrections by the robot would contribute to targeting precision. We present a solution that balances the requirement of minimal target geometric miss and the consistency of global body orientation by combining both spinal alignment and tumor fiducial tracking techniques.

3.2 Material and Methods

3.2.1 Fiducial Tracking and Tumor Orientation

The use of fiducial markers with CyberKnife was originally introduced for spinal radiosurgery [1]. Its application was successful; the system's precision is widely acknowledged [2, 3]. The fiducial tracking algorithm has demonstrated its accuracy and precision throughout the early experience of CyberKnife spinal radiosurgery. This method was naturally adopted when the application of CyberKnife was extended to body radiosurgery for soft tissue tumors. One should remember that the success of fiducial-based spinal radiosurgery lies in the fact that fiducial arrays deform very little because they are implanted into the rather rigid spinal bony structures; thus, the geometry of the fiducial array reconstructed by the imaging system can be reliably and accurately correlated to the tumor orientation. The same, however, does not hold true for soft-tissue tumors, for which fiducials are placed directly inside or adjacent to the tumor. Studies have shown that soft-tissue tumors in anatomic sites like the lung, liver or pancreas can easily deform. This deformation can range from a few millimeters up to 1 cm [4–6]. A fiducial array will inevitably be deformed when implanted in such an environment. This naturally means the movement of the implanted fiducials may not be coherent. Under such conditions, the orientations (translation and rotation) derived from the reconstructed geometry of the fiducial array will not be the complete representation of the tumor spatial orientation. This can be demonstrated by the following example.

Three fiducials were implanted in a tumor, forming an equilateral triangle with a distance of 2 cm between any two fiducials. The fiducial array was used to define the treatment geometry. The fiducial tracking procedure was then carried out and the fiducial array was aligned to minimize the offsets in all six degrees of freedom to < 0.2 (mm or degrees). One of the three fiducials was then offset by 2 mm in the direction shown in Figure 3.1. The image tracking following this modification failed, giving a message of "Rigid body constraint failure". Upon increasing the "Rigid Body Tolerance" to 2.5 mm, the tracking was processed and 5 degrees of rotation were reported. Here, the 2-mm offset of one fiducial represents a local deformation. Under such conditions, the calculated rotations do not reflect reality, and the corrections based on such information would have caused a geometric miss.

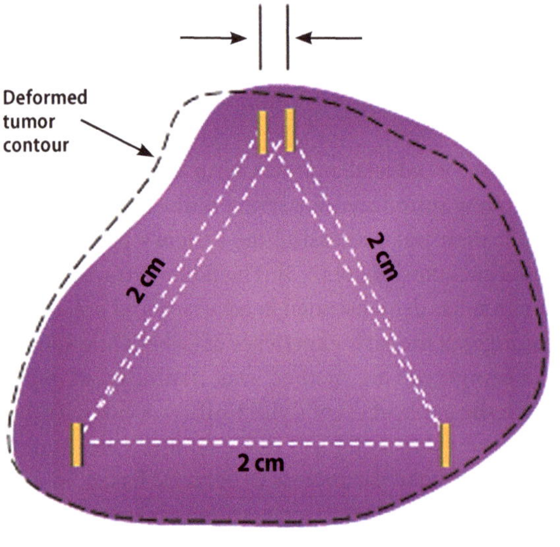

Fig. 3.1 A fiducial array in an equilateral triangle formation (not to scale). A local tumor deformation causes the top fiducial to shift by 2 mm. The CyberKnife tracking algorithm would report 5 degrees of rotation.

Furthermore, changes of tumor position, caused either by local displacement or deformation, are often inconsistent with global body movement. When the tracking module returns three rotational parameters, the system assumes the patient's whole body has rotated by those values. This can fail to reflect physical reality. A correction of beam geometry following such instruction would cause dosimetric error, since the source-to-surface distance (SSD) and the depth of calculation would vary. Such a situation is illustrated in Figure 3.2. Although the use of a large number of non-coplanar beams may "wash out" such an effect from each individual beam, the overall effect should still be carefully examined.

In light of the above, it should be acknowledged that increasing the number of fiducial markers may not always be advantageous in cases of easily deformed tumors. When multiple fiducials are used for these tumors, the rotational parameters generated by the tracking system should be interpreted with caution.

3.2.2 Global Alignment and Local Tumor Tracking

Since a deformed fiducial array can offer little meaningful rotational information, there is no obvious need to meet the theoretical requirement of at least three fiducials for 6D tracking when deciding the number of fiducials to be used. Moreover, fiducials implanted in distal surrounding deformable tissue (such as normal lung tissue) may result in additional error in projecting the "CT center" (the center of image tracking), and therefore the target location itself. This leads to a new strategy of implanting a minimal number of fiducials (often one) in deformable soft-tissue tumors, and using only the translational parameters for tracking. When a single fiducial is used, it should always be implanted at or near the center of the tumor mass. Placing fiducials in adjacent normal lung tissue should be avoided whenever possible.

However, if rotational information from fiducial(s) is to be discarded, global patient alignment would become indispensable to ensure accurate treatment de-

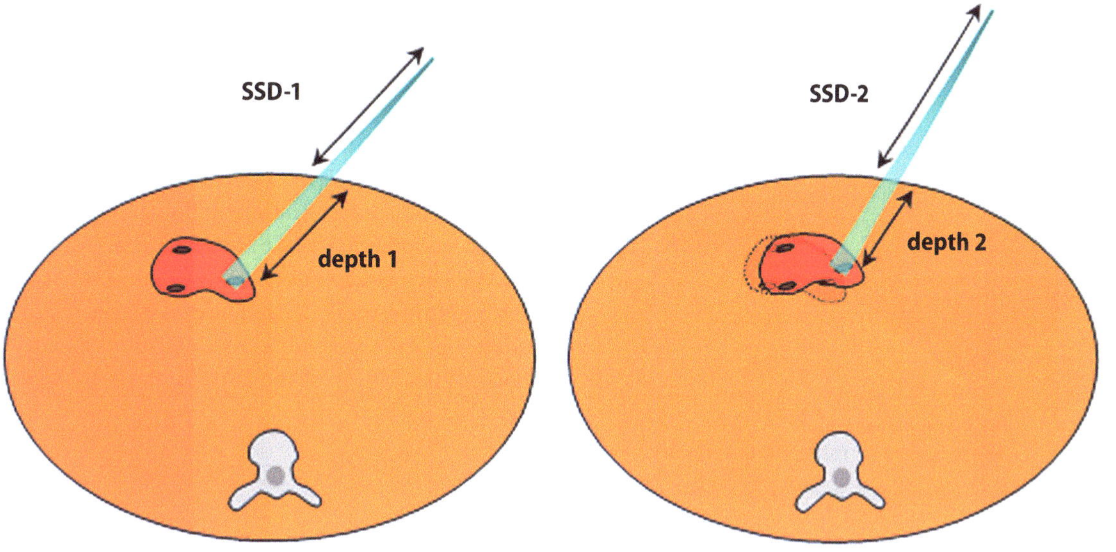

Fig. 3.2 Incoherent tumor-body movement/orientation causes variation in dosimetric parameters such as SSD and depth.

livery. This is apparent especially when a single fiducial is used for target tracking, since the tracking algorithm would not distinguish one position from another, not even between supine and prone patient positioning. Using spinal bony structure for global patient alignment is a natural choice. A new methodology was thus formulated for tracking soft-tissue tumors: achieving global patient alignment followed by translational local tumor tracking using a reduced number (one or two) of fiducials implanted within the tumor mass [7].

In our early implementation of this methodology, global alignment was achieved by implanting three or four fiducials in the spinal processes at appropriately selected levels. This was later replaced by a newly-developed technology, called skeletal structure tracking (SST) or Xsight™ (Accuray Incorporated, Sunnyvale, CA) Spine Tracking System [8, 9]. Xsight, using a unique non-rigid image registration algorithm, was designed for SRS of spinal tumors. It provides accurate spinal orientation information with six degrees of freedom, and thus eliminates the use of fiducials for spinal tracking.

3.2.3 Treatment Margin Determination

If the rotational parameters are to be excluded in tumor tracking, the possibility of geometric miss needs to be addressed. The problem can be reduced to the issue of determining an additional treatment margin. The following analysis provides a general guideline.

Assume a fiducial is implanted inside the tumor and is set as the image tracking center or CT Center C, as shown in Figure 3.3. Consider a point P at the edge of the tumor. A rotation of θ degrees would result in a linear displacement of Δs, which can be determined by the following simple relation (for a small value of θ):

$$\Delta s = R \cdot \theta$$

where R is the distance from P to C, and θ is in the unit of radians. Table 3.1 presents linear displacements for a range of values of R and θ.

Since the direction of the displacement Δs is always tangential to R, a sharp contour would be subjected to a greater degree of spatial displacement. On the other extreme, a spherical tumor with the CT center located at the center of the tumor would suffer no such spatial offset. This implies that the location of the fiducial has an effect on the rotation-induced spatial displacement, and that such displacement would not be uniform around the tumor boundary. However, with a well-centered fiducial placement, an outward expansion by the projected maximum value of Δs would suffice to account for

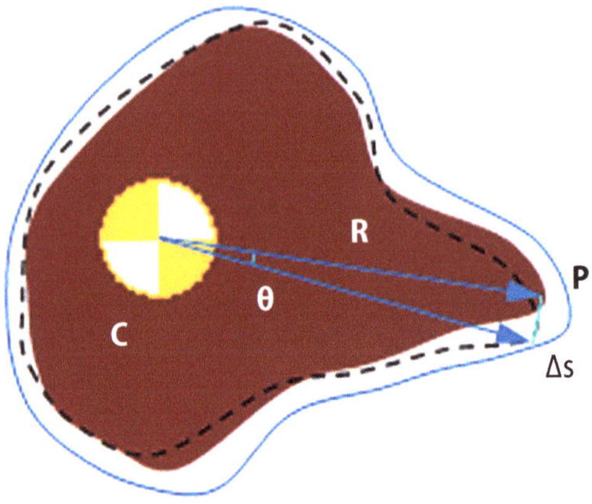

Fig. 3.3 Linear displacement caused by rotational movement.

Table 3.1 Linear Displacement Δs (mm) vs Rotation θ & Radius R

R (mm)	Rotation θ (degree)									
	0.5	1.0	1.5	2.0	2.5	3.0	3.5	4.0	4.5	5.0
5	0.04	0.09	0.13	0.17	0.22	0.26	0.31	0.35	0.39	0.44
10	0.09	0.17	0.26	0.35	0.44	0.52	0.61	0.70	0.79	0.87
15	0.13	0.26	0.39	0.52	0.65	0.79	0.92	1.05	1.18	1.31
20	0.17	0.35	0.52	0.70	0.87	1.05	1.22	1.40	1.57	1.74
25	0.22	0.44	0.65	0.87	1.09	1.31	1.53	1.74	1.96	2.18
30	0.26	0.52	0.79	1.05	1.31	1.57	1.83	2.09	2.36	2.62
35	0.31	0.61	0.92	1.22	1.53	1.83	2.14	2.44	2.75	3.05
40	0.35	0.70	1.05	1.40	1.75	2.09	2.44	2.79	3.14	3.49
45	0.39	0.79	1.18	1.57	1.96	2.36	2.75	3.14	3.53	3.93
50	0.44	0.87	1.31	1.75	2.18	2.62	3.05	3.49	3.93	4.36

the geometric uncertainty due to rotational movement. Our experience indicates that after global patient alignment, the detectable local tumor rotations are usually within 3 degrees, and unlikely to exceed 5 degrees. Using this as a guideline and taking into consideration fiducial location and tumor shape, an additional margin beyond the usual tumor margin (for microscopic disease and systemic uncertainty) can be determined to account for the rotational error. Using lung tumor SRS as an example, we are now typically using a total of 8-mm expansion around the gross tumor volume (GTV) as the planning target volume (PTV) for primary lung tumors, and a 5-mm expansion around the GTV for metastatic lung tumors. It is worth noting that very often after 3D expansion sharp edges of the GTV are significantly smoothed, thus making tumor rotation less critical in affecting the accuracy of dose delivery.

The determination of PTV in radiotherapy is a well-recognized decisive factor directly affecting treatment outcomes. The methodology of defining PTV in radiotherapy has been systemically studied [10, 11]. The principles behind the definition of the PTV in radiosurgery depend heavily on the characteristics of each individual SRS delivery system. A thoughtful scrutiny of the issue is naturally called for.

3.2.4 Treatment Planning and the CT Center Selection

For all image tracking modes, the greatest accuracy can be achieved when the region of tracking is near the center of the imaging system, because the central region of the imaging system is the primary focus of the system calibration.

Following this logic, a separate alignment plan is generated using either the fiducial mode or the Xsight mode. When the fiducial mode is used for the alignment plan, only the spinal fiducials are defined, and the CT center falls on the center of the spinal fiducial array. When Xsight is used, the CT center is customarily defined at the anterior aspect of the spinal canal.

The actual dose delivery plan is done with only the tumor fiducial(s) being defined and with the CT center at the center of the tumor fiducial(s). To minimize dosimetric errors caused by the inevitable variation in the SSD and the depth of calculation during treatment delivery (Fig. 3.2), special effort is made to maximize the isotropic beam configuration. To ensure a good conformal plan with satisfactory dose fall-off, the total number of beams used is usually around 100.

3.2.5 Treatment Delivery

With the alignment plan and the dose delivery plan at hand, the translational shifts in x,y,z directions from the setup position to the treatment position can be calculated by subtracting the coordinates of the CT centers of the two plans.

Treatment delivery is carried out in two steps. The alignment plan for the global setup is first loaded. The patient is then aligned to the spinal bony structures, either with the spinal fiducials or Xsight. Following a satisfactory global alignment, the dose delivery plan is then loaded. The pre-calculated shifts between the alignment CT center and the treatment CT center are entered to drive the automatic couch to the treatment position. Treatment is then carried out with fiducial/Synchrony® (Accuray Incorporated, Sunnyvale, CA) tracking. As explained, rotational corrections are omitted during treatment delivery. Figure 3.4 illustrates this two-step process in a case of primary lung tumor treatment. For that case, the Xsight tracking mode was used for the global patient alignment.

3.3 Results and Discussion

Since 2004, more than 250 patients with soft-tissue tumors have been treated at the CyberKnife Center of Miami using the target-tracking strategy described here, with favorable clinical outcome [12, 13]. For the first 100 patients, spinal fiducials were used for the spinal-global alignment before the Xsight tracking system was introduced. For most of the approxi-

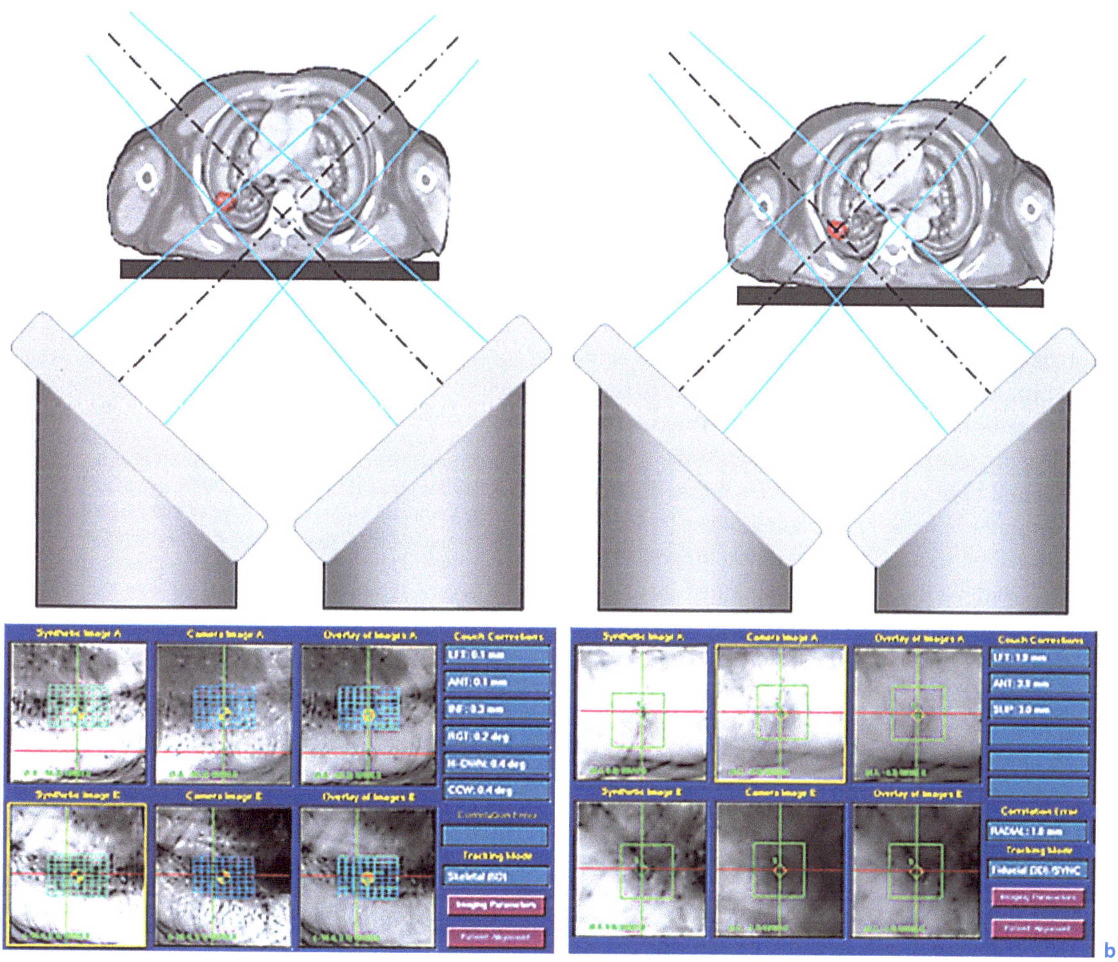

a b

Fig. 3.4a,b. Global alignment and local tumor tracking for a primary lung tumor (*shown in red*). **a** Global alignment using Xsight spinal tracking. **b** Treatment with the new CT center on the tumor fiducial mately 250 patients, a single fiducial was used for tumor tracking.

It is important to realize that the use of the multiple-fiducial tracking technique for soft-tissue tumors presents a different scope of challenges than the spinal application. When tracking a soft-tissue tumor with multiple fiducials, the operators often need to relax the tracking parameters, such as the rigid body tolerance, in order to make the orientation parameters obtainable. One should be conscious of the fact that when these conditions are changed, the margin of tracking error increases, and the interpretation of the rotational parameters becomes difficult. Multiple fiducials may be necessary for some larger target volumes, such as the prostate gland. Caution should be exercised in such situations while interpreting reported orientation parameters.

The method described here requires a minimal number of fiducials (often one). This is advantageous both because complications due to fiducial placement can be reduced and because the tracking process during treatment is significantly eased. For our lung patients, we have experienced a much lower rate of pneumothorax caused by fiducial implantation.

When a single fiducial is used, local fiducial migration is a potentially serious problem. An interval of 5–7 days between the time of fiducial placement and the time of treatment CT acquisition is our standard protocol. This allows the fiducial time to settle into a fixed tissue location, typically within scar tissue. During planning CT acquisition, if the fiducial were found to be outside the tumor mass, the fiducial would have to be re-implanted. Post-treatment CT showed no sign of local migration for any of our patients with a single fiducial implanted inside the tumor mass.

We believe that real-time tumor tracking is important in treating tumors that move with respiration. However, 6D tracking may not be advantageous at the present time for the reasons we have discussed. The ultimate solution of soft-tissue tumor SRS would involve the implementation of 4D CT-acquisition with an adaptive dose calculation method, and a real-time target tracking technique capable of synchronizing the real-time patient position and the 4D planning image set. In the absence of a true 4D planning and treatment delivery system, a combined global patient alignment and translational local tumor tracking technique is a practical solution for minimizing both geometric and dosimetric misses.

The current two-step/two-CT center procedure might be further simplified by integrating the process into the existing planning and treatment delivery software.

References

1. Ryu SI, Chang SD, Kim DH, *et al.* Image-guided hypofractionated stereotactic radiosurgery to spinal lesions. *Neurosurgery* 2001; 49:838–846.
2. Murphy MJ. Fiducial-based targeting accuracy for external-beam radiotherapy. *Med Phys* 2002;29:334–344.
3. Wu X, Main W, Schwade GJ. Measurement of total clinical accuracy of an image-guided radiosurgery system. *Med. Phys.* 2004; 31:1901.
4. Balter JM, Brock KK, Litzenberg DW, *et al.* Daily targeting of intrahepatic tumors for radiotherapy. *Int J Radiat Oncol Biol Phys* 2002; 52:266–271.
5. Balter JM, Dawson LA, Kazanjian S, *et al.* Determination of ventilatory liver movement via radiographic evaluation of diaphragm position. *Int J Radiat Oncol Biol Phys* 2001; 51:267–270.
6. Loo BW, Thorndyke BR, Maxim PG, *et al.* Determining margin for target deformation and rotation in respiratory motion-tracked stereotactic radiosurgery of pancreatic cancer. *Int. J. Rad. Oncol. Biol. Phys.* 2005; 63:S31.
7. Wu X, Sosa O, Yang J, *et al.* Practical target tracking strategy for soft-tissue tumor SRS with Cyberknife. Whole Body Radiosurgery Symposium. Washington, DC; 2005.
8. Fu D, Kuduvalli G. Enhancing skeletal features in digitally reconstructed radiographs. In: Reinhardt JM, Pluim JP, editors. Proceedings of SPIE. Vol 6144: International Society for Optical Engineering; 2006. pp. 846–851.
9. Fu D, Kuduvalli G, Maurer CR, *et al.* 3D target localization using 2D local displacements of skeletal structures in orthogonal X-ray images for image-guided spinal radiosurgery. *International Journal of Computer Assisted Radiology and Surgery (Suppl 1)* 2006:198–200.
10. Langer MP, Papiez L, Spirydovich S, *et al.* The need for rotational margins in intensity-modulated radiotherapy and a new method for planning target volume design. *Int J Radiat Oncol Biol Phys* 2005; 63:1592–1603.
11. Papiez L, Langer M. On probabilistically defined margins in radiation therapy. *Phys Med Biol* 2006; 51:3921–3939.
12. Brown WT, Perman M, Wu X, *et al.* Image-guided robotic stereotatic radiosurgery for treatment of lung tumors. In: Mould RF, Bucholz RD, Gagnon GJ, *et al.*, editors. Robotic Radiosurgery. Vol I. Sunnyvale, CA: The Cyberknife Society Press; 2005. pp. 255–268.
13. Perman M, Bellairs EE, Wu X, *et al.* Cancer of the pancreas with special reference to epidemiology & radiosurgery. In: Mould RF, Bucholz RD, Gagnon GJ, *et al.*, editors. Robotic Radiosurgery. Vol I. Sunnyvale, CA: The Cyberknife Society Press; 2005. pp. 301–314.

Section 2

Thoraco-Abdominal Dosimetry, Radiobiology, and Imaging

The Radiobiological Rationale for Hypofractionation of Lung and Liver Tumors

4

Alan Alfieri, Jill Rossinow, Madhur Garg, Shalom Kalnicki, and Chandan Guha

CONTENTS

As far as the laws of mathematics refer to reality,
they are not certain;
and as far as they are certain,
they do not refer to reality.

Albert Einstein,
"Geometry and Experience", January 27, 1921

4.1 Introduction

The radiobiological concepts of intrinsic radiation sensitivity, oxygenation, and dose-volume effects have been reasonably delineated in the context of conventional radiotherapy (RT). Yet, for circumstances in which large doses are delivered in single-fraction or hypofractionated regimens, these intrinsic radiobiological concepts are relatively poorly understood. Stereotactic radiosurgery (SRS) is a radical departure from the current RT approach in which large fields, cone downs, and protracted therapies are used for normal tissue preservation and to maximize the therapeutic ratio. SRS is the precise, highly focused delivery of radiation beams to lesions whereby only a fraction of the total dose is received by surrounding normal tissues. The usage of SRS is currently expanding well beyond its roots as an ablative tool for thalamotomies, arteriovenous malformations, and cranial vault tumors. Hence, widely believed dogmas concerning the tolerance of critical structures to conventionally fractionated doses, such as the dose-volume effect, total dose, and time (latency) dependency, have to be reevaluated for hypofractionated radiation therapy.

Do improvements in therapeutic ratio with hypofractionated radiation require changing radiobiological paradigms? The answer to this question must reflect a convergence of rationales from conventional therapeutic approaches combined with the realization that optimal therapeutic strategies will be based on delivering the maximum dose tolerated by late-responding normal tissue volumes, while reducing the treatment time compatible with tissue tolerance of acutely reacting tissues.

4.2 Therapeutic Ratio: Tumor Control Probability versus Normal Tissue Sequelae

The basic tenet of radiation oncology is to eradicate tumor cells while minimizing radiation-induced late tissue toxicities. Ideally, all tumor cells are rendered nonviable, and toxicities are nonexistent. In the worst case scenario, many dividing tumor cells survive and long-lasting side effects are severe or even fatal. Clearly, this latter scenario is unacceptable. In reality, outcomes fall along this spectrum, and variations in treatment planning (including total delivered dose, dose per fraction, and volume of tissue treated) are carefully evaluated to achieve an acceptable balance. This balance can be graphically expressed by plotting the Tumor Control Probability (TCP) and the Normal Tissue Complication Probability (NTCP) as a function of dose (Fig. 4.1). Mathematically, the ratio of the TCP and the NTCP is known as the Therapeutic Ratio (TR). For the TCP to be clinically meaningful (i.e., to attain local tumor control), there should be no viable tumor cells remaining, which requires that the volume of tissue treated include all tumor cells and that the effective dose be high enough to overcome the capacity of tumor cells to repair radiation-induced DNA damage. The NTCP, however, tends to increase with increasing dose and treatment volume, making preservation of the TR a challenge. It is important to note that this conundrum exists, especially when treating lesions near critical structures where achieving a therapeutic effect is likely to result in normal tissue toxicity. Thus, if the tumor is relatively radioresistant, or if the surrounding normal tissues are sensitive to radiation, then the TCP and NTCP (curves T and C, respectively, in Figure 4.1) will be very close to one another and could even overlap, yielding a TR ~ 1. Historically, in the treatment of cancer, the clinical objective has been to separate the TCP and NTCP by exploiting biological differences between normal and malignant cells such that the probability of damage to the tumor is greater than the probability of damage to normal tissue at an isoeffective dose. SRS promises to achieve a high TR by sculpting the radiation dose to the tumor and limiting the dose to surrounding critical structures. This is accomplished through dose delivery using multiple coplanar or non-coplanar beams and possibly intensity modulation.

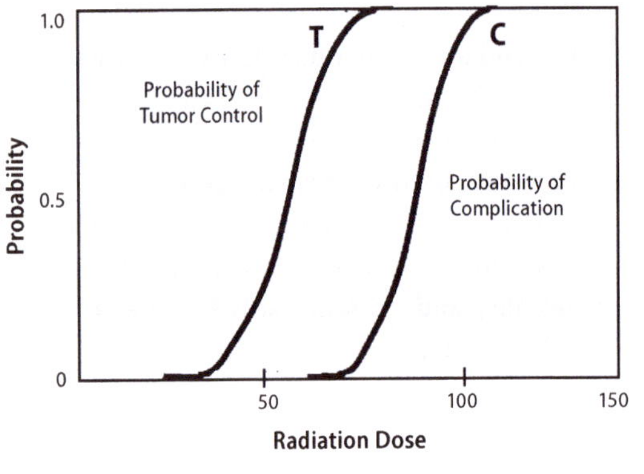

Fig. 4.1 Therapeutic Ratio. Curve T represents the probability of tumor control and curve C represents the probability of normal tissue complications. In ideal conditions, T is to the left of C such that the RT dose with 100% tumor control probability results in 0% complications. If the tumor is radioresistant (curve T shifts to the right) or if normal tissue is radiosensitive and complication arises after exposure to lower doses of RT (curve C shifts to the left) then the TR is lowered. Examples of the latter include high radiosensitivity of normal organs, such as, the liver, lung, intestine, and kidney. Since the tumor does not contribute to normal tissue physiological function, SRS promises to increase the TR by sculpting high RT dose to the tumor with very little exposure of the normal tissue.

4.3 Evolution of Dose Fractionation

The evolution of dose fractionation for normal tissue preservation and reduced NTCP was pioneered in the early part of last century by Coutard [1], Regaud [2], Baclesse [3], and Miescher [4]. It was not until Strandquist compiled data on recurrences of skin cancer and normal tissue (skin) complications that isoeffective doses could be estimated for comparing tumor control with skin reactions following different regimens of radiotherapy [5]. Thus, fractionation evolved as an effort to minimize acute and late normal tissue reactions without affecting tumor control. Dose fractionation within limits was completely dependent upon total dose and overall treatment time, unknowingly (at the time) allowing the physician to exploit the physiological processes of tumor reoxygenation, redistribution, reassortment, repopulation, and damage repair.

4.4 The Linear-Quadratic Model

The relationship between mammalian cell survival and radiation dose can be described by a "survival curve" which is plotted as the surviving fraction (SF) on a logarithmic scale and radiation dose (D) on the linear axis. This survival curve is most usually approximated by the Linear Quadratic (LQ) model, $SF = \exp(-\alpha D - \beta D^2)$, in which α and β are parameters which describe the radiosensitivity of the cells.

The ratio α/β describes the non-linearity of the curve when plotted on a log-normal scale. This ratio (α/β) is typically larger for early responding normal tissues and many tumors, and is typically smaller for late responding normal tissues. The α/β ratios obtained from multifraction preclinical and clinical data for early and late responding tissues and tumors have been compiled over the last decade by numerous investigators. The resulting efforts have been transposed by Withers et al. [7] into a functional mathematical expression of the LQ model that accounts only for the repair capacity of the tissue treated. The LQ model of cell survival can be used to calculate 'iso-effective' dose regimens such that multiple combinations of dose per fraction, d, and number of fractions, n, give a constant surviving fraction of cells for a specific α/β ratio. Perhaps the most commonly used iso-effect calculation method is Biologically Effective Dose (BED), which is defined as $BED = nd\,[1 + d/(\alpha/\beta)]$ and has the unit Gy.

BED has become a useful tool for many clinicians employing both dose modifications and changes in fractionation patterns when comparing alternative treatment protocols. The LQ model and the estimations of α/β ratio of normal tissues have given us a strong radiobiological rationale for predictions concerning early and late radiation-induced sequelae with conventional therapy. A fundamental change in the radiation approach, using technologies such as SRS and conformal techniques should effectively change the current LQ modeling concept for normal tissue damage as an "overestimate" of the biological response. This may be explained by inhomogeneities in dose distribution and the complexities of tissue architecture in situ (serially versus non-serially arranged), where the stromal-epithelial and cell-to-cell interactions are also known to affect radiation responsiveness.

4.5 Biologically Effective Dose of Radiation for a Tumor is Substantially Increased with Hypofractionation

Although large fraction sizes will effectively kill tumor cells, they also overwhelm normal tissue repair mechanisms, resulting in severe late adverse effects. It has been well established, initially by empirical evaluation and later by clinical and experimental data, that damage to late-responding tissues increases with increasing fraction size. The conventional solution has been to reduce fraction size and deliver a biologically effective total radiation dose in a more protracted regimen. This dose must account not only for fraction size and number of fractions (total dose), but it must also account for the inherent sensitivity of the cell type to a specific fraction size, and the cell type's ability to repair sublethal damage between fractions, as well as other radiobiological parameters, such as repopulation, reoxygenation, and redistribution. From the BED equation, it can be seen that for the same total dose, as schedules become more protracted, the BED decreases, requiring additional fractions to be delivered to maintain the BED. For example, the same total dose of 40 Gy delivered as 20 fractions of 2 Gy, 10 fractions of 4 Gy, or 4 fractions of 10 Gy will have corresponding BEDs of 48 Gy, 56 Gy and 80 Gy, respectively (Fig. 4.2). The highest dose/fraction (4 fractions of 10 Gy) would effectively be a "67% more effective" dose than 20 fractions of 2 Gy. It can also be seen that the sensitivity of BED to change in fraction size is determined by the 'fractionation factor' F, given by $F = [1 + d/(\alpha/\beta)]$, and therefore the effect is greater for tissues with low α/β ratios, i.e., late-responding tissues. The same 40 Gy delivered in 20, 2-Gy fractions to a tissue with an α/β ratio of 3 results in a BED of 67 Gy, as opposed to 48 Gy for an early-responding tissue with an α/β ratio of 10. Therefore, reducing the fractional dose will reduce the BED for normal late-responding tissues, allowing for the repair of sublethal damage between fractions and reduction of late toxicities. Smaller fractions permit the surviving tumor cells, which are typically less sensitive to changes in fractionation (assuming they are associated with a large α/β ratio), to be maintained with only a modest increase in the total dose. Furthermore, protraction

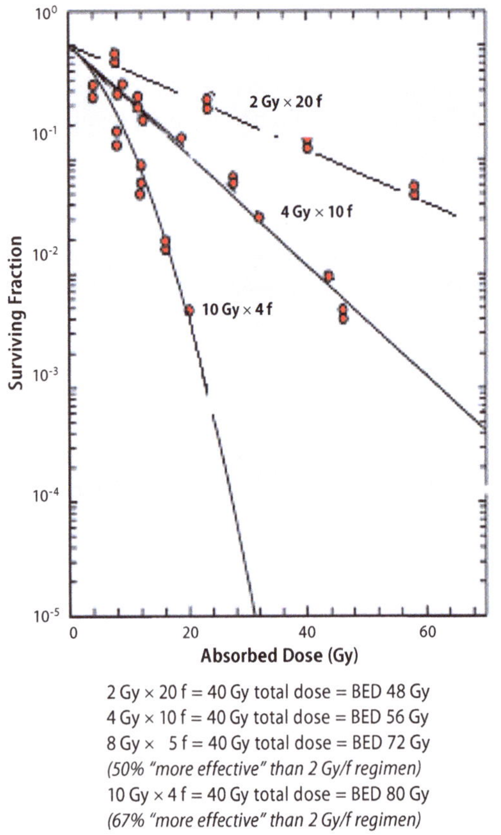

Fig. 4.2 Biological Effective Dose (BED) of tumor is subtantially increased with hypofractionation. BED = nd [1 + d / (β / α)], where "n" is the number of fractions, "d" is the dose per fraction, and [1 + d / (β/α)] is "f," the fractionation factor. The same total dose of 40 Gy will have very different BED values based upon dose per fraction. BED dose calculations for 2 Gy × 20 f = 48 Gy, 4 Gy × 10 fractions = 56 Gy and 10 Gy × 4 f = 80 Gy for the tumor with a/b estimated as 10. This demonstrates that 8 Gy × 5 f has higher BED dose.

may also allow for reoxygenation and reassortment of the SF of cancer cells to occur between fractions, increasing the potential for cell kill during the next RT fraction. Thus, the probability of normal late tissue complications is decreased, while the probability of tumor control is preserved, which ultimately increases the TR. This exploitation of the differences in early- and late-responding tissues forms the basis for mainstream clinical RT, and its delivery of multiple, small fractions to large treatment fields. The protracted therapy time, however, might enable tumor cells that are intrinsically radioresistant to repair radiation-induced damage and undergo accelerated repopulation, resulting in recurrence of the primary tumor.

Protracted fractionation is diametrically opposed to the hypofractionation approach of SRS. In order for SRS to be applied safely and effectively in cancer treatment, the tolerance of normal tissues must be respected. Little is known, however, about the true tolerance of critical normal structures with SRS and to what extent the existing body of data generated from fractionated treatment can be applied to SRS. Much of what we do know about the tolerance of normal structures to SRS comes from animal studies, which show that the histologic response of central nervous system (CNS) tissues depends upon the radiation dose, the time elapsed after treatment, and the treatment volume. As dose increases, the latency period for both radiological and histological changes shortens. As treatment volume increases, the sensitivity of tissues to radiation-induced damage increases. In addition, changes in different tissues become evident at different doses, with vascular damage occurring at lower doses, and with a shorter latency period, than most late reacting tissues.

4.6 Normal Tissue Response to Irradiation

Normal tissue response to ionizing radiation is categorized as either early or late. These very different responses to the same radiation exposure result from inherent differences in cell cycle kinetics and DNA repair capacities within these tissues. Early responding tissues exhibit their response to radiation relatively soon after exposure, because their cell cycle times are shorter. Conversely, cells which have a longer cycle time, or which have a small percentage of cells in the DNA-synthesis phase, will manifest damage much later, often long after treatment is completed. Normal tissues of the spinal cord, brain, kidney, lung, pancreas and liver, which cycle slowly (unless these tissues are surgically compromised), fall into the late-responding category. Certain tumor types, such as prostate adenocarcinoma, malignant melanoma, and meningioma also behave like late-responding tissues. Late toxicities associated with fibrosis in the lung, pancreas, and veno-occlusive disease in the liver that contribute to loss of function, may not be observed for several months after treatment. Early-responding normal tissues, however, which normally include skin, jejunum, colon and testes, and for the most part, tumors will generally manifest radiation-induced damage well before conventional fractionation is completed. Most tumors also manifest radiation damage like an early responding normal tissue. Hence, the early onset tissue reactions associated with desquamation, mucositis, pancreatitis and pneumonitis, which are often observed as interstitial fluid loss, sloughing, microvascular damage and regenerative replacement, can also be noted in the pancreas and lung. The early radiation-induced reactions observed in these organs indicate that a distinct cellular population is the target of acute response to radiation via direct effects of irradiation on cell cycling and early loss of clonogens or the indirect effects via cytokine release.

In addition to differences in cell cycle kinetics, the repair of sublethal injury continues for a longer time in late-responding tissues [10–12]. For late-responding tissues, the contribution of single event killing is minor compared to early responding tissues for the same dose. For early responding tissues, the SF α component shoulder is sharper and the dose range is larger, when compared to late responding tissues, reflecting the ability to repair damage over a wide range of doses. Here, the contribution of multiple killing events does not fully occur until higher doses are reached. For example, the percentage of cells surviving a single dose of 2 Gy (if $\alpha = 0.3\ Gy^{-1}$ and $\beta = 0.03\ Gy^{-1}$) would calculate as 49% ($SF = e^{-(0.3 \cdot 2) - (0.03 \cdot 4)}$). The survival curves that are representative of these tissues are less smooth than the acutely responding tissues, indicating that the survival rate decreases rapidly with increasing dose per fraction in late-responding tissues (Fig. 4.3). Such repairable damage is referred to as "sublethal cell damage." As indicated earlier, SRS could reduce these normal tissue complications by limiting the dose to surrounding critical structures of early- and late-responding tissues and sculpting the radiation dose to the tumor using multiple beams and intensity modulation.

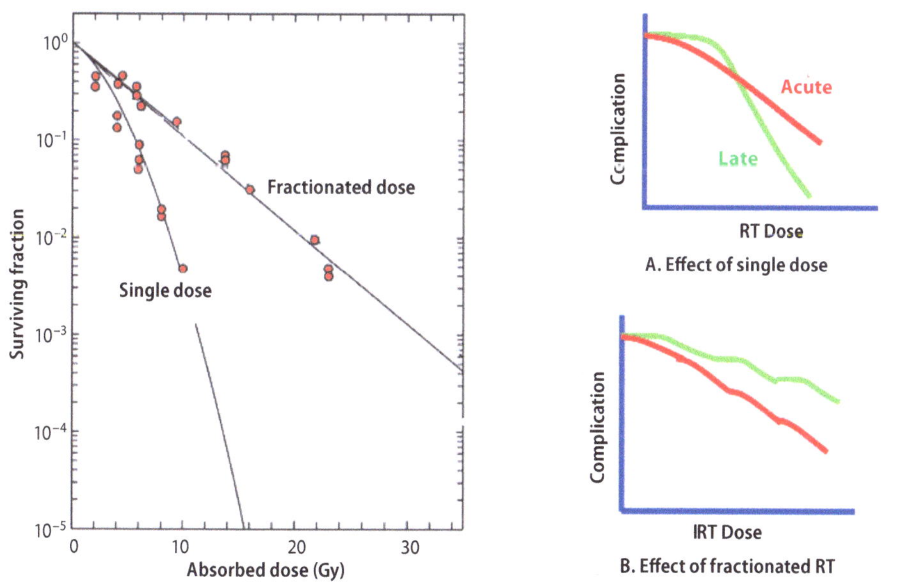

Fig. 4.3 Basis of Fractionation. *Acute effects* of radiation occur relatively soon after exposure because the cell cycle times are shorter for early responding tissues. Acute effects depend upon fraction size and overall treatment time. While prolonging treatment time spares acute effects, little sparing occurs for late effects. *Late effects* depend upon fraction size and total dose. The dose response curves that are representative of these tissues are less smooth than the acutely responding tissues, indicating that the survival rate decreases rapidly with increasing dose per fraction in late-responding tissues. Fractionation has a large impact on late effects.

4.7 Dose-Response Relationships of Normal Tissues: Volume Effect in RT

The tolerance dose (TDn/y) that results in an n percent incidence of complications after y years depends on the cell's clonogenic ability to maintain a sufficient number of mature cells for tissue organization and organ function. In this regard, the response of a tissue or organ to radiation depends on the organizational capacity of the tissue, the kinetics of the cell population that is the principle component of the tissue, and the inherent sensitivities of the individual cells making up the tissue. This is referred to as the Functional Sub-Units (FSU) of an organ/tissue, which contains a relatively constant number of clonogens and which can be repopulated by as little as a single surviving clonogen. FSUs are discrete, anatomically delineated cells whose relationship to tissue function is obvious because they represent the "critical and functional volume" of organs with an organizational and spatial arrangement that is well delineated. Therefore, the survival of structurally defined FSUs depends on the survival of one or more clonogenic cells within them, and tissue/organ survival in turn depends on the number and radiosensitivity of these clonogens. To fully assess the response of varying tissues to ionizing radiation, it is equally important to fully understand and appreciate tissue organization. Most normal tissues are *non-serially arranged* with regards to radiation-induced damage as they demonstrate a graded dose response with respect to the irradiated volumes. Examples of such organs include liver, kidney and lung. The large reserve capacity and increased tolerance to partial volume irradiation are due to the *parallel organization* of the FSU of these organs, e.g., liver acini, lung aveoli or the glomeruli and nephrons. Small volumes of these organs can be treated to very high doses without compromising the functional capacity of the respective organs. The apparent lack of a volume effect at lower volumes of radiation coverage is because of proliferative repair and regeneration that occurs from surviving FSUs scattered throughout the treatment volume and from the nonirradiated volume. Serially arranged FSUs, however, are arranged like the "links of a chain" and, therefore, the elimination of any one results in the measurable probability of a complication. A typical example of this volume effect, for example, is the spinal cord, in which the loss of any one FSU subunit will impact on the conduction of the nerve signals and thereby affect the entirety of the organ, resulting in myelopathy, independent of the status of other subunits within the same tissue. Interestingly, stem cells and surviving clonogens both have been found to migrate from one FSU to another, allowing repopulation of a depleted FSU. An example of this is the epidermal layer of skin or the crypts of the jejunum, from which cells can migrate to repopulate radiation-depleted neighboring crypts.

Thus, early and late radiation-induced tissue effects depend on the organization of the particular organs FSUs, the treatment volume of the FSU, the total dose, and the dose/fraction. In addition, the functional reserve of an organ will also depend upon its basal physiological state. In many cancers, such as lung and hepatocellular carcinomas, the organ is diseased and has poor physiological reserve. Since primary liver cancer usually arises in cirrhotic liver infected with viral hepatitis, the anatomical volume of the liver will underestimate the functional reserve of the organ and therefore, the dose-volume histogram would underestimate the functional reserve of the liver. A similar situation may be encountered in a patient with primary lung cancer who is a chronic smoker and has underlying lung disease, such as chronic obstructive pulmonary disease.

4.8 Heterogeneity of Radiation Response in Normal and Tumor Tissues: A Role for Hypofractionation

Tissues are a mixture of a primary functional cell type, the parenchymal cell type, and the stroma, which consists of various nonparenchymal cell types, such as the endothelial cells, macrophages, fibroblasts, and resident lymphocytes. The capacity to repair radiation damage in normal tissues, such as muscle, liver, neurons, bone and cartilage, depends upon the extent of radiation-induced depletion of parenchymal cells, regeneration or repopulation of parenchymal and microvascular cells after radiation injury, and the modulatory effects of cytokines released after exposure to irradiation. Tissue regeneration may involve the proliferation and differentiation of stem cells that reside in or migrate to injured tissue in response to radiation-induced damage and stress signals. The severity of radiation-induced microvascular damage may promote focal regions of ischemia, which in turn may increase the depletion of parenchymal cells and tissue stem/progenitors. Thus, the effect of radiation depends upon the heterogeneity of response to various target cells in the normal tissues.

Similarly, tumor tissues are not just a clonal population of tumor cells, but they consist of complex multicellular elements that comprise the tumor microenvironment. The progression of a tumor and the ability of a treatment modality to cure the tumor depends upon the genetic signature of the tumor cells, the response of those cells to intercellular signals from the stroma, and the stromal response to therapy. Although tumors arise initially as a monoclonal population, the inherent genetic instability of tumor cells during tumor progression results in a heterogeneous mixture of tumor cells with varying levels of radiation sensitivity. Thus, the existence of tumor progenitor clonogens or stem cells, which are postulated to be inherently more resistant to chemotherapy and RT, points towards mechanisms of local and systemic recurrence of tumor after therapy. The classical model of tumor radiobiology assumes that the success of RT depends upon the ability of radiation to deplete the number of tumor stem cells and progenitor clonogens. Nevertheless, it is becoming clear that the tumor-stromal interactions modulate the curative effects of radiation. For example, Dewhirst and colleagues have demonstrated that bursts of reactive oxygen species generated from the waves of hypoxia/reoxygenation during daily fractions of conventional fractionated RT, increase the translation of HIF-1 mRNA transcripts stored in specialized cytosolic stress granules of hypoxic tumor cells [13]. HIF-1 is a transcription factor that augments the expression of vascular endothelial growth factor (VEGF) and other angiogenic factors which protect the tumor microvasculature from lethal effects of radiation and promote angiogenesis. While HIF-1 confers radioresistance to the tumor endothelium, it regulates the expression of genes that enhance tumor cell radiosensitivity by increasing the apoptotic potential, proliferation rates, and ATP metabolism of tumor cells [14]. Interestingly, inhibition of HIF-1 results in regression of tumor microvasculature and significant enhancement of tumor growth delay [14].

The interaction of the tumor microvasculature with tumor cells modulates the repair of radiation-induced damage and determines the probability of tumor control by RT. Recent studies by Garcia-Varros et al. revealed the earliest tissue damage indicator was a rapid wave of endothelial apoptosis at 1–6 hours when murine tumors were exposed to a single dose of 15–20 Gy [15]. Furthermore, endothelial cell apoptosis and microvascular dysfunction contributed significantly to the tumoricidal effects of RT. *These studies suggest that a large single fraction of RT would bypass the effects of radiation-induced, HIF-1-mediated release of VEGF and its protective effects on tumor vasculature seen during conventional fractionated RT.* The lethal effects of a large, single fraction of RT, however, also applies to radiation damage in normal tissues. For example, radiation has been postulated to target intestinal stem cells within the crypts of Lieberkuhn to initiate the lethal GI syndrome. Paris et al. demonstrated that microvascular endothelial apoptosis is the primary lesion leading to stem cell dysfunction at doses between 12–16 Gy [16]. These investigators further demonstrated that the inhibition of endothelial cell apoptosis pharmacologically by intravenous basic fibroblast growth factor (bFGF) or genetically by deletion of the acid sphingomyelinase gene, prevented radiation-induced crypt damage, organ failure, and death from the GI syndrome. As the dose per fraction increased to 18 Gy, however, radioresistance in the jejunal endothelial cells could not protect crypt damage and organ failure [17], indicating a radiation dose response that switched the primary target from the GI endothelium cells to intestinal stem cells. The authors demonstrate that in normal intestine, ATM, a serine-threonine kinase that is activated following exposure to radiation, prevents the activation of ceramide synthase in intestinal crypt cells [17]. ATM is the apical kinase that plays a central role in the DNA damage surveillance pathway triggered by radiation injury [18, 19]. As such, ATM promotes cell cycle arrest, activates proteins that participate in DNA repair, and induces or suppresses apoptosis in a tissue-specific manner. Depending upon cell type, ATM deficiency can result in exquisite enhancement of radiosensitivity, while other tissues are unaffected or even protected from irradiation [20–23]. Thus, in intestinal crypt cells, ATM confers radioprotection by inhibiting the ceramide synthase that catalyzes the synthesis of ceramide, a second messenger that induces apoptotic cell death after a variety of stresses, including radiation [24, 25]. At higher doses of radiation (>18 Gy), the defense mechanism initiated by ATM is overwhelmed and ceramide-induced apoptosis occurs in the crypt progenitor cells (Fig. 4.4). The switching/reordering of molecular targets with a radiation dose threshold could be exploited to increase the therapeutic ratio of RT [26]. For example, inhibition of ATM protein expression by antisense ATM gene therapy resulted in enhancement of the intrinsic tumor radiosensitivity in glioblastomas [27] and prostate tumor cells [28]. Similarly, exposure of cells to KU-55933, a small molecule inhibitor of the ATM kinase, resulted in a significant sensitization to both the cytotoxic effects of ionizing radiation and to the DNA double-strand break-inducing chemotherapeutic agents etoposide, doxorubicin, and camptothecin [29]. Conversely, administration of growth factors and cytokines, such as, bFGF [30, 31], keratinocyte growth factor (KGF) [32], or interleukin-11 (IL-11) [33] could prevent radiation-induced target cell loss in normal tissues, such as intestinal endothelial cells and crypt clonogens.

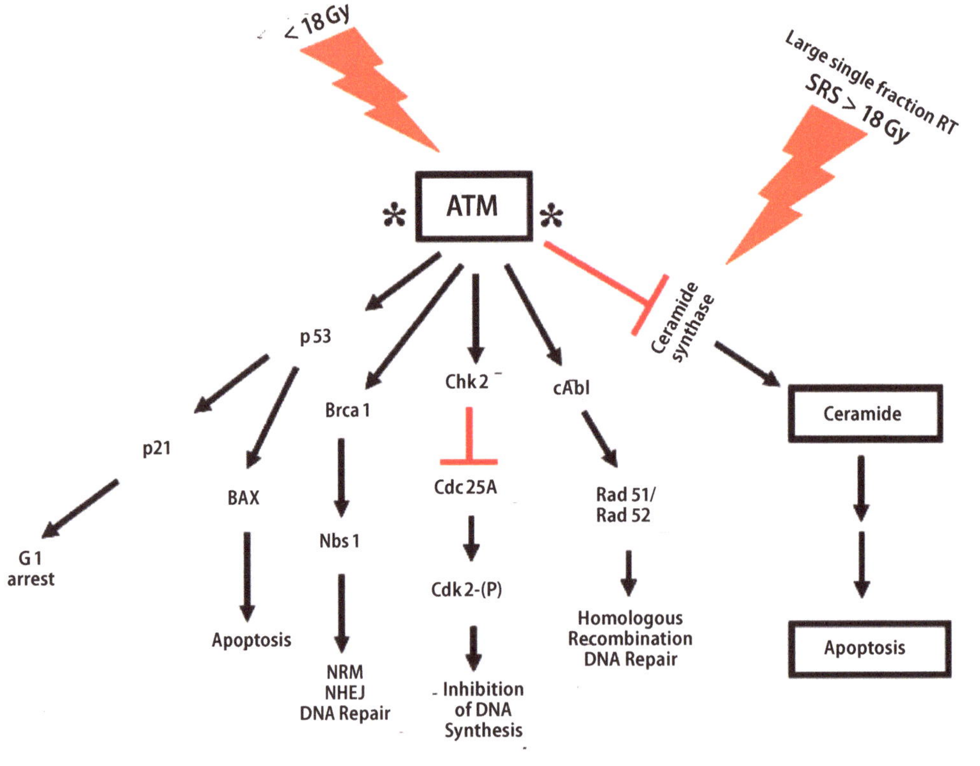

Fig. 4.4 Rationale for Hypofractionation. Conventional fractions of RT induce the DNA damage surveillance pathway by activating ATM. ATM, a serine-threonine kinase, phosphorylates a battery of proteins, some of which are shown. ATM-mediated phosphorylation events induce cell cycle arrest, inhibition of DNA synthesis, induce DNA repair, and promote survival of irradiated cells. Large single-fraction RT (>18 Gy) appears to overwhelm the radioprotective functions of ATM and induce cell death in stem cells. Thus, large single fractions of RT delivered by SRS may bypass the ATM protective functions and be more effective in killing tumor cells.

4.9 Clinical Aspects of Hypofractionation

4.9.1 Liver

The liver is usually regarded as a prototypical example of a parallel FSU organ. The major limitation of whole liver irradiation is induction of potentially lethal radiation-induced liver disease (RILD), which may develop when more than 30–35 Gy radiation is administered in conventional daily fractionations of 1.8–2 Gy [34, 35]. RILD is a clinical syndrome of anicteric hepatomegaly, ascites, and elevation of alkaline phosphatase more than the transaminases [35]. In patients with underlying cirrhosis, viral hepatitis, and primary liver cancer, however, transaminases may be elevated along with induction of reactive viral hepatitis with 2 weeks of initiation of RT [36]. The histological hallmark of RILD is veno-occlusive disease (VOD) and the formation of platelet lakes in the hepatic central veins, marked venous congestion of the central portion of the liver lobule, and eventual atrophy of hepatocytes and fibrosis/sclerosis in the pericentral region of the liver lobule. Although the underlying mechanism of the radiation injury is not clearly understood, the postulated central dogma has been injury to hepatic sinusoidal endothelial cells as the primary event in RILD. Interestingly, although hepatic VOD is not induced in irradiated rodent livers, perivenous atrophy of hepatocytes has been noted in rodent models of RILD [37]. Further investigations have demonstrated that similar to humans, the murine and rodent microvascular endothelial cells are sensitive to radiation injury [31, 38]. Thus, we postulate that RILD is induced by a combination of radiation injury to the hepatic parenchymal cells and nonparenchymal cells, such as sinusoidal endothelial cells. The constellation of RILD symptoms are then modulated by activation of the stellate cells and immune effector cells that secrete cytokines, such as transforming growth factor-β1 and β3 [39]. In support of our hypothesis, we have demonstrated that transplantation of hepatocytes following high-dose liver irradiation and hepatic resection ameliorates the histopathological manifestation of RILD and improves survival in rats [37]. This is the first description of parenchymal cell replacement modulating the late effects of RT in rodent livers. Furthermore, our experiments demonstrated that the irradiated host hepatocytes could be completely replaced by the donor cells, suggesting a role for hepatic irradiation as a preparative regimen for liver repopulation by transplanted hepatocytes [37, 40–42]. This might open new avenues for the application of hypofractionated SRS in preparative regimens of liver cell transplantation as an alternative to orthotopic organ transplantation [43].

Recently, advances in 3D conformal RT planning using non-axial beams and normal tissue complication probability (NTCP) modeling have allowed researchers to escalate RT doses in patients with unresectable hepatobiliary cancers, after exclusion of radiographically "normal" liver from the treatment field. [44, 45] Radiation doses to parts of the liver were escalated to levels as high as 90 Gy (median 58.5 Gy, range 28.5–90 Gy), in combination with concurrent chemotherapy, to obtain local tumor control in patients with unresectable hepatobiliary cancers [46, 47]. The Lyman NTCP model has been used to estimate the volume effect of normal tissue toxicity to radiation [48, 49]. This model predicts a sigmoid relationship of dose-volume effect to the probability of a normal tissue complication. The Lyman model is dependent upon three parameters: TD50, the radiation dose associated with a 50% probability of toxicity; m, the slope of the dose response curve at TD50; and n, a volume effect parameter (range 0–1). Based on a review by Emami et al. [50], Burman and colleagues estimated the Lyman parameters of liver RT: TD50 – 45 Gy; m = 0.15; and n = 0.32 [51]. The volume effect was underestimated in this report. Investigators at the University of Michigan calculated the Lyman parameters based upon data from 204 patients and estimated the volume effect to be much higher with n = 0.97 [52, 53]. Based upon this study, the tolerance dose of partially irradiating one-third and two-thirds of liver with a 5% risk of developing RILD is >100 Gy and 54 Gy, respectively, for patients with metastatic disease and 93 Gy and 47 Gy, respectively, for patients with primary liver cancer (Fig. 4.5). Mean liver dose, primary liver cancers, bromodeoxyuridine (BudR) chemotherapy, and male sex were statistically significant factors in the induction of RILD [52]. There are no data regarding what the optimal α/β for the liver is. Since RILD is a late effect of RT, the α/β is assumed to be between 2

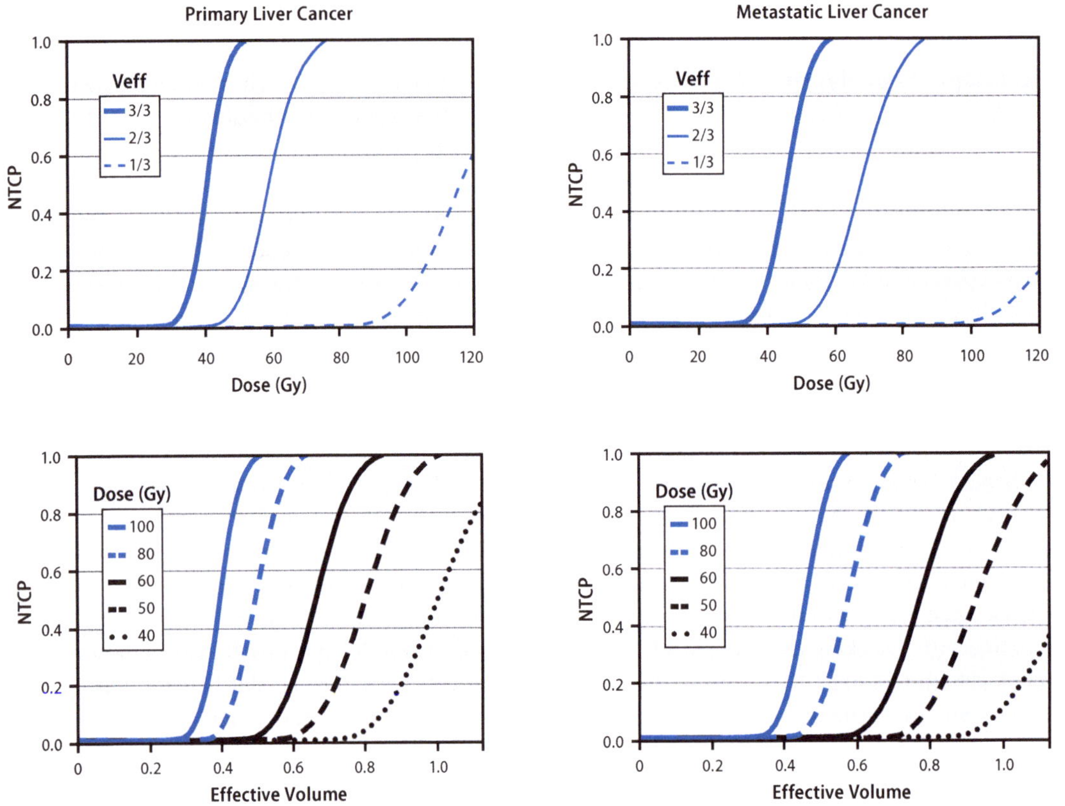

Fig. 4.5 Analysis of radiation-induced liver disease using the Lyman NTCP model [52]. Top row, NTCP versus radiation dose as a function of the volume irradiated uniformly (V_{eff}). Bottom row, NTCP versus V_{eff} as a function of reference dose (prescribed dose normalized to 1.5 Gy bid).

and 3 and has been used as such in the NTCP model corrections for dose per fraction for liver RT. The Michigan model appears to underestimate the dose threshold and the volume effect in patients with primary liver cancer with underlying cirrhosis. A recent report estimated the hepatic tolerance dose with 5% risk of RILD in patients with primary liver cancer to be 21 Gy and 6 Gy, for Child-Pugh A and B patients, respectively [54]. Hypofractionated SRS has also been used to treat focal liver cancers [55–57]. The fractionation schemes varied from 7.7 to 45 Gy per fraction, for 1 to 4 fractions to a total dose of 7.7 to 45 Gy. Herfarth et al. reported no RILD in patients treated with one radiation fraction of 14 to 26 Gy [56]. In patients with metastatic liver cancer, RILD never developed when they received SRS using dose constraints whereby 50% of the liver did not receive more than 15 Gy in 3 fractions (or 7 Gy in 1 fraction) and 30% of the liver did not receive more than 21 Gy in 3 fractions (or 12 Gy in 1 fraction) [53]. Thus, the NTCP models can describe the partial liver volume tolerance to radiation and prospectively assign the prescribed radiation dose to tumor, while maintaining the same risk of RILD. The current NTCP models for RILD do not allow for spatial dose-volume data. In patients with lung and liver tumors associated with underlying pathology, such as viral hepatitis, diffuse cirrhosis or obstructive pulmonary dysfunction, the probability of normal tissue toxicity is higher after radiation than the NTCP predictive risk, although the curve fits could be matched to these physiological variables. Prospective imaging studies with MRI or MR spectroscopy has to be conducted to determine whether these imaging modalities can estimate and detect the functional liver volume spatially for use in the NTCP-based treatment planning during SRS. In the Michigan studies, serial CT scans demonstrated atrophy in the irradiated liver lobes and hypertrophy of the untreated liver with maximal effect seen 2–3 months after RT [58]. More recently, a function has been defined as the "mean fraction-size equivalent dose" which can be adjusted to the volume treated with a normal tissue complication probability (NTCP) and has similarity to the BED model adjusted to radiation dose/fraction as:

$$MeanFED^{f_s}_{\alpha/\beta} = \sum_{n=1}^{N_b} [FED^{f_s}_{\alpha/\beta}]_i V_i$$

While these models will clearly have to be validated with carefully designed clinical trials, it has now become abundantly clear that if the effective irradiated liver volume is very low (<25%), then very high-doses in the 70–80 Gy range using conformal techniques could be used to safely treat liver tumors, provided liver function is intact.

4.9.2 Lung

Similar to the liver, the lung is a tissue with a parallel FSU architecture affected by the volume and dose/fraction of RT. There are two distinct syndromes induced by RT: pneumonitis and lung fibrosis. Radiation-induced pneumonitis is an early response that develops within 12 weeks of RT. Radiation-induced lung fibrosis, on the other hand, is a late response that develops slowly over several months or years. The lung is among the most sensitive of late-responding organs, although because of the structural organization of the FSU (the alveoli), the toxicity is dose-limiting only when a large volume is irradiated. The development of radiation pneumonitis depends upon treatment-related factors, such as total radiation dose, dose per fraction, volume of lung irradiated, and use of concurrent chemotherapy [59, 60]. Other patient-related pre-existing conditions, such as history of smoking, chronic obstructive pulmonary disease, poor lung function, and certain genetic signatures, such as loss of heterozygosity at the mannose 6-phosphate insulin-like growth factor 2 receptor [61], predisposes patients to the development of pneumonitis. The early syndrome of radiation pneumonitis consists of mild dry cough, progressive dyspnea, and low-grade fever which usually resolve after common doses of antibiotics and steroids. Abnormality in pulmonary function tests is seen more frequently than symptoms. Radiological changes in computerized tomography (CT) or single-photon-emission computed tomography (SPECT) are seen essentially in all patients. "Late" pulmonary toxicity occurs several months to years after RT as progressive chronic dyspnea associated with fibrosis of the irradiated lung which is unresponsive to steroids (mechanisms of radiation-induced fibrosis are reviewed in [62, 63]). Both acute and late effects appear to be a result of parenchymal cell loss, damage to lung microvasculature, and the modulatory effects of a cascade of inflammatory cytokines that are released in response to the radiation injury (detailed mechanisms of radiation-induced lung injury are reviewed in Marks et al. [59] and Tsoutsou et al. [60]). The plasma level of transforming growth factor beta (TGFβ) has been used as a predictor of radiation-induced lung injury [64, 65]. Several dosimetric parameters have also been developed to predict radiation-induced lung injury (reviewed in Marks et al. [66] and Miller et al. [67]). As expected, higher dose/fraction size and total dose were shown to be associated with symptomatic toxicity. Parameters such as the percent of lung volume receiving greater than 20 Gy (V_{20}), the mean lung dose, and the NTCP have been related to the incidence of clinically significant radiation lung damage [68–70]. Another study reported the effect of radiation dose on the risk of postoperative lung injury in esophageal cancer patients [71]. In that study, DVH and dose-mass histograms (DMHs) for the whole lung were used to fit Lyman NTCP models with the following estimated values (with 95% confidence intervals in parentheses): n = 1.85 (0.04, infinity), m = 0.55 (0.22, 1.02), and TD50 = 17.5 Gy (9.4 Gy, 102 Gy). Interestingly, the absolute volume of lung receiving <5 Gy, along with mean lung dose and effective dose were predictive of the radiation-induced lung injury risk. These parameters are most useful in comparing competing treatment plans for assessing relative risk of lung injury. It is believed, however, that the spatial configuration of the irradiated volume may also have functional significance. Thus, in some studies treatment of the lung lower lobe has been suggested to contribute to pulmonary dysfunction more than the upper lobe [72]. This could be secondary to the presence of a greater pulmonary reserve, underlying lung pathology, or preferential perfusion in the lower lobe of the lung. An imaging modality to define functional lung volume is required. Preliminary reports suggest that the dose distribution within perfusion-SPECT-defined lung volume is more predictive for lung injury than CT volumes [73, 74]. The diffusion capacity of carbon monoxide decreases when the radiation dose to the lung volume exceeds 13 Gy, indicating that lung tissue is exquisitely sensitive to radiation damage [75]. Since the lung FSU is dead and nonfunctional with a very low threshold of RT dose (13 Gy), it is preferred to treat a minimal lung volume with high doses than it is to treat a large lung volume with a low dose.

The dose-volume relationship for radiation-induced lung injury has low α/β ratios (<3 Gy), with data compiled for conventional fractionation. Conventional fractionated therapies (2 Gy/daily fraction for 5 weeks with boost/cone down for 1–2 additional weeks to a 70 Gy total dose) have normal tissue BED doses and tumor BEDs of 116 Gy and 84 Gy, respectively. While such treatment is intended to be curative, this BED dose is substantially lower than the BED doses currently being reported [76] in patients treated with three-dimensional conformal hypofractionated high-dose radiotherapy for small pulmonary lesions at 10 Gy/fraction to 50 and 70 Gy total dose (100 Gy and 140 Gy, tumor BED's respectively). Wada et al. reported the treatment of primary non-small-cell lung cancers (NSCLC) using high single doses (20–30 Gy) with excellent 3-year tumor control probability and less than 3% of the patients developing Grade 3 pneumonitis [55]. Of considerable interest in these early hypofractionation studies were the normal tissue responses demonstrating that the diaphragmatic lobes of the lung appeared to be more sensitive to radiation-induced damage than the apical lobes [77]. Furthermore, these investigators used mean lung dose, V_7 and V_{10} as predictors for lung injury. Since there is a dose-response relationship for NSCLC, dose escalation studies are warranted. Timmerman et al. performed a dose escalation study starting at a three-fraction regimen of 8 Gy per fraction and escalated it to 20 Gy per fraction (total dose = 60 Gy; BED = 180 Gy) without achieving maximal tolerated dose [78]. While these preliminary studies are promising, one word of caution is to not use large single fractions of SRS for the treatment of tumors near large airways [79]. The airway behaves as a serially functioning tissue, and acute hemorrhage or late fibrosis and collapse of the airway leads to downstream atelectasis and fibrosis. Since the V_{20} values for SRS treatment of small lung tumors are usually between 3–10%, the incidence of pneumonitis is low for such treatments.

4.10 Summary and Conclusions

Many of the contemporary clinical procedures used in radiation therapy are based on earlier clinical practices and empirical observations. At present, the cure rate for many cancers leaves an inordinate amount of room for improvement in standard therapies and opens the possibility for changes to the central dogmas for therapeutic techniques and technologies. In the past, modifications in fractionation regimens and concerns for therapeutic ratios between normal and malignancy were a dominant focus for designing new treatment regimens. SRS is a radical departure from the strategy of using large fields and protracted therapies for normal tissue preservation; only a fraction of the total (high, hypofractionated) dosage is received by the surrounding normal tissues. If a tumoricidal dose of radiation can be fully circumscribed to the gross tumor volume with only nominal radiation dose delivered to neighboring normal tissue volumes, then SRS could rival surgical excision for the control of primary tumors.

References

1. Coutard H. Principles of x-ray therapy of malignant diseases. *Lancet* 1934; 2:1–8.
2. Regaud C, Ferroux R. Discordance des effects de rayons X, d'une part dans le testicule, par le fractioment de la dose. *C R Soc Biol* 1927; 97:431–434.
3. Baclesse F. Hyperfractionation. *Am J Roentgenol Radium Ther Nucl Med* 1964; 91:32–36.
4. Miescher G. Erfolge der karzinombehandlung an der Dermatologischen Klink Zurich. Einzeitige Hochstdosis and Fraktionierte Behandlung. *Strahlentherapie* 1934; 49:65–81.
5. Strandqvist M. Studien uber die cumulative Wirkung der Rontgenstrahlen bei Fraktionierung. Erfahrungen aus dem Radiumhemmet an 280 Haut und Lippenkarzinomen. *Acta Radiol.* 1944;55 (Suppl):1–300.

6. Ellis F. Dose, time and fractionation: a clinical hypothesis. *Clin Radiol* 1969; 20:1-7.
7. Withers HR, Thames HD, Jr., Peters LJ. A new isoeffect curve for change in dose per fraction. *Radiother Oncol* 1983; 1:187-191.
8. Puck TT, Markus PI. Action of x-rays on mammalian cells. *J Exp Med* 1956; 103:653-666.
9. Bentzen SM. Preventing or reducing late side effects of radiation therapy: radiobiology meets molecular pathology. *Nat Rev Cancer* 2006; 6:702-713.
10. Bentzen SM, Saunders MI, Dische S. Repair halftimes estimated from observations of treatment-related morbidity after CHART or conventional radiotherapy in head and neck cancer. *Radiother Oncol* 1999; 53:219-226.
11. Orton CG. High-dose-rate brachytherapy may be radiobiologically superior to low-dose rate due to slow repair of late-responding normal tissue cells. *Int J Radiat Oncol Biol Phys* 2001; 49:183-189.
12. Guttenberger R, Thames HD, Ang KK. Is the experience with CHART compatible with experimental data? A new model of repair kinetics and computer simulations. *Radiother Oncol* 1992; 25:280-286.
13. Moeller BJ, Cao Y, Li CY, *et al.* Radiation activates HIF-1 to regulate vascular radiosensitivity in tumors: role of reoxygenation, free radicals, and stress granules. *Cancer Cell* 2004; 5:429-441.
14. Moeller BJ, Dreher MR, Rabbani ZN, *et al.* Pleiotropic effects of HIF-1 blockade on tumor radiosensitivity. *Cancer Cell* 2005; 8:99-110.
15. Garcia-Barros M, Paris F, Cordon-Cardo C, *et al.* Tumor response to radiotherapy regulated by endothelial cell apoptosis. *Science* 2003; 300:1155-1159.
16. Paris F, Fuks Z, Kang A, *et al.* Endothelial apoptosis as the primary lesion initiating intestinal radiation damage in mice. *Science* 2001; 293:293-297.
17. Ch'ang HJ, Maj JG, Paris F, *et al.* ATM regulates target switching to escalating doses of radiation in the intestines. *Nat Med* 2005; 11:484-490.
18. Lavin MF, Birrell G, Chen P, *et al.* ATM signaling and genomic stability in response to DNA damage. *Mutat Res* 2005; 569:123-132.
19. Lavin MF, Delia D, Chessa L. ATM and the DNA damage response. Workshop on ataxia-telangiectasia and related syndromes. *EMBO Rep* 2006; 7:154-160.
20. Savitsky K, Bar-Shira A, Gilad S, *et al.* A single ataxia telangiectasia gene with a product similar to PI-3 kinase [see comments]. *Science* 1995; 268:1749-1753.
21. Elson A, Wang Y, Daugherty CJ, *et al.* Pleiotropic defects in ataxia-telangiectasia protein-deficient mice. *Proceedings of the National Academy of Sciences of the United States of America* 1996; 93:13084-13089.
22. Herzog KH, Chong MJ, Kapsetaki M, *et al.* Requirement for Atm in ionizing radiation-induced cell death in the developing central nervous system. *Science* 1998; 280:1089-1091.
23. Heyer BS, MacAuley A, Behrendtsen O, *et al.* Hypersensitivity to DNA damage leads to increased apoptosis during early mouse development. *Genes Dev* 2000;14:2072-2084.
24. Haimovitz-Friedman A. Radiation-induced signal transduction and stress response. *Radiat Res* 1998; 150-S102-108.
25. Liao WC, Haimovitz-Friedman A, Persaud RS, *et al.* Ataxia telangiectasia-mutated gene product inhibits DNA damage-induced apoptosis via ceramide synthase. *J Biol Chem* 1999; 274:17908-17917.
26. Weichselbaum R. Radiation's outer limits. *Nat Med* 2005; 11:477-478.
27. Guha C, Guha U, Tribius S, *et al.* Antisense ATM gene therapy: a strategy to increase the radiosensitivity of human tumors. *Gene Ther* 2000; 7:852-858.
28. Fan Z, Chakravarty P, Alfieri A, *et al.* Adenovirus-mediated antisense ATM gene transfer sensitizes prostate cancer cells to radiation. *Cancer Gene Ther* 2000; 7:1307-1314.
29. Hickson I, Zhao Y, Richardson CJ, *et al.* Identification and characterization of a novel and specific inhibitor of the ataxia-telangiectasia mutated kinase ATM. *Cancer Res* 2004; 64:9152-9159.
30. Fuks Z, Persaud RS, Alfieri A, *et al.* Basic fibroblast growth factor protects endothelial cells against radiation-induced programmed cell death in vitro and in vivo. *Cancer Res* 1994; 54:2582-2590.
31. Maj JG, Paris F, Haimovitz-Friedman A, *et al.* Microvascular function regulates intestinal crypt response to radiation. *Cancer Res* 2003; 63:4338-4341.
32. Potten CS, O'Shea JA, Farrell CL, *et al.* The effects of repeated doses of keratinocyte growth factor on cell proliferation in the cellular hierarchy of the crypts of the murine small intestine. *Cell Growth Differ* 2001; 12:265-275.
33. Du XX, Doerschuk CM, Orazi A, *et al.* A bone marrow stromal-derived growth factor, interleukin-11, stimulates recovery of small intestinal mucosal cells after cytoablative therapy. *Blood* 1994; 83:33-37.
34. Bearman SI. The syndrome of hepatic veno-occlusive disease after marrow transplantation. *Blood* 1995; 85:3005-3020.
35. Lawrence TS, Robertson JM, Anscher MS, *et al.* Hepatic toxicity resulting from cancer treatment. *International Journal of Radiation Oncology, Biology, Physics* 1995; 31:1237-1248.
36. Cheng JC, Wu JK, Lee PC, *et al.* Biologic susceptibility of hepatocellular carcinoma patients treated with radiotherapy to radiation-induced liver disease. *Int J Radiat Oncol Biol Phys* 2004; 60:1502-1509.
37. Guha C, Sharma A, Gupta S, *et al.* Amelioration of radiation-induced liver damage in partially hepatectomized rats by hepatocyte transplantation. *Cancer Res* 1999; 59:5871-5874.
38. Guha C. Unpublished findings.
39. Seong J, Kim SH, Chung EJ, *et al.* Early alteration in TGF-beta mRNA expression in irradiated rat liver. *Int J Radiat Oncol Biol Phys* 2000; 46:639-643.
40. Guha C, Parashar B, Deb NJ, *et al.* Liver irradiation: a potential preparative regimen for hepatocyte transplantation. *Int J Radiat Oncol Biol Phys* 2001; 49:451-457.
41. Guha C, Parashar B, Roy-Chowdhury N, *et al.* Complete long-term normalization of serum bilirubin levels in Gunn rats after hepatocyte transplantation following partial hepatectomy and liver irradiation. *Hepatology* 1999; 30:108.
42. Guha C, Yamanouchi K, Jiang J, *et al.* Feasibility of Hepatocyte Transplantation-Based Therapies for Primary Hyperoxalurias. *Am J Nephrol* 2005; 25:161-170.
43. Zhou H, Yamanouchi K, Liu L, *et al.* A new paradigm for tissue regeneration: preparative irradiation for cell-based

therapies as an alternative to organ transplantation. *Int J Radiat Oncol Biol Phys* 2006; 66:S98–S99.

44. Lawrence TS, Dworzanin LM, Walker-Andrews SC, *et al.* Treatment of cancers involving the liver and porta hepatis with external beam irradiation and intraarterial hepatic fluorodeoxyuridine. *International Journal of Radiation Oncology, Biology, Physics* 1991; 20:555–561.
45. Robertson JM, Lawrence TS, Dworzanin LM, *et al.* Treatment of primary hepatobiliary cancers with conformal radiation therapy and regional chemotherapy. *Journal of Clinical Oncology* 1993; 11:1286–1293.
46. Robertson JM, Lawrence TS, Andrews JC, *et al.* Long-term results of hepatic artery fluorodeoxyuridine and conformal radiation therapy for primary hepatobiliary cancers. *International Journal of Radiation Oncology, Biology, Physics* 1997; 37:325–330.
47. Dawson LA, McGinn CJ, Normolle D, *et al.* Escalated focal liver radiation and concurrent hepatic artery fluorodeoxyuridine for unresectable intrahepatic malignancies. *J Clin Oncol* 2000; 18:2210–2218.
48. Lyman JT. Complication probability as assessed from dose-volume histograms. *Radiat Res Suppl* 1985; 8:S13–19.
49. Lyman JT. Normal tissue complication probabilities: variable dose per fraction. *Int J Radiat Oncol Biol Phys* 1992; 22:247–250.
50. Emami B, Lyman J, Brown A, *et al.* Tolerance of normal tissue to therapeutic irradiation. *Int J Radiat Oncol Biol Phys* 1991; 21:109–122.
51. Burman C, Kutcher GJ, Emami B, *et al.* Fitting of normal tissue tolerance data to an analytic function. *Int J Radiat Oncol Biol Phys* 1991; 21:123–135.
52. Dawson LA, Normolle D, Balter JM, *et al.* Analysis of radiation-induced liver disease using the Lyman NTCP model. *Int J Radiat Oncol Biol Phys* 2002; 53:810–821.
53. Dawson LA, Ten Haken RK. Partial volume tolerance of the liver to radiation. *Semin Radiat Oncol* 2005; 15:279–283.
54. Xu ZY, Liang SX, Zhu J, *et al.* Prediction of radiation-induced liver disease by Lyman normal-tissue complication probability model in three-dimensional conformal radiation therapy for primary liver carcinoma. *Int J Radiat Oncol Biol Phys* 2006; 65:189–195.
55. Wada H, Takai Y, Nemoto K, *et al.* Univariate analysis of factors correlated with tumor control probability of three-dimensional conformal hypofractionated high-dose radiotherapy for small pulmonary or hepatic tumors. *Int J Radiat Oncol Biol Phys* 2004; 58:1114–1120.
56. Herfarth KK, Debus J, Lohr F, *et al.* Stereotactic single-dose radiation therapy of liver tumors: results of a phase I/II trial. *J Clin Oncol* 2001; 19:164–170.
57. Blomgren H, Lax I, Naslund I, *et al.* Stereotactic high dose fraction radiation therapy of extracranial tumors using an accelerator. Clinical experience of the first thirty-one patients. *Acta Oncol* 1995; 34:861–870.
58. Yamasaki SA, Marn CS, Francis IR, *et al.* High-dose localized radiation therapy for treatment of hepatic malignant tumors: CT findings and their relation to radiation hepatitis. *AJR. American Journal of Roentgenology* 1995; 165:79–84.
59. Marks LB, Yu X, Vujaskovic Z, *et al.* Radiation-induced lung injury. *Semin Radiat Oncol* 2003; 13:333–345.
60. Tsoutsou PG, Koukourakis MI. Radiation pneumonitis and fibrosis: Mechanisms underlying its pathogenesis and implications for future research. *Int J Radiat Oncol Biol Phys* 2006; 66:1281–1293.
61. Kong FM, Anscher MS, Sporn TA, *et al.* Loss of heterozygosity at the mannose 6-phosphate insulin-like growth factor 2 receptor (M6P/IGF2R) locus predisposes patients to radiation-induced lung injury. *Int J Radiat Oncol Biol Phys* 2001; 49:35–41.
62. O'Sullivan B, Levin W. Late radiation-related fibrosis: pathogenesis, manifestations, and current management. *Semin Radiat Oncol* 2003; 13:274–289.
63. Rodemann HP, Bamberg M. Cellular basis of radiation-induced fibrosis. *Radiotherapy & Oncology* 1995; 35:83–90.
64. Anscher MS, Murase T, Prescott DM, *et al.* Changes in plasma TGF beta levels during pulmonary radiotherapy as a predictor of the risk of developing radiation pneumonitis. *Int J Radiat Oncol Biol Phys* 1994; 30:671–676.
65. Anscher MS, Kong FM, Andrews K, *et al.* Plasma transforming growth factor beta1 as a predictor of radiation pneumonitis. *Int J Radiat Oncol Biol Phys* 1998; 41:1029–1035.
66. Marks LB. Dosimetric predictors of radiation-induced lung injury. *Int J Radiat Oncol Biol Phys* 2002; 54:313–316.
67. Miller KL, Shafman TD, Marks LB. A practical approach to pulmonary risk assessment in the radiotherapy of lung cancer. *Semin Radiat Oncol* 2004; 14:298–307.
68. Kwa SL, Lebesque JV, Theuws JC, *et al.* Radiation pneumonitis as a function of mean lung dose: an analysis of pooled data of 540 patients. *Int J Radiat Oncol Biol Phys* 1998; 42:1–9.
69. Graham MV, Purdy JA, Emami B, *et al.* Clinical dose-volume histogram analysis for pneumonitis after 3D treatment for non-small cell lung cancer (NSCLC). *Int J Radiat Oncol Biol Phys* 1999; 45:323–329.
70. Bradley J, Graham MV, Winter K, *et al.* Toxicity and outcome results of RTOG 9311: a phase I-II dose-escalation study using three-dimensional conformal radiotherapy in patients with inoperable non-small-cell lung carcinoma. *Int J Radiat Oncol Biol Phys* 2005; 61:318–328.
71. Tucker SL, Liu HH, Wang S, *et al.* Dose-volume modeling of the risk of postoperative pulmonary complications among esophageal cancer patients treated with concurrent chemoradiotherapy followed by surgery. *Int J Radiat Oncol Biol Phys* 2006; 66:754–761.
72. Yorke ED, Jackson A, Rosenzweig KE, *et al.* Dose-volume factors contributing to the incidence of radiation pneumonitis in non-small-cell lung cancer patients treated with three-dimensional conformal radiation therapy. *Int J Radiat Oncol Biol Phys* 2002; 54:329–339.
73. Seppenwoolde Y, Muller SH, Theuws JC, *et al.* Radiation dose-effect relations and local recovery in perfusion for patients with non-small-cell lung cancer. *Int J Radiat Oncol Biol Phys* 2000; 47:681–690.
74. Lind PA, Marks LB, Hollis D, *et al.* Receiver operating characteristic curves to assess predictors of radiation-induced symptomatic lung injury. *Int J Radiat Oncol Biol Phys* 2002; 54:340–347.
75. Gopal R, Tucker SL, Komaki R, *et al.* The relationship between local dose and loss of function for irradiated lung. *Int J Radiat Oncol Biol Phys* 2003; 56:106–113.
76. Kyas I, Hof H, Debus J, *et al.* Prediction of radiation-induced changes in the lung after stereotactic radiosurgery of non-small-cell lung cancer. *Int J Radiat Oncol Biol Phys* 2006.

77. Xia T, Li H, Sun Q, *et al.* Promising clinical outcome of stereotactic body radiation therapy for patients with inoperable Stage I/II non-small-cell lung cancer. *Int J Radiat Oncol Biol Phys* 2006; 66:117–125.
78. Timmerman R, Papiez L, McGarry R, *et al.* Extracranial stereotactic radioablation: results of a phase I study in medically inoperable stage I non-small cell lung cancer. *Chest* 2003; 124:1946–1955.
79. Kavanagh BD, McGarry RC, Timmerman RD. Extracranial radiosurgery (stereotactic body radiation therapy) for oligometastases. *Semin Radiat Oncol* 2006; 16:77–84.

^{18}FDG-PET/CT for Treatment Planning and Follow-Up

5

Medhat M. Osman, Allison Wall, Matthew D. Miller, Nghi Nguyen, Dana A. Oliver, Richard D. Bucholz, and Bruce J. Walz

CONTENTS

5.1 Introduction

Stereotactic radiosurgery is well-established in the treatment of intracranial lesions, but its use in extracranial lesions is relatively new. Stereotactic radiation delivery technologies such as the CyberKnife® (Accuray Incorporated, Sunnyvale, CA) are being used with increasing frequency to treat lesions outside the cranium, such as in the spine, thorax, and abdomen; lesions that were not previously amenable to radiosurgical treatment due to inherent motion. The critical role of imaging technologies in radiosurgery, both for planning treatments and assessing their effects, makes an understanding of developments in these technologies imperative for radiosurgery professionals. In this chapter, we review how positron emission tomography (PET) is combined with computed tomography (CT) to enhance disease staging, radiosurgery and radiotherapy planning, and follow-up.

5.2 Cancer Imaging

Modern imaging technologies visualize various aspects of cancer in a non-invasive way by taking advantage of different anatomic, molecular, and functional attributes common to malignant cells. In the last few years, we witnessed a great deal of advancements in anatomic imaging modalities such as CT and magnetic resonance imaging (MRI). Based on its fast image acquisition and high spatial resolution, CT is particularly well suited for use in daily management of cancer patients. CT, however, has its limitations including difficulty diagnosing malignancy in a normal-sized lymph node, difficulty distinguishing between an enlarged lymph node due to malignancy or benign disease, and difficulty differentiating disease recurrence from post-therapeutic changes.

Whole-body (WB) PET imaging with [^{18}F]2-fluoro-2-deoxy-D-glucose (FDG) is a functional imaging modality that addresses many of the limitations of CT. FDG-PET detects the accelerated glucose metabolism common in malignancy. ^{18}F-FDG PET imaging is used in the diagnosis, staging, and follow-up of many cancers with accuracies ranging from 80% to 90% [1]. Because of the high accuracy of FDG-PET for molecular imaging of disease biology, the U.S. Food and Drug Administration (FDA) has approved FDG-PET for all cancers. In addition, the Centers for Medicare and Medicaid Services (CMS) have approved Medicare reimbursement of FDG-PET imaging for different types of cancer. FDG-PET, however, has limitations. Increased FDG-avidity is not limited to malignancy, which can result in false positives.

On the other hand not all malignancies are FDG-avid, which can produce false negatives. In addition, whole-body FDG-PET is a slow imaging technique, requiring two hours for a whole body (WB) field of view (FOV) as compared to less than a minute for the same FOV using a multi-slice CT scanner. Furthermore, while FDG-PET facilitates lesion identification, localization of such lesions can be complicated due to the lack of anatomical landmarks.

5.3 PET/CT Image Fusion

Fusion of PET and CT images enhances the inherent clinical potential of both techniques and provides knowledge that is greater than the sum of information provided by each modality alone. Fusion of PET and CT images can be performed at three levels: visual, software, and hardware fusion. In traditional visual fusion, a physician compares separate PET and CT images viewed next to each other when available, i.e., fusion takes place in the physician's head. Visual fusion has often been considered sufficient, however, clinical practice proved this technique to be sub-optimal and often unsuccessful, underscoring the need for a formal fusion mechanism. The concept of "anatometabolic" software fusion of PET and CT studies using fiducial and anatomical landmarks, was first introduced over 15 years ago [2]. Since then, several sophisticated software fusion algorithms have been developed and validated, particularly in the brain. Software fusion is often logistically challenging owing to differences in the patient's positioning when imaged by two different modalities on two different occasions. The alignment process is labor intensive and far from routine at most medical centers. This situation changed dramatically with the recent introduction of hardware fusion in 1998 using the combined PET/CT scanner. PET/CT systems can acquire anatomical and functional information in the same examination at nearly the same time [3]. In such systems, CT data can be used for PET attenuation correction (AC) instead of the traditionally used germanium-68 [4]. It is safe to say that the most exciting development in PET is the emergence of combined PET/CT imaging devices (Fig. 5.1). The combination of form (CT) and function (PET) has several advantages [5]. First, biological and anatomical WB can be performed in one examination. Second, because of limited patient motion, due to the almost simultaneous acquisition of PET and CT images, optimal fusion of biological and anatomical images can be achieved. Finally, the combination of anatomical landmarks provided by CT greatly facilitates the assignment of PET depicted biological abnormalities to specific anatomical structures.

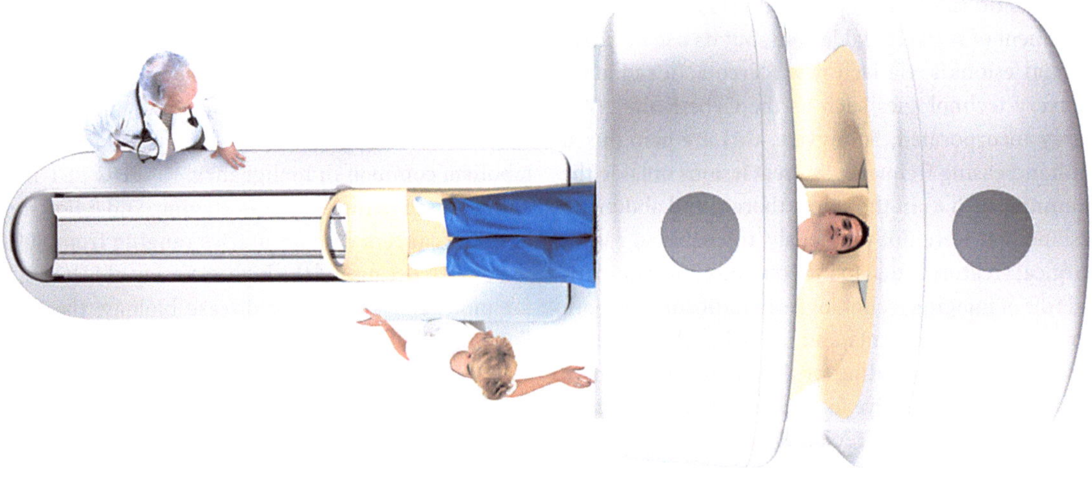

Fig. 5.1 GEMINI TF at Saint Louis University Hospital; a 64-slice PET/CT scanner with TruFlight technology from Philips Medical Systems.

5.4 Clinical Data Supporting PET/CT

PET/CT is still in its infancy; however, several studies published over the last few years demonstrated that PET/CT transforms image fusion from primarily a research tool to everyday clinical practice. In addition, these studies proved that PET/CT has a higher diagnostic accuracy than PET or CT alone or visually correlated PET and CT. There are more important data on the use of PET/CT in lung cancer than any other type of malignancy. One study evaluated the impact of PET/CT over PET in the localization and the confidence of diagnosis in patients with non-small cell lung cancer (NSCLC) [5]. PET/CT led to a 32% reduction in the number of "probable and equivocal" lesions and a 41% increase in the number of "definite" localizations. Another prospective study assessed the diagnostic accuracy of PET/CT in 50 patients with NSCLC [6]. Integrated PET/CT provided information beyond that provided by conventional visual correlation of PET and CT in 41% of the patients. Additional information included precise localization of lymph nodes, identification of chest wall infiltration, correct differentiation between inflammation and malignancy, and localization of distant metastases. Integrated PET/CT had better diagnostic accuracy than PET alone, CT alone, or visual correlation of PET and CT. A more recent prospective study assessed the value of PET/CT in the diagnosis and clinical management of suspected recurrent NSCLC in 42 patients [7]. The sensitivity, specificity, and positive and negative predictive values of PET/CT for diagnosis of recurrence were 96%, 82%, 89%, and 93% compared with 96%, 53%, 75%, and 90%, respectively, for PET. Furthermore, PET/CT changed the PET lesion classification in 52% of patients, by determining the precise localization of sites with increased FDG uptake. In addition, PET/CT changed the management of 29% of patients by eliminating previously planned diagnostic procedures, by initiating a previously unplanned treatment option, or by inducing a change in the planned therapeutic approach.

The added value of PET/CT over PET alone and CT alone is not limited to NSCLC. A recent prospective study assessed the clinical performance of a combined PET/CT system in patients with various types of cancers [8]. In 204 patients with 586 suspicious lesions, PET/CT provided additional information over the separate interpretation of PET and CT in 99 patients (49%) with 178 sites (30%). As expected, PET/CT precisely defined the anatomic location of malignant FDG uptake in 6%, and it lead to retrospective lesion detection on PET or CT in 8%. Moreover, PET/CT improved characterization of equivocal lesions as definitely benign in 10% of sites and as definitely malignant in 5% of sites. The results of PET/CT had an impact on the management of 28 patients (14%), obviated the need for further evaluations in 5 patients, guided further diagnostic procedures in 7 patients, and assisted in planning therapy for 16 patients.

Data comparing PET/CT to other imaging modalities are not limited to PET and/or CT. One prospective study compared the staging accuracies of both WB PET/CT and WB MRI for different malignant diseases [9]. In 98 patients, the TNM stage was correctly determined in 77% with PET/CT and in 54% with MRI. Moreover, 12% of patients scanned with PET/CT resulted in management changes as compared to 2% with MRI. Authors of this work advocated the use of PET/CT as a first-line imaging modality for WB tumor staging, restaging, and post therapy response assessment in different types of cancer.

5.5 PET and PET/CT in Radiation Therapy

Conventional staging of patients scheduled to receive radiation therapy (RT) is typically based on anatomic imaging modalities such as CT or MRI. Such anatomic staging, however, may be inadequate and/or inaccurate, resulting into several problems. First, inaccurate identification of macroscopic tumors may result in suboptimal treatment volumes in some patients and unnecessarily large volumes in others. Second, interobserver variability may be caused by difficulty identifying tumor boundaries due to atelectasis or pleural effusion. Similar problems exist in using anatomic imaging modalities in assessing the response to RT. Many centers are beginning to adopt ^{18}F-FDG PET before RT, for more accurate staging, and after RT, for better assessment of response to RT. Incorporating PET in RT planning typically has an impact in the decision-making process prior to RT, and results in changes in therapy in about 25% of patients.

Initial data suggest that PET imaging is useful in RT planning (RTP) based on the concept of biologic tumor volume (BTV) as opposed to gross tumor volume (GTV) determined by traditional CT planning. The planning target volume can be altered in one of two ways. First, the PET images may show more extensive local disease or distant metastases which necessitate enlargement of the planned radiation field or, in extreme cases, termination of radiotherapy. Second, the PET scan may identify a smaller volume of viable tumor than suspected by CT planning thus demanding a reduction in the size of the radiation field so that nearby normal tissue is spared. Studies incorporating PET data using ^{18}F-FDG images have shown significant alteration in tumor volume coverage in approximately 30–60% of the cases [10–13]. A recent study evaluated the delineation of GTV and clinical target volume (CTV) in patients with extracranial malignancies [14]. In this study, PET altered the GTV or CTV in 44% of studied patients. Methods for incorporating PET into the RTP include visual, side-by-side comparisons, image overlays, direct fusion of PET and CT images, and PET/CT simulation. In PET/CT, in addition to other staging information performed prior to the initiation of RT, the CT data from the PET/CT exam can be used for RTP, provided the CT data are properly acquired. Moreover, when PET data are used in addition to CT data for planning, there is more consistent agreement among observers on how to delineate GTV [15]. PET/CT has also demonstrated usefulness in post-treatment restaging since clinically relevant treatment responses are often not apparent on morphological imaging studies [16, 17].

Nevertheless, controversies still remain in the PET literature regarding how to define the region of interest (ROI) needed for tumor volume delineation as well as response assessment [18].

5.6 Added Value of PET/CT in CyberKnife Stereotactic Radiosurgery: Initial Experience

Our university was one of few academic institutions in the country to procure a state-of-the-art CyberKnife frameless stereotactic radiosurgery delivery system. In addition, we had a 16-slice PET/CT scanner, which has recently been replaced by the nation's first 64-slice PET/CT scanner with True Flight technology. Although PET/CT is in rapid dissemination and plays a growing role in radiation therapy, there is a gap in the literature regarding the added value of PET/CT in CyberKnife treatment. The presence of both devices in the same facility at our institution gave us a unique opportunity to address this gap by systematically studying the added value of PET/CT in patients undergoing CyberKnife therapy [19].

Our initial experience with PET/CT scanning prior to CyberKnife therapy in 24 patients showed that value was added by providing clinically useful staging and treatment response data. Approximately 60% of patients studied had no change to known sites of disease when the PET/CT was compared to other imaging modalities previously employed. In the last group, confirmation of local disease was useful as it indicated that treatment with CyberKnife, a local treatment, was appropriate. The remaining 40% of the patients had a change of stage when the PET/CT was compared to other imaging studies. See Figure 5.2 for an instance in which upstaging occurred and Figure 5.3 for an example of downstaging.

This is similar to recently published literature in which PET/CT frequently provided statistically significant improvements over PET or CT alone in staging and restaging of different cancers [20]. In our study, however, far more of our patients were downstaged instead of upstaged, likely reflecting physician patient selection and extensive workup before CyberKnife evaluation of this heavily pretreated population. More accurate staging can influence the decision to proceed with CyberKnife by improving patient selection.

Of the 24 treated patients, 16 (67%) patients with 20 lesions had PET/CT scans before and after the CyberKnife treatment. For these patients, we compared the volume (in cm^3) and mean standardized uptake value (SUV) max before and after CyberKnife treatment. There was a statistically significant difference in the mean pre-treatment volume (17.5 cm^3) compared to the mean post-treatment volume (8.7 cm^3), $p=0.003$. Additionally, a statistically significant difference in metabolic response was found when pre-treatment max SUV (5.9) was compared to post-treatment max SUV (3.0; $p=0.001$). After CyberKnife therapy, nine patients (56%) had continued response

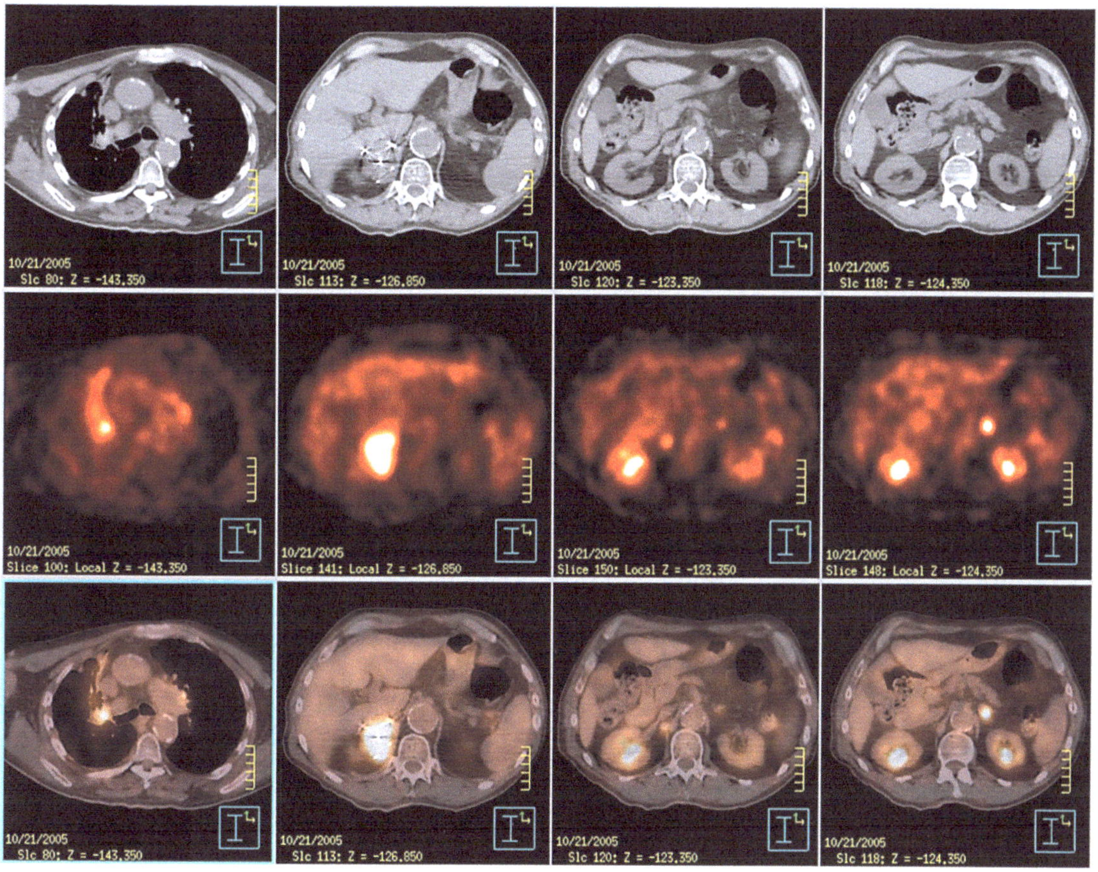

Fig. 5.2 These images are from a 70-year-old male with recurrent small-cell lung cancer evaluated for CyberKnife treatment of an isolated right adrenal lesion. PET/CT showed previously unsuspected mediastinal, left adrenal, and retroperitoneal disease.

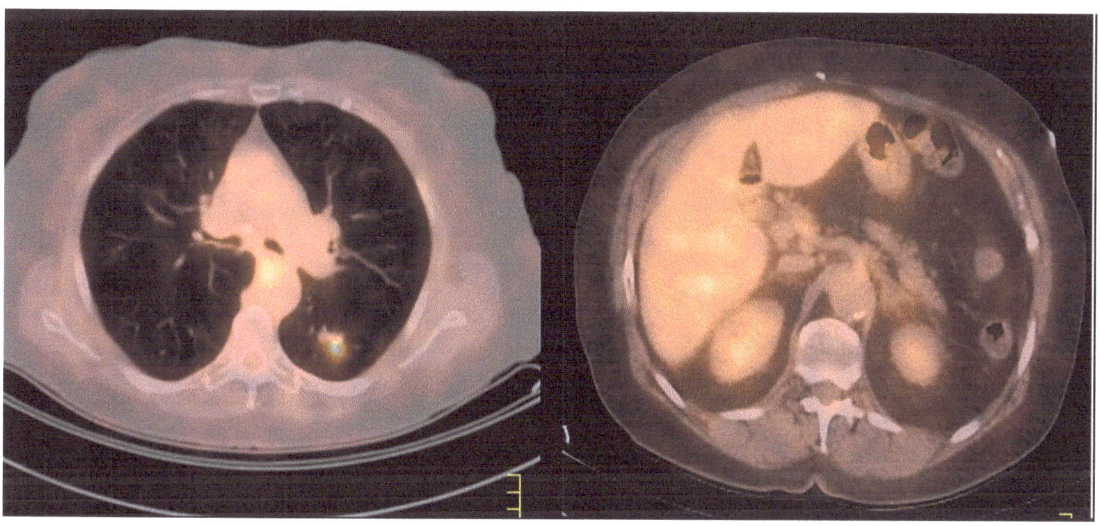

Fig. 5.3 A 73-year-old female with a new lower left lobe lesion and enlarged left adrenal gland on CT scan. PET/CT showed the enlarged adrenal gland to be non-FDG avid, thus benign.

at last imaging. Seven patients (44%) had failed therapy based on post-treatment imaging (2 local failures, 2 with distant metastases, and 3 failed both locally and with distant metastases).

Inflammatory changes can persist for several months after radiation therapy; inflammatory changes can cause false positives in imaging and be mistaken for progressive disease. Therefore, a 3–6 month wait is typically recommended before using PET/CT imaging in assessing response to RT. The proper timing of post-CyberKnife PET/CT imaging, however, is not known at this point. One of our patients (Fig. 5.4) had a prior left pneumonectomy, CyberKnife to a solitary lesion in the right upper lobe, and four post-treatment PET/CT scans from day 46 to day 251. After an initial response,

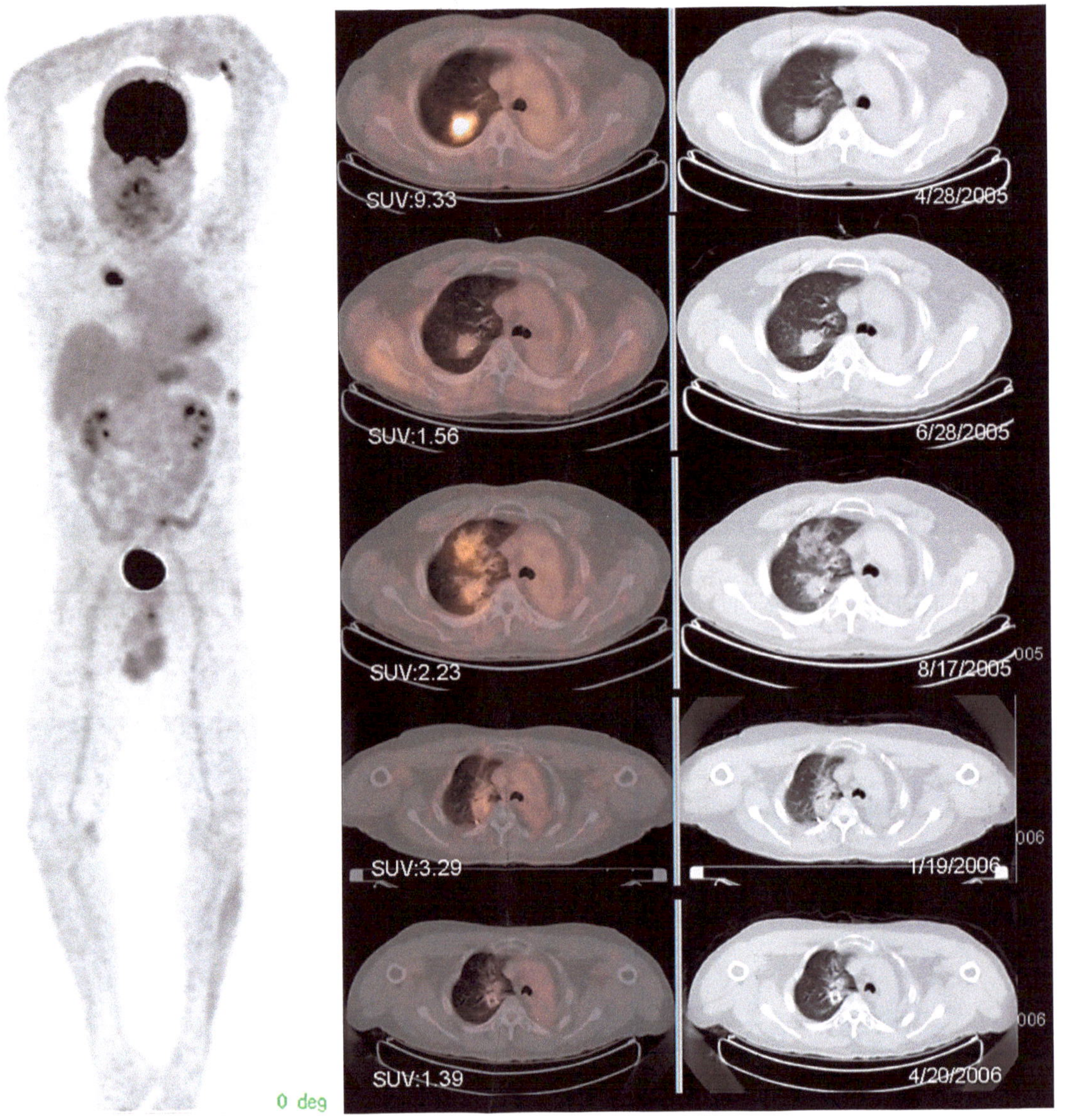

Fig. 5.4 These images are from a 67-year-old male with history of left lung squamous cell carcinoma, status post left pneumonectomy in 2001. In April 2005 he was diagnosed with the same cancer in the right lung and was treated with CyberKnife. Post-therapy PET/CT scan in June 2005 showed decrease in metabolic activity consistent with good response to therapy. Follow-up scans in August 2005 and January 2006 revealed increased metabolic activity which it was thought may have represented disease progression; however, a subsequent scan in April 2006 showed decrease in metabolic activity without any interval therapy, which is consistent with improving post-radiation changes.

measured by volume and mean SUV decrease, the lesion grew larger and more intensely FDG-avid in the setting of pneumonitis peaking at day 186 post-treatment. Surprisingly, the lesion began to regress in size and FDG-avidity by day 251 with no further therapy. This suggests that a delayed period of inflammation could be mistaken for failure unless followed closely with serial PET/CT scans. The optimal way to answer the question of when this phenomenon peaks after CyberKnife radiotherapy should be determined by a PET/CT scan at approximately 30 days post-treatment and at regular intervals in multiple patients. The ongoing prospective study, at our institution, will attempt to answer this interesting question by requiring a PET/CT 30–60 days after completion of CyberKnife and monthly thereafter. The change in volume and mean SUV over time and incorporation of PET/CT imaging data in radiation treatment planning for CyberKnife also are planned in subsequent prospective analyses.

5.7 Conclusion

The migration from PET to PET/CT is inevitable and the combined PET/CT designs will continue to benefit from recent advancements in instrumentation and protocol development. PET/CT continues to play an evolving role in the management of patients undergoing stereotactic radiosurgery by providing more accurate staging, precise tumor volume definition, and earlier assessment of response to therapy.

References

1. Czernin J, Phelps ME. Positron emission tomography scanning: current and future applications. *Annu Rev Med* 2002; 53:89–112.
2. Wahl RL, Quint LE, Cieslak RD, *et al.* "Anatometabolic" tumor imaging: fusion of FDG PET with CT or MRI to localize foci of increased activity. *J Nucl Med* 1993; 34:1190–1197.
3. Beyer T, Townsend DW, Brun T, *et al.* A combined PET/CT scanner for clinical oncology. *J Nucl Med* 2000; 41:1369–1379.
4. Akhurst T, Chisin R. Hybrid PET/CT machines: optimized PET machines for the new millennium? *J Nucl Med* 2000;41:961–963.
5. Osman M, Cohade C, Leal J, *et al.* Direct comparison of FDG-PET and PET-CT imaging in staging and restaging patients with lung cancer. *Journal of Nuclear Medicine* 2002; 43:151P.
6. Lardinois D, Weder W, Hany TF, *et al.* Staging of non-small-cell lung cancer with integrated positron-emission tomography and computed tomography. *N Engl J Med* 2003; 348:2500–2507.
7. Keidar Z, Haim N, Guralnik L, *et al.* PET/CT using 18F-FDG in suspected lung cancer recurrence: diagnostic value and impact on patient management. *J Nucl Med* 2004; 45:1640–1646.
8. Bar-Shalom R, Yefremov N, Guralnik L, *et al.* Clinical performance of PET/CT in evaluation of cancer: additional value for diagnostic imaging and patient management. *J Nucl Med* 2003; 44:1200–1209.
9. Antoch G, Vogt FM, Freudenberg LS, *et al.* Whole-body dual-modality PET/CT and whole-body MRI for tumor staging in oncology. *Jama* 2003; 290:3199–3206.
10. Bradley JD, Perez CA, Dehdashti F, *et al.* Implementing biologic target volumes in radiation treatment planning for non-small cell lung cancer. *J Nucl Med* 2004;45 Suppl 1:96S–101S.
11. Erdi YE, Rosenzweig K, Erdi AK, *et al.* Radiotherapy treatment planning for patients with non-small cell lung cancer using positron emission tomography (PET). *Radiother Oncol* 2002; 62:51–60.
12. Kiffer JD, Berlangieri SU, Scott AM, *et al.* The contribution of 18F-fluoro-2-deoxy-glucose positron emission tomographic imaging to radiotherapy planning in lung cancer. *Lung Cancer* 1998; 19:167–177.
13. Nestle U, Walter K, Schmidt S, *et al.* 18F-deoxyglucose positron emission tomography (FDG-PET) for the planning of radiotherapy in lung cancer: high impact in patients with atelectasis. *Int J Radiat Oncol Biol Phys* 1999; 44:593–597.
14. Brianzoni E, Rossi G, Ancidei S, *et al.* Radiotherapy planning: PET/CT scanner performances in the definition of gross tumour volume and clinical target volume. *Eur J Nucl Med Mol Imaging* 2005; 32:1392–1399.
15. Ciernik IF, Dizendorf E, Baumert BG, *et al.* Radiation treatment planning with an integrated positron emission and computer tomography (PET/CT): a feasibility study. *Int J Radiat Oncol Biol Phys* 2003; 57:853–863.
16. Juweid ME, Cheson BD. Positron-emission tomography and assessment of cancer therapy. *N Engl J Med* 2006; 354:496–507.
17. Mac Manus MP, Hicks RJ, Matthews JP, *et al.* Metabolic (FDG-PET) response after radical radiotherapy/chemoradiotherapy for non-small cell lung cancer correlates with patterns of failure. *Lung Cancer* 2005; 49:95–108.
18. Krak NC, Boellaard R, Hoekstra OS, *et al.* Effects of ROI definition and reconstruction method on quantitative outcome and applicability in a response monitoring trial. *Eur J Nucl Med Mol Imaging* 2005; 32:294–301.
19. Miller M, Nguyen N, Oliver D, *et al.* Added Value of 18F FDG-PET/CT in CyberKnife Stereotactic Radiosurgery. 5th Annual CyberKnife Users' Meeting. Carlsbad, CA; 2006.
20. Czernin J, Allen-Auerbach M, Schelbert HR. Improvements in Cancer Staging with PET/CT: Literature-Based Evidence as of September 2006. *J Nucl Med* 2007; 48 Suppl 1:78S–88S.

Treatment Planning of Abdominal Lesions

6

Charles L. Lee

Contents

6.1 Introduction

The CyberKnife® (Accuray Incorporated, Sunnyvale, CA) was initially used to treat lesions of the central nervous system (CNS), but in recent years the most rapid increases in utilization have been in the treatment of soft tissue lesions outside the skull and spine. Soft tissue tumors may be broadly classified as lung, abdominal, or pelvic tumors. Like lung tumors, abdominal tumors are sufficiently close to the diaphragm that they normally require correction for motion due to breathing, but the large number of critical organs in the abdomen raises treatment issues that are common to pelvic lesions that move little with respiration, such as prostate adenocarcinoma.

The combination of breathing motion and high organ density has limited the development of stereotactic radiosurgery and hypofractionated radiotherapy for abdominal tumors. Large planning target volumes (PTVs) are a consequence of the appreciable tumor margins needed to account for breathing using either breath holding or gating techniques. The size of these targets restricts the prescription dose for lesions close to high-risk organs, such as the duodenum and kidneys. CyberKnife fiducial tracking throughout treatment, in combination with Synchrony® (Accuray Incorporated, Sunnyvale, CA) continuous breathing motion tracking, reduces PTV volumes [1]. In addition, non-isocentric beam delivery and inverse treatment planning limits hot spots so that abdominal radiosurgery and hypofractionated radiotherapy are readily achievable.

The range of tumor locations that have been treated with CyberKnife in the abdomen is impressive and continues to expand. Metastases from other primary cancers are treated in the liver, as well as primary hepatocellular carcinoma. CyberKnife treatment of late stage pancreatic adenocarcinoma in conjunction with chemotherapeutic agents has been described using both single-fraction and hypofractionated approaches [2–4]. The improved radiosensitivity of renal cell carcinoma to hypofractionated radiation delivery in spinal metastases has been described [5, 6], but only recently have primary renal cell treatments in the kidneys been reported [7]. Several sites have described the treatment of very large (greater than 500 cc) soft tissue sarcomas, many of retroperitoneal origin, and anecdotal evidence suggests similar increases in radiosensitivity for these tumors [8, 9]. Metastases may arise in many other parts of the abdomen, including gall bladder, duodenum, spleen, stomach, and lymph nodes; these may also be amenable to CyberKnife treatment.

These abdominal targets raise a number of issues concerning treatment planning and delivery that must be addressed before a safe and efficacious CyberKnife treatment can be offered. In this chapter I describe some of these issues and suggest means of dealing with them based on my experience with the CyberKnife system. Some of these suggestions are based on features found in Multiplan® (Accuray Incorporated, Sunnyvale, CA) version 1.5.2 and may need to be altered for future versions of the software. Note that, although many of the clinical concerns that are outlined below are common to all technologies used for stereotactic hypofractionated delivery of radiation to moving targets, the methods used to address them are specific to the CyberKnife. More general information on this topic may be found in other published sources [10].

6.2 Pre-Treatment Planning Considerations

6.2.1 Imaging and Fusion

Contouring and treatment planning for abdominal CyberKnife cases may be enhanced by fusing additional imaging studies to the main CT image set and through judicious choices during the CT scanning process. Blurring will arise in abdominal MRI scans unless breath holding is employed. While this approach may yield useful information, several effective alternatives have achieved more widespread use.

In recent years, positron emission tomography (PET) has become a standard clinical tool for identifying metabolically active regions of the body and consequently helping to distinguish regions of rapid cell growth from surrounding structures that are radiographically similar. All locations within the abdomen are amenable to PET imaging for fusion to CT, and while breath holding is not normally used for PET scans, their limited resolution (generally no better than 3 mm slice thickness) means that this is generally not a problem.

The automatic image registration tool in MultiPlan works well when fusing CT and PET scans, especially when these images are obtained as part of a single study. Starting points for automatic registration are best seen in the sagittal and coronal views on the PET scan, rather than the default 3D views. Careful examination of CT/PET registration is crucial to contouring, and in some cases, manual registration at a specific anatomic location is preferable to the automated global registration that does not account for soft tissue deformations. The colorwash overlay mode available in MultiPlan is excellent for CT and PET displays, both when checking the quality of registration and when contouring. Switching to the PET-only display and adjusting the window and level to display only extremely hot regions before using the colorwash overlay mode will maximize the effectiveness of this modality for image registration (Fig. 6.1). Care must be taken to avoid mistaking common areas of high radiopharmaceutical uptake for active tumor, especially near the heart, kidneys, and major vessels.

Intravenous (IV) iodine contrast is highly desirable in abdominal CT scans for differentiating the array of critical soft tissues. For patients with an iodine allergy, a PET scan becomes more important. The timing between the IV injection and CT acquisition is crucial to maximizing uptake in the abdomen, especially if a PET scan is not used for contouring. Oral contrast administered just before imaging is also helpful to distinguish stomach and duodenum from surrounding tissues, and ingesting additional contrast several hours prior to the CT study can opacify the lower gastrointestinal (GI) tract if desired. Treatment planning should use a homogeneous electron density model (chosen in the Setup page of the Plan tab in MultiPlan) when oral contrast is present in the CT scan.

Breath holding during the CT study will remove the motion blurring that will occur if the scan is obtained using normal respiration. The patient's breath may be held at maximum inhalation, exhalation, or a midpoint between these extremes; patient training will maximize the reproducibility of this technique. Fusing a breathing-blurred PET scan to a CT scan taken during breath holding will lead to mismatch that must be carefully evaluated through manual

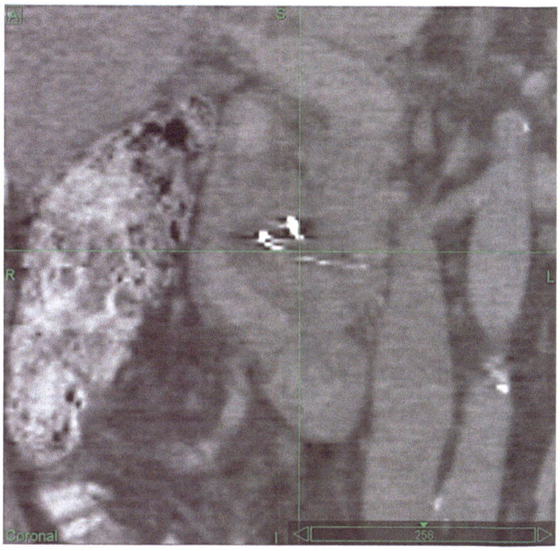

a

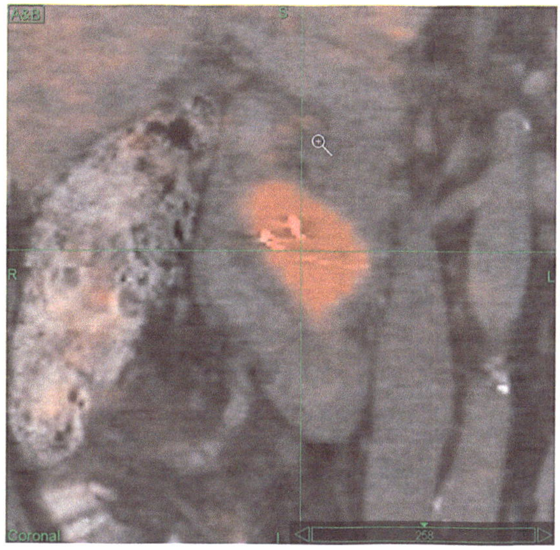

b

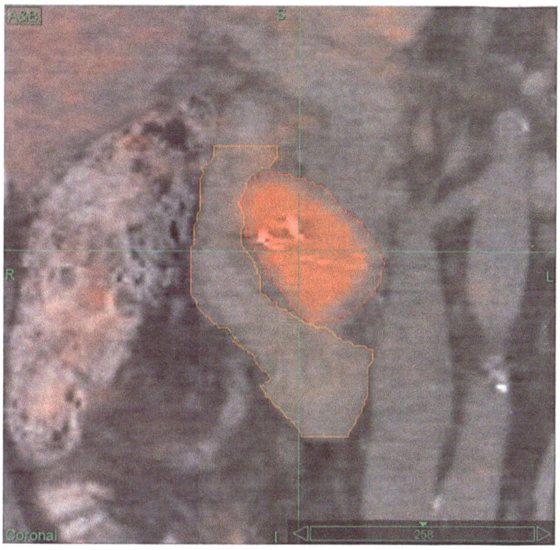

c

Fig. 6.1a–c. Target delineation with CT/PET fusion. With CT only (**a**), the melanoma metastasis to the patient's pancreatic head is difficult to distinguish from surrounding soft tissues. A colorwash overlay of the PET image (**b**) clearly identifies the region of hypermetabolic activity. This aids in contouring both target and adjacent duodenum (**c**).

registration of the images. It may also be useful to fuse two CT studies together in order to evaluate any changes to the tumor shape at different parts of the breathing cycle and construct a patient-specific PTV to account for this deformation.

For lesions in the upper abdomen, arms placed across the chest may interfere with certain oblique beams, while keeping the arms above the patient's head may be difficult to maintain during the length of time necessary for a typical CyberKnife treatment. Many centers treat abdominal CyberKnife patients with arms at their side, but care must be exercised to minimize the use of treatment beams passing through them. In addition, the CT scanner's field of view (FOV) can cut through the patient's body on one or more slices. The planning software disables beams that enter through portions of the body cut off by the top and bottom slices of the CT scan, but beams entering through the regions attenuated by the FOV are turned off inconsistently (Fig. 6.2). Under no circumstances should treatment beams pass through any part of the body cut off by the scanner FOV; this may be accomplished by contouring a structure that outlines the cut-off portion of the scan and disabling beams passing through this structure. If the patient's size requires that part of the body will be cut off by the FOV, adjust the patient position so that beams will not be turned off on the same side of the body as the tumor. For example, the entire right side of the patient should be intact for liver lesions;

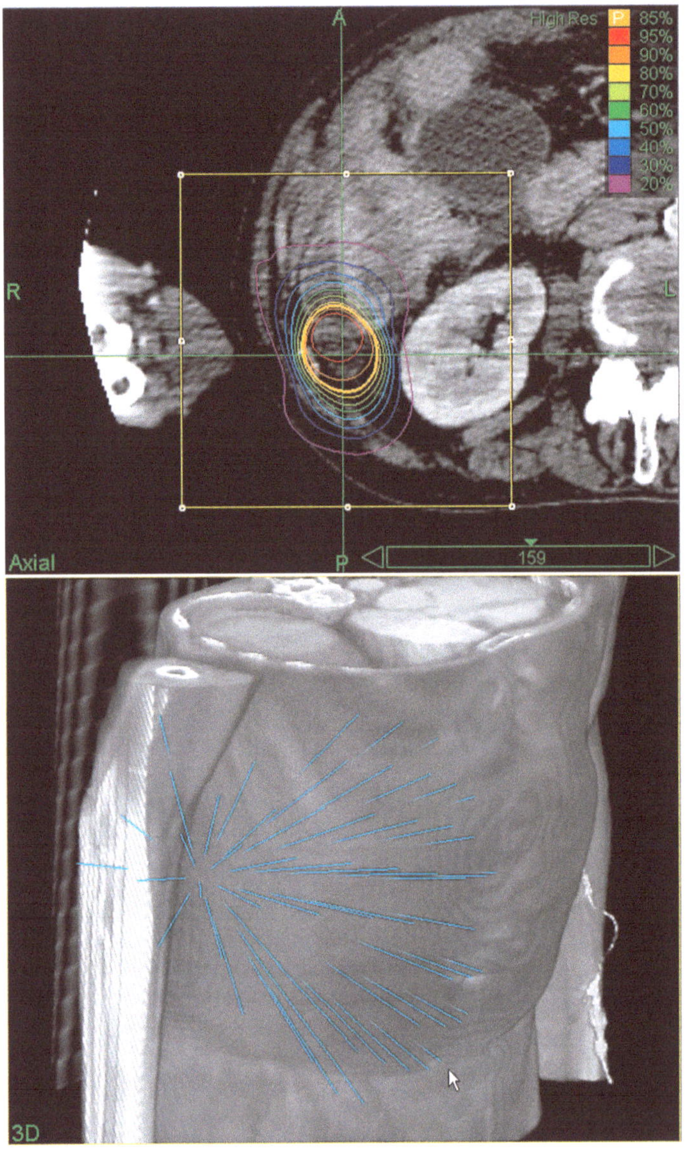

Fig. 6.2 Body cut off by CT scanner field of view. The right arm placed by the patient's side for simulation (and subsequent treatment) is cut off by the FOV (top image). As shown in the 3D image on the bottom, the isocentric treatment of this posterior liver metastasis will include two beams passing through this region (in which tissue depth is incorrect). Contouring the intersection of the FOV with tissue and disabling beams passing through this structure will eliminate this problem.

this may require the FOV to attenuate the left side of larger patients' bodies.

6.2.2 Fiducial Placement

While CyberKnife treatments for skull, spine, and even lung are progressing towards sub-millimeter targeting accuracy using anatomical landmarks, abdominal targets are still targeted and tracked using implanted fiducials. Gold fiducials are used instead of stainless steel screws, and the number of fiducials depends on the number of lesions, lesion size, and tracking goals. In determining the appropriate number of fiducials to use for a given patient, a balance must be struck between the advantages of fewer fiducials (greater distance between fiducials and easier to distinguish fiducials) and the disadvantages (less accuracy and less stability). Adding more and more fiducials will not automatically improve tracking; in fact, placing fiducials in extremely close proximity may make it impossible to lock onto any of them. If surgical clips are also present, differentiating between fiducials and clips is challenging and may work better if the patient is placed on the CyberKnife couch prior to treatment planning and diagnostic set-up X-ray images are obtained for comparison to MultiPlan-generated digitally reconstructed radiographs (DRRs).

Translations and rotations determined from fiducial tracking are susceptible to small positional variations in individual markers, e.g., organ deformation, and large variations, e.g. migration. Deformation is a significant concern in the liver, for which the extremely elastic tissue and position directly below the diaphragm are likely to lead to significant difficulties in tracking. Multiple lesions in the liver, treated in a single plan, present a particular challenge with no ideal solution. The physicians and physicist must anticipate this difficulty during the treatment planning stage by either increasing PTV sizes to account for this inherent uncertainty or making the plan less sensitive to extreme rotational changes, e.g., using isocentric beams.

It has been pointed out (Wu et al. [11]) that significant patient rotations (greater than a few degrees) can lead to large changes in the source-to-surface distance (SSD) for treatment beams when compared to the CT image used for planning. For this reason, some CyberKnife sites have decided to use fiducial or Xsight™ (Accuray Incorporated, Sunnyvale, CA) plans at the beginning of treatment for global rotational positioning of the patient, followed by translation-only tracking during the actual treatment delivery, using the fiducials in or near the tumor [12]. This is the same approach used by the new Xsight Lung™ tumor tracking method, but with the Xsight and tumor tracking portions of treatment separated into two separate plans. The tumor, and fiducials, may be rotated relative to its position in the CT, so appropriate margins are added to generate a PTV that is expected to provide full tumor coverage throughout treatment. For abdominal cases, this approach facilitates treatment and Synchrony modeling which may reduce a patient's total treatment time. It is also expected that the SSD for each treatment beam will closely match the treatment plan. The loss of information about the tumor's rotational behavior, however, makes this approach less stereotactic and closer to a variation of intensity modulated radiotherapy (IMRT).

6.2.3 Contouring

Careful contouring techniques are essential to assuring high target conformality in all dimensions. Because all contouring is performed in a single slice direction (usually on axial slices), it is easy to contour carefully on a slice-by-slice basis and neglect the overall three-dimensional shapes that are formed. The physician or physicist contouring for CyberKnife should always be mindful of the axial, sagittal and coronal views of the contoured structures. The automatic interpolation tool in MultiPlan provides an effective way to assure three-dimensional smoothness. An effective contouring approach will choose a contouring interval and initially draw structures only on those slices, e.g., contour the tumor only on even slice numbers, or contour the spinal canal on every fifth slice. The interpolated slices then have a starting point for contouring that is likely to need little or no adjustment.

Margins for creating PTVs are at the planner's discretion, and while many CNS CyberKnife treatments target the gross tumor volume (GTV) without the addition of a margin, it is recommended that at least 3–5 mm margins be used in the abdomen. The additional targeting inaccuracy that arises from Synchrony tracking accounts for most of the margin, and soft tissue deformability requires a few more millimeters. Recall that fused CT scans at maximum inhalation and exhalation may be used to create anisotropic margins that account for patient-specific tumor deformation.

The contour dilation tool in MultiPlan can create jagged contours, so that any PTV created by dilating the GTV will be less smooth and resulting plans may be less conformal. The PTV may be smoothed using the contour interpolation tool. Following PTV generation, move slice-by-slice through the structure and delete contours on even slices only. Note that with interpolation turned on, the deleted contours will be replaced by interpolated slices that are smoother. Once all even-slice contours are deleted, pass back through the structure and solidify each interpolated contour. Repeat the process but delete only odd-numbered slices. For complex shapes, this approach changes the PTV shape and is ineffective, but a large class of PTVs (and other structures created by dilation of another structure) will benefit from this technique. This approach adds to the time for contouring, but the potential improvement in plan quality makes it worthwhile.

The critical structures that need to be contoured differ from case-to-case, but most abdominal CyberKnife treatments will consider the liver, stomach, duodenum, spinal cord, and both kidneys as critical structures. Other potential structures to consider: adrenal glands, gall bladder (if not included with liver), colon, small bowel, and major vessels like the aorta, vena cava, or mesenteric artery. A distinction should be made between contouring the spinal cord, the spinal canal, or dural sac. Contouring the canal is fast and straightforward, but it may overestimate the dose to the spinal cord. It is also extremely important to contour any structures that will either be damaged by radiation (pacemakers and defibrillators for upper abdomen) or interfere with beam delivery (prostheses or pain pumps).

6.3 Treatment Planning Considerations

6.3.1 Tuning Structures

As with IMRT, inverse planning is based only on weighted constraints specified in certain volumes of interest (VOIs). For regions without constraints, the optimization algorithm accepts any dose levels without penalty. Very low maximum dose constraints with a high weight in one part of the body can lead to dose "streaking" away from this area, usually in unpredictable ways. Tuning structures provide a mechanism to control these effects, but different geometric forms provide different types of control over dose delivery.

A tuning structure may simply be a polygon drawn on one side of the tumor. For example, pushing dose posteriorly away from the anterior small bowel when treating the pancreas. Alternatively, another tuning structure may trace the outer edge of the patient. For example, pushing dose away from the skin for lateral liver metastases. While these tuning structures are useful, they can suffer from the same problem as critical structures and push dose in unanticipated directions.

One solution to this dilemma is to create a new shell structure in another VOI by specifying the inner gap and shell thickness in millimeters. If no inner gap is specified, the entire region surrounding the target is restricted isotropically, and a series of these shells may be specified, each with a progressively smaller maximum dose, to control dose conformality and limit dose streaking. One limitation of this method is that hard constraints will generally prevent simplex optimization from achieving solutions, but soft constraints will allow some dose to exceed the dose specification (with a penalty that is minimized). If the inner gap is nonzero and the shell thickness is small, the shell may be used as a near-absolute limit on dose. This approach can effectively force the system to create plans with high conformality. That is, by surrounding a target with a thin shell (inner gap and thickness of 2 mm) and specifying a hard-weight constraint with maximum dose equal to the minimum target dose, the resulting prescription isodose line should be highly conformal. This approach of creating thin shells with large inner gaps can also be used to constrain low isodose lines, whose distributions are often unpredictable.

Point constraints are useful when the overall direction of optimization is acceptable but a handful of specific locations are not behaving as desired. This may be due to the absence of dose grid calculation points at these locations. Another cause may be beam targeting entirely towards the periphery of the lesion, leading to a cold spot in the center. Point constraints are ideal for resolving these issues. Using point constraints as the primary method to shape isodoses, however, is inefficient because repairing hot or cold spots in this manner often produces new defects, which in turn require new constraints that create still more areas of concern.

The lack of dose volume histogram (DVH) constraints may require creativity on the part of the treatment planner to achieve certain dose distributions in critical structures. For example, the physician may be willing to accept a maximum dose of 15 Gy to one of the kidneys provided no more than 20% receives in excess of 8 Gy. Specifying a maximum kidney dose of 15 Gy may result in most of the organ receiving dose near this limit; on the other hand, specifying a dose closer to 8 Gy is overly restrictive and may limit some of the optimization algorithm's options. As another example, the physician may be willing to sacrifice coverage of a tumor in the pancreatic head in order to keep the maximum dose to the adjacent duodenum below 50% of the prescription dose. Specifying the prescribed doses to tumor and duodenum, along with giving the duodenum a high weight compared to tumor, will push dose away from duodenum but may reduce coverage on the other side of the tumor as well.

Both of these examples may be solved by dividing structures into smaller parts, each with different constraints. In particular, the duodenum may be copied into a dummy critical structure which is dilated by 5 to 10 mm. The tumor may then be divided into two sub-structures, near-duodenum tumor and bulk tumor; the former is defined as the intersection of the expanded duodenum and tumor, while the bulk tumor is the remainder (Fig. 6.3). Near-duodenal tumor can have a low weight to allow dose to push away from the duodenum while the bulk tumor uses high weights on the minimum dose to ensure good coverage of the majority of the tumor.

6.3.2 Dose Grid

Plan optimization is considered "low resolution" and occurs on a grid uniformly distributed along each axis of the calculation box. Resizing the box adjusts the density of calculation points in each

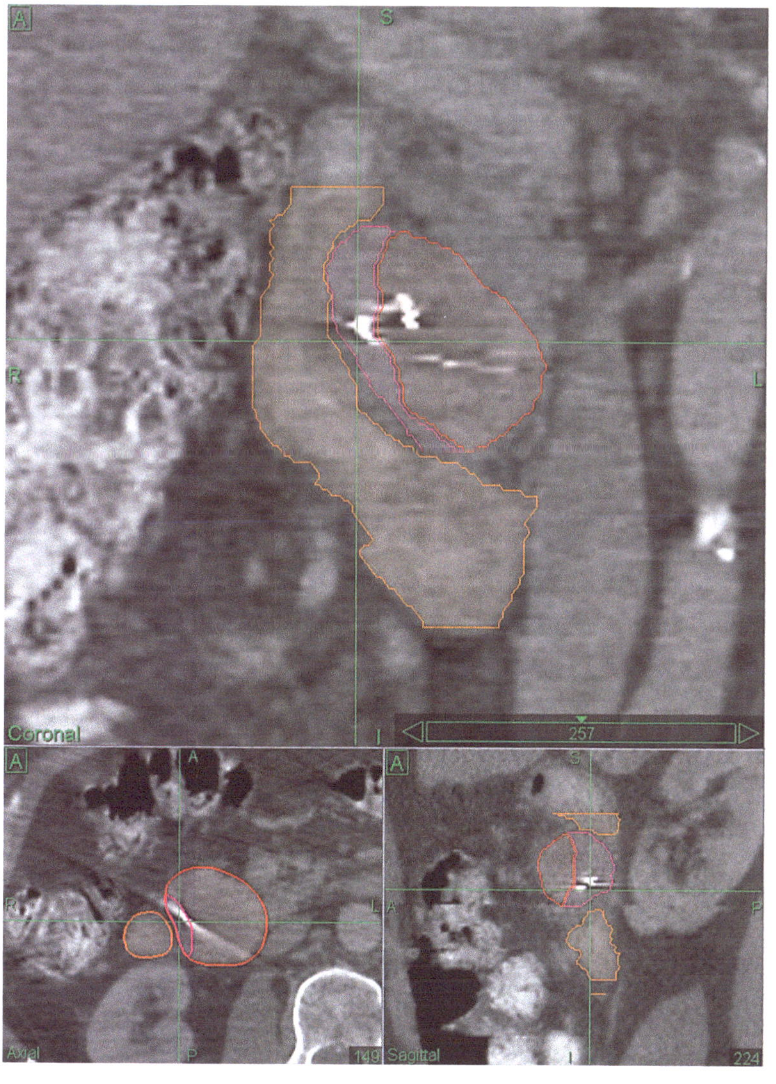

Fig. 6.3 Dividing the target into two regions. The pancreatic lesion shown here has been divided into two regions, the bulk tumor (*in red*) and near-tumor (*in purple*). The near-tumor is that part of the lesion located within a 10-mm radius of the duodenum (*in orange*). The duodenum was anisotropically expanded in the left, right, anterior, and posterior directions by 10 mm. The intersection of the target contour with this expanded contour formed the near-tumor, and the bulk tumor was created by copying all near-tumor contours as cavities into a new contour set for the original lesion. Different constraints may be applied to each target during optimization, and beams may be directed to only one of the two structures.

dimension, but the number of points does not change. Shrinking the box to just cover the tumor will provide many calculation points within the tumor and adjacent critical surfaces and encourage high conformality, but the lack of information outside the grid can produce unanticipated hot spots away from the tumor. Expanding the box to fill the entire abdomen allows the planner to prevent hot spots at the expense of too few calculation points within the tumor and consequentially poor conformality. The planner must strike a balance between these extremes when producing a treatment plan. Regardless of the grid size used for optimization, after saving the final plan the grid should be expanded to include the entire abdomen and a high resolution calculation used to check for hot spots away from the target.

6.4 Isocentric Treatment Planning

In isocentric CyberKnife treatments, all beams are directed towards a location within the patient to create an ellipsoidal volume of high dose. The operator specifies this location, as well as the size of the collimator. By default, all beams are of equal weight. Isocentric treatments deliver all dose to the same geometric point, so that many fewer monitor units (MU) are necessary to achieve a given prescription dose than with non-isocentric beams. This will significantly reduce the patient's time on the treatment couch, which in some instances may be of high clinical importance. Unfortunately, there is little user control over the shape of isocentric dose distributions, so that only spherical or ellipsoidal targets may be treated in this manner and the dose distributions will be less conformal. MultiPlan has two tools to increase the range of tumors that may be treated isocentrically: conformal isocentric planning and dose re-targeting. Multiple isocenters would also be an option to control the shape of isocentric distributions.

Once the location of an isocenter has been chosen, the user has the option to adjust the individual beam weights by specifying dose constraints for iterative or simplex optimization similar to non-isocentric inverse planning. MultiPlan will keep the beams isocentric but attempt to meet the constraints by adjusting each beam weight, effectively stretching the dose distribution. For nearly spherical lesions this approach can be highly effective, but the isodose curves are limited to convex shapes. Figure 6.4 shows isodose distributions for two colorectal metastases to the liver treated in a single plan, one lesion treated isocentrically and the other non-isocentrically.

On the Physics page under the Settings tab, a tool has been created to optimally center isocentric distributions on a target. This tool was designed for ball-cube end-to-end tests for CyberKnife quality assurance, but it can be used in patient treatment planning to increase the percentage of tumor covered by a dose distribution through iteratively retargeting the treatment isocenter.

While many early-stage peripheral lung tumors lend themselves to isocentric treatments, appropriate cases in the abdomen will generally be limited to certain liver metastases. However, some tumors in the pancreatic head may be sufficiently distant from the duodenum to permit isocentric delivery, and small round tumors may be amenable to this method. Because dose calculations for isocentric treatments are rapid compared to non-isocentric treatments, it is worthwhile to initially consider this possibility before considering the more complex task of non-isocentric inverse planning.

6.5 Non-Isocentric Treatment Planning

Unlike conventional IMRT, in which the planner chooses gantry angles and the number of beams but does not directly adjust the pencil beam intensities, CyberKnife inverse planning uses fixed paths and node positions but requires the user to choose the size of the one or more collimators for the treatment. Choosing the appropriate collimator (or collimators) for each plan becomes easier with experience, but the unique aspects of each target shape and prescription requirements for surrounding critical structures make the adoption of any hard-and-fast rules impossible.

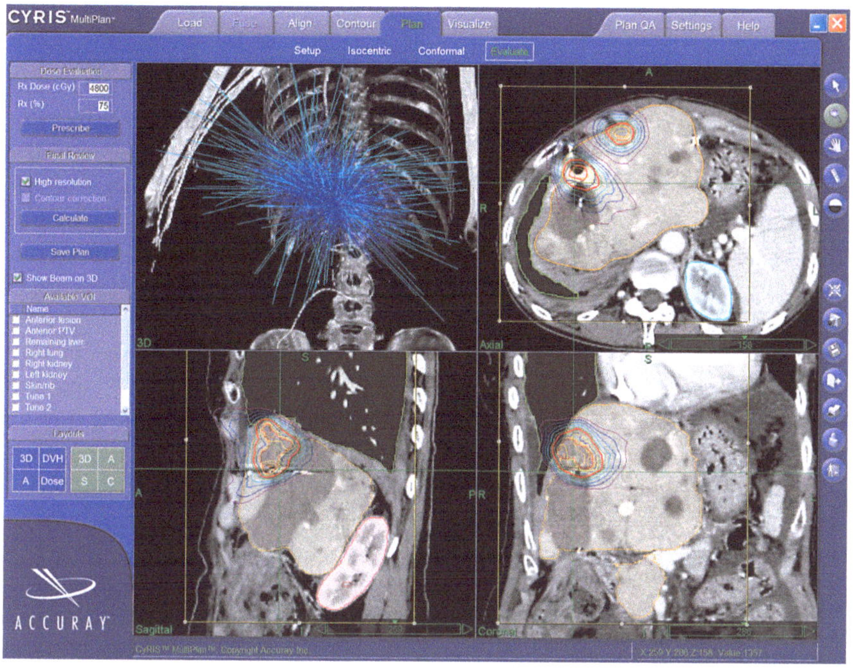

Fig. 6.4 A combined treatment using both isocentric and non-isocentric beams. Here, two symptomatic colorectal liver metastases were treated in a single plan. The anterior lesion is nearly spherical and was treated with isocentric beams, indicated by its uniform isodose distribution. The posterior lesion is concave and therefore not amenable to isocentric treatment. CyberKnife treatment planning is clearly capable of mixing these targeting methodologies within a single plan.

In general, the collimator should be small enough to shape the prescription isodose conformally around the target contours, but large enough that the number of nonzero beams and monitor units is not excessive and treatment delivery times are reasonable. A common rule of thumb suggests this balance is possible using a collimator diameter 2/3 to 3/4 of the target diameter. If the target is elongated or very concave, this rule will not apply and the user is limited to a collimator no larger than the target's smallest dimension. It is recommended that once an acceptable treatment plan has been generated using a given collimator, the plan should be re-optimized using the same constraints with the next-largest collimator; often, this second plan will also be acceptable but with fewer nonzero beams and monitor units.

Directing multiple collimators at a target can provide the advantages of both large and small collimators, but the planner should beware that the monitor units required for these plans are normally dominated by the smallest collimator, and indiscriminate use of very small collimators can lead to excessive treatment times. If multiple collimators are used for generation three (G3) CyberKnife systems, each component of the three-path sets should use only a single collimator. If all collimators are applied to the entire path – which is the only option for generation four (G4) models – the increase in robot travel time may extend the length of treatment more than the total monitor units implies.

For plans treating more than one lesion, the planner can use a different collimator for each target or direct more than one collimator to each. This provides a wide range of options for planning, but the user should remember that the longer treatment time needed to treat multiple lesions may be further extended by using more than one collimator. For multiple liver lesions, it is recommended that each path is only directed at a single lesion. This is due to the extreme elasticity of the organ, which makes it difficult to track all implanted fiducials at the same time. If all beams for a path are directed at one lesion, only the nearest fiducials may be used to generate the Synchrony tracking model. The model can then be re-generated for the next path, i.e., the next lesion.

The large number of critical structures in any abdominal radiosurgery plan makes it imperative that the surgeon and radiation oncologist prioritize their treatment goals and communicate them to the physicist. These goals may include, but are not limited to the following:

- The patient's condition may necessitate a short treatment time, even at the expense of conformality or critical structure sparing
- Concerns about vessels passing through the tumor may require high dose homogeneity to minimize hot spots
- Near-perfect prescription dose conformality along the border between the target and a critical structure, such as a pancreatic tumor adhered to the duodenal wall, may be the physician's primary concern (and consequently limit the prescription dose)
- Prior external beam radiotherapy may limit the volume of soft tissue (including small bowel) able to receive 20% to 40% of prescription dose
- A prior nephrectomy could make the dose distribution in the remaining kidney the dominant concern of the planner

Translating these clinical concerns into appropriate inverse planning constraints involves many different adjustable parameters and hence the solution space is enormous. This makes a systematic approach to planning essential, but trial and error are almost always needed to assure all clinical requirements are met. Initially employing iterative optimization permits the user to observe the step-by-step adjustment of the isodose curves and dose volume histograms (DVHs), pause the algorithm if it is not converging towards a workable solution, change constraints and resume calculations. In most cases, this algorithm rapidly creates a dose distribution close to the final result, but many iterations may be required to reduce the number of nonzero beams and monitor units to a workable level.

If the constraints determined with this approach are applied to a simplex optimization, a solution is possible in less time using fewer beams and monitor units. The simplex algorithm, however, is not guaranteed to produce a solution and may fail if the constraints are too rigid, while the iterative algorithm will always produce some solution. Another approach begins with a simplex optimization followed by an iterative adjustment.

At this time, no large-scale clinical studies of acute and long-term side effects from radiosurgery and hypofractionated radiotherapy are found in the literature, so the clinical team is restricted to extrapolations from traditional fractionation schedules and experience. The linear-quadratic (LQ) cell survival model provides a starting point for these extrapolations until clinical data are available. Table 6.1 presents normal tissue tolerances for key abdominal structures, in terms of TD 5/5 volumes, based on the analysis of Emami et al. [13]. It must be emphasized that this table is not based on clinical data for radiosurgery or hypofractionated radiotherapy of abdominal tissues, but it may provide the clinical team with a starting point for evaluating CyberKnife dose distributions and DVHs.

6.6 Additional Considerations

Creating a treatment plan for the CyberKnife is only the first step prior to setting up the patient for treatment and delivering the radiation. Patient comfort is more important for CyberKnife treatments than for conventional radiotherapy treatments due to the extended treatment times – a comfortable patient will remain still for a longer period and reduce unexpected emergency stops (ESTOPs) and losses of the Synchrony model. Clearance of the robot is also of prime importance, since treatments of the lower abdomen may leave the patient's head extended closer to the robot than normal and require the radiation therapist's attention for safe dose delivery.

Identifying fiducials implanted in soft tissue tumors of the abdomen may be difficult without adjusting tracking parameters. For example, patients with prior surgery may have clips in or near the field of fiducials, requiring an increase of the contrast thresholds to prevent the tracking algorithm from choosing them over the real fiducials. In addition, the fiducials may be difficult to identify in large patients, even by visual inspection, without increasing the voltage, current and exposure times of the di-

Table 6.1 LQ model extrapolation of critical structure dose limits. Structures of interest to abdominal radiosurgery are listed along with damage endpoint and partial volume restrictions for radiotherapy limits based on conventional daily doses. Both conventional and hypofractionated limits are total doses. Radiotherapy limits are taken from Emami et al. [13].

Structure	Volume/Point	Radiotherapy limits (Gy)	Hypofractionated limits (Gy)*			Endpoint
			1 fraction	3 fractions	5 fractions	
Spinal cord	D_{max}	45	13.3	21.4	26.2	Myelitis necrosis
Liver	Highest 1/3	50	14.4	23.3	28.6	Liver failure
	Highest 2/3	35	11	17.5	21.2	
	Total volume	30	9.8	15.5	18.7	
Kidney	Highest 2/3	30	9.8	15.5	18.7	Clinical nephritis
	Total volume	23	8.1	12.6	15	
Small bowel	Total volume	40	12.1	19.4	23.7	Obstruction/ perforation/ fistula/ulceration
Large bowel	Highest 1/3	55	15.5	25.1	31.1	
	Total volume	45	13.3	21.4	26.2	
Stomach	Highest 1/3	60	16.6	27	33.4	
	Highest 2/3	55	15.5	25.1	31.1	
	Total volume	50	14.4	23.3	28.6	
Heart	Highest 1/3	60	16.6	27	33.4	Pericarditis
	Highest 2/3	45	13.3	21.4	26.2	
	Total volume	40	12.1	19.4	23.7	
Cauda equina	D_{max}	60	16.6	27	33.4	Nerve damage

*Assumes $\alpha/\beta = 3$ Gy and 25 fraction radiotherapy course; neglects repair time

agnostic X-ray sources. In general, the physicist or therapist must change the tracking parameters to either encourage the algorithm to look in the correct location for the fiducials, or discourage it from considering other objects that may be mistaken for fiducials. The treatment team must also balance the need for translational and rotational targeting accuracy with the needs of the patient; constant interruption of the treatment when tracking five implanted fiducials may not be justifiable if uninterrupted treatment occurs when only three fiducials are followed.

The remarkable targeting accuracy that may be achieved with Synchrony tracking makes it possible to create smaller PTVs than with other radiotherapy systems, but this advantage to CyberKnife radiosurgery evaporates if the quality of the Synchrony model is poor. Throughout the course of treatment, the therapist must carefully examine both diagnostic images immediately after they are acquired in order to assure that the tracking algorithm has correctly identified the fiducials. It is also critical that all portions of the breathing cycle are sampled prior to the start of treatment, including narrow peaks in the cycle that may be difficult to incorporate.

It is also possible to treat without Synchrony in two important but common situations. If the Synchrony model shows a flat response for all three translational directions, the fiducials in the tumor are not appreciably moving with respiration and Synchrony is unnecessary; if there is no problem tracking the patient's breathing, it may remain on, but for patients with chronic cough, sleep apnea, or other breathing irregularities, disabling Synchrony may greatly accelerate treatment. Another possible difficulty may arise if the Synchrony model is lost while one or both diagnostic cameras are blocked by the linear accelerator or robot; in this case, continuation of the treatment with Synchrony will require aborting the treatment, returning the robot to perch, re-establishing the fiducial tracking and Synchrony models, and following the treatment path around the patient to the original stopping point. When this happens near the end of the path, this process may add up to 30 minutes to the patient's treatment. However, the treatment may be continued from the point of interruption with Synchrony disabled if minimal dose delivery is required before the cameras are unblocked. A small portion of the treatment will inadequately target the moving tumor, but the physician may decide that adding 30 minutes to the treatment may be a worse option.

6.7 Future Directions

The observations and suggestions in the previous sections should give the new user a strong basis for an abdominal CyberKnife program and provide experienced users with another perspective and perhaps some useful new techniques for their own programs. This field will continue to grow, ever more rapidly, and involve newer techniques for improved treatments and clinical outcomes.

At press time, in fact, a new version of MultiPlan (2.0) has been unveiled that improves significantly on the already impressive capabilities of the current release. The new tools that will impact abdominal CyberKnife treatments include, but are not limited to, treatment planning using time-averaged contours from four-dimensional (4D) CT scans; a fine tune menu that gives the user greater control over individual treatment beams; the ability to limit the total monitor units delivered per node; and improved fusion tools. The user will have the ability to fuse up to three moving images to a single fixed image, and fusion of coronal and sagittal MRI images sets will be possible.

References

1. Casamassima F, Cavedon C, Francescon P, *et al.* Use of motion tracking in stereotactic body radiotherapy: Evaluation of uncertainty in off-target dose distribution and optimization strategies. *Acta Oncol* 2006; 45:943–947.
2. Koong AC, Christofferson E, Le QT, *et al.* Phase II study to assess the efficacy of conventionally fractionated radiotherapy followed by a stereotactic radiosurgery boost in patients with locally advanced pancreatic cancer. *Int J Radiat Oncol Biol Phys* 2005; 63:320–323.
3. Koong AC, Le QT, Ho A, *et al.* Phase I study of stereotactic radiosurgery in patients with locally advanced pancreatic cancer. *Int J Radiat Oncol Biol Phys* 2004; 58:1017–1021.
4. Hoffelt C, Didolkar MS, Brenner MJ. Results of stereotactic radiosurgery (CyberKnife®) in uresectable and locally recurrent adenocarcinoma of the pancreas. CyberKnife Society Users' Meeting. Carlsbad, CA; 2006.
5. Gerszten PC, Burton SA, Ozhasoglu C, *et al.* Stereotactic radiosurgery for spinal metastases from renal cell carcinoma. *J Neurosurg Spine* 2005; 3:288–295.
6. Levine AM, Brenner M, Coleman C, *et al.* CyberKnife stereotactic radiosurgery for the treatment of renal cell carcinoma metastases. CyberKnife Society Users' Meeting. Carlsbad, CA; 2006.
7. McNeely L, Taylor D, McIntyre M. Image-guided radiosurgery of renal and perinephric malignancies. CyberKnife Society Users' Meeting. Carlsbad, CA; 2006.
8. Rosenthal H. Stereotactic radiosurgery in the treatment of soft tissue sarcomas. CyberKnife Society Users' Meeting. Palm Springs, CA; 2007.
9. Levine AM, Brenner M, Coleman C, *et al.* CyberKnife stereotactic radiosurgery for the treatment of sarcomas of the spine. CyberKnife Society Users' Meeting. Palm Springs, CA; 2007.
10. Papiez L, Timmerman R, DesRosiers C, *et al.* Extracranial stereotactic radioablation: physical principles. *Acta Oncol* 2003; 42:882–894.
11. Wu X, Fu D, Yang J, *et al.* Patient alignment and target tracking in radiosurgery of soft-tissue tumors using combined fiducial and skeletal structures tracking techniques CyberKnife Society Users' Meeting. Carlsbad, CA; 2006.
12. Lee C, Brenner M, Coleman C, *et al.* An alternative approach to soft tissue fiducial tracking at Sinai Hospital of Baltimore CyberKnife Society Users' Meeting. Carlsbad, CA; 2006.
13. Emami B, Purdy JA, Manolis J, *et al.* Three-dimensional treatment planning for lung cancer. *Int J Radiat Oncol Biol Phys* 1991; 21:217–227.

Use of a Target Complexity Index in Radiosurgical Plan Evaluation

7

Gregory J. Gagnon, Walter Jean, Sonja Dieterich,
Huaying Ji, and Donald A. McRae

CONTENTS

7.1 Introduction

There is a great need for reliable tools for radiosurgery plan evaluation. With the increasing sophistication of radiosurgical treatment planning systems, the radiosurgeon finds a more challenging treatment environment and feels the need to utilize more sophisticated methods of plan evaluation. Ultimately, these methods carry the promise of assisting treatment decisions, or even replacing some of them with consistent, reliable, and verifiable measures of probable treatment success. Among the most common are indices of homogeneity, dose uniformity across the target area, and conformity, the shaping of the radiation dose to the target area. These can often be expressed as simple ratios of treatment target and normal tissue volumes receiving certain radiation doses, although more complicated forms exist. The importance of these tools lies in their rendering of complex concepts into simple values, allowing either more sophisticated additions to multiple clinical treatment parameters, or more simplification of a limited treatment parameter set to a limited metric. These numeric simplifications can assist in plan evaluation or be used as a basis for optimization algorithms. They may also be useful for predicting local control and complications from radiotherapy or radiosurgery treatment. Reliable measures of homogeneity and conformity of dose distribution in relation to the target can also allow comparison of results between medical centers or between different radiosurgical technologies.

Although complex syntheses of the pertinent treatment parameters into evaluation measures are possible, it is desirable to have measures that are both easily calculable and generalizable across various treatment techniques and plans. Several measures of conformity, for example, have been described in the literature. An incomplete list is seen in Table 7.1. These are generally ratios of simple volumes. They are useful measures but may not include enough information to be broadly applicable or reliable.

Ideally, these measures should be immune to target size and target shape considerations. The measures in Table 7.1 are not immune from target size – smaller targets are typically associated with larger indices of conformity. This is seen in simplified form in Figure 7.1. This is explained by the fact that the proportional increase in volume associated with a fixed margin becomes greater as the target volume

Table 7.1 Commonly Used Conformity Indices [4–12]. An incomplete listing of commonly used conformity indices. For these indices, DTV_{min} is the minimum isodose covering the target volume, RI is the reference isodose, V_{RI} is the prescription isodose volume, or the volume encompassed in the prescription isodose surface, TV is the target volume, TV_{RI} is the target volume covered by the reference isodose, V_{RI} is the volume of the reference isodose, PITV is the prescription isodose surface and is identical to V_{RI}, CTV is clinical target volume, V is the volume that receives a dose higher than or equal to the minimum tumor dose, V_0 is the tumor volume, PTC is the percentage of tumor covered, V_x is × Gray volume, T_{PIV} is the target volume within the planning isodose volume (PIV), LV is the lesion volume, HTV is the healthy tissue volume, LV_{RI} is the lesion volume covered by the prescription isodose, HTV_{RI} is the healthy tissue volume covered by the prescription isodose, and TIV is the target isodose volume.

RTOG (Nedzi et al. 1993) [11]	$Coverage = \frac{DTV_{min}}{RI}$
RTOG (Nedzi et al. 1993) [11]	$Conformity = \frac{V_{RI}}{TV}$
Lomax (Lomax et al. 2001) [9]	$CI = \frac{TV_{RI}}{V_{RI}}$
van't Riet (van't Riet et al. 1997) [12]	$CN = \frac{TV_{RI}^2}{TV \cdot V_{RI}}$
Kubo (Kubo et al.1999) [8]	$PITV_{ratio} = \frac{PITV}{CTV}$
Ma (Ma et al. 1999) [10]	$TVR = \frac{V}{V_o}$
Borden (Borden et al. 2000) [5]	$QF_x = \frac{PTC \cdot TV}{V_x}$
Paddick (Paddick et al. 2000) [4]	$NCI = \frac{TV_{PIV}^2}{TV \cdot PIV}$
Dejean (Dejean et al. 2001) [6]	$CI_{RTOG} = \frac{LV_{RI}}{LV} + \frac{HTV_{RI}}{LV}$
Gibbs (Gibbs et al. 2002) [7]	$ACI = \frac{TIV \cdot PIV}{TV^2}$

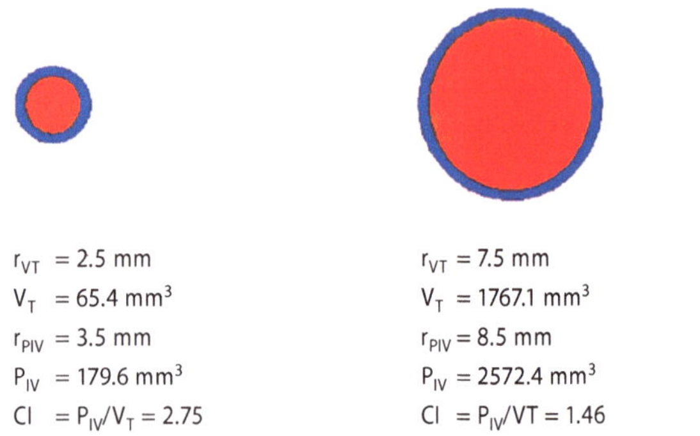

Fig. 7.1 A simplified case: Two spherical targets (*VT*, in red) and two spherical planning isodose volumes (*PIV*, in blue) which encompass the targets evenly with a 1 mm margin. Despite having similar treatment margins, there is a large difference in the RTOG conformity index (*CI*).

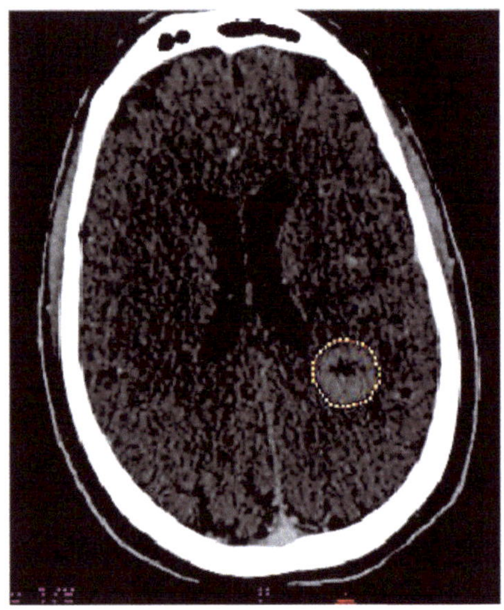

Fig. 7.2 A solitary brain metastasis.

decreases. Dependence of conformity measures on target size is a significant limitation that can be corrected by considering target size as an independent variable.

Target shape is another variable that affects conformity measures. This variable is more difficult to assess and to correct for, mainly because we have no measures of shape complexity. Figures 7.2 and 7.3 show typical radiosurgery cases. We believe that most radiosurgeons would consider the target in Figure 7.2, a spherical brain metastasis, to be less complex than the target shown in Figure 7.3 which has a more complex shape and has surrounding dose-limiting structures. Therefore, it would be easier to achieve a good conformity measure for the target in Figure 7.2 than the target in Figure 7.3.

Another important variable affecting the conformity index is the prescription isodose line. In many measures of conformity, an isodose can be selected

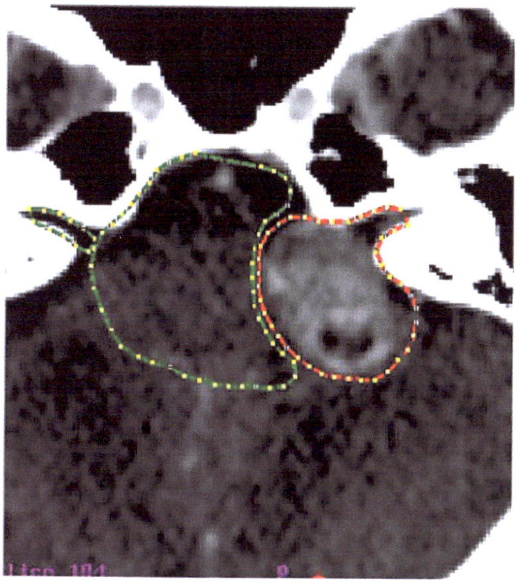

Fig. 7.3 A large vestibular schwannoma.

to give any conformity value within a broad range, although often at the expense of target coverage. Clearly, conformity measures may vary widely depending on target volume, target shape, proximal critical structures, or prescription isodose. In addition, intuitively, it seems reasonable to believe that prescription isodose conformity with simple target shapes is easier to achieve than with complex shapes.

A measure of shape complexity is desirable and could extend the applicability of these measures of conformity. Such a complexity measure should be easy to calculate, should be rotationally and size invariant, and should be normalized to 1.0 in the idealized case. We describe a simple measure of shape complexity (CpI) based on the ratio of surface area to volume. Our idealized case is the sphere. The sphere is intuitively the simplest target volume and has the quality of minimizing the surface area to volume ratio. Any deviation from sphericity will increase this ratio. It is this ratio, the surface area to volume ratio, which we will use as our measure of complexity (Fig. 7.4).

7.2 Derivation of the Shape Complexity Parameter

A reasonable shape complexity parameter should be size invariant as well as rotationally invariant. We begin with the ratio S/V, where S is surface area and V is volume of the target. Starting from the sphere:

$$\frac{S_{Sphere}}{V_{Sphere}} = \frac{4\pi r^2}{\frac{4}{3}\pi r^3}$$

In order to accomplish size invariance, it is necessary to remove the contribution of the radius r to the ratio. This can be accomplished by squaring the denominator and cubing the numerator, effectively canceling-out r. This gives:

$$\frac{(S_{Sphere})^3}{(V_{Sphere})^2} = 36\pi$$

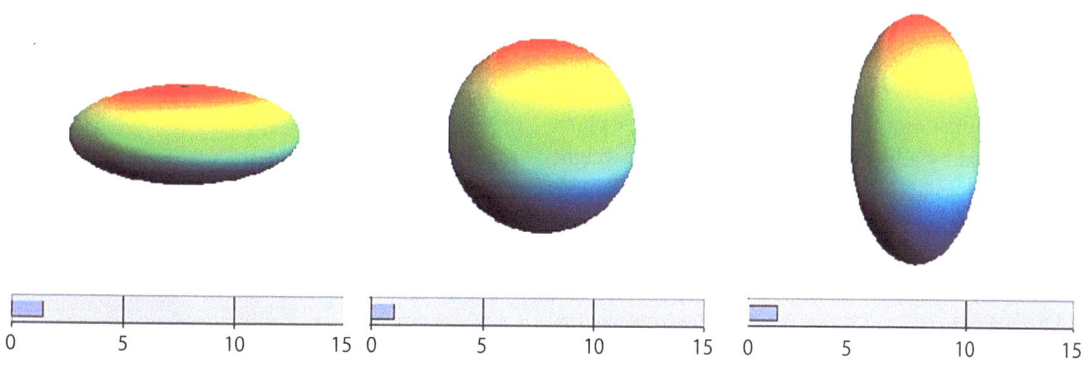

Fig. 7.4 An oblate spheroid, a sphere, and a prolate spheroid. The CpI is displayed for each shape in the bar graph under the shape.

We can now define the shape complexity measure for the sphere as:

$$CpI = \frac{(S_{Sphere})^3}{36\pi(V_{Sphere})^2} = 1$$

This generalizes to:

$$CpI = \frac{S^3}{36\pi V^2}$$

which is our measure of complexity for an arbitrary shape of volume V and surface area S.

7.2.1 Dose-Planning Studies with the Complexity Index

We considered the simplified case of spheroids, objects similar in shape to spheres but the radius along one axis is either larger than along the other axes (a prolate spheroid), or shorter than along the other two axes (an oblate spheroid). We chose these shapes because simple formulas for calculating surface area and volume are available, and because we can adjust one variable, the unequal radius, to get a variety of shapes to assess.

A dose planning study was completed to explore the utility of this measure in a phantom. A series of prolate spheroids was constructed within a CT-scanned head phantom with axes of 1:1:1, 1:1:2, 1:1:4, 1:1:6 and 1:1:8. These were scaled such that the volume included in all spheroids was equal and they were aligned along the y-axis (toward the cranial vertex). Planning was performed with the SGI-based CyberKnife® (Accuray Incorporated, Sunnyvale, CA) treatment planning system using 5 mm collimators. The planning system uses a non-coplanar, non-isocentric geometry, and optimizes the weights of over 1200 beams using a linear-programming (Simplex) algorithm. Ten inverse plans were created by adjusting the maximal beam weight for each spheroid and the best plan was chosen. The isodose covering 100% of the target volume was used for calculation of the RTOG conformity Index (CI) and this was the determinant of the best plans. A clear relationship between CI and the complexity measure, CpI, was observed as seen in Figure 7.5.

7.2.2 Clinical Evaluation of the Complexity Index

Before considering clinical data, it is necessary to clarify one important point. While the volume of a target is relatively insensitive to the scale used to measure it, the same cannot be said of the surface area. Small perturbations of surface area will add to the variability already present in conformity indices, which is undesirable. For example, by contouring the target volume on axial slices and recreating the volume as a stack of axial contours, differences in slice thickness or in contouring slice-by-slice will cause a variable surface area for similar shapes. This is similar to the question asked in fractal mathematics of "How long is the coast of Britain?" [1]. It is clear that the length of the coast increases as the scale to measure it decreases becoming, in theory, infinite at infinitely small scales. To correct for this effect, it is necessary to smooth or standardize the surface area calculation. In this study, this is accomplished by mapping 362 near equally-spaced vectors of a unit sphere to the target surfaces. Each surface is simplified in this manner to a 362 vertex shape and triangulation of this surface yields a consistent surface area calculation method.

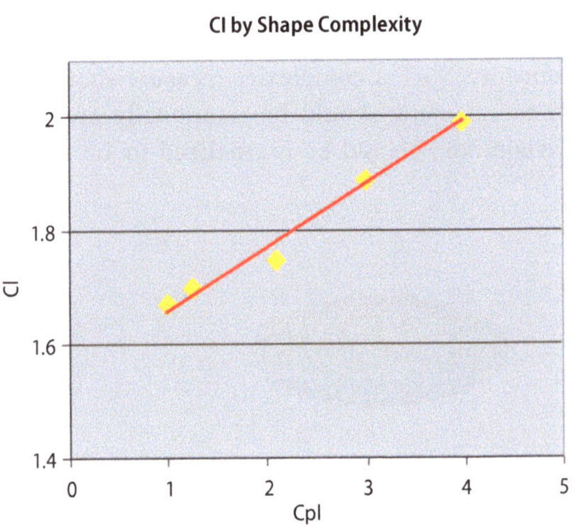

Fig. 7.5 The RTOG conformity index (CI) as a function of the complexity Index (CpI). A linear fit is seen in red.

7.3 Patient Data

A total of 374 CpI values were calculated for a variety of intracranial and extracranial targets, all treated at Georgetown University Hospital's Radiosurgery Center. A mean value of 2.037 and a median of 1.705 were found (range 1.09–8.50). Univariate analysis showed that CI inversely correlated with tumor volume, V_T ($p = 0.049$). The New Conformity Index (NCI) by Paddick et al. [2–4] inversely correlated with V_T ($p = 0.008$). The CpI correlated with CI ($p = 0.001$) but not with NCI ($p = 0.566$). The inverse correlation between target volume and conformity is expected, but it also dilutes the apparent strength of the association between CpI and conformity. One attempt to control for the effect of target size is to group the results by target size. Figure 7.6 shows such a group, body targets larger than 4 cm^3 in size. A regression line and equation are also provided.

Multiple regressions were also performed as an alternate method to control for target size. With the multiple regressions, CI was significantly predicted by CpI ($p < 0.001$) and VT ($p = 0.046$). The derived regression equation, $CI = 1.532 + 0.113CpI - 0.0103VT$ was significant with an F-Ratio of 9.377 ($p < 0.001$). The NCI was not predicted by CpI ($p = 0.408$) or VT ($p = 0.272$) and the regression equation had an F-Ratio 0.809 ($p = 0.446$). Interestingly, the linear regression equation from the phantom study of spheroids mentioned earlier, $CI = 1.5534 + 0.1092CpI$, is very similar to that of the clinical data on the 374 patient targets, giving some confidence that the measure is robust.

Another interesting result of these regressions is that the CpI correlates well with the CI but poorly with the NCI of Paddick [2–4]. The reasons for these results are not entirely clear.

It is possible to use these regressions to account for the effects of target volume and shape complexity on CI, reducing the variability in this measure. Determination of a reference CI for a given target volume and CpI will also allow a determination of whether the CI obtained in a plan is in fact adequate. In addition, determination of the regression equations of CpI on CI for different radiosurgery systems will allow a more accurate determination of conformal capabilities of these systems, as some systems might outperform others only at high complexity levels. A better system would be expected to have a shallower slope on the regression equation. We are attempting such a project but currently have no results.

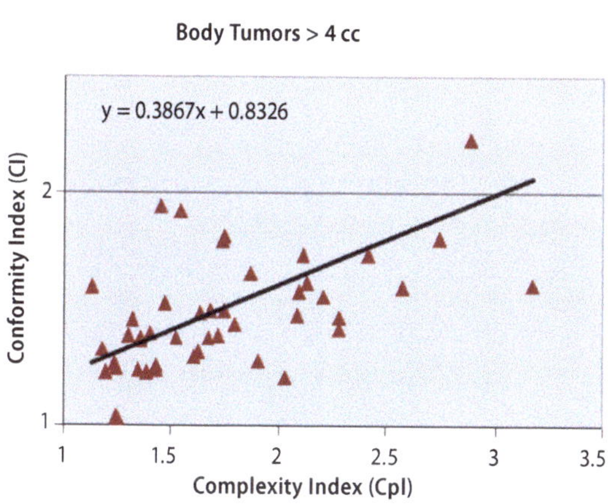

Fig. 7.6 The conformity index as a function of complexity index for treated patients with targets larger than 4 cm^3. A fitted regression line and regression equation are shown.

7.4 Conclusions

A simple complexity index was derived for a more accurate assessment of radiosurgical or radiotherapy plan evaluation. This measure could improve the use of conformity indices by decreasing their variability. An assessment of the target complexity, before treatment and planning decisions are made, may also guide these decisions and improve workflow by proper allocation of planning resources based on complexity. Finally, comparisons between different radiosurgical devices may be enhanced by the use of this measure.

References

1. Mandelbrot B. How Long Is the Coast of Britain? Statistical Self-Similarity and Fractional. *Science* 1967; 156:636–638.
2. Nakamura JL, Pirzkall A, Carol MP, *et al.* Comparison of intensity-modulated radiosurgery with gamma knife radiosurgery for challenging skull base lesions. *Int J Radiat Oncol Biol Phys* 2003; 55:99–109.

3. Nakamura JL, Verhey LJ, Smith V, *et al.* Dose conformity of gamma knife radiosurgery and risk factors for complications. *Int J Radiat Oncol Biol Phys* 2001;51:1313–1319.
4. Paddick I. A simple scoring ratio to index the conformity of radiosurgical treatment plans. Technical note. *J Neurosurg* 2000; 93 Suppl 3:219–222.
5. Borden JA, Mahajan A, Tsai JS. A quality factor to compare the dosimetry of gamma knife radiosurgery and intensity-modulated radiation therapy quantitatively as a function of target volume and shape. Technical note. *J Neurosurg* 2000; 93 Suppl 3:228–232.
6. Dejean C, Lefkopoulos D, Nataf F, *et al.* [Radiosurgery of cerebral arteriovenous malformations: methodology of the dose prescription and its implications for dosimetric results. Retrospective study of 408 cases]. *Neurochirurgie* 2001; 47:239–245.
7. Gibbs IC. Frameless image-guided intracranial and extracranial radiosurgery using the Cyberknife robotic system. *Cancer Radiother* 2006; 10:283–287.
8. Kubo HD, Wilder RB, Pappas CT. Impact of collimator leaf width on stereotactic radiosurgery and 3D conformal radiotherapy treatment plans. *Int J Radiat Oncol Biol Phys* 1999; 44:937–945.
9. Lomax NJ, Scheib SG. Quantifying the degree of conformity in radiosurgery treatment planning. *Int J Radiat Oncol Biol Phys* 2003; 55:1409–1419.
10. Ma L, Xia P, Verhey LJ, *et al.* A dosimetric comparison of fan-beam intensity modulated radiotherapy with Gamma Knife stereotactic radiosurgery for treating intermediate intracranial lesions. *Int J Radiat Oncol Biol Phys* 1999; 45:1325–1330.
11. Nedzi LA, Kooy HM, Alexander E, 3rd, *et al.* Dynamic field shaping for stereotactic radiosurgery: a modeling study. *Int J Radiat Oncol Biol Phys* 1993; 25:859–869.
12. van't Riet A, Mak AC, Moerland MA, *et al.* A conformation number to quantify the degree of conformality in brachytherapy and external beam irradiation: application to the prostate. *Int J Radiat Oncol Biol Phys* 1997; 37:731–736.

Dose Escalation with the Frameless Respiratory Tumor Tracking System of the CyberKnife for Early Stage Non-Small-Cell Lung Cancer

8

Joost Jan Nuyttens, Jean-Briac Prévost, Mischa S. Hoogeman, and Peter C. Levendag

CONTENTS

8.1 Abstract

We used the respiratory movement tracking system of the CyberKnife®, called Synchrony® (Accuray Incorporated, Sunnyvale, CA), to develop dose plans delivering 45 Gy (3 times 15 Gy) for the treatment of early stage non-small cell lung cancer (NSCLC). Characteristics of those plans were compared with plans developed for 3-Dimensional conformal radiotherapy (3D-CRT) administering 60 Gy (20 times 3 Gy) based on a slow CT. Ten patients with Stage I NSCLC previously treated with 3D-CRT were replanned with the CyberKnife treatment planning system. In the 3D-CRT plan, the planning target volume (PTV) equaled the gross tumor volume $(GTV)_{slow}$ + 15 mm. In the CyberKnife plan, the PTV equaled the GTV + 8 mm. The physical dose of both treatment plans was converted into the normalized total dose using the linear quadratic model with an α/β_{tumor} = 10 Gy and $\alpha/\beta_{organs\ at\ risk(OAR)}$ = 3 Gy. The mean doses administered to the PTV with the CyberKnife and 3D-CRT were 115.8 Gy and 66 Gy, respectively ($p < 0.0001$). The mean V_{20} of the CyberKnife and 3D-CRT plan was 8.2% and 6.8%, respectively ($p=0.124$). Both plans respected the constraints of the other organs at risk (OAR). In this context the CyberKnife can administer a much higher biological dose than 3D-CRT without increasing the dose (V_{20}) to the lungs.

8.2 Introduction

Medically inoperable patients with non-small cell lung cancer (NSCLC) are often treated by conventional radiotherapy with doses of 60–66 Gy. This treatment yields 5-year survival rates of 6-30%, with local control rates of 40–70% [1–6]. The poor local control rates could be a consequence of insufficient dose administration [7–11] and/or a geographical miss due to variable target motion [12, 13]. Attempts have been made to escalate this dose, but the motion of lung tumors necessitated the use of large treatment fields resulting in high overall doses to the lungs. For that reason, several techniques, such as immobilizing the patient [14] or respiratory gating [15], were developed to reduce the field size. In this chapter we briefly review dose escalation in 3D-CRT and describe a comparison of characteristics of treatment plans from 3D-CRT and CyberKnife to determine the viability of dose escalation with the CyberKnife.

8.3 Dose Escalation

Dose escalation has been proven to result in better local control. A Radiation Therapy Oncology Group (RTOG) study [16] randomized patients into 4 treatment arms receiving total doses of 40 Gy in 20 fractions, 50 Gy in 25 fractions, 60 Gy in 30 fractions, or a split course consisting of 4 Gy per day for 5 days followed by a 3-week interruption before a second course of 20 Gy in 5 fractions. Improved local control on chest X-ray was seen with increased dose. Nevertheless, the dose needed to achieve a high level of local control and survival was substantial [17]: A "standard" total dose of 60 Gy, administered in 30 fractions of 2 Gy, yields an estimated progression-free survival after 30 months of 16%. Using stereotactic radiotherapy a total dose of 60 Gy, delivered in 3 fractions of 20 Gy, is estimated to achieve a progression-free survival after 30 months of > 99%. Three fractions of 20 Gy is equivalent to a total dose of 150 Gy given in 2 Gy per fraction (α/β_{tumor}=10 Gy).

Higher doses, however, resulted in an increase in toxicity. To predict radiation pneumonitis, dose volume histograms were used. Graham et al. [18] found that the V_{20} (the volume of normal lung that received a total dose of 20 Gy or more) with daily standard fractionation was related to severe or life-threatening pulmonary toxicity. The results of the RTOG 9311 indicated that radiation doses with fractions of 2.15 Gy could safely be escalated using 3D conformal techniques to 83.8 Gy for patients with V_{20} values of < 25% and to 77.4 Gy for patients with V_{20} values between 25% and 36%. The 90.3-Gy dose level was too toxic, resulting in dose-related deaths in 2 patients [19]. Other researchers found that a mean lung dose (MLD) of 19 Gy predicts a 20% risk of developing radiation pneumonitis [20]. A dose escalation study using MLD reported that the radiation dose could safely be escalated to 87.8 Gy if MLD was below 12 Gy and to 81.0 Gy if MLD was below 24 Gy [21]. Another dose escalation study found symptomatic bronchial stenosis occurred at a rate of 4% at a dose of approximately 74 Gy and 25% at a dose of 86 Gy [22]. These dose escalation studies included mainly larger tumors (Stage II and III), resulting in irradiation of a large amount of the lungs and a higher probability of toxicity. The treatment of smaller tumors is feasible if the MLD is below 24 Gy, and recent studies with conventional fractionation were able to increase the dose up to 92–103 Gy with 5-year local control of 50% [23].

New techniques such as intensity-modulated radiotherapy (IMRT) and stereotactic body radiotherapy (SBRT) have been tested in NSCLC. IMRT, however, was found not to offer additional benefits over 3D-CRT in node-negative patients [24]. SBRT with a reduction of the margin around the tumor is now the main tool for dose escalation in early-stage NSCLC [25–27].

8.4 Margin Reduction with the CyberKnife

According to the ICRU 50 and 62 report recommendations [28], extra margins to the target volume are added because the tumor and patient move during treatment. An internal margin is added to account for all variations in size, shape, and position of the target in reference to the patient's coordinate system using anatomic reference points. A set-up margin is added to account for all uncertainties in patient-beam positioning in reference to the treatment machine's coordinate system. To deliver high doses with SBRT, however, a reduction of the margin around the tumor is necessary. Several methods have been developed to reduce the set-up margin, the internal margin, or both. A reduction in the set-up margin can be achieved by immobilizing the patient with a non-invasive body frame [14, 29] or by image-guided positioning [30–32]. A reduction of the internal margin can be realized with the use of a 4-Dimensional CT scan. The use of an abdominal pressure belt [33] minimizes respiratory motion and allows the internal margin to be reduced. Other methods to reduce the respiratory motion are deep inspiration breath hold [34–36] and gating [15]. Synchrony, the respiratory tumor tracking system of CyberKnife, automatically compensates for the tumor motion by continuously adjusting the position of the beam [37]. Therefore, this system reduces the set-up as well as the internal margin.

At the Erasmus Medical Center, the planning target volume (PTV) for CyberKnife treatment plans was compared with the PTV of 3D-CRT plans in 10 patients with Stage I NSCLC. For the 3D-CRT plan, a breath hold CT of the entire thorax was used to define the GTV_{plan}. A slow CT scan [38] was performed to capture the tumor motion. The breath hold and slow CT scan were matched on bony anatomy. A GTV_{slow} was defined by the delineation of the tumor on the slow CT scan and was expanded by 5 mm (residual motion) to create the internal target volume (ITV). If the GTV_{plan} was positioned outside the ITV then the ITV was adapted to fully encompass the GTV_{plan}. The PTV was created by adding 5 mm (microscopic extension) and 5 mm (set-up margin) to the ITV (Fig. 8.1a). For the CyberKnife plan, the $GTV_{plan,}$ as used for the 3D-CRT plan, was expanded by 5 mm (microscopic extension) to define the CTV. For the PTV, 3 millimeters were symmetrically added to the CTV to take the inaccuracy of the linear model of the Synchrony into account (Fig. 8.1b) [39]. The motion information from the slow CT was not used in the CyberKnife plan considering the system's ability to track the lung tumor after the insertion of markers. The mean PTV in the CyberKnife plan and 3D-CRT was 53.7 cc and 132.7 cc, respectively (Table 8.1). Due to the Synchrony system of the CyberKnife, a safe volume reduction of approximately 60% was feasible.

Table 8.1 The volume of the PTV for 3D-CRT and CyberKnife.

Patient	3D-CRT (cc)	CyberKnife (cc)	% (CyberKnife/3D-CRT)
1	258.2	50.6	19.6
2	107.6	43.2	40.1
3	136.4	58.7	43.0
4	115.1	46.9	40.8
5	228.5	111.8	48.9
6	91.4	70.2	76.9
7	90.4	46.2	51.1
8	75.8	30.8	40.6
9	94.2	32.8	34.8
10	129.1	46.2	35.7
Mean	132.7	53.7	43.2 $p=0.0009$

8.5 Dose Escalation with the CyberKnife

Ten consecutive patients with Stage I NSCLC, who had been previously treated with 3D-CRT incorporating tumor motion with a slow CT, were replanned with the CyberKnife treatment planning. The 3D-CRT technique involved a three-dimensional, coplanar isocentric technique. A dose of 60 Gy (3 Gy/fraction) was prescribed according to ICRU 50 recommendations. The dose to the spinal cord was limited to a maximum of 50 Gy for all patients. All patients were planned on a linear accelerator using a 6 MV photon beam. The 3D beam angles, weights, and wedge angles were determined to ensure adequate protection of normal tissues and homogeneous coverage of the PTV. The CyberKnife plan was done with the On-Target treatment planning system allowing inverse planning, non-isocentric and non-coplanar radiation delivery. The total dose of 45 Gy (15 Gy/fraction)

a

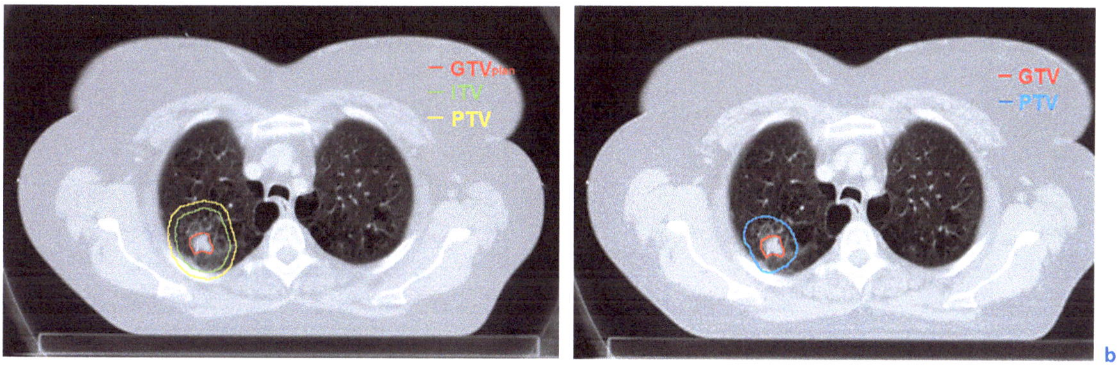

b

Fig. 8.1a,b. An example of the GTV and PTV definition in one patient for 3D-CRT (**a**) and CyberKnife planning (**b**).

was always prescribed to the 80% isodose line. An example of the dose distribution for one patient with 3D-CRT and CyberKnife planning is given in Figure 8.2a and 8.2b.

To enable comparison of the treatment plans, the dose volume histograms (DVHs) of the physical dose distribution were converted to the normalized total dose (NTD) using the linear quadratic model with an α/β_{OAR} of 3 Gy and an α/β_{tumor} of 10 Gy [40]. In the remainder of this chapter all the doses are reported in NTD. To assess the radiobiologic implications of normal lung irradiation, the mean normalized total lung dose [41] (MNTLD, the mean lung dose normalized in 2 Gy fractions) and the $V_{20,NTD}$ (volume of lung irradiated to doses above 20 Gy, normalized in 2 Gy fractions) [18], were computed. The lung volume was defined based on the CT, excluding the PTV [42]. The MNTLD and the mean dose to organs at risk were approximated by the area under the cumulative DVH (converted to NTD). The PTVs for both techniques were derived as described above.

The mean minimal dose, 93 Gy, administered to the PTV with the CyberKnife plan was statistically significantly higher than the dose of 61 Gy delivered by the 3D-CRT plan (p< 0.0001, Table 8.2). The average mean dose delivered to the PTV with the CyberKnife plan and 3D-CRT treatment plan was 115.8 Gy and 66.0 Gy, respectively (Table 8.2). The CyberKnife plan administered a 75.3% higher mean dose to the PTV than the 3D-CRT plan. The maximum dose administered by both plans is shown in Table 8.2. The DVH of the lungs showed a similar low-dose region of the CyberKnife plan and the 3D-CRT plan (Fig. 8.3). The mean $V_{20,NTD}$ of the CyberKnife plan, 8.2%, was not statistically different from the 6.8% of the 3D-CRT (p=0.124). The mean lung dose (MLD) delivered by the CyberKnife plan and 3D-CRT treatment plan was on average 9.8 Gy and 6.5 Gy, respectively.

The doses administered to the other organs at risk were below tolerance levels for both techniques. The mean of the maximum dose delivered to the spinal cord by the 3D-CRT and CyberKnife plan was 8.6 Gy and 9.9 Gy, respectively (p=0.615). The mean of the maximum dose administered to the esophagus by the 3D-CRT was 10.5 Gy, which is statistically significantly higher than the 5.9 Gy delivered with the CyberKnife plan (p=0.028). There was no statistical difference between the dose of 19.8 Gy administered to the heart by the CyberKnife and the 26.6 Gy delivered with the 3D-CRT (p=0.32).

Table 8.2 Dose to the PTV for 3D-CRT and CyberKnife

	3D-CRT (range)	CyberKnife (range)	P-value	Percent Dose (CyberKnife/3D-CRT)
Mean of the min dose (Gy)	61 (60–62)	93 (88–96)	< 0.0001	151
Mean of the max dose (Gy)	68 (67–71)	135	< 0.0001	198
Mean of the dose (Gy)	66.0 (65.6–67.9)	115.8 (112.9–117.7)	< 0.0001	175

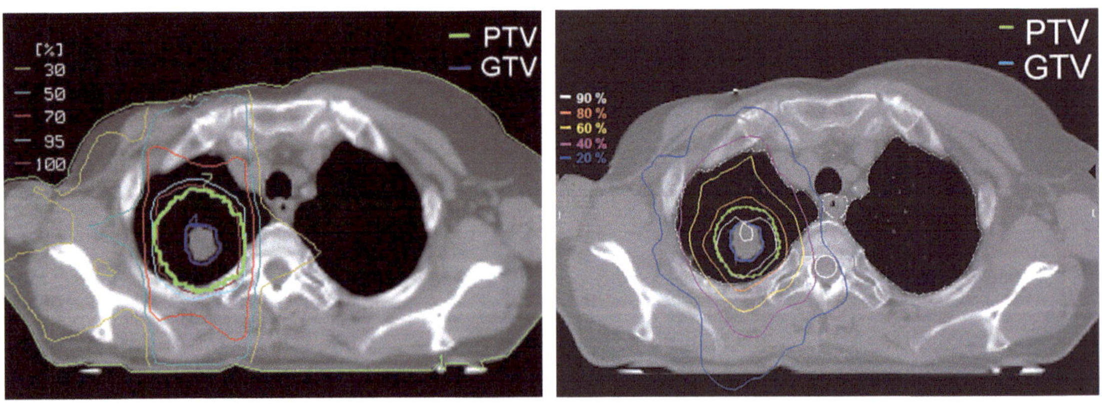

a b

Fig. 8.2a,b. An example of the 3D-CRT (**a**) and CyberKnife (**b**) dose distribution for the same patient.

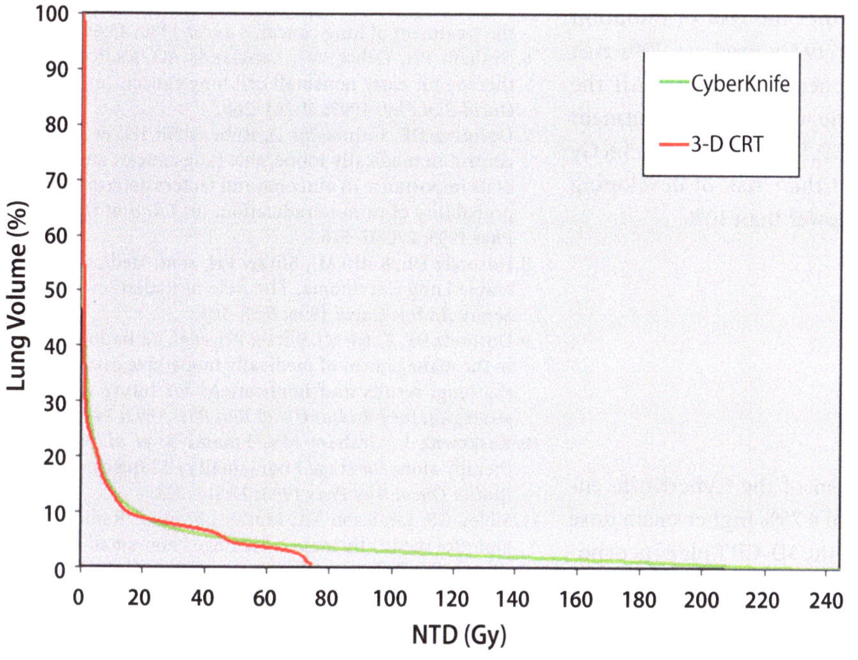

Fig. 8.3 The DVH of the lung for 3D-CRT and CyberKnife planning.

8.6 Discussion

In this planning study, we converted the physical dose to normalized dose, an estimate of the biological dose delivered under altered fractionation schemes, in order to compare the dosimetric parameters of both plans. The CyberKnife plan administered a mean dose to the PTV of 116 Gy, 75% higher than the mean dose administered with the 3D-CRT plan. The CyberKnife plan administered on average a minimal dose (93 Gy) that was 51% higher than the dose (61 Gy) delivered by the 3D-CRT plan. Despite these large differences, the overall dose to the lung did not differ between the two treatment modalities. These outcomes could be achieved by using the tumor tracking system (Synchrony) of the CyberKnife. With this tumor tracking system we are able to reduce the required safety margin and consequently the volume of the PTV. In the CyberKnife plan, the mean PTV was 53.74 cc, significantly smaller than the mean PTV of the 3D-CRT plan of 132.7 cc, defined on a slow CT. The 56.8% reduction of the PTV demonstrates the superiority of the Synchrony system in comparison to the 3D-CRT using a slow CT and an off-line set-up protocol to reduce the systematic set-up error.

The ability to track the moving tumor makes it safe to use an internal margin of only 3 mm, which contributes to the key advantage of this system; the ability to administer higher biological doses compared to conventional methods. The 3-mm margin takes into account the inaccuracy of the linear treatment model of the Synchrony, which was derived from a study performed by Accuray in collaboration with several CyberKnife clinical users [39]. They analyzed the data files of 14 Synchrony treatments containing 510 intra-treatment model errors. The overall mean errors were 1.4 mm with a standard deviation of 1.0 mm; 90% of all errors were less than 3.0 mm. The linear model, describing the relationship between internal fiducial and external optical marker positions, is accurate for most patient breathing patterns. At present, the Synchrony system also uses non-linear models to model the tumor trajectory.

As normal lung is the major organ at risk in radiotherapy for NSCLC, dose escalation to the tumor is restricted by the tolerance of the surrounding normal lung tissue. Graham et al. reported the risk for radiation pneumonitis to be zero when the V_{20} was less than 22% [18]. With CyberKnife treatment planning, a V_{20} of 8.2% was obtained, indicating that risk for radiation pneumonitis is very low. The MLD is another parameter to predict the risk of radiation pneumonitis. An MLD of 19 Gy predicts 20% risk of developing radiation pneumonitis [20]. All the patients planned with the CyberKnife treatment planning system had an MLD ranging from 5.6 Gy to 15.8 Gy, indicating that their risk of developing radiation pneumonitis is lower than 10%.

8.7 Conclusion

The tumor tracking system of the CyberKnife enables the administration of a 75% higher mean dose than the dose delivered by the 3D-CRT plan incorporating tumor motion based on a slow CT. The dose to the lungs or other organs at risk with the CyberKnife treatment planning was not increased compared to the 3D-CRT. Based on the correlation between dose, local control and overall survival in radiation treatment for NSCLC [1, 2, 4, 6, 8] the poor outcomes of conventional radiotherapy treatment schedules are expected to be improved upon by the administration of higher biological doses with the CyberKnife. This will be investigated in a prospective trial in which a dose of 3 times 20 Gy will be delivered to Stage I NSCLC patients when achievable by the parameters of the organs at risk.

References

1. Werner-Wasik M, Scott C, Cox JD, *et al.* Recursive partitioning analysis of 1999 Radiation Therapy Oncology Group (RTOG) patients with locally-advanced non-small-cell lung cancer (LA-NSCLC): identification of five groups with different survival. *Int J Radiat Oncol Biol Phys* 2000; 48:1475–1482.
2. Krol AD, Aussems P, Noordijk EM, *et al.* Local irradiation alone for peripheral stage I lung cancer: could we omit the elective regional nodal irradiation? *Int J Radiat Oncol Biol Phys* 1996; 34:297–302.
3. Sandler HM, Curran WJ, Jr., Turrisi AT, 3rd. The influence of tumor size and pre-treatment staging on outcome following radiation therapy alone for stage I non-small cell lung cancer. *Int J Radiat Oncol Biol Phys* 1990; 19:9–13.
4. Cooper JD, Pearson G, Todd TR, *et al.* Radiotherapy alone for patients with operable carcinoma of the lung. *Chest* 1985; 87:289–292.
5. Coy P, Kennelly GM. The role of curative radiotherapy in the treatment of lung cancer. *Cancer* 1980; 45:698–702.
6. Graham PH, Gebski VJ, Langlands AO. Radical radiotherapy for early nonsmall cell lung cancer. *Int J Radiat Oncol Biol Phys* 1995; 31:261–266.
7. Dosoretz DE, Galmarini D, Rubenstein JH, *et al.* Local control in medically inoperable lung cancer: an analysis of its importance in outcome and factors determining the probability of tumor eradication. *Int J Radiat Oncol Biol Phys* 1993; 27:507–516.
8. Dosoretz DE, Katin MJ, Blitzer PH, *et al.* Medically Inoperable Lung Carcinoma: The Role of Radiation Therapy. *Semin Radiat Oncol* 1996; 6:98–104.
9. Dosoretz DE, Katin MJ, Blitzer PH, *et al.* Radiation therapy in the management of medically inoperable carcinoma of the lung: results and implications for future treatment strategies. *Int J Radiat Oncol Biol Phys* 1992; 24:3–9.
10. Kaskowitz L, Graham MV, Emami B, *et al.* Radiation therapy alone for stage I non-small cell lung cancer. *Int J Radiat Oncol Biol Phys* 1993; 27:517–523.
11. Sibley GS, Jamieson TA, Marks LB, *et al.* Radiotherapy alone for medically inoperable stage I non-small-cell lung cancer: the Duke experience. *Int J Radiat Oncol Biol Phys* 1998; 40:149–154.
12. Seppenwoolde Y, Shirato H, Kitamura K, *et al.* Precise and real-time measurement of 3D tumor motion in lung due to breathing and heartbeat, measured during radiotherapy. *Int J Radiat Oncol Biol Phys* 2002; 53:822–834.
13. Shirato H, Seppenwoolde Y, Kitamura K, *et al.* Intrafractional tumor motion: lung and liver. *Semin Radiat Oncol* 2004; 14:10–18.
14. Wulf J, Hadinger U, Oppitz U, *et al.* Stereotactic radiotherapy of extracranial targets: CT-simulation and accuracy of treatment in the stereotactic body frame. *Radiother Oncol* 2000; 57:225–236.
15. Shirato H, Shimizu S, Kunieda T, *et al.* Physical aspects of a real-time tumor-tracking system for gated radiotherapy. *Int J Radiat Oncol Biol Phys* 2000; 48:1187–1195.
16. Perez CA, Stanley K, Rubin P, *et al.* A prospective randomized study of various irradiation doses and fractionation schedules in the treatment of inoperable non-oat-cell carcinoma of the lung. Preliminary report by the Radiation Therapy Oncology Group. *Cancer* 1980; 45:2744–2753.
17. Fowler JF, Tome, W.A., Welsh, J.S. On the Estimation of Required Doses In Stereotactic Body Radiation Therapy. In: Kavanagh BD, Timmerman, R.D., editor. Stereotactic Body Radiation Therapy. Philadelphia: Lippippincott, Williams and Wilkins; 2006. pp. 7–14.
18. Graham MV, Purdy JA, Emami B, *et al.* Clinical dose-volume histogram analysis for pneumonitis after 3D treatment for non-small cell lung cancer (NSCLC). *Int J Radiat Oncol Biol Phys* 1999; 45:323–329.
19. Bradley J, Graham MV, Winter K, *et al.* Toxicity and outcome results of RTOG 9311: a phase I-II dose-escalation study using three-dimensional conformal radiotherapy in patients with inoperable non-small-cell lung carcinoma. *Int J Radiat Oncol Biol Phys* 2005; 61:318–328.
20. De Jaeger K, Hoogeman MS, Engelsman M, *et al.* Incorporating an improved dose-calculation algorithm in conformal radiotherapy of lung cancer: re-evaluation of dose in normal lung tissue. *Radiother Oncol* 2003; 69:1–10.
21. Belderbos JS, De Jaeger K, Heemsbergen WD, *et al.* First results of a phase I/II dose escalation trial in non-small

cell lung cancer using three-dimensional conformal radiotherapy. *Radiother Oncol* 2003; 66:119–126.

22. Miller KL, Shafman TD, Anscher MS, *et al.* Bronchial stenosis: an underreported complication of high-dose external beam radiotherapy for lung cancer? *Int J Radiat Oncol Biol Phys* 2005; 61:64–69.

23. Kong FM, Ten Haken RK, Schipper MJ, *et al.* High-dose radiation improved local tumor control and overall survival in patients with inoperable/unresectable non-small-cell lung cancer: long-term results of a radiation dose escalation study. *Int J Radiat Oncol Biol Phys* 2005; 63:324–333.

24. Grills IS, Yan D, Martinez AA, *et al.* Potential for reduced toxicity and dose escalation in the treatment of inoperable non-small-cell lung cancer: a comparison of intensity-modulated radiation therapy (IMRT), 3D conformal radiation, and elective nodal irradiation. *Int J Radiat Oncol Biol Phys* 2003; 57:875–890.

25. McGarry RC, Papiez L, Williams M, *et al.* Stereotactic body radiation therapy of early-stage non-small-cell lung carcinoma: phase I study. *Int J Radiat Oncol Biol Phys* 2005; 63:1010–1015.

26. Timmerman R, Papiez L, McGarry R, *et al.* Extracranial stereotactic radioablation: results of a phase I study in medically inoperable stage I non-small cell lung cancer. *Chest* 2003; 124:1946–1955.

27. Nyman J, Johansson KA, Hulten U. Stereotactic hypofractionated radiotherapy for stage I non-small cell lung cancer–mature results for medically inoperable patients. *Lung Cancer* 2006; 51:97–103.

28. Purdy JA. Current ICRU definitions of volumes: limitations and future directions. *Semin Radiat Oncol* 2004; 14:27–40.

29. Wulf J, Hadinger U, Oppitz U, *et al.* Impact of target reproducibility on tumor dose in stereotactic radiotherapy of targets in the lung and liver. *Radiother Oncol* 2003; 66:141–150.

30. Ford EC, Chang J, Mueller K, *et al.* Cone-beam CT with megavoltage beams and an amorphous silicon electronic portal imaging device: potential for verification of radiotherapy of lung cancer. *Med Phys* 2002; 29:2913–2924.

31. Sonke JJ, Zijp L, Remeijer P, *et al.* Respiratory correlated cone beam CT. *Med Phys* 2005; 32:1176–1186.

32. Kupelian PA, Ramsey C, Meeks SL, *et al.* Serial megavoltage CT imaging during external beam radiotherapy for non-small-cell lung cancer: observations on tumor regression during treatment. *Int J Radiat Oncol Biol Phys* 2005; 63:1024–1028.

33. Negoro Y, Nagata Y, Aoki T, *et al.* The effectiveness of an immobilization device in conformal radiotherapy for lung tumor: reduction of respiratory tumor movement and evaluation of the daily setup accuracy. *Int J Radiat Oncol Biol Phys* 2001; 50:889–898.

34. Hanley J, Debois mm, Mah D, *et al.* Deep inspiration breath-hold technique for lung tumors: the potential value of target immobilization and reduced lung density in dose escalation. *Int J Radiat Oncol Biol Phys* 1999; 45:603–611.

35. Mah D, Hanley J, Rosenzweig KE, *et al.* Technical aspects of the deep inspiration breath-hold technique in the treatment of thoracic cancer. *Int J Radiat Oncol Biol Phys* 2000; 48:1175–1185.

36. Rosenzweig KE, Hanley J, Mah D, *et al.* The deep inspiration breath-hold technique in the treatment of inoperable non-small-cell lung cancer. *Int J Radiat Oncol Biol Phys* 2000; 48:81–87.

37. Schweikard A, Shiomi H, Adler J. Respiration tracking in radiosurgery. *Med Phys* 2004; 31:2738–2741.

38. Lagerwaard FJ, Van Sornsen de Koste JR, Nijssen-Visser MR, *et al.* Multiple „slow" CT scans for incorporating lung tumor mobility in radiotherapy planning. *Int J Radiat Oncol Biol Phys* 2001; 51:932–937.

39. Kilby W, Main WM, Dieterich S, *et al.* In-vivo assessment of the Synchrony respiratory motion tracking system accuray. CyberKnife Users Meeting. Half Moon Bay, CA; 2005.

40. Lebesque JV, Keus RB. The simultaneous boost technique: the concept of relative normalized total dose. *Radiother Oncol* 1991; 22:45–55.

41. Kwa SL, Lebesque JV, Theuws JC, *et al.* Radiation pneumonitis as a function of mean lung dose: an analysis of pooled data of 540 patients. *Int J Radiat Oncol Biol Phys* 1998; 42:1–9.

42. Seppenwoolde Y, Lebesque JV, de Jaeger K, *et al.* Comparing different NTCP models that predict the incidence of radiation pneumonitis. Normal tissue complication probability. *Int J Radiat Oncol Biol Phys* 2003; 55:724–735.

Section 3

Thoraco-Abdominal Fiducial Placement Strategies

Percutaneous Placement of Fiducial Markers for Thoracic Malignancies

9

Filip Banovac, Donald McRae, Sonja Dieterich, Kenneth Wong, Lisa Dias, and Thomas Chang

CONTENTS

9.1 Abstract

Image-guided placement of fiducial markers is in some ways an extension of percutaneous procedures such as needle biopsy of lung pathology, which are native to most interventional radiology practices. To that extent, learning the procedure is not difficult for those who are familiar with the basic principles of image-guided lung nodule biopsy. However, there are significant modifications in the procedure that are necessary in order to ensure appropriate placement and distribution of the fiducial markers. Proper positioning of fiducial markers in specific geometric configurations is essential for accurate targeting of the nodule. This chapter focuses on the principles of CT-guided percutaneous placement of fiducial markers. For the most part, this procedure is performed on a consultative basis by interventional radiologists, physicians who specialize in minimally invasive image-guided therapy. Special considerations for patient selection, pre-procedural preparation, techniques, and post-procedural care are explained.

9.2 Introduction

The insertion of fiducial markers into and around thoracic pathology has become increasingly important as the uses of CyberKnife® (Accuray Incorporated, Sunnyvale, CA) technology have expanded to treat lung neoplasms. Initially, the CyberKnife was used for stereotactic radiosurgery of intracranial tumors and spine [1–3]. Clinical applications evolved once the algorithms for tracking targets subject to respiratory-induced motion became more robust [4, 5]. CyberKnife radiosurgery based on implantable fiducial targeting in extracranial sites became more common, including treatment of neoplasms in the prostate [6], kidney [7], and thorax [8]. This necessitated development of new techniques that allow reliable implantation of fiducial markers in or near tumors. For tumors within the thorax, more specifically the lung, CT-guided percutaneous placement was adopted for more peripheral lesions [8], and bronchoscopic approaches [9] were utilized for more central lesions.

Proper distribution of fiducial markers is important for two reasons. First, the CyberKnife has to identify the fiducial markers as separate and distinct points in the 45-degree orthogonal treatment imaging planes. Second, the aggregate of fiducial marker distribution has to define a volume of tissue. If the fiducial markers overlap or are too close together, CyberKnife treatment planning and treatment execution become a challenge. Basic principles and techniques applied to CT-guided fiducial insertion described here are based on the experience in our institution. Although we find these techniques to be clinically acceptable, they are probably not all-encompassing and new methods will be evolving as the aggregate experience in this field expands.

9.3 Principles of Fiducial Placement from the Tracking Perspective

There are two separate issues that must be addressed to successfully track tumors moving with respiration. These are the general alignment of the patient into the desired treatment position and the real-time motion tracking of the tumor with Synchrony® (Accuray Incorporated, Sunnyvale, CA) while the radiation beam is on.

It is difficult to obtain reliable translational and rotational information if only a single fiducial is placed in soft tissue because of its mutability and movement with respiration. Further, with only one fiducial it can be difficult to determine whether fiducial migration has occurred between the planning CT and CyberKnife treatment. One fiducial, however, is better than two or more poorly placed, closely spaced fiducials because they are hard to distinguish as separate points on the planning CT and by the targeting system. A well-placed and spaced geometric array of fiducials alleviates these concerns and allows the CyberKnife planning and treatment system to work with Synchrony to track tumor volume movement in all directions and accurately account for respiratory movement as well as small movements of the patient.

It has become apparent over our four year experience of delivering CyberKnife treatments to soft tissues that fiducial placement is the most important part of achieving accurate targeting with the CyberKnife.

9.3.1 Fiducial Distribution

Fiducials should be placed in or near the lesion in such a way that they move with the target. Ideally, they should all be at least 2 cm apart, with no more than two fiducials in a co-linear relationship to one another. In small lesions, this interfiducial distance can be reduced to 1 cm. When implanting fiducials for tumors of larger volumes, all fiducials should be within 12 cm of each other in any dimension (preferably ≤ 10 cm). When technically feasible, it is generally preferred not to place fiducials in an axial plane containing materials that cause large CT artifacts.

9.3.2 45-Degree Lateral Oblique Projections

Figure 9.1 shows the geometry of the X-ray projections for fiducial tracking. It is of primary impor-

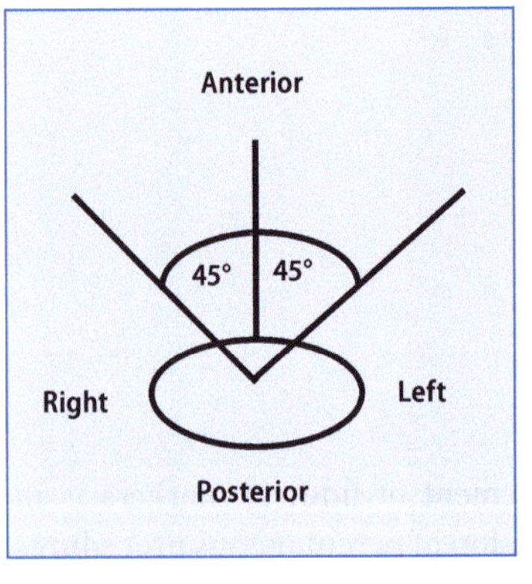

Fig. 9.1 The angles used for imaging of the fiducials on the CyberKnife are ± 45-degrees Right and Left Lateral Obliques.

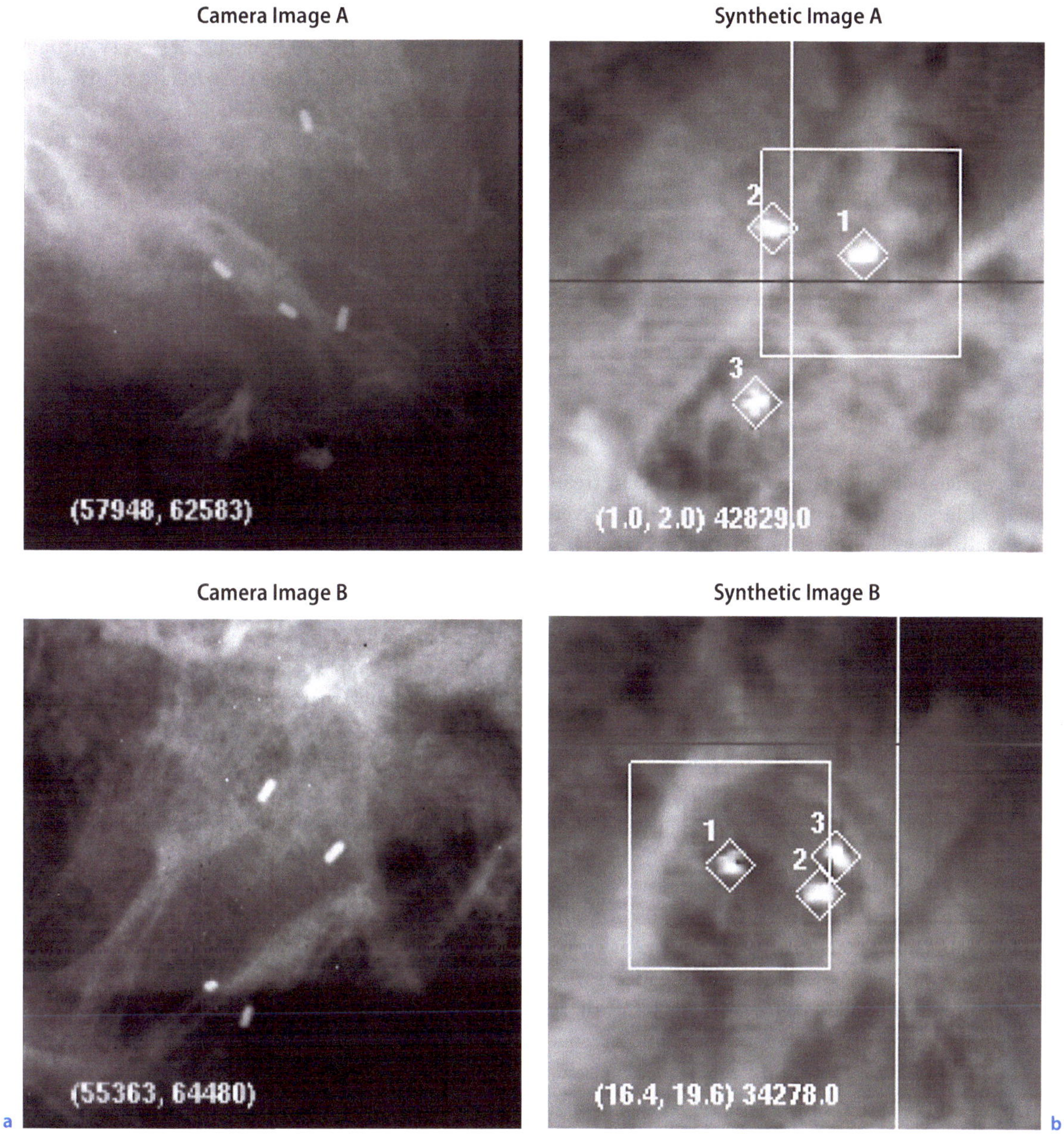

Fig. 9.2a,b. Orthogonal 45 degree views as seen during treatment tracking of **a** 4 well-placed gold seed fiducials in soft tissue; **b** 3 trackable fiducials where #2 and #3 in the lower image are pushing the limit of appearing too close together in that projection.

tance to recognize that the aim of proper fiducial placement is to ensure that the fiducial array is visible in both of these 45-degree projections. Utilizing the CT scout views at the 45-degree oblique angles or being able to mentally re-construct these views by the clinician from observation of the CT slice during implantation is valuable for envisioning the evolving fiducial array. Fiducials must not appear to overlap when viewed at these 45-degree angles and preferably should appear at least 1 cm apart. If two fiducials must be placed in or near the same axial plane, one must place them at different left-right or superior-inferior positions to separate them in space in the 45-degree views. Two good examples of tracking images with these desired angles are shown (Fig. 9.2). If there is metallic hardware, place the fiducials so that they are not obstructed while also maintaining the desired angles.

In some cases, full 6D localization and/or tracking (3 translational directions and 3 rotational angles) is desired. If so, a minimum of three properly-placed fiducials is required. To achieve this, we recommend implanting at least four fiducials. The rules for 45-degree visualization above must be followed.

9.4 Patient Selection

Appropriate selection of patients for CyberKnife treatment of lung nodules and masses is complex. Given the availability of surgical options, radiation therapy options, and chemotherapeutic options [10–13], all backed by years of experience and sound data on efficacy and outcomes, the role of CyberKnife in the treatment of lung neoplasms is still evolving. Studies are emerging, however, showing the benefits of stereotactic radiosurgery for the thoracic malignancies [8, 14, 15], and without a doubt, the role of CyberKnife in this realm will expand in the coming years.

For intended fiducial placement, most patients are referred to interventional radiology by their medical oncologists, radiation oncologists, or thoracic surgeons after a thorough evaluation of available treatment options for the given patient's specific disease and stage. The majority of patients with lung nodules will undergo an evaluation for primary resection, appropriate chemotherapy, and/or conventional radiation therapy. In cases where surgery is not an option due to expected morbidity or extent of disease, CyberKnife and other local ablative techniques [16] are emerging as viable options for treatment. Occasionally, patients also do not wish to undergo more invasive therapeutic options, and choose CyberKnife as a less invasive alternative.

9.4.1 Indications

The invariable indication for placement of fiducial markers for CyberKnife therapy of lung neoplasms is the presence of a neoplastic mass in the lung or thorax. It is essential to either have a cytopathologic confirmation of the malignancy or a very strong clinical suspicion of disease. There are many situations where a moderate clinical suspicion exists and interventional radiologists are consulted for a biopsy of a suspected target prior to placement of fiducial markers. Later in this chapter we elaborate on some of the practical approaches that allow both the biopsy and the fiducial marker placement to be performed simultaneously or as separate procedures.

9.4.2 Contraindications

There are relatively few absolute contraindications to percutaneous placement of fiducial markers. There are many clinical situations, however, where relative contraindications exist, and additional precautions and pre-procedural steps have to be taken to reduce the incidence of complications. Absolute contraindications include uncorrectable coagulopathy and severe respiratory compromise that would endanger the life of the patient if an additional superimposed procedure-related complication occurred (e.g., pneumothorax).

Most coagulopathies are correctable by discontinuation of anti-coagulant medications or transfusion of fresh frozen plasma or platelets. In our practice, serologic confirmation of prothrombin time (PT), activated partial thromboplastin time (aPTT), and international normalized ratio (INR) are deemed acceptable if the patient has obtained a serologic confirmation within three weeks prior to the planned procedure date. If these patients are not on chronic anticoagulant therapy, we have little reason to believe that a severe coagulopathy would develop in the interim period. Unless there are other confounding factors, we use PT of 16 seconds, aPTT of 40 seconds, and INR of 1.6 as the upper limits of acceptable values for the fiducial placement procedure to proceed. We prefer that the platelet count be at least 50,000/uL.

The presence of severe respiratory compromise as an absolute contraindication poses a more formidable challenge. At baseline, many of the patients who have lung nodules also have poor respiratory function. Many are or have been tobacco smokers, and many have concomitant chronic pulmonary lung disease. In our center, severe respiratory compromise is a common comorbid condition in patients referred for CyberKnife fiducial placement. We evaluate the risks of potential complication for each individual case. Certainly, pneumothorax or minor to moderate pulmonary hemorrhage is not an insurmountable problem in most cases. Patients with poor pulmonary function, however, could have catastrophic consequences because of their reduced tolerance for additional pulmonary insults.

Relative contraindications to the procedure include the need for positive pressure ventilation, lack of patient cooperation, uncontrollable cough in the patient, pulmonary hypertension, local bullous disease near the intended site, and presence of highly vascular lesions. It is probably best to consider the risk to potential benefit in each individual case with relative contraindications. In most cases, these patients are not surgical candidates and have no realistic chance for cure unless some therapy is initiated. For patients with a solitary lesion with potentially curative radiosurgery, the benefit of fiducial marker placement is far greater than in a patient with multifocal metastatic disease where radiosurgery is being considered as a palliative measure. In our center we do not have absolute cut-off values for pulmonary function testing that would exclude the patient from candidacy for fiducial marker placement.

9.4.3 Pre-Procedure Evaluation

The evaluation starts with the initial consultation to interventional radiology for CT-guided percutaneous fiducial marker placement. The referring physician supplies the demographic information along with imaging illustrating the mass or nodule to be treated. The consulting interventional radiologist examines all the available imaging information, most commonly CT scans of the thorax, or occasionally PET-CT scans or magnetic resonance images, to determine the size, position, and imaging characteristics of the mass in question.

The location of the mass or nodule, the proximity of the nodule to critical organs and vascular structures, and the interposition of organs or ribs, are all considered when determining the feasibility of the procedure. All of these factors, plus theoretical modeling of ideal patient positioning for the procedure, are considered to allow for the safest approach to the nodule or mass in question. If the procedure is considered technically feasible, the patient is scheduled for fiducial marker placement.

Virtually all fiducial marker placement procedures are performed on an outpatient basis. In our practice, the patient is called several days in advance by our nursing staff in order to give general pre-procedural instructions. These include discontinuation of anticoagulant therapy at least 48–72 hours prior to the procedure if this is clinically allowable. Some examples of medications that we routinely screen for in our pretreatment history and discontinue include anticoagulants such as Warfarin and enoxaparin (Lovenox), and drugs that interfere with platelet function such as clopidogrel (Plavix) and aspirin. Patients are also instructed not to eat after midnight the day before the procedure.

The patient is evaluated in an interventional radiology examination room where a detailed and complete history of present illness is obtained and a physical exam is performed. This history is focused on assessing information regarding all known allergies, past surgeries, previous lung biopsies, and any ensuing complications of past procedures. Baseline oxygen requirements are noted and any need for supplemental oxygen therapy is assessed at this time. Baseline vital signs and pulse oximetry are routinely obtained and recorded.

9.4.4 Informed Consent

Informed written consent is obtained prior to every procedure. If a concomitant biopsy is planned, consent for the biopsy is also obtained. The risks of the procedure are reviewed, particularly the risk of pneumothorax, pulmonary hemorrhage, and procedure related infection.

9.5 Techniques

9.5.1 Patient Positioning

Patients are positioned on a CT scanner with CT-fluoroscopy capability in either supine or prone position, or less commonly in a lateral decubitus position. The positioning is based on the location of the tumor and the interposition of organs or structures in the projected needle path. The availability of CT-fluoroscopy in the scanner adds substantial flexibility during this procedure, allowing for frequent manipulations of and small adjustments to the needle trajectory.

9.5.2 Conscious Sedation and Monitoring

We achieve conscious sedation of the patient by using a combination of fentanyl and midazolam (Versed). We give 50 micrograms of intravenous fentanyl and 1 mg of intravenous midazolam initially, and titrate as appropriate throughout the procedure. The patient's pulse oximetry and blood pressure are monitored continuously.

9.5.3 Tools for Fiducial Placement

There are several manufacturers of the gold fiducial markers that are suitable for CyberKnife radiosurgery. In our practice, we find that gold fiducial markers (Alpha-Omega Services, Inc., Bellflower, CA) are well suited for lung and thoracic applications. Their small outer diameter is ideal for placement through the shaft of the needle trocar. Other fiducial markers may also be used, but pre-loaded systems are generally avoided in the lung because they would require multiple transgressions of the pleura. With a coaxial system, the pleura is typically only crossed once.

The needle systems that have a large enough channel to allow fiducial marker delivery include most 18 gauge needles such as the Chiba needle (Cook Inc., Bloomington, IN), or the needle preferred in our practice, the 19 gauge Bard TruGuide coaxial biopsy needle (Bard Peripheral Vascular, Inc., Tempe, AZ). The Bard TruGuide needle comes in three lengths: 19.8 cm, 13.8 cm, and 8 cm. The 19 gauge needle has a theoretical advantage over the 18 gauge systems because pneumothorax rates were reported to be lower with 19 gauge systems during lung biopsy [17]. Certainly, there are other needles on the market that would be suitable for this purpose. Any needle that is to be used should be tested before the procedure in order to ensure that the fiducial markers can easily pass through the trocar shaft.

9.5.4 Pre-Procedural Imaging

After the patient is adequately sedated and comfortable, initial localizing scans are performed. Selective portions of the thorax are imaged to allow clear and unobstructed visualization of the target lesion. Volumetric acquisition with image reconstruction to 5-mm slices is the preferred method in our center. The path for needle insertion is then chosen and the initial skin entry point is determined and marked.

9.5.5 Sterile Field Preparation

The skin is cleaned with chlorhexidine gluconate (or another appropriate antiseptic solution), and then draped with a sterile cover. The radiologist wears a lead apron, covered by a sterile gown and gloves, mask, and a protective hat. The skin entry point is anesthetized with Lidocaine 1% through a 21 gauge hypodermic needle. The skin tract is then anesthetized to the level of the pleura with a longer needle. Care is taken not to transgress the pleura with the hypodermic needle as this carries a separate risk of pneumothorax.

9.5.6 Fiducial Insertion Technique

The 19 gauge needle is inserted while the trajectory and depth are frequently checked with CT-fluoroscopy. Small adjustments in needle trajectory can be achieved by applying appropriate torque to the shaft of the needle.

We generally deliver a total of four gold fiducials for any given lesion. As explained above, four well-placed fiducial markers provide an optimal target for the treatment planning system. The fiducial markers do not necessarily have to be in the lesion of interest. In fact, it is completely appropriate to place the fiducial markers around the lesion of interest. Not touching the tumor directly has a theoretical advantage in preventing any possible seeding of malignant cells in the tract of the needle.

After CT-fluoroscopy confirms the desired location of the needle tip, the needle stylette is removed and a single gold fiducial marker is deposited through the hub into the shaft of the needle (Fig. 9.3). The stylette is then used to push the fiducial marker through the shaft of the needle and introduce it into the lung parenchyma. The needle is then withdrawn a few centimeters and a new trajectory is chosen. After redirecting the needle four times and delivering four fiducial markers, the needle is removed.

As a matter of practice, we avoid pulling the needle completely out of the lung. This would neces-

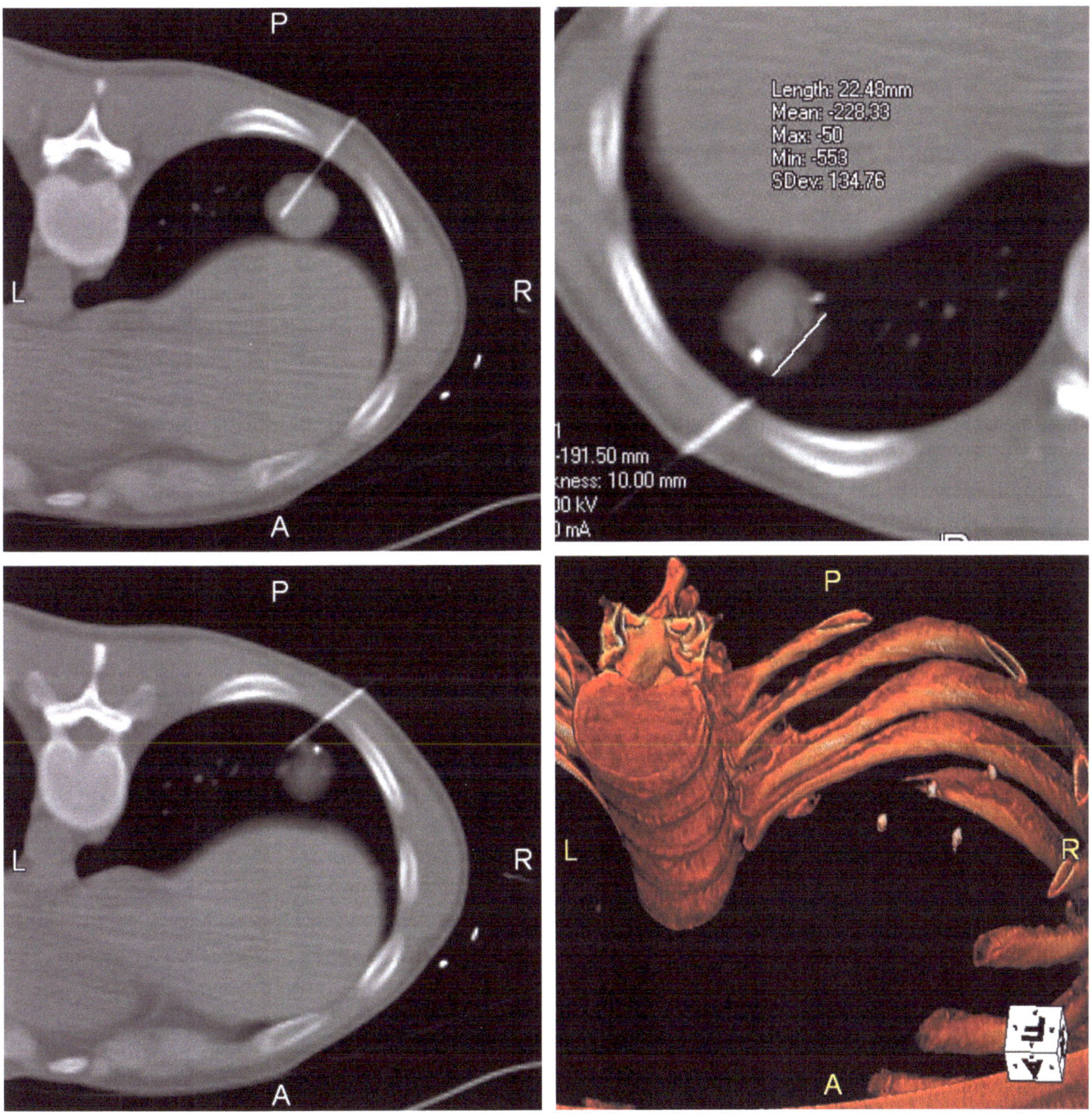

Fig. 9.3. a Axial CT image shows the 19 gauge delivery needle in the distal edge of the tumor. A single gold fiducial is deposited in this location. b The first two fiducials are placed more than 2 cm apart to optimize their discrete visualization during respiratory tracking. The best approach would have been to place these fiducials at slightly different axial slice levels. c An axial CT image shows a different pass of the needle after depositing two fiducials. The needle was retracted before reinsertion but the pleura was not crossed. d Three-dimensional reconstruction of the final fiducial distribution of four fiducials deposited around this lung nodule.

sitate crossing the pleura with a separate puncture, with an accompanying increased risk of complications. It is believed that factors associated with an increased risk of pneumothorax during lung biopsy include depth of lesion and presence of emphysema [18, 19]. Although we have no formal study to confirm the same for the fiducial placement procedure, we routinely chose shortest trajectories to the target. We also avoid patient positions that would necessitate crossing more than one lung fissure whenever possible.

9.5.7 Concomitant Biopsy and Fiducial Placement

Occasionally, cytopathologic confirmation is needed to establish that the given lesion is malignant. Certainly, one approach is to perform the biopsy in one visit and fiducial placement on a separate occasion but this would expose the patient to two separate procedures, each with their own set of risks. We established that our pneumothorax rate of 33% for fiducial placement alone [20] is similar to the pneumothorax rate during lung biopsy established by a very large multicenter study that included 124 centers in Japan. In that study, data were collected from 9783 biopsies, with a pneumothorax rate of 35% [21].

The preferred approach at our institution is to do the biopsy and fiducial placement at the same time. The biopsy is performed first. We use a 19 gauge Temno Coaxial Biopsy System (Allegiance Healthcare, McGraw Park, IL). The guiding 19 gauge trocar is inserted just proximal to the lesion using CT-fluoroscopic guidance. At this point, a fine needle aspiration is usually performed coaxially using a 21 or 22 gauge Chiba needle and the specimen is given to the cytopathologist who is present in the exam area. If presence of malignant cells is established, the same 19 gauge guiding trocar can be used to deliver the fiducials in and around the lesion. We have found that this combined method of biopsy and fiducial placement has a higher chance of pneumothorax [20] than performing fiducial placement alone. Our study was retrospective, however, and we did not prospectively establish the pneumothorax rate if the two procedures were to be done in sequence on separate visits.

9.6 Post-Procedure Care

A CT scan of the thorax is obtained immediately after fiducial marker placement, while the patient is still in the scanner. If the patient has a large or symptomatic pneumothorax, a chest tube is inserted. If the patient has a small pneumothorax, a chest radiograph is obtained (to be used as a baseline for comparison to a later chest radiograph). If there is no pneumothorax, a chest radiograph is generally not needed at that time.

The patient is observed for approximately three to four hours after fiducial marker placement, with the vital signs and oxygen saturations monitored periodically. A chest radiograph is obtained at the end of this period to rule out a new (or expanding) pneumothorax. If the patient complains of shortness of breath or chest pain, a chest radiograph should be obtained sooner. The patient should have minimal discomfort after fiducial placement, and any pain that is experienced should be adequately treated with acetaminophen or a non-steroidal anti-inflammatory drug. If the pain is more severe, particularly if it is pleuritic in nature, a pneumothorax should be suspected as a possible underlying etiology. It is advisable to keep the patient fasting during the observation period, in case the patient requires chest tube placement.

If the patient's follow-up chest radiograph reveals no pneumothorax, the patient can be discharged. The patient should be advised to go to the nearest emergency room if he or she becomes short of breath or experiences pleuritic chest pain. On rare occasions a pneumothorax can develop even after the 4 hour observation period.

If the chest radiograph reveals a small, stable pneumothorax and the patient is asymptomatic, the patient can generally be discharged, with a repeat chest radiograph obtained the next day on an outpatient basis. Alternatively, if the patient lives alone or lives far away from any medical facility, it is safer to observe the patient overnight.

If the chest radiograph reveals a large pneumothorax, an expanding pneumothorax, or a pneumothorax of any size in a symptomatic patient, a chest tube should be inserted. We place most of our chest tubes with image guidance, using either fluoroscopy (for large pneumothoraces) or computed tomography (for small pneumothoraces). We generally use a modification of the Seldinger technique [22], with an introducer needle inserted into the pleural cavity under image guidance, placing a guide wire through the needle into the pleural cavity, followed by serial dilatations over the wire, and finally insertion of a drainage catheter over the wire. The chest tube should be positioned anteriorly if possible because air will tend to redistribute to this location. We place our chest tubes to low wall suction using a Pleur-Evac in order to facilitate lung re-expansion. In certain clinical scenarios, however, it is also appropriate to attach the chest tube to a Heimlich valve.

If a chest tube is placed, we will admit the patient for overnight observation. A repeat chest radiograph will be obtained the following morning, and if the results are satisfactory, we will clamp the chest tube and obtain a repeat chest radiograph in 1 hour. If the patient develops any symptoms with the chest tube clamped, it should be connected to low wall suction. If the repeat chest radiograph demonstrates no pneumothorax or a small stable pneumothorax in an asymptomatic patient, then the chest tube can be removed. The patient can be observed for one to two hours after chest tube removal for development of symptoms. If the patient remains asymptomatic, he/she is discharged with instructions to return to the emergency room with development of any new symptoms.

9.6.1 Considerations after Implantation

For soft tissue implantations, a minimum of one week should be allowed to permit the fiducials to become fixed in the tissue and to ensure that the post-treatment edema has resolved completely. Shortening this time interval may result in fiducial migration and/or position changes between the planning CT and treatment. If this occurs, a new planning CT must be obtained. Another consideration is that the patient must be comfortable enough after implantation to be able to remain in a still position during the CT planning session (up to 1 hour), because this position must be replicated during treatment.

9.7 Conclusion

Fiducial placement is an essential step in CyberKnife treatment of lung neoplasms. Our experience thus far indicates that it is a clinically feasible procedure. Pneumothorax and to a lesser extent, pulmonary hemorrhage are the most common complications of transthoracic fiducial placement. Our experience on complications in 48 patients that were a part of the recent study suggests a pneumothorax rate of 33%; however, thoracostomy placement was only needed in 6 out of 48 patients (12.5%) [20]. Misplacement of fiducials can severely compromise treatment planning and accuracy of CyberKnife beam delivery. The technical principles of transthoracic fiducial placement are still evolving and the approach may still be modified in the future.

References

1. Ryu SI, Chang SD, Kim DH, *et al.* Image-guided hypofractionated stereotactic radiosurgery to spinal lesions. *Neurosurgery* 2001; 49:838–846.
2. Harada K, Nishizaki T, Adachi N, *et al.* Pediatric acoustic schwannoma showing rapid regrowth with high proliferative activity. *Childs Nerv Syst* 2000; 16:134–137.
3. Chang SD, Murphy M, Geis P, *et al.* Clinical experience with image-guided robotic radiosurgery (the Cyberknife) in the treatment of brain and spinal cord tumors. *Neurol Med Chir (Tokyo)* 1998; 38:780 783.
4. Schweikard A, Glosser G, Bodduluri M, *et al.* Robotic motion compensation for respiratory movement during radiosurgery. *Comput Aided Surg* 2000; 5:263–277.
5. Schweikard A, Shiomi H, Adler J. Respiration tracking in radiosurgery. *Med Phys* 2004; 31:2738–2741.
6. King CR, Lehmann J, Adler JR, *et al.* CyberKnife radiotherapy for localized prostate cancer: rationale and technical feasibility. *Technol Cancer Res Treat* 2003; 2:25–30.
7. Ponsky LE, Crownover RL, Rosen MJ, *et al.* Initial evaluation of Cyberknife technology for extracorporeal renal tissue ablation. *Urology* 2003; 61:498–501.
8. Whyte RI, Crownover R, Murphy MJ, *et al.* Stereotactic radiosurgery for lung tumors: preliminary report of a phase I trial. *Ann Thorac Surg* 2003; 75:1097–1101.
9. Reichner CA, Collins BT, Gagnon GJ, *et al.* The Placement of Gold Fiducials for CyberKnife Stereotactic Radiosurgery Using a Modified Transbronchial Needle Aspiration Technique. *Journal of Bronchology* 2005; 12:193–195.
10. Zierhut D, Bettscheider C, Schubert K, *et al.* Radiation therapy of stage I and II non-small cell lung cancer (NSCLC). *Lung Cancer* 2001; 34 Suppl 3:S39–43.

11. Kupelian PA, Komaki R, Allen P. Prognostic factors in the treatment of node-negative nonsmall cell lung carcinoma with radiotherapy alone. *Int J Radiat Oncol Biol Phys* 1996; 36:607–613.
12. Landreneau RJ, Sugarbaker DJ, Mack MJ, *et al.* Wedge resection versus lobectomy for stage I (T1 N0 M0) non-small-cell lung cancer. *J Thorac Cardiovasc Surg* 1997; 113:691-698; discussion 698–700.
13. Ginsberg RJ, Rubinstein LV. Randomized trial of lobectomy versus limited resection for T1 N0 non-small cell lung cancer. Lung Cancer Study Group. *Ann Thorac Surg* 1995; 60:615–622; discussion 622–613.
14. Wulf J, Haedinger U, Oppitz U, *et al.* Stereotactic radiotherapy for primary lung cancer and pulmonary metastases: a noninvasive treatment approach in medically inoperable patients. *Int J Radiat Oncol Biol Phys* 2004; 60:186–196.
15. Blomgren H, Lax I, Naslund I, *et al.* Stereotactic high dose fraction radiation therapy of extracranial tumors using an accelerator. Clinical experience of the first thirty-one patients. *Acta Oncol* 1995; 34:861–870.
16. Steinke K, Sewell PE, Dupuy D, *et al.* Pulmonary radiofrequency ablation–an international study survey. *Anticancer Res* 2004; 24:339–343.
17. Geraghty PR, Kee ST, McFarlane G, *et al.* CT-guided transthoracic needle aspiration biopsy of pulmonary nodules: needle size and pneumothorax rate. *Radiology* 2003; 229:475–481.
18. Kazerooni EA, Lim FT, Mikhail A, *et al.* Risk of pneumothorax in CT-guided transthoracic needle aspiration biopsy of the lung. *Radiology* 1996; 198:371–375.
19. Topal U, Berkman YM. Effect of needle tract bleeding on occurrence of pneumothorax after transthoracic needle biopsy. *Eur J Radiol* 2005; 53:495–499.
20. Yousefi S, Chang T, Collins B, *et al.* Thoracic Fiducial Placement with CT guidance for the Purpose of CyberKnife Radiotherapy. Cyberknife Users' Meeting. Carlsbad, CA; 2006.
21. Tomiyama N, Yasuhara Y, Nakajima Y, *et al.* CT-guided needle biopsy of lung lesions: A survey of severe complication based on 9783 biopsies in Japan. *Eur J Radiol* 2006.
22. Seldinger SI. Catheter replacement of the needle in percutaneous arteriography; a new technique. *Acta Radiol* 1953; 39:368–376.

Use of a Blood Patch Technique to Prevent Pneumothorax During Insertion of Lung Fiducials

10

James W. Blalock, Clinton A. Medbery, Astrid E. Morrison, Marianne M. Young, Julie Hensley, and Vikki Harriet

CONTENTS

10.1 Introduction

The use of the CyberKnife® (Accuray Incorporated, Sunnyvale, CA) for radiosurgical treatment of lung tumors [1] typically requires placement of one or more gold fiducials for tracking during treatment. Possible exceptions to this requirement are tumors involving the posterior chest wall, which may be localized with Xsight™ (Accuray Incorporated, Sunnyvale, CA) based on spinal anatomy. Placement of gold fiducial seeds has been associated with post-placement pneumothorax at rates of 25–40% [1]. Some of these patients will require treatment by chest tube placement, and all of them will require repeated physician exams and radiological follow-up. In addition, some patients are refused consideration for radiosurgery for fear they will be unable to tolerate a potential pneumothorax. Pneumothorax prevention, therefore, has become a major priority in CyberKnife radiosurgery for lung cancer.

10.2 Methods

Two major methods have been developed to address this problem. The first method, described by the group at Georgetown University, uses a coaxial needle system to allow bronchoscopic placement [2]. This method is optimally employed for more centrally located tumors, and becomes somewhat cumbersome for placement of 3–4 fiducials. It has been reported to result in no pneumothorax risk but some risk of vascular embolization, thus far without serious consequences. The second method is the use of a blood patch technique during percutaneous seed placement. This technique is most suitable for peripherally placed lesions and can result in the placement of 4 fiducials with two needle sticks. It is this latter technique that will be described here.

In the blood patch technique (Table 10.1), the patient is generally sedated prior to the procedure, and is then positioned on the CT couch as if for a biopsy. A 10-cc syringe of the patient's blood is obtained from their IV site and set aside on a sterile field. A CT scan is obtained and used to plan the entrance point

Table 10.1 Key Points of Blood Patch Technique

1.	CT is performed and lesion location is noted
2.	Seed placement is planned
3.	Needle with seeds placed and position confirmed using CT
4.	Fiducials deployed, needle left in lung parenchyma
5.	Syringe of patient's clotted blood attached to fiducial needle

(marked on the skin), the angle of approach, and the depth to which the needle should be inserted. Once the entry has been planned, the needle is inserted, using markers on the needle to gauge depth. With the needle in its initial position, another CT scan (Fig. 10.1) is obtained through the area of interest to verify needle position.

Needle placement is confirmed and, if correct, the fiducials are deployed (Figs. 10.2 and 10.3). We generally place two fiducials above and two below the tumor. With large tumors, placement within the tumor may be possible, but such placement within small tumors often results in obscured tumor borders from the CT artifact produced by the gold seeds. The fiducials are deployed by withdrawing the needle while holding the stylet stable, in a fashion similar to that used for prostate seed implants. Once the second seed is deployed, and before the needle is withdrawn, the syringe containing the patient's clotted blood is attached to the needle (Fig. 10.4). The hub does not accommodate direct attachment to the needle, so a blunt needle of the type usually used for withdrawing fluids from vials is used as an intermediary connection. The patient's blood is then injected while the fiducial needle is withdrawn. This sterile "blood patch" with the patient's blood provides a barrier by which pneumothorax complications are substantially minimized.

We have worked with Med-Tec (Orange City, IA) to design shorter needles (Figs. 10.4 and 10.5) to hold the gold seeds, thus making it possible to have the needle go through the CT scanner with less risk of hitting the gantry, causing possible displacement or lung injury. With large bore CT scanners, this may be unnecessary.

It is important to note that the blood can cause an artifact possibly resembling tumor on CT scans, so caution should be taken when contouring tumor volumes.

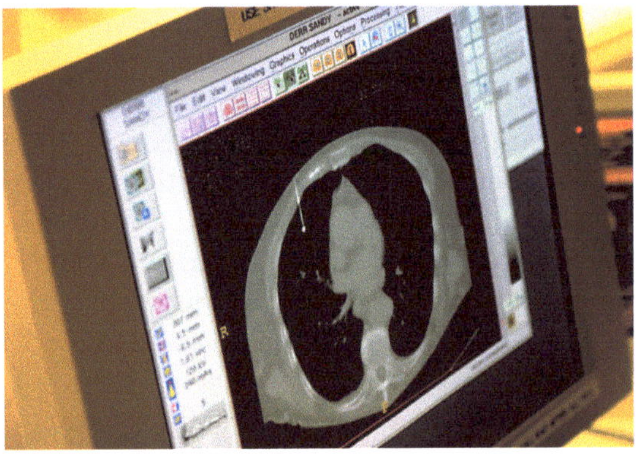
Fig. 10.1 CT scan confirming initial needle placement.

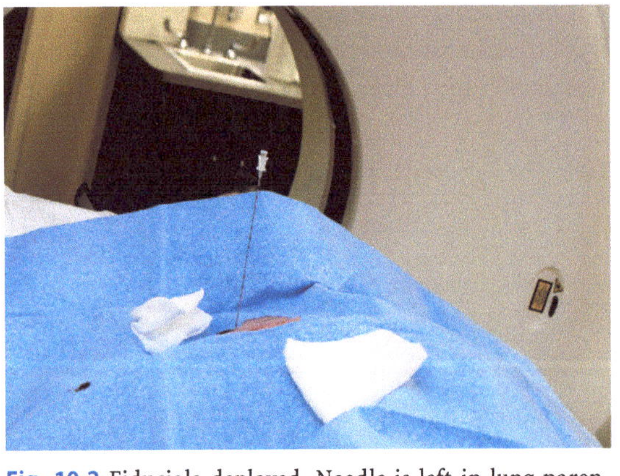
Fig. 10.2 Fiducials deployed. Needle is left in lung parenchyma.

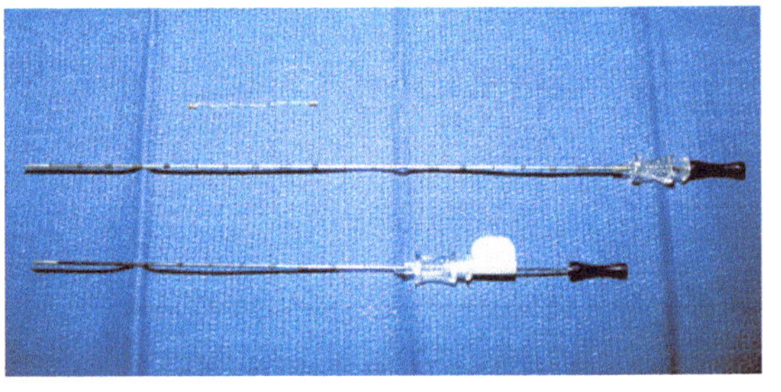
Fig. 10.3 Long and short needle; seeds with spacers.

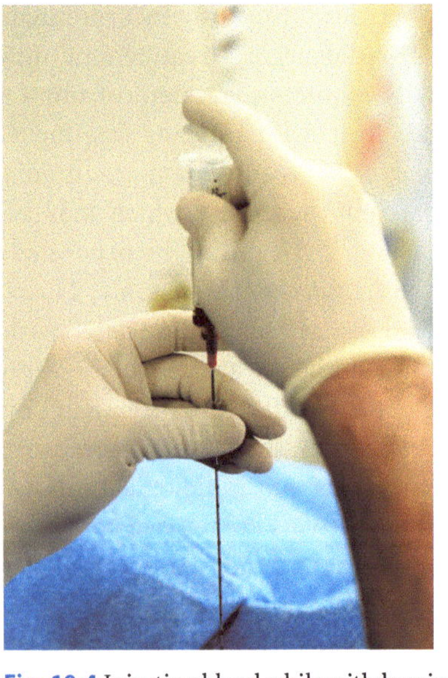
Fig. 10.4 Injecting blood while withdrawing fiducial needle.

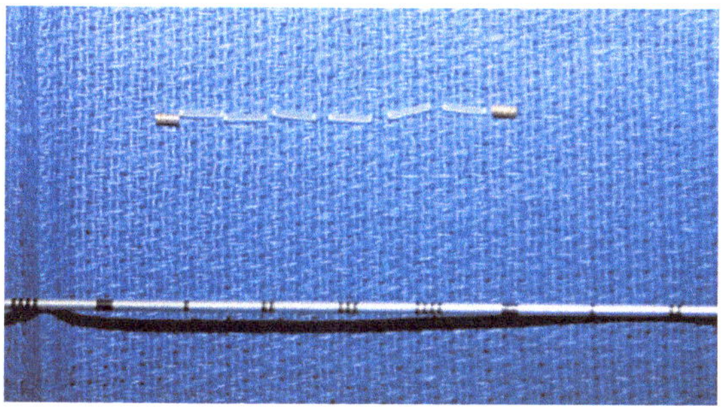

Fig. 10.5 Close-up of seeds with spacers and needle showing centimeter markings.

10.3 Discussion

In our study of 21 patients (22 fiducial placements consisting of 4 seeds each), use of a blood patch technique resulted in a substantial reduction in the incidence of pneumothorax. In a group of 12 patients for which the blood patch technique was not used, we performed a total of 13 seed placement procedures (one patient had bilateral tumors). Of these 13 fiducial placements, 4 (31%) resulted in pneumothorax large enough to require treatment. Two of these were delayed by one or more days. Once the blood patch technique was instituted, there were no pneumothoraces in the next 9 patients. This reduction in risk translates into considerable improvement in cost and safety of CyberKnife radiosurgery for lung cancers.

10.4 Conclusion

While no method of fiducial placement can be considered absolutely safe, use of a blood patch technique allows placement of fiducials in peripheral lung locations with greatly reduced risk. Use of a bronchoscopic technique can do the same for more centrally located tumors [2]. Evidence continues to mount showing the efficacy of stereotactic hypofractionated radiation delivery for lung tumors, with tumor control rates that exceed 70% in some studies [3–6]. In this setting CyberKnife radiosurgery stands poised to play a major role in lung cancer treatment, perhaps even becoming an alternative to surgery for early stage lung cancers. With continued positive experience, we may be able to start offering this treatment even to very high-risk patients, such as those who have previously undergone a contralateral pneumonectomy. When the International Association for the Study of Lung Cancer institutes its randomized trial between surgery and radiosurgery next year, CyberKnife will be uniquely positioned to participate.

References

1. Whyte RI, Crownover R, Murphy MJ, *et al.* Stereotactic radiosurgery for lung tumors: preliminary report of a phase I trial. *Ann Thorac Surg* 2003; 75:1097–1101.
2. Reichner CA, Collins BT, Gagnon GJ, *et al.* The placement of gold fiducials for CyberKnife stereotactic radiosurgery using a modified transbronchial needle aspiration technique. *Journal of Bronchology* 2005; 12:193–195.
3. McGarry RC, Papiez L, Williams M, *et al.* Stereotactic body radiation therapy of early-stage non-small-cell lung carcinoma: phase I study. *Int J Radiat Oncol Biol Phys* 2005; 63:1010–1015.
4. Nagata Y, Takayama K, Matsuo Y, *et al.* Clinical outcomes of a phase I/II study of 48 Gy of stereotactic body radiotherapy in 4 fractions for primary lung cancer using a stereotactic body frame. *Int J Radiat Oncol Biol Phys* 2005; 63:1427–1431.
5. Onishi H, Araki T, Shirato H, *et al.* Stereotactic hypofractionated high-dose irradiation for stage I nonsmall cell lung carcinoma: clinical outcomes in 245 subjects in a Japanese multiinstitutional study. *Cancer* 2004; 101:1623–1631.
6. Timmerman R, McGarry R, Yiannoutsos C, *et al.* Excessive toxicity when treating central tumors in a phase II study of stereotactic body radiation therapy for medically inoperable early-stage lung cancer. *J Clin Oncol* 2006; 24:4833–4839.

Thoracic Fiducial Placement via Flexible Bronchoscopy

11

Eric D. Anderson, Brian T. Collins, Gregory J. Gagnon,
Sean P. Collins, Timothy Mahoney, Filip Banovac, Carlos Jamis-Dow,
Shakun Malik, and Cristina A. Reichner

CONTENTS

11.1 Abstract

CyberKnife® (Accuray Incorporated, Sunnyvale, CA) Frameless Image-Guided Radiosurgery [1, 2] with the Synchrony® Motion Tracking Module (Accuray Incorporated) [3, 4] is now available for the treatment of thoracic malignancies. Gold fiducial markers are required for the treatment planning and tracking of the tumor during each treatment. Fiducials have traditionally been placed percutaneously under CT-guidance. This chapter describes our center's experience placing fiducials using flexible bronchoscopy with a transbronchial aspiration needle (TBNA).

11.2 Introduction

The Synchrony Motion Tracking Module provides accurate and effective tracking of tumor movement throughout the respiratory cycle and provides compensation for the observed movement. To optimize tumor identification, the Synchrony module currently requires the placement of radio-opaque markers called fiducials, which can be visualized on plain film radiographs acquired during CyberKnife treatment. Traditionally, thoracic fiducial placement has been performed percutaneously under computed tomography (CT) guidance and has been associated with a 20–40% incidence of pneumothorax. Efforts to overcome this risk have led to increased adoption of alternative techniques, such as bronchoscopic methods of fiducial placement.

Fiducial markers are typically composed of gold and measure 3 mm in diameter (Fig. 11.1). Gold fiducials are placed either within the tumor or in close approximation to the tumor. Optimal treatment requires 3 to 5 well-placed fiducials to allow for assessment of both directional and rotational tumor movement.

Thoracic fiducial placement requires the placement of fiducial markers within the lung and thus poses significant risks to patients, such as pneumothorax, hemorrhage, hemoptysis, and migration of the markers. Patients undergoing CyberKnife stereotactic radiosurgery for thoracic malignancies are usually high-risk patients with multiple comorbidities that may exclude them from conventional therapy. Often they are not surgical candidates due to severe chronic obstructive pulmonary disease (COPD), coronary artery disease, or prior surgery and thus are more likely to experience deleterious effects from additional risks posed to them.

At our institution the risk of pneumothorax after percutaneous CT-guided fiducial placement has been 35%. This risk is increased if a biopsy is obtained at the time of fiducial placement. Twenty-four percent of the patients who developed a pneumothorax required thoracostomy tube insertion. In addition, 24% of patients have developed hemorrhage seen on CT images, 4% have developed hemoptysis, and 5% had fiducials dropped into the pleural space during insertion [5].

Flexible bronchoscopy may be performed under conscious sedation or general anesthesia and is a common, safe, and accurate technique used to biopsy peripheral lung masses as well as mediastinal lymphadenopathy. The risks of flexible bronchoscopy include: reaction to the anesthetics, hypoxemia, bronchospasm, and bleeding (1–4%) after transbronchial biopsy. The incidence of pneumothorax with a transbronchial biopsy is 1–4%. No cases of pneumothorax following transbronchial needle aspiration (TBNA) of the mediastinum have been reported in the literature [6]. Single cases of hemomediastinum and pneumomediastinum have been reported with TBNA [7, 8].

We hypothesized that placing fiducials through flexible bronchoscopy using a modified transbronchial needle aspiration technique would be a safe, accurate, and lasting method for marking tumors in the mediastinum and larger masses in the parenchyma of the lung.

11.3 Methods

All patients referred to Georgetown University Hospital for CyberKnife stereotactic radiosurgery treatment of thoracic malignancies, were reviewed by a multidisciplinary thoracic oncology team. Patients were evaluated by a radiation oncologist, pulmonologist, and radiologist to determine if the intrathoracic tumor would be more amenable to fiducial placement via flexible bronchoscopy or CT-guided percutaneous techniques. Patients with centrally located or larger peripheral tumors (> 5 cm) were selected for fiducial placement via flexible bronchoscopy. Informed consent was obtained from all patients and collection of data was approved by the Institutional Review Board.

Video flexible bronchoscopy (Pentax Medical Company, Montvale, NJ) was performed under conscious sedation (fentanyl/midazolam or propofol) or general anesthesia at the discretion of the anesthesiologist. Patients who underwent general anesthesia were intubated with an endotracheal tube or a laryngeal mask airway. Two percent lidocaine was used for topical anesthesia of the airways.

Each sterilized gold fiducial (item no. 351-1; Best Medical International, Inc., Springfield, VA) was placed in the 19-gauge needle of a 19/21-gauge Wang transbronchial needle (C. R. Bard Inc., Billerica, MA) with the 21-gauge needle retracted (Fig. 11.2). The needle tip was then dipped in sterile surgical

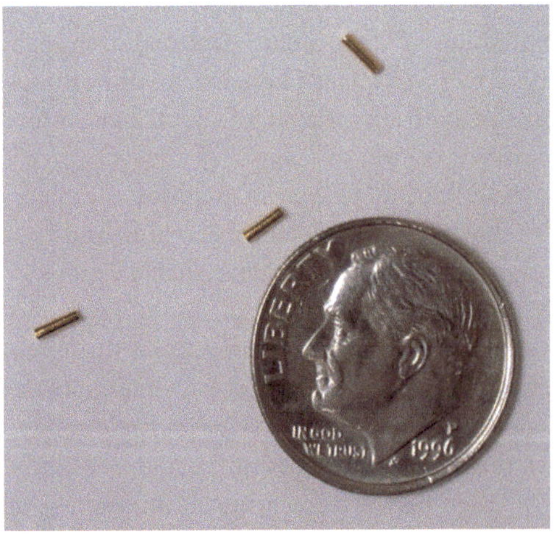

Fig. 11.1 Gold fiducials.

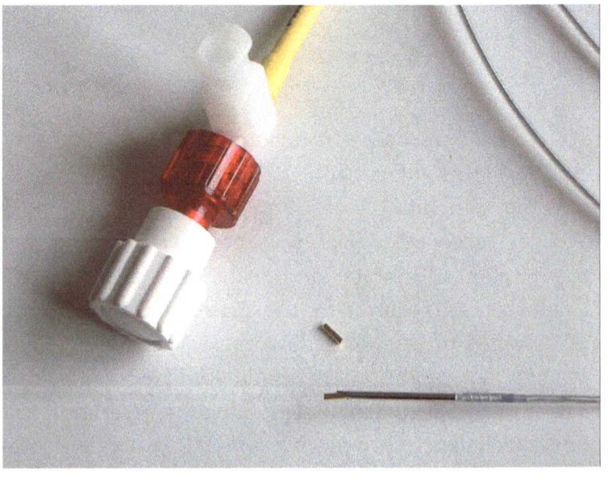

Fig. 11.2 Loading of gold fiducials into a 19/21 gauge TBNA needle.

lubricant (Surgilube; Fougera, Melville, NY) to improve the fiducial's adherence to the needle. Keeping the 21-gauge needle retracted, the 19-gauge fiducial-loaded needle was then retracted into the sheath and the sheath passed through a flexible bronchoscope. At the desired location, the 19-gauge needle was extended and inserted into the tumor through a jabbing method [7] using fluoroscopic guidance. The 21-gauge needle was tightened, deploying the fiducial (Fig. 11.3). Leaving the 21-gauge needle extended, the needle was then withdrawn from the tissue. The technique was repeated until all markers were placed.

Fiducials were positioned into or near the lung tumor. Careful planning was performed with the assistance of the radiation oncologist. Three to four markers were placed approximately 2 cm apart from each other and mindful of the 45 degree angle of the fluoroscopy units used in the CyberKnife suite. Care was taken to avoid great vessels in the mediastinum and hilum during placement. Portable chest radiographs were obtained after all procedures to rule out pneumothorax and confirm the position of the fiducials (Fig. 11.4) [9].

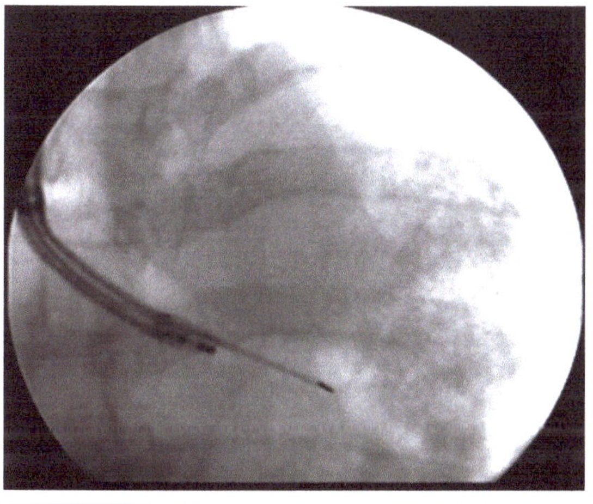
a

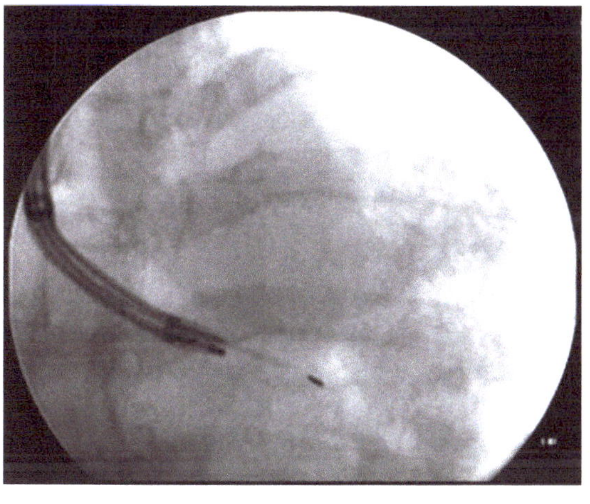
b

Fig. 11.3a,b. The TBNA needle is extended into the target and fluoroscopy is used to visualize deployment of the fiducial.

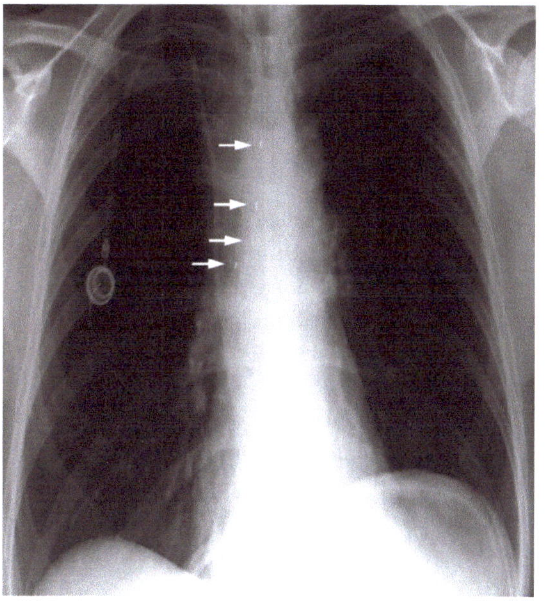

Fig. 11.4 Post procedure chest X-ray demonstrating mediastinal fiducial placement.

Patients underwent planning CT scan 7–10 days after fiducial placement to allow for fixation of fiducials. Data collected included patient demographics, number and location of fiducials placed, and complications associated with their placement.

11.4 Results

Between July 2004 and August 2006 a total of 32 patients underwent fiducial placement via flexible bronchoscopy. One patient underwent a second procedure to obtain additional tissue for pathology and because one fiducial had migrated out from a paratracheal location. Sixty-seven percent of patients were female and 87% were Caucasian. The age range was 35–81 years (mean age 59 years).

The most common diagnosis was non-small cell lung cancer (58%). Other diagnoses included metastatic disease to chest (38%) and one patient was diagnosed with small cell lung cancer at the time of bronchoscopy. The main reason for choosing CyberKnife therapy was previous radiotherapy to the chest. Other reasons included: suboptimal anatomic location of tumor for treatment by conventional irradiation because of adjacent critical structures, lack of response to prior conventional irradiation, severe COPD or coronary artery disease, and prior lung resection.

A total of 124 fiducials were successfully inserted (average of 3.9 fiducials per target lesion). Fifty-two were placed in the mediastinum, 19 in the hila, and 53 in the lung parenchyma (Table 11.1). Sixty-nine percent of patients underwent general anesthesia and 31% conscious sedation.

During the bronchoscopies 25 fiducials were dropped in the airways prior to insertion, 18 were removed with bronchopulmonary-coated disposable biopsy forceps (C. R. Bard Inc., Billerica, MA), two were suctioned, three were coughed out, and two were not retrieved. The latter were not seen on post-procedure chest X-rays (Fig. 11.5a and b).

Complications included one fiducial migration after insertion as noted above. One patient with severe COPD, who underwent endotracheal intubation, developed bronchospasm requiring mechanical ventilation for 48 hours. There were no pneumothoraces or significant bleeding.

Table 11.1 Location of Fiducials Placed via Bronchoscopy

32 Patients	124 Fiducials placed
Average fiducials per tumor	3.9
Locations	
Paratracheal area	15 (station 2 and 4)
Subcarinal	21 (station 7)
Left mainstem	5
Left hilum	12 (station 11L)
Right mainstem	2
Right hilum	7
Right bronchus intermedius	9 (station 11R)
Right upper lobe (RUL)	6
Right middle lobe (RML)	4
Right lower lobe (RLL)	7
Left upper lobe (LUL)	20
Left lower lobe (LLL)	16

Three fiducials in two patients embolized during insertion via the pulmonary artery without adverse clinical consequence. The embolizations were immediately visible under fluoroscopy. The first occurred during placement of a fiducial in a subcarinal location. The other two occurred in a single patient in whom fiducials were placed in a left parahilar mass (Fig. 11.6).

11.5 Discussion

CyberKnife frameless image-guided radiosurgery with the Synchrony motion tracking module is now available for the treatment of thoracic malignancies. Gold fiducial markers are required for the treatment planning and tracking of the tumor during each treatment. Fiducials have traditionally been placed percutaneously under CT-guidance. Flexible bronchoscopy may be used to safely and accurately place fiducials for centrally located and larger peripheral tumors in the chest.

Tumors in the periphery should be greater than 1 cm in order to be visible under fluoroscopy. Peripheral lesions not visible under fluoroscopy should be considered for CT-guidance. At least three fiducials were placed in close proximity to each target lesion. Additional markers were often placed to account for any poorly positioned markers and to assure that tracking could be performed accurately. Mindful of the 45 degree orthogonal plain film imaging used in the CyberKnife suite to avoid overlap of the markers, fiducials were placed 2 cm apart from each other.

The complications of bronchoscopic fiducial placement include the risks inherent to bronchoscopy such as bronchospasm and respiratory failure. This patient population will be particularly at risk as many have severe lung disease already. Pneumothorax and hemorrhage have not occurred and should be a rare complication of bronchoscopic insertion. Placement in more peripheral locations may increase the likelihood of pneumothorax.

Fiducials dropped during the bronchoscopy may easily be removed with biopsy forceps. If not able to be located, dropped fiducials are usually coughed out immediately after the procedure. The develop-

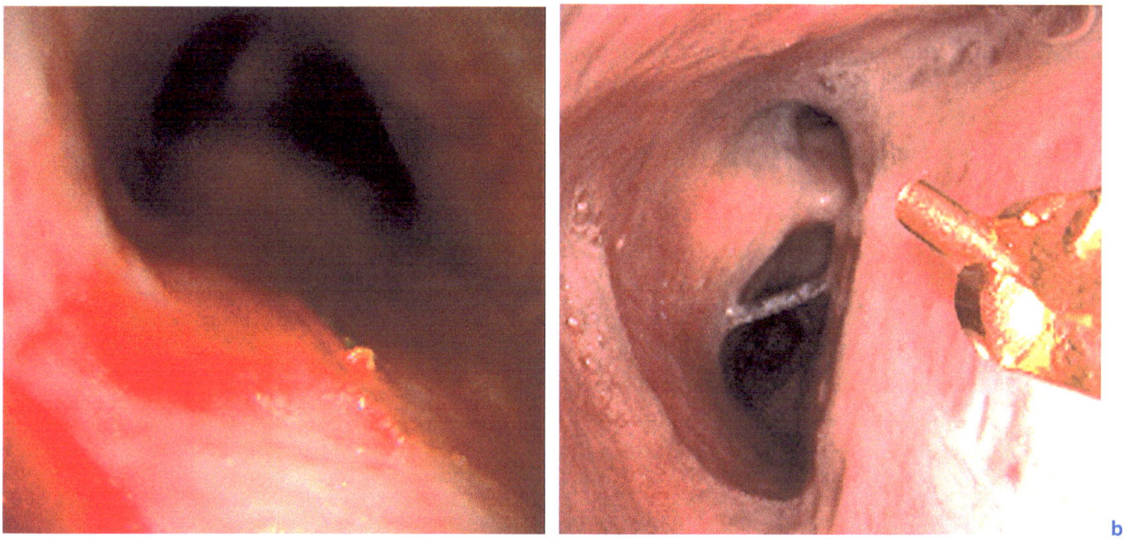

Fig. 11.5. a Dropped fiducial in left mainstem bronchus. **b** Retrieval of fiducial using biopsy forceps

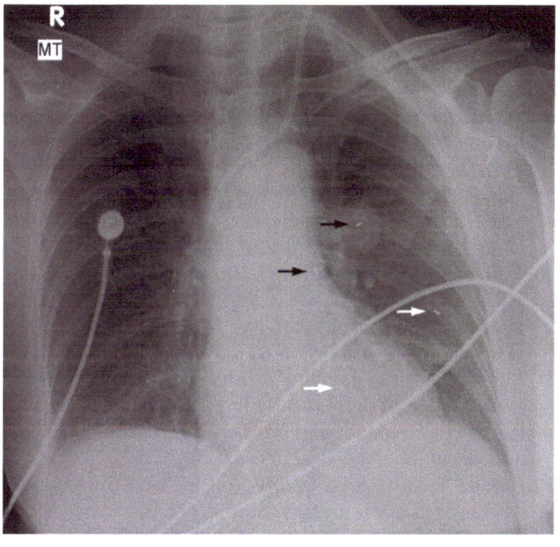

Fig. 11.6 Chest X-ray demonstrating left hilar nodule with fiducial placement. Two fiducials were observed in left lower lobe after embolization via left pulmonary artery (*white arrows*).

ment of a transbronchial needle deployment device might help decrease the incidence of this minor complication. Only a single fiducial, placed early in our experience, migrated out into the airway and was expectorated prior to treatment. It is possible that fiducials may migrate after implantation. Nevertheless, this has been infrequent and has not been a clinical problem secondary to the placement of multiple fiducials and the ability of the CyberKnife system to recognize fiducial migration.

The most important complication of bronchoscopic fiducial placement was embolization into the pulmonary artery. Unlike traditional TBNA, no blood return can be seen in the catheter if inadvertent puncture of a vessel occurs during insertion. Careful planning is essential to avoid puncture of great vessels in the mediastinum, especially those of the systemic circulation where systemic embolization could result in a cerebrovascular accident or peripheral emboli. Endobronchial ultrasound, which allows for visualization of the major vascular structures in the mediastinum at the time of bronchoscopy, may be utilized in the future to reduce the risk of such an occurrence.

Insertion of fiducials into parenchymal tumors may result in embolization as the airways run adjacent to the pulmonary vasculature. This complication has not yet been reported after CT-guided placement of fiducials, but seems possible. Embolization into the pulmonary artery, however, appears to be of low risk. Neither of our patients in whom this occurred developed late sequelae of fiducial embolization. Radioactive seeds of similar dimensions are commonly used to treat early stage prostate cancer and have been known to embolize to the pulmonary arteries. It is estimated that 1% of prostate seeds migrate to the lungs and have not been associated with clinical symptoms. These patients did not require systemic anticoagulation and were not found to have an increased incidence of thromboembolic disease [10, 11].

An alternative method of bronchoscopic fiducial placement has been described [12–14], using a wire to push the fiducial through a plastic catheter extended through the bronchoscope working channel. In this method, however, markers were pushed into the distal airways, but not implanted into tissue. As a result, a large number (25–35%) of fiducials became dislodged prior to the planning CT scan and treatment. This complication was more common when placing fiducials in upper lobe tumors and centrally located tumors.

The use of electromagnetic guidance and virtual bronchoscopy may allow for more accurate placement of fiducials in smaller peripheral tumors. These techniques allow for CT reconstruction to plan a guided passage of the fiducial into a more precise location in the lung. Registration between the virtual bronchoscope and the video bronchoscope is accomplished and then a sheath is extended through the bronchoscope channel and positioned using a trackable probe under electromagnetic guidance. The fiducial needle may then be placed through the sheath. This technology may improve accuracy of bronchoscopic fiducial placement, especially for those lesions located in the periphery of the lung [15].

11.6 Conclusion

Flexible bronchoscopy using a modified transbronchial aspiration needle technique appears to be a safe method for placement of fiducials for mediastinal and centrally located thoracic tumors as a precursor to CyberKnife stereotactic radiosurgery. At our institution, flexible bronchoscopy is the preferred method for insertion of fiducials in mediastinal, hilar, and larger peripheral tumors. Percutaneous CT-guidance remains the preferred method of insertion for smaller peripheral nodules.

References

1. Adler JR, Jr., Murphy MJ, Chang SD, *et al.* Image-guided robotic radiosurgery. *Neurosurgery* 1999; 44:1299–1306; discussion 1306–1297.
2. Kuo JS, Yu C, Petrovich Z, *et al.* The CyberKnife stereotactic radiosurgery system: description, installation, and an initial evaluation of use and functionality. *Neurosurgery* 2003; 53:1235–1239; discussion 1239.
3. Coste-Manière È, Olender, D., Kilby, W., Schulz, R.A. Robotic Whole Body Stereotactic Radiosurgery: Clinical Advantages of the CyberKnife Integrated System. *Medical Robotics and Computer Assisted Surgery* 2005; 1:28–39.
4. Schweikard A, Shiomi H, Adler J. Respiration tracking in radiosurgery. *Med Phys* 2004; 31:2738–2741.
5. Reichner CA, Collins, B.T., Gagnon, G.J., Malik, S., Jamis-Dow, C., Anderson, E.D. Comparison of Fiducial Placement for Cyberknife Stereotactic Radiosurgery using CT-Guidance or Flexible Bronchoscopy. *Chest* 2005; 128:162–163S.
6. Wang K-P, Mehta, A., Turner,.Jr. J.F. Transbronchial Needle Aspiration for Cytology and Histology Specimens. In: Flexible Bronchoscopy: Malden: Blackwell Publishing; 2004. pp. 117–137.
7. Dasgupta A, Mehta AC. Transbronchial needle aspiration. An underused diagnostic technique. *Clin Chest Med* 1999;20:39–51.
8. Harrow EM, Oldenburg FA, Jr., Lingenfelter MS, *et al.* Transbronchial needle aspiration in clinical practice. A five-year experience. *Chest* 1989; 96:1268–1272.
9. Reichner CA, Collins, B.T., Gagnon, G.J., Malik, S., Jamis-Dow, C., Anderson, E.D. The Placement of Gold Fiducials for CyberKnife Stereotactic Radiosurgery Using a Modified Transbronchial Needle Aspiration Technique. *Journal of Bronchology* 2005; 12:193–195.
10. Ankem MK, DeCarvalho VS, Harangozo AM, *et al.* Implications of radioactive seed migration to the lungs after prostate brachytherapy. *Urology* 2002; 59:555–559.
11. Nag S, Vivekanandam S, Martinez-Monge R. Pulmonary embolization of permanently implanted radioactive palladium-103 seeds for carcinoma of the prostate. *Int J Radiat Oncol Biol Phys* 1997; 39:667–670.
12. Harada T, Shirato H, Ogura S, *et al.* Real-time tumor-tracking radiation therapy for lung carcinoma by the aid of insertion of a gold marker using bronchofiberscopy. *Cancer* 2002; 95:1720–1727.
13. Imura M, Yamazaki K, Shirato H, *et al.* Insertion and fixation of fiducial markers for setup and tracking of lung tumors in radiotherapy. *Int J Radiat Oncol Biol Phys* 2005; 63:1442–1447.
14. Morice A, Jimenez, L., Eapen, G.A., Nelson, C., Stevens, C.W., Starkschall, G. Bronchoscopic implantation of gold fiducials for estimating lung tumor motion during gated radiation therapy. *Chest* 2005; 128:163S.
15. Schwarz Y, Greif J, Becker HD, *et al.* Real-time electromagnetic navigation bronchoscopy to peripheral lung lesions using overlaid CT images: the first human study. *Chest* 2006; 129:988–994.

Endoscopic Ultrasound-Guided Fiducial Placement for CyberKnife Radiosurgery

12

Aline Charabaty-Pishvaian, Richard Desi, and Nadim Haddad

CONTENTS

12.1 Introduction

The Synchrony® (Accuray Incorporated, Sunnyvale, CA) Respiratory Tracking System allows the CyberKnife® (Accuray Incorporated, Sunnyvale, CA) to deliver precisely high-dose radiation to targets that move with respiration. Tracking of moving tumors currently requires the use of implanted radiographic markers (or fiducials) as reference points [1, 7]. Several reports describe the placement of fiducials through surgical or percutaneous methods under image guidance (ultrasonography or computed tomography) in the prostate, spine, and lungs [2, 3, 6–8]. Fiducial implantation surgery can be invasive, and the percutaneous approach carries risks and limitations, especially when the lesions are in a deep location such as the posterior mediastinum or the abdomen. Innovations in the fiducial implantation procedure may improve its accuracy and reduce its risks, ultimately enhancing comfort for the patient. We recently described a new approach to placing fiducials in these deep structures using endoscopic ultrasound (EUS) [9]. This chapter briefly describes this approach.

EUS employs an endoscope with an ultrasound and a Doppler function incorporated into its tip. The ultrasound allows precise imaging of the different layers of the GI tract and beyond, into regions such as the posterior mediastinum and celiac area, and imaging of neighboring structures such as the pancreas, the gallbladder, and the left lobe of the liver [10, 11]. The Doppler function allows the visualization of vessels and helps differentiate venous from arterial vessels. Hence, EUS has been used as an imaging modality in the diagnosis and staging of tumors in this region. Linear-array endoscopes can also be used for invasive diagnostic and therapeutic procedures whereby a needle is introduced in the endoscope working channel and advanced into the area of interest under ultrasound visualization [10]. Fine needle aspiration of the tumor or injection of therapeutic substances into a lesion can then be performed [12–14]. The Doppler function verifies that there are no intervening vessels between the tumor and the needle. Real-time sonographic visualization of the needle being introduced into the lesion and the Doppler function both enhance the safety and accuracy of EUS-guided procedures. These same properties make placement of fiducial markers by EUS guidance an attractive approach for tumors in the posterior mediastinum and the abdomen.

12.2 Methods

12.2.1 Patient Selection

CyberKnife radiosurgery outside of the central nervous system is typically used in patients for whom conventional radiation therapy is contraindicated. Patients with a tumor located in the same field of a prior radiation treatment cannot undergo further conventional radiation therapy because of the potential for serious injury to the area. CyberKnife treatment is also indicated for intra-abdominal lesions because of the concern for radiation injury to the surrounding organs such as the bowel and the liver. Finally, CyberKnife is of particular interest in radiating intrathoracic lesions in patients with limited lung capacity (patients with COPD, emphysema, or with previous radiation injury to the lungs) who would not tolerate further damage to their lungs.

EUS-guided placement of fiducials should be considered in patients with a radiosensitive tumor located within the scanning field of the EUS probe, i.e., within 5 cm of the GI tract. Contraindications to EUS-guided fiducial placement include the inability of the patient to tolerate sedation for the procedure, coagulopathy (INR > 1.5 or platelet count < 50,000/cmm), and pregnancy.

12.2.2 Equipment and Procedure

The procedure is performed on an outpatient basis, in the endoscopy suite in a room where fluoroscopy is available. The patient is kept NPO after midnight the day before the procedure. For the procedure, the patient receives intravenous sedation, typically midazolam and fentanyl, or propofol. It is also recommended that patients receive prophylactic antibiotics, such as ciprofloxacin, on the day of the procedure and for three days afterward. A linear-array echoendoscope (Pentax, Orangeburg, NY) is introduced into the patient's mouth and advanced into the upper GI tract under direct endoscopic visualization. Using the ultrasound function of the echoendoscope, the intestinal or extraintestinal tumor is localized. The Doppler function is then used to verify that no vessel is present between the needle and the target. The tip of a 19-gauge needle (MEDI-Globe, Achenmuhle, Germany, or Sonotip II, Wilson-Cook, Winston-Salem, NC) is inserted into the lesion under EUS guidance (Fig. 12.1). The stylet inside the needle channel is removed and a fiducial is manually inserted into the needle lumen. The stylet is then reintroduced in the needle channel to push the fiducial through the channel until it enters the target tissue. The position of the fiducial should be verified by EUS (where it appears as a bright hyperechoic linear structure) and by fluoroscopy (Figs. 12.2 and 12.3).

This should be repeated to place 3 to 6 fiducials around the target area. The optimum angle and distance between the fiducials should also be monitored under real-time ultrasonography and fluoroscopy. The goal is to keep a minimum distance of 2 cm and an angle of at least 15 degrees between two fiducials in order to get accurate fiducial tracking by the CyberKnife system [15]. The size of the fiducials needs to fulfill two requirements: the fiducial should fit in the needle channel and it should be detectable by the CyberKnife system. We find that fiducials from Best Medical International (Springfield, VA) with a length of 3 or 5 mm and a diameter of 0.8 mm best fulfill these criteria (Fig. 12.4). An initial attempt to

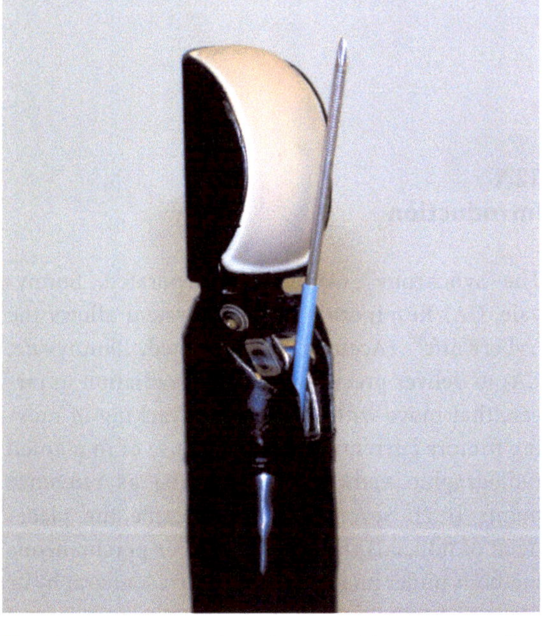

Fig. 12.1 Linear-array echoendoscope with 19-gauge needle.

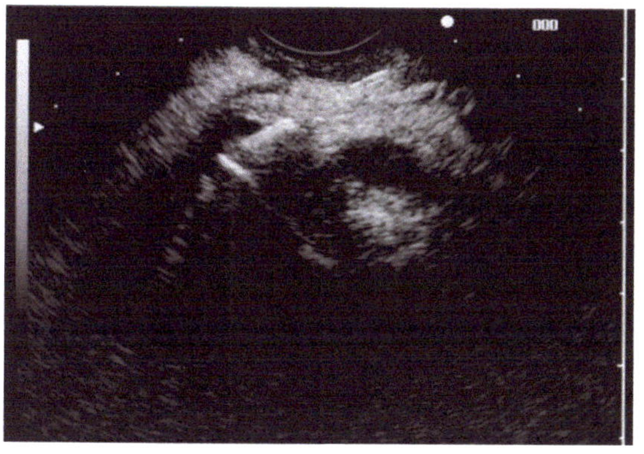

Fig. 12.2 EUS image with placement of gold fiducial.

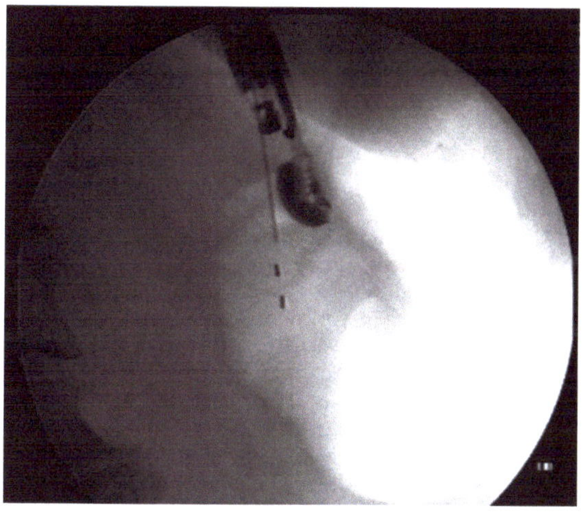

Fig. 12.3 Appearance of fiducials as seen with fluoroscopy.

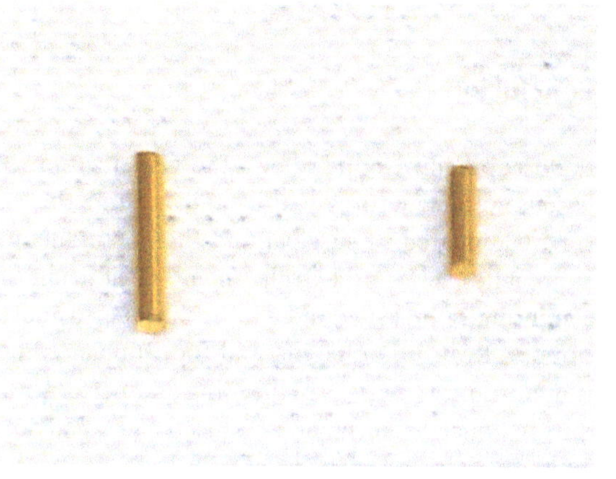

Fig. 12.4 A 5-mm and 3-mm gold fiducial.

place 5 mm fiducials should be made because this is the size approved for use with the CyberKnife system. However, in instances where the 5 mm fiducial cannot be passed into the needle for technical reasons, then a 3-mm fiducial can be used. For example, the tip of the EUS scope may be acutely bent in order to visualize a tumor, making the insertion of a long 5 mm fiducial beyond the bending point difficult to do. A smaller, 3-mm-long fiducial can be easier to use in these cases.

12.2.3 Technical Issues

Every attempt should be made to keep the EUS scope straight, to facilitate the passage of 5 mm fiducials; as mentioned above, if necessary a 3-mm fiducial may be used. Another potential difficulty is the use of a 19-gauge needle. This is a large needle that can be difficult to penetrate hard tissue, such as pancreatic tumors. Finally, when marking lesions in the mediastinum, it can be difficult to achieve the optimal angle and distance between fiducials because of the nature of the mediastinal space.

12.3 Results

Pishvaian et al. reported that fiducials were successfully implanted near tumors throughout the abdomen and mediastinum in 11 of 13 patients. The procedure failed in 2 patients, in one because the progress and alignment of the endoscope was impeded by a pancreatic tumor obstructing the gastric outlet, and in the other because the aorta was in the path of the needle. The authors reported one infectious complication out of the 13 patients in the study [9]. The infection resolved with antibiotic treatment. Potential complications of the procedure not observed by Pishvaian et al. include the risks of sedation on the cardiovascular and respiratory systems, and the risk of infection, bleeding, and perforation of the GI tract.

12.4 Conclusion

Endoscopic ultrasound allows the visualization of lesions and structures within and around the GI tract. Compared to surgical and the percutaneous approaches, EUS-guided placement of fiducials is a minimally invasive and safe technique to mark tumors. Real-time imaging and the Doppler function minimize the risks of bleeding and perforation during fiducial placement. EUS has the great advantage of accessing lesions located deep in the posterior mediastinum and in the abdomen that would otherwise be difficult to reach. Hence, EUS-guided fiducial placement gives patients the possibility of undergoing CyberKnife treatment when conventional radiation therapy is not a good option.

References

1. Adler JR, Jr., Chang SD, Murphy MJ, *et al.* The Cyberknife: a frameless robotic system for radiosurgery. *Stereotact Funct Neurosurg* 1997; 69:124–128.
2. Gerszten PC, Ozhasoglu C, Burton SA, *et al.* CyberKnife frameless single-fraction stereotactic radiosurgery for benign tumors of the spine. *Neurosurg Focus* 2003; 14: e16.
3. King CR, Lehmann J, Adler JR, *et al.* CyberKnife radiotherapy for localized prostate cancer: rationale and technical feasibility. *Technol Cancer Res Treat* 2003; 2:25–30.
4. Kuo JS, Yu C, Petrovich Z, *et al.* The CyberKnife stereotactic radiosurgery system: description, installation, and an initial evaluation of use and functionality. *Neurosurgery* 2003; 53:1235–1239; discussion 1239.
5. Lax I, Blomgren H, Naslund I, *et al.* Stereotactic radiotherapy of malignancies in the abdomen. Methodological aspects. *Acta Oncol* 1994; 33:677–683.
6. Welsh JS, Berta C, Borzillary S, *et al.* Fiducial markers implanted during prostate brachytherapy for guiding conformal external beam radiation therapy. *Technol Cancer Res Treat* 2004; 3:359–364.
7. Whyte RI, Crownover R, Murphy MJ, *et al.* Stereotactic radiosurgery for lung tumors: preliminary report of a phase I trial. *Ann Thorac Surg* 2003; 75:1097–1101.
8. Shirato H, Harada T, Harabayashi T, *et al.* Feasibility of insertion/implantation of 2.0-mm-diameter gold internal fiducial markers for precise setup and real-time tumor tracking in radiotherapy. *Int J Radiat Oncol Biol Phys* 2003; 56:240–247.
9. Pishvaian AC, Collins B, Gagnon G, *et al.* EUS-guided fiducial placement for CyberKnife radiotherapy of mediastinal and abdominal malignancies. *Gastrointest Endosc* 2006; 64:412–417.
10. Erickson RA. EUS-guided FNA. *Gastrointest Endosc* 2004; 60:267–279.
11. Fusaroli P, Caletti G. Endoscopic ultrasonography. *Endoscopy* 2005; 37:1–7.
12. Chang KJ, Nguyen PT, Thompson JA, *et al.* Phase I clinical trial of allogeneic mixed lymphocyte culture (cytoimplant) delivered by endoscopic ultrasound-guided fine-needle injection in patients with advanced pancreatic carcinoma. *Cancer* 2000; 88:1325–1335.
13. Gan SI, Thompson cc, Lauwers Gy, *et al.* Ethanol lavage of pancreatic cystic lesions: initial pilot study. *Gastrointest Endosc* 2005; 61:746–752.
14. Gunaratnam NT, Sarma AV, Norton ID, *et al.* A prospective study of EUS-guided celiac plexus neurolysis for pancreatic cancer pain. *Gastrointest Endosc* 2001; 54:316–324.
15. Olender D. Fiducials for target localization. In: Heilbrun MP, editor. CyberKnife radiosurgery: a practical guide. Sunnyvale, CA: CyberKnife Society; 2003. pp. 80–94.

Section 4

Lung Tumor Treatment – Techniques and Experiment

Stereotactic Body Radiation Therapy in the Treatment of Early Stage Non-Small Cell Lung Cancer

13

Ronald C. McGarry

CONTENTS

13.1 Introduction

Lung cancer is a significant cause of morbidity and mortality in the United States, with an estimated 173,000 new lung cancer cases and 156,300 deaths due to lung cancer in 2003 [1]. Eighty-five percent of these patients are diagnosed with non-small cell lung cancer (NSCLC), of which 15–20% will have early stage disease [2]. Treatment options for early stage NSCLC depend on patient factors, such as pulmonary reserve, to determine if a patient is a surgical candidate or not. Typically, standard "cut-off" medical guidelines for surgical resection of NSCLC include the following: baseline forced expiratory volume in 1 s (FEV_1) < 40% of predicted, post-operative predicted FEV_1 < 30%, severely reduced diffusion capacity, baseline hypoxemia and/or hypercapnia, and exercise oxygen consumption < 50% predicted. For the relatively healthy patient, the traditional treatment of choice for Stage I (T1-2, N0, M0) NSCLC usually consists of radical surgery (lobectomy, pneumonectomy) resulting in an approximate 60% to 80% 5-year overall survival [3–5].

For patients who are considered high risk for surgery, treatment options include limited or sublobar resection, conventionally fractionated local radiation therapy, radiofrequency ablation, and stereotactic body radiation therapy (SBRT). Here we briefly review the indications, goals, and outcomes of surgery and conventional radiation therapy followed by a detailed review of the literature and technical challenges of SBRT.

13.2 Surgery

The goal of surgical resection is to obtain a complete resection of the primary tumor. For those patients with NSCLC who are considered surgical candidates, surgery with either lobectomy or pneumonectomy, leaving clear margins, is considered the gold standard. Although mediastinal lymph node dissection at the time of surgery is mandatory, the degree of dissection is controversial. Lymph node dissection plays a role in staging of patients, but its therapeutic role is uncertain. In most surgical series, pathological staging of clinically staged patients results in upstaging of Stage I patients in approximately 20–30% of cases, although positron emitted tomography (PET) scanning is helping identify those patients with nodal and other metastatic disease prior to surgery [6, 7]. In an analysis of 598 surgical patients, 560 of whom received mediastinal lymph node dissection, 5-year survival was 82% for pT1N0M0 patients and 68% for pT2N0M0 patients. Local or regional recurrence was low (7%), but systemic relapse was much more common, occurring in up to 20% of patients [6].

For patients with lesser pulmonary reserve, limited resection (commonly known as wedge or segmental resection) may be considered a substitute for lobectomy, but local failure with this procedure is considered suboptimal by the Lung Cancer Study Group [7]. Few randomized studies have examined the role of limited surgery. The Lung Cancer Study Group examined 276 patients with early stage (T1N0M0) NSCLC randomly assigned to limited-resection vs. lobectomy groups. There were no significant differences for all stratification variables, selected prognostic factors, perioperative morbidity, mortality, or late pulmonary function. There was a 75% increase in overall recurrence rates in patients who underwent limited resection, including a tripling of local recurrence rate. Overall death rate increased by 30% and the rate of death with cancer rate increased by 50% relative to patients undergoing lobectomy. The Group concluded that limited pulmonary resection does not improve perioperative morbidity, mortality, or late postoperative pulmonary function. Because of the higher death rate and locoregional recurrence rate after limited resection, lobectomy is the preferred procedure for patients with peripheral T1N0 NSCLC.

13.3 Radiation Therapy for Medically Inoperable Stage I NSCLC

The alternative treatment for medically inoperable patients with Stage I NSCLC has been radiation therapy, typically delivered with standard fractionation (180–200 cGy per day, five days per week, to a total of approximately 6000 cGy) to a relatively small volume of lung. While this approach is reasonably well tolerated, the 5-year overall survival is 30%, only half of that for patients receiving surgical resection [8, 9].

As several studies seemed to show a dose-response relationship [10–14], it was hypothesized that greater local control could be gained if higher doses of radiation were given. SBRT, or extracranial stereotactic radiation therapy, is rapidly becoming the method of choice for dose escalation of early stage NSCLC. The goal of SBRT is to deliver hypofractionated radiation therapy in several large doses via multiple beam geometries to the defined target volume, with a rapid dose fall-off to minimize the amounts of normal tissue receiving significant integral radiation dose. Challenges to achieving these treatment goals and gaining good local tumor control include dealing with the respiratory motion of the tumor, assuring conformality of the dose to the target volume along with rapid fall-off of dose outside the target volume, and defining of the target volume to encompass gross and microscopic disease. An important secondary challenge concerns the issue of monitoring tumor response and detecting treatment failure.

13.4 Management of Respiratory Movement

Several techniques for targeting lung cancers and controlling respiration are being investigated. With each technique some degree of patient immobilization is necessary to ensure reproducible target localization. *Active breathing control* couples the patient to a spirometer-like device used in several different ways. The patient can be coached, receive verbal cues through headphones, or watch a monitor displaying the respiratory cycle to control their respiration. Most commonly, during deep inspiration, a form of "beam-on" gating can be used, either electronically by the LINAC software or manually, as the therapist watches the breath hold on a monitor. *Four-dimensional radiation therapy* uses prospective collection of CT slices during various phases of respiration for treatment and gating of the linear accelerator beam-on time to treat only during a defined portion of the respiratory cycle. *Chasing* techniques, described at recent meetings by Uematsu's group, use a multileaf collimator to "track" the cancer and move the field as defined by the multileaves with respiration. A commercially available approach to chasing the tumor with respiration is the Synchrony® Respiratory Tracking System developed for the CyberKnife® (Accuray Incorporated, Sunnyvale, CA). This system is described more completely in several of the chapters in this volume.

The development of frame-based treatments of intra-thoracic and intra-abdominal lesions is generally attributed to researchers at the Karolinska Hospital in Stockholm, Sweden [15, 16]. In 1994, Lax et al. [16] described an extracranial frame that incorporated a fiducial stereotactic coordinate system along its side panels. To decrease respiratory excursion, an abdominal press was employed that forced the patient to perform relatively more chest wall breathing than diaphragmatic breathing. A formal verification of reproducibility showed that target motion was reduced to within 0.5 cm in the axial plane and 1.0 cm in the caudal/cephalad direction. At the time of simulation, the degree of abdominal compression is determined by watching the tumor under fluoroscopy and compressing to patient tolerance or until respiratory tumor movement is at a minimum.

13.5 Target Definition for SBRT

The object of SBRT is to define closely the tumor target so that the volume of normal lung that is irradiated is minimized. Prophylactic coverage of nodal regions has not been attempted in most studies because tissue/volume constraints limit dose that can be delivered without undue toxicity. Nevertheless, caution must be exercised to avoid a geographic miss. Giraud et al. [17] examined 70 primary NSCLC surgical resection specimens and marked the apparent border between the tumor and adjacent lung parenchyma by naked eye. Careful histopathologic examination for microscopic extension (ME) was performed. They concluded that ME of lung cancers averaged 2.69 mm for adenocarcinomas and 1.48 mm for squamous cell carcinomas. From these data it was reasoned that to cover 95% of ME, 8 mm margins are needed for adenocarcinomas and 6 mm margins are required for squamous cell tumors. It is apparent that one must take into account ME extension to limit the risk of a geographic miss in SBRT.

As performed in the Indiana University setting [18, 19], all patients with histologically proven lung cancers have a high-quality PET-CT to confirm as much as possible that their cancers are truly early stage. Following CT-simulation with contrast in the Elekta stereotactic body frame (Elekta Oncology, Norcross, GA), targets are defined. The selection of CT contrast level is important, and contouring of lung tumors should always be done in typical lung windows. The gross tumor volume (GTV) is outlined and ground glass opacities/atelectasis in the vicinity of the tumor are not included. No specific clinical target volume is defined, so margins are added to the GTV to create the planning target volume (PTV). Margins of 0.5 cm in the axial plane and 1.0 cm in the superior/inferior plane are added to allow for some degree of target motion.

13.6 Selection of Dose/Fractionation

Conventionally fractionated radiation has been the standard of care for Stage I medically inoperable patients for many years. Stage I lung cancer masses are commonly treated to doses of 66–70 Gy in the absence of chemotherapy. High local recurrence rates prompted studies of dose escalation in an attempt to improve results. Thus, the rationale for SBRT is to escalate dose to cancers while maintaining toxicity at acceptable levels.

Studies of outcomes following definitive radiotherapy for medically inoperable NSCLC have shown that outcomes depend critically on both tumor size and radiation dose. For example, Sandler et al. [20] reviewed 77 patients with Stage I lung cancer treated to a median dose of 60 Gy. They reported a 22% actuarial disease-free survival. The 3-year disease-free survival for patients with tumors less than 3 cm, 3–6 cm, and greater than 6 cm was 30%, 17%, and 0%, respectively. Patients in this study were treated from 1970 through 1987, an era that spanned the development of CT and PET scanning. Nevertheless, the survival differences related to tumor size are striking.

A dose response was detected by Sibley et al. [14], who reviewed the results of 10 studies on treating medically inoperable early stage NSCLC with radiotherapy. All patients received megavoltage radiotherapy to doses >55 Gy and median doses of 60–66 Gy. Grade 3–5 complications occurred in fewer than 2% of patients. Patients in these studies generally had a 15% median 5-year survival with 25% dying of intercurrent disease and 30% dying of metastatic disease. The 5-year cause-specific survival was 32%. A pattern-of-failure analysis showed that 42% of failures were local-only and 38% were distant-only. Improved survival correlated with local control and approached significance for higher radiotherapy dose ($p=0.07$) and larger treated fields ($p=0.08$).

Most studies of dose escalation have included patients with higher stage disease. Due to the high risk of local recurrence, several studies have undertaken dose escalation using different strategies to maintain toxicity at acceptable levels. Most strategies try to equate dose/fractionation schemes by employing linear quadratic normalization (biological effective dose, or BED). This is somewhat difficult since it does not take into account the duration of treatment in hypofractionated schemes. Applying the linear quadratic equation ($BED=nd(1+d/\alpha/\beta)$ where n=the number of fractions, d=the dose/fraction and α/β ratio of 10 for acute reacting tissue and tumor cells), 70 Gy will have a BED of approximately 84 Gy. Using a mathematical model, Martel et al. [21] predicted that for NSCLC patients, the dose to achieve significant probability of tumor control may be at least 84 Gy for longer (>30 months) local progression-free survival. RTOG 9311 was a multi-institutional, dose-escalation trial in which 179 patients were treated with 3D radiotherapy [22]. At each dose, patients were stratified depending on the percentage of the total lung volume that received greater than 20 Gy (V_{20}, an indicator of risk for pneumonitis; see below). Patients with a $V_{20} < 25\%$ (Group 1) received 70.9 Gy in 33 fractions, 77.4 Gy in 36 fractions, 83.8 Gy in 39 fractions, or 90.3 Gy in 42 fractions. Patients with a V_{20} of 25–36% (Group 2) received doses of 70.9 Gy and 77.4 Gy, successively. The radiation dose was safely escalated using 3D conformal techniques and fraction sizes of 2.15 Gy to 83.8 Gy for patients in Group 1 and to 77.4 Gy for patients in Group 2. The 90.3-Gy dose level was too toxic, resulting in dose-related deaths in 2 patients. Local nodal failure occurred in <10% of patients. This study showed that for patients receiving 3D conformal radiation therapy (3D-CRT) alone, doses of 83.8 Gy were tolerable, with excess mortality observed at 90.3 Gy. When concurrent chemotherapy and 3D-CRT are used, the maximum tolerated dose is reduced to about 70 to 74 Gy [23]. Nonetheless, it has not been prospectively demonstrated that dose escalation produces superior outcomes, and careful planning is needed to avoid toxicities such as pneumonitis, esophagitis and stricture, in addition to compromise of other dose limiting organs in the field.

Willner et al. [24] retrospectively examined the influence of total dose and tumor volumes on local control and survival in primary radiotherapy of NSCLC. They concluded that there is a dose effect on local control and survival with doses of at least 70 Gy (standard fractionation), and that tumors with volumes greater than or equal to 100 cc may require higher doses.

It stands to reason and evidence suggests that improved local control may result from higher local radiation doses. Dose escalation may be accomplished by delivering higher doses to restricted volumes in fewer fractions, i.e., hypofractionated stereotactic radiation therapy. In an analysis of various hypofractionated stereotactic radiation schemes, Wulf et al. [25] reported that for optimal control of Stage I lung cancer, a BED of >100 Gy is required. It is apparent that our current conventional treatment regimens may be responsible for the high local recurrence rates, but when one considers the dose-limiting organs in the chest, dose escalation becomes difficult.

13.7 Stereotactic Body Radiation Therapy: Clinical Trials

In a pioneering study of patients with Stage Ia/Ib (NSCLC), Uematsu et al. [26] treated 50 patients (21 medically inoperable, the rest refused surgery) to tumor doses of 50–60 Gy in 5–10 fractions using a frameless stereotactic technique (SRT). Tumor sizes ranged from 0.8 to 5.0 cm and were treated with 2.0 cm margins. Eighteen patients had received

40–60 Gy in 20–33 fractions before SRT. At a median follow-up of 36 months, 94% showed no local progression, the 3-year overall survival rate was 66%, and survival in the medically fit patients (those who refused surgery) was 86%. The 3-year cause-specific survival was 88%. Two minor bone fractures occurred and 6 patients reported pleural pain.

In another report [27], 22 patients with Stage I NSCLC (13 Stage Ia, 9 Stage Ib) were treated with SBRT. Patients were not immobilized in specific devices, but coverage of the tumors was evaluated by CT scan at full inspiration, expiration, and free breathing to ensure the GTV was covered in all phases of the respiratory cycle. Treatment volumes were defined as CTV + 1.0 cm for a PTV that encompassed the average respiratory movement observed under fluoroscopy. Dose to the CTV was prescribed to the 80% line and patients received 60 Gy in 8 fractions. After a patient experienced a tracheobronchial ulcer, the dose was reduced to 48 Gy in 8 fractions for target volumes involving the lobar or more central bronchus. Median follow-up was 24 months. The rate of local control at 2–6 months was 94%. Cancer-specific and relapse-free survival was reported at 94% and 71% at 12 months and 73% and 67% at 24 months, respectively. No other clinically significant toxicities and no significant decline in pulmonary function tests (PFTs) were reported.

Cheung et al. [28] treated 33 medically inoperable patients with Stage Ia/Ib NSCLC with SBRT without prophylactic nodal irradiation. Standard radiotherapy planning was utilized without specific immobilization. Tumor extent and position was determined using conventional planning with a 1.5 cm margin around the GTV symmetrically. Patients received a total of 48 Gy in 12 Gy fractions. Overall survival was 80% at one year and 46% at 2 years with a median survival of 22.6 months. Cause-specific survival was 89.8% at one year and 54.1% at 2 years. Recurrence-free survival was 66.4% at one year and 40% at 2 years. Reported toxicities included dermatitis (30.3%) and late cutaneous/subcutaneous fibrosis (24.2%). Significantly poorer outcomes were obtained when tumors were treated with margins of less than 2.0 cm.

Hiraoka et al. [29] reported a retrospective analysis of 241 patients from 13 Japanese institutions. This report focused on solitary NSCLC tumors less than 4 cm. Primary lung cancers were eligible if they were solitary, histologically confirmed tumors less than 4 cm that were inoperable (or the patient refused surgery) and not near the spinal cord, patients did not require oxygen support, and patient ECOG performance status was equal to or less than 2. A total dose of 48 Gy was delivered in four fractions over 2 weeks in most patients. No Grade II toxicities were observed in the spinal cord, bronchus, pulmonary artery, or esophagus. Overall 3-year survival for patients with Stage IA and IB disease was 87% and 80%, respectively. No cancers recurred during the 3–50 month follow-up. Regional lymph node recurrence developed in 1 patient and distant metastases developed in 4 patients. The local recurrence rate was 20% when the BED was less than 100 Gy and 6.5% when the BED was over 100 Gy. Overall survival at 3-years was 42% at BED < 100 Gy and 46% at BED > 100 Gy. A high 3-year survival rate (91%) was obtained in operable patients that refused surgery. This fractionation regimen is currently being used in a Japan-wide protocol to assess outcomes in a multi-institutional trial.

A Phase I dose escalation trial was undertaken at Indiana University [18, 19]. Patients with Stage IA and IB (up to 7.0 cm) NSCLC who were declared medically inoperable were treated. Eligible patients had clinically Staged T1 or T2 (≤7 cm), N0, and M0 biopsy-confirmed NSCLC. All patients had co-morbid medical problems that precluded lobectomy and Karnofsky performance status from 60-100. Based on the early Karolinska data, total dose was given in three fractions, with escalation beginning at 800 cGy per fraction prescribed to the 80% line covering a minimum of 95% of the PTV. All treatments used a minimum of 7 non-coplanar beams. Dose limiting toxicity was not reached, but dose escalation was stopped for Stage IA tumors at 2000 cGy per fraction (6000 cGy total dose, given in three fractions). For the Stage IB cancers (3.0–7.0 cm), a separate dose escalation was continued to 2400 cGy/fraction (7200 cGy), which was felt to be limiting based on appearance of pneumonitis. In a follow-up report from the Indiana University Phase I study [18] patterns of failure were evaluated. It was found that 9 local failures occurred at doses less than 1600 cGy per fraction, while only 1 patient recurred locally at higher doses. For all Stage I patients (N=47), 4 had recurrence distally, 3 had recurrence regionally, and 7 had both regional and distant recurrence. The conclusion was that SBRT is a well tolerated and safe treatment option for early stage lung cancers.

A prospective Phase II trial of this treatment at Indiana University has completed accrual with Stage IA receiving 6000 cGy and Stage IB receiving 6600 cGy delivered in three fractions [30]. Eligibility was the same as the completed Phase I trial. All 70 patients completed all three fractions of SBRT. Preliminary results [30] show the three-month major response rate (complete and partial) was 60%. Fibrotic changes and atelectasis occurred commonly after 6 months making local assessment difficult. Seventeen patients had local enlargement of radiographic changes in the vicinity of the treated tumor prompting a PET scan. After a median follow-up of 17.5 months (range 0.6–44.2 months), a total of three patients experienced a local recurrence of cancer. Actuarial local control at two years is 95%. Altogether, 28 patients died of cancer or co-morbid illnesses; 6/70 patients were deemed by the Data Monitoring committee to have "possibly" died due to treatment with causes of death including pneumonia, respiratory decline (questionable in already compromised patient), hemoptysis (one patient with local recurrence), and pericardial effusion. Kaplan-Meier estimates indicate a median overall survival of 32.6 months and actuarial 2-year overall survival of 54.7%. Grade 3–4 toxicity occurred in a total of 14 patients including pneumonia in 4 patients, pulmonary function test (PFT) decline in 3 patients, pleural effusion in 2 patients, and apnea, vocal cord palsy, skin burn, esophagitis, and hemoptysis each in one patient at a median of 7.5 months post-therapy.

Patients treated for tumors in the peripheral lung had 2-year freedom from severe toxicity of 83% compared to only 54% for patients with perihilar/central tumors. We have concluded that very high rates of local control are achieved with this SBRT regimen in this population of medically inoperable patients with Stage I NSCLC. Local tumor recurrence and toxicity occur late after this treatment. Concern has been expressed regarding higher incidence of toxicity for tumors near central airways and caution should be exercised with these patients. Based on this trial, a multi-institution Phase II trial conducted by the RTOG (Radiation Therapy Oncology Group 0236) has completed accrual. While the stereotactic radiation was well tolerated, a minority of patients had what appears to be partial lung atelectasis downstream from the treated areas. Most of the atelectasis has been clinically insignificant; nevertheless, a decision was made in the development of the RTOG trial to designate a "safety zone" of 2 cm around the major airways which the primary tumor cannot encroach upon.

In summary, the data to support SBRT as an option in the curative management of early stage NSCLC is mounting. Further studies including healthy patients with early stage NSCLC are needed to assess properly not only long-term survival, but also long-term toxicities. At this point, toxicities in very frail patients seem to be minimal and SBRT for the medically inoperable patient is a reasonable treatment option.

References

1. Jemal A, Siegel R, Ward E, *et al.* Cancer statistics, 2006. *CA Cancer J Clin* 2006; 56:106–130.
2. Brown cc, Kessler LG. Projections of lung cancer mortality in the United States: 1985-2025. *J Natl Cancer Inst* 1988; 80:43–51.
3. AJCC Cancer Staging Manual, 6th Edition. New York: Springer-Verlag; 2002.
4. Annual Smoking attributable mortality, years of potential life lost and economic costs – United States,1997–2001. *MMWR Morb Mortal Wkly Rep* 2005; 54:625–628.
5. Travis WD, Colby TV, Corrin B, *et al.*, editors. Histologic typing of lung and pleural tumors: World Health Organization Histological Classification of Tumors. 3rd ed. Berlin: Springer-Verlag; 1999.
6. Martini N, Bains MS, Burt ME, *et al.* Incidence of local recurrence and second primary tumors in resected stage I lung cancer. *J Thorac Cardiovasc Surg* 1995; 109:120–129.
7. Ginsberg RJ, Rubinstein LV. Randomized trial of lobectomy versus limited resection for T1 N0 non-small cell lung cancer. Lung Cancer Study Group. *Ann Thorac Surg* 1995; 60:615–622; discussion 622–613.
8. Chen M, Hayman JA, Ten Haken RK, *et al.* Long-term results of high-dose conformal radiotherapy for patients with medically inoperable T1-3N0 non-small-cell lung cancer: is low incidence of regional failure due to incidental nodal irradiation? *Int J Radiat Oncol Biol Phys* 2006; 64:120–126.
9. Qiao X, Tullgren O, Lax I, *et al.* The role of radiotherapy in treatment of stage I non-small cell lung cancer. *Lung Cancer* 2003; 41:1–11.
10. Cox JD, Azarnia N, Byhardt RW, *et al.* A randomized phase I/II trial of hyperfractionated radiation therapy with total doses of 60.0 Gy to 79.2 Gy: possible survival benefit with greater than or equal to 69.6 Gy in favorable patients with Radiation Therapy Oncology Group stage III non-small-cell lung carcinoma: report of Radiation Therapy Oncology Group 83-11. *J Clin Oncol* 1990; 8:1543–1555.

11. Dosoretz DE, Galmarini D, Rubenstein JH, *et al.* Local control in medically inoperable lung cancer: an analysis of its importance in outcome and factors determining the probability of tumor eradication. *Int J Radiat Oncol Biol Phys* 1993; 27:507–516.
12. Dosoretz DE, Katin MJ, Blitzer PH, *et al.* Radiation therapy in the management of medically inoperable carcinoma of the lung: results and implications for future treatment strategies. *Int J Radiat Oncol Biol Phys* 1992; 24:3–9.
13. Sibley GS, Jamieson TA, Marks LB, *et al.* Radiotherapy alone for medically inoperable stage I non-small-cell lung cancer: the Duke experience. *Int J Radiat Oncol Biol Phys* 1998; 40:149–154.
14. Sibley GS. Radiotherapy for patients with medically inoperable Stage I nonsmall cell lung carcinoma: smaller volumes and higher doses–a review. *Cancer* 1998; 82:433–438.
15. Blomgren H, Lax I, Naslund I, *et al.* Stereotactic high dose fraction radiation therapy of extracranial tumors using an accelerator. Clinical experience of the first thirty-one patients. *Acta Oncol* 1995; 34:861–870.
16. Lax I, Blomgren H, Naslund I, *et al.* Stereotactic radiotherapy of malignancies in the abdomen. Methodological aspects. *Acta Oncol* 1994; 33:677–683.
17. Giraud P, Antoine M, Larrouy A, *et al.* Evaluation of microscopic tumor extension in non-small-cell lung cancer for three-dimensional conformal radiotherapy planning. *Int J Radiat Oncol Biol Phys* 2000; 48:1015–1024.
18. McGarry RC, Papiez L, Williams M, *et al.* Stereotactic body radiation therapy of early-stage non-small-cell lung carcinoma: phase I study. *Int J Radiat Oncol Biol Phys* 2005; 63:1010–1015.
19. Timmerman R, Papiez L, McGarry R, *et al.* Extracranial stereotactic radioablation: results of a phase I study in medically inoperable stage I non-small cell lung cancer. *Chest* 2003; 124:1946–1955.
20. Sandler HM, Curran WJ, Jr., Turrisi AT, III. The influence of tumor size and pre-treatment staging on outcome following radiation therapy alone for stage I non-small cell lung cancer. *Int J Radiat Oncol Biol Phys* 1990;19:9-13.
21. Martel MK, Ten Haken RK, Hazuka MB, *et al.* Estimation of tumor control probability model parameters from 3-D dose distributions of non-small cell lung cancer patients. *Lung Cancer* 1999; 24:31–37.
22. Bradley J, Graham MV, Winter K, *et al.* Toxicity and outcome results of RTOG 9311: a phase I-II dose-escalation study using three-dimensional conformal radiotherapy in patients with inoperable non-small-cell lung carcinoma. *Int J Radiat Oncol Biol Phys* 2005; 61:318–328.
23. Bradley J. A review of radiation dose escalation trials for non-small cell lung cancer within the Radiation Therapy Oncology Group. *Semin Oncol* 2005; 32:S111–113.
24. Willner J, Baier K, Caragiani E, *et al.* Dose, volume, and tumor control prediction in primary radiotherapy of non-small-cell lung cancer. *Int J Radiat Oncol Biol Phys* 2002; 52:382–389.
25. Wulf J, Baier K, Mueller G, *et al.* Dose-response in stereotactic irradiation of lung tumors. *Radiother Oncol* 2005; 77:83–87.
26. Uematsu M, Shioda A, Suda A, *et al.* Computed tomography-guided frameless stereotactic radiotherapy for stage I non-small cell lung cancer: a 5-year experience. *Int J Radiat Oncol Biol Phys* 2001; 51:666–670.
27. Fukumoto S, Shirato H, Shimzu S, *et al.* Small-volume image-guided radiotherapy using hypofractionated, coplanar, and noncoplanar multiple fields for patients with inoperable Stage I nonsmall cell lung carcinomas. *Cancer* 2002; 95:1546–1553.
28. Cheung PC, Mackillop WJ, Dixon P, *et al.* Involved-field radiotherapy alone for early-stage non-small-cell lung cancer. *Int J Radiat Oncol Biol Phys* 2000; 48:703–710.
29. Hiraoka M, Nagata Y. Stereotactic body radiation therapy for early-stage non-small-cell lung cancer: the Japanese experience. *Int J Clin Oncol* 2004; 9:352–355.
30. Timmerman R, McGarry R, Yiannoutsos C, *et al.* Excessive toxicity when treating central tumors in a phase II study of stereotactic body radiation therapy for medically inoperable early-stage lung cancer. *J Clin Oncol* 2006; 24:4833–4839.

Dosimetric Considerations for Stereotactic Body Radiotherapy Delivery in the Lung

14

Lech Papiez

CONTENTS

14.1 Introduction

In the performance of stereotactic body radiotherapy (SBRT), one must take into consideration not only the biology of the cancer being treated, but the physics and dosimetry behind delivery of the radiation dose. In a sense, investigation of the clinical application has begun before the basic research has been performed. Investigation of the biology of large-fraction radiotherapy has only started, but due to its potential for significant toxicity, extremely stringent conditions must be applied after due consideration. Here we examine some of the issues.

14.2 Conditions for Dose

The proper conduct of hypofractionated SBRT requires both precise dose targeting and precise dose shaping. In particular, dose distribution around the target should show evidence of isotropic and rapid fall-off to normal surrounding tissues [1–6]. The realization of these dose distribution properties in SBRT treatments takes place when:

- The dose isosurface, defined by the collection of points within the body that are exposed to a given level of prescribed dose, closely approximates the delineated surface of the target. Hot spots of the dose distribution (points in tissue where the highest doses are delivered) are located inside the target and the tissue exposed to the prescribed dose that falls outside of the target is confined to regions near the target boundary.
- The dose distribution shows rapid fall-off from the tumor to healthy tissue isotropically in all directions.
- The dose delivered to the target is characterized by a non-uniform distribution within the volume of the tumor, with the highest dose delivered to the section of the tumor where hypoxic cells reside.

The assessment of dose shaping in SBRT is important for the evaluation of the applied treatment quality. This evaluation should provide information about the target coverage by the prescription dose, the proximity of the dose cloud wrapping around the target, and the steepness and uniformity of the gradient of dose decrease away from the target. Some traditional functions that are used in radiation therapy for measuring quality of dose distributions are useful but not completely adequate for these purposes. Common measures of dose distribution quality include the radiation conformality index [7, 8] and the characteristic parameters of dose-volume histograms (DVH), e.g., the V_{20}, the proportion of an organ exposed to 20 Gy or more per course of treatment [9, 10]. DVH parameters that are particularly useful for these assessments include the volume covered by 95% of the prescribed dose, the sharpness of the volume decrease from full volume to zero with increased dose (equivalent to uniformity of dose over target), and so forth. None of these indexes, however, characterizes in adequate detail critical aspects of SBRT dosimetry.

The radiation conformality index may indicate that the volume of the planning target volume (PTV) is the same as the volume of tissue exposed to a given level of dose; however, it does not specify how much the volumes overlap each other. V_{20} may tell us how much of the lung (liver, rectum, etc.) is protected from a threshold level dose but it does not identify if the portion of lung exposed to large dose is close to or far from the tumor. Generally, the DVH functions give detailed information on how large portions of organs are exposed to different levels of dose but do not show where in the organs these different levels of dose are located.

14.3 Measures of SBRT Dose Quality

To design appropriate measures for SBRT treatment planning one needs to introduce functions that naturally verify the desired properties of SBRT dose distributions. In the first-order approximation, SBRT dose quality can be measured by adherence to constraints given in the form of a table (Table 14.1). The tabled factors for evaluation of SBRT dosimetry scrutinize three main features of dose quality: (a) target coverage, (b) high-dose "spillage," and (c) low-dose "spillage."

With regard to target coverage, it is established practice in SBRT [11–15] to normalize the dose (i.e., set to 100%) relative to the center of the mass of the PTV, which is generally identified with the isocenter. With such normalization, the edge of the PTV is usually required to be covered by 60–90% of the prescription dose. The prescription dose is given to the isodose encompassing at least 95% of the PTV volume. Furthermore, it is required that 99% of the PTV volume gets at least 90% of the prescription dose. Considerable dose heterogeneity is accepted.

Table 14.1 Target Volume Scoring

Target Volume Scoring	
No Variation	**Major Variation**
95% of the PTV volume covered by 60 Gy (20Gy/fraction) ***and***	Failure to achieve *No Variation*
99% of the PTV volume covered by 54 Gy (18 Gy/fraction) ***and***	
The dose at the Normalization point (100% prescription point, i.e., COM_{PTV}) should be ≥66 Gy (22 Gy/fraction) and ≤100 Gy (33 Gy/fraction)	

The higher-than-prescribed dose is, as a rule, manipulated to occur within the gross tumor volume (GTV).

Any high-dose "spillage" into normal tissue should be in close proximity to the target. More precisely, any dose greater than 105% of the prescription dose (not the maximum dose) should occur within the PTV. With regard to high-dose spillage volume, it is clearly beneficial to keep the conformality ratio (volume of the prescription isodose coverage over the PTV volume) close to 1. As with high-dose spillage, low-dose spillage is evaluated by criteria relating to location and volume.

For the location of low-dose spillage to be acceptable, the dose to any point 2 cm away from the PTV surface (D_{2cm}) must be below a limit that is a function of PTV volume as shown in Table 14.2. The

Table 14.2 Critical Organ Dose-Volume Limits. This table lists maximum dose limits to a point or volume within several critical organs. For protocol purposes, any doses to the organs at risk (OAR) in excess of the specified dose by more than 2.5% will be scored as a minor deviation. Any doses to the OAR in excess of the specified dose by more than 5% will be considered a major deviation.

Organ	Volume	Dose (cGy)
Spinal Cord	Any point	18 Gy (6 Gy per fraction)
Esophagus	Any point	27 Gy (9 Gy per fraction)
Ipsilateral Brachial Plexus	Any point	24 Gy (8 Gy per fraction)
Heart	Any point	30 Gy (10 Gy per fraction)
Trachea and Ipsilateral Bronchus	Any point	30 Gy (10 Gy per fraction)
Whole Lung (Right & Left)	(See Table 14.3)	(See Table 14.3)

volume of low-dose spillage must fit within a low-dose conformality ratio defined as the ratio of the 50% isodose coverage to the PTV. This parameter, called R50%, is also a function of the PTV volume as shown in Table 14.3.

These evaluations, target coverage, high-dose "spillage," and low-dose "spillage", coupled with a respect for normal tissue dose-volume absolute limits, must be satisfied in most circumstances for a plan to be used in treating a patient with SBRT. It should be stressed that the criteria and parameters for SBRT delivery discussed above, and explicitly quoted in Tables 14.1, 14.2 and 14.3, have been derived mostly from lung treatments. Therefore, they are recommended for treatments that are modeled after the RTOG 0236 SBRT study, but they should only be treated as general guidelines for treatment of other sites with SBRT techniques.

More comprehensive than these tabled constraints on dose distribution are measures designated as "dose allocation histograms" (DAH) [16, 17]. DAH functions are designed to provide a simple and natural measure of dose anisotropy and dose gradient away from the target (Figs. 14.1, 14.2, and 14.3). The intuitive justification for the introduction of such functions is the idea of estimating the unbalanced decrease in exposure to radiation of successive shells of tissue moving outward from the target (Fig. 14.1). This estimate would allow verification of how uniformly shells of tissue adjacent to the target are exposed to desirable, large doses and how uniformly shells far away from the target are spared. To this end, numerical representations of deviation from the balanced exposure decrease away from the spherical target, for differently planned shapes of prescribed dose, are calculated and displayed in Figures 14.2 and 14.3. The need for high dose to shells in contact with the target arises from the need to sterilize microscopic extension of the tumor in normal tissue, extension which is not identified in imaging [18]. The other reason for high dose inside shells near the target is that margins used to accommodate motion and positional error may fail, with some probability, during therapy delivery. In other words, one always has to accept the possibility that the capacity to contain the tumor within prior defined margins may fail due to limited capabilities of registering the motion of the target and the imperfections of motion-correcting tools. On the other hand, there is no reason to believe that deposition of high doses in tissue far from the target would benefit the treat-

Table 14.3 Conformality of prescription dose. The ratio of the prescription isodose volume to the PTV must be no greater than that indicated in the table. The ratio of the volume of 50% of the prescription dose (10 Gy per fraction = 30 Gy total) isodose to the volume of the PTV must be no greater than $R_{50\%}$ where $R_{50\%}$ is given by the table. The maximum absolute dose to any point 2 cm from the surface of the PTV in any direction must be no greater than D_{2cm} where D_{2cm} is given by the table. The fraction of the entire lung (right and left together minus the GTV and the major bronchi) that receives a total dose of 20 Gy or more must be no greater than V_{20} where V_{20} is given by the table. Note: For values of PTV dimension or volume not specified, linear interpolation between table entries is required.

Maximum PTV Dimension (cm)	Ratio of Prescription Isodose Volume to the PTV – Deviation		Ratio of 50% Prescription Isodose Volume to the PTV, $R_{50\%}$ – Deviation		Maximum Dose 2 cm from PTV in any Direction, D_{2cm} (Gy) – Deviation		Percent of Lung receiving 20 Gy total or more, V_{20} (%) – Deviation		PTV Volume (cc)
	none	minor	none	minor	none	minor	none	minor	
2.0	<1.2	1.2–1.4	<3.9	3.9–4.1	<28.1	28.1–30.1	<10	10–15	1.8
2.5	<1.2	1.2–1.4	<3.9	3.9–4.1	<28.1	28.1–30.1	<10	10–15	3.8
3.0	<1.2	1.2–1.4	<3.9	3.9–4.1	<28.1	28.1–30.1	<10	10–15	7.4
3.5	<1.2	1.2–1.4	<3.9	3.9–4.1	<28.1	28.1–30.1	<10	10–15	13.2
4.0	<1.2	1.2–1.4	<3.8	3.8–4.0	<30.4	30.4–32.4	<10	10–15	21.9
4.5	<1.2	1.2–1.4	<3.7	3.7–3.9	<32.7	32.7–34.7	<10	10–15	33.8
5.0	<1.2	1.2–1.4	<3.6	3.6–3.8	<35.1	35.1–37.1	<10	10–15	49.6
5.5	<1.2	1.2–1.4	<3.5	3.5–3.7	<37.4	37.4–41.7	<10	10–15	69.9
6.0	<1.2	1.2–1.4	<3.3	3.3–3.5	<39.7	39.7–41.7	<10	10–15	95.1
6.5	<1.2	1.2–1.4	<3.1	3.1–3.3	<42.0	42.0–44.0	<10	10–15	125.8
7.0	<1.2	1.2–1.4	<2.9	2.9–3.1	<44.3	44.3–46.3	<10	10–15	162.6

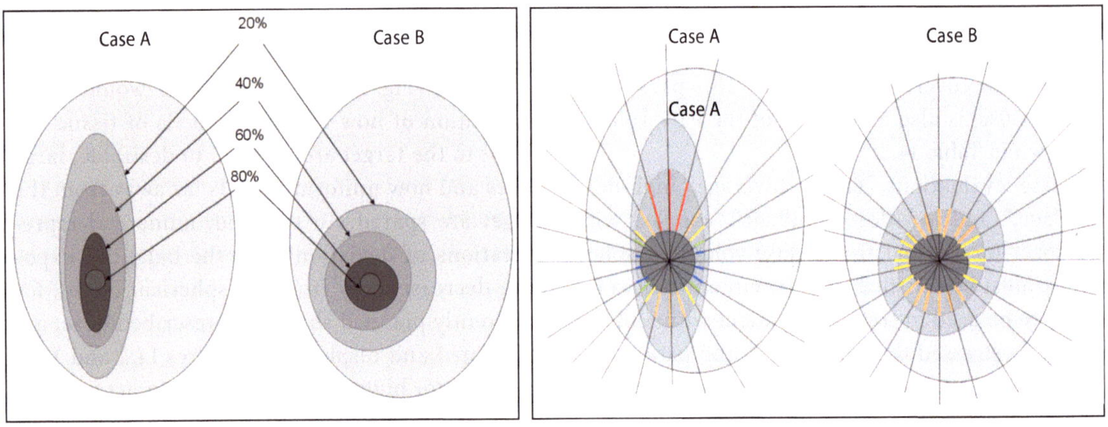

Fig. 14.1a,b. Illustration of the inadequacy of DVH as a precise measure of dose allocation in the sensitive organ. **a** Two large ovals (cases A and B) each represent a lung of the same volume. In each lung, identical targets are located in the same position relative to the lung volume. Targets in both lungs are irradiated and isodose lines illustrating the same levels of irradiation (80%, 60%, 40%, and 20%, respectively) are displayed as two sequences of converging ellipses. The areas enclosed by ellipses of the same level are equal. The DVH determining volumes of lung exposed to 80%, 60%, 40% and 20%, respectively, are identical in both cases even though the quality of dose distributions in both cases are diverse. In particular, if the isotropic decrease of dose away from the target is most desirable from the therapeutic point of view, the dose distribution in case B is superior to dose distribution in case A. **b** The dose distribution for two lungs (cases A and B of panel **a**). The sequence of radial lines diverging from the center of each target are drawn and distances along these radial lines measuring separation between isolines of 40% and 60% are shown. The variation of distances is clearly larger for case A, where greater anisotropy of dose distribution within the lung is visible, than in case B.

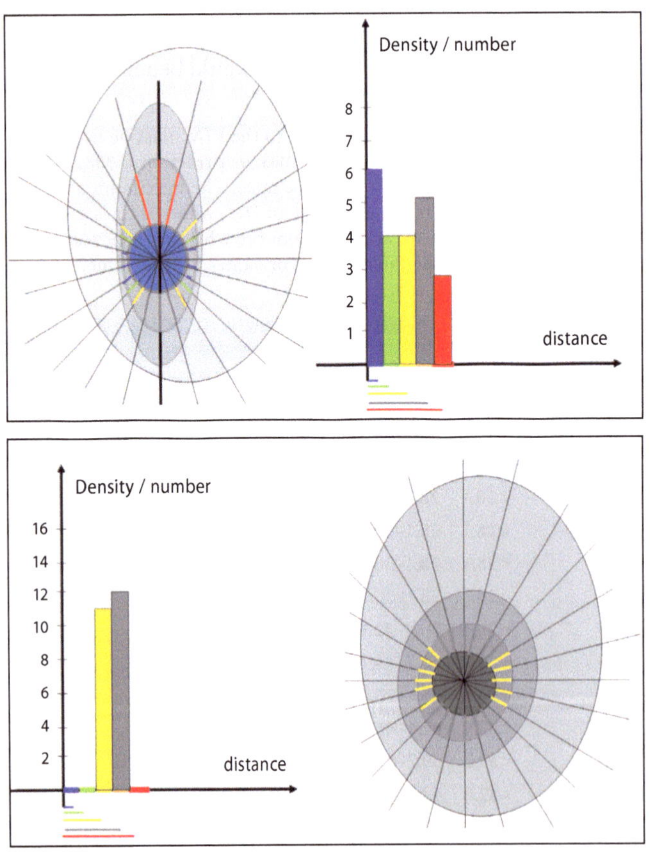

Fig. 14.2a,b. Quantification of the distance distribution between isolines of 40% and 60% created in Figure 14.1b. The graph illustrates the correlation between the spread of distances between these isolines and dose distribution symmetry. **a** The figure shows the distribution of distances for case A of lung irradiation. Notice the almost uniform histogram for distances between isolines of 40% and 60%. This indicates the asymmetric behavior of the dose distribution around the target. **b** The figure shows the distribution of distances for case B of lung irradiation. Notice the concentration of the number of distances between isolines of 40% and 60% around the given length. This indicates the symmetric behavior of the dose gradient away from the target.

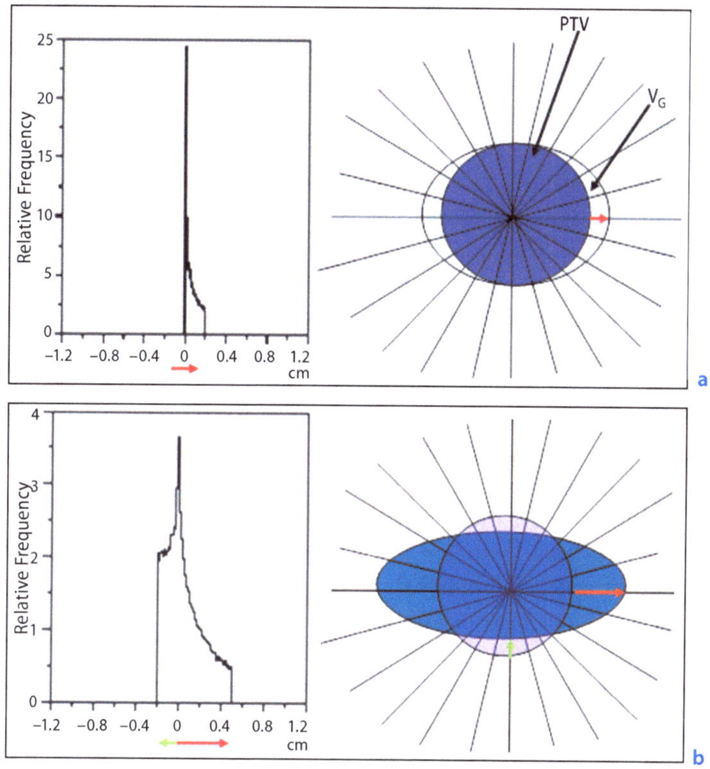

Fig. 14.3. **a** The DAH function for spherical PTV with radius 1 cm and ellipsoid surface of the volume defined by the prescribed dose. This volume has the same center as PTV and its principal axes have lengths of 2.0 cm, 2.0 cm and 2.4 cm. The graph displays compressed information about the level of conformity between PTV and prescribed dose volume. In the case under consideration, the density function of distances between the surface of the prescribed dose and the surface of the PTV is a decreasing function (i.e. its values decrease with increasing distance x between surfaces) that falls abruptly to zero when the value of x reaches 0.2 cm. **b** The DAH function for spherical PTV with radius 1 cm and ellipsoid surface of the volume defined by the prescribed dose (with the same center as PTV and principal axes equal to 1.6 cm, 1.6 cm and 4.0 cm). The graph displays compressed information about the level of conformity between PTV and the volume defined by prescribed dose. In the case under consideration, the density function of distances between the surface defined by the prescribed isodose and the surface of PTV is an increasing function on the interval (-0.2, 0) and is a decreasing function on the interval (0, 0.5). The shape of this function indicates that part of the PTV is outside of the volume defined by the prescribed dose and also that part of the volume defined by prescribed dose is located outside of the PTV. Displayed properties of DAH function indicate that the ideal conformity between two structures/volumes is achieved when density of DAH is concentrated at distance x=0 (i.e., DAH is a Dirac's delta function). Moreover, the distribution of densities that DAH displays over x axis (axis of distances between structures) over points x away from 0 provides a simple measure of departure from the ideal conformity between two structures/volumes.

ment (Fig. 14.1). Thus, it is logical to expect that tissue in shells far removed from the target should be exposed to low doses only.

14.4 Treatment Planning for SBRT

As we have stressed before, the most important factor responsible for limiting the toxicity of SBRT treatment is the minimization of the volume of tissue exposed to high dose levels and localization of this volume inside, or in close proximity to, the target. Thus, it is important to keep beam margins as small as possible when transforming GTV (or CTV) into PTV [1–6, 11, 12, 17, 19, 20]. Nevertheless, these margins have to account, at a reasonable confidence level, for all motions of targeted tissue plus uncertainties inherent in patient positioning and error in the target recognition [21]. Radial extensions of margins for SBRT planning (i.e., lengths of intervals connecting GTV and PTV surfaces along a particular radius emanating from GTV center) are most commonly chosen to be equal to 0.5 cm along any radius pointing away from the target center in any direction in transverse planes and equal to 1.0 cm along all parallel lines perpendicular to the axial plane [2–6, 11]. Margins may need to be increased if equipment has not been critically evaluated or for patients whose target identification and/or target motion are difficult to assess. Margins may also be decreased for individual cases in accordance with the physician's judgment, provided care is taken to insure that set-up accuracy and the means of motion compensation justify shrinking the periphery of the PTV exposed to high dose.

A specific instance when the decrease in margins, particularly in the direction perpendicular to axial plane, may be considered is for treatments involving utilization of real-time tracking such as the Synchrony® (Accuray Incorporated, Sunnyvale, CA) respiratory tracking. Still, careful analysis and phantom measurement of moving target dosimetry is recommended before the decision is made to decrease the margin in the inferior/superior direction, particularly when frameless delivery is practiced. There are two potential sources of error in delivery relative to information provided by a priori data and feedback from treatment monitoring. The actual, not surrogate, motion-based information about target position is provided only intermittently during therapy. For cases of irregular patient breathing, the error in tracking may be not trivial and actually quite difficult to assess. The other source of error is related to dosimetry calculations. The homogeneous algorithms for dose calculations in lung tumors generally overestimate the dose at inferior and superior regions of the PTV. The heterogeneous dose calculations that correct only for density changes for the primary component of the beam lead generally to larger overestimation of doses in inferior/superior regions than homogeneous dose calculations. Thus, errors in both inaccurate tracking and inaccurate dose calculation would generally enhance each other and may lead to local failures if appropriately large margins in inferior/superior direction are not prescribed.

It is clear from a direct analysis of a cumulative contribution to error in SBRT delivery that typically used margins are not large enough to ensure that the prescribed dose will be delivered to the entire GTV with a probability of 1.0 [21]. This perceptible inadequacy of dose shaping in SBRT is resolved by the deliberate shaping of the dose in the vicinity of the target so that it compensates for the margin's failure. Following the conditions discussed earlier, dose in all directions away from PTV assures that the target volume is immersed in the shell of the high-dose region (70–90% of prescribed dose) with thickness of approximately 1 cm. In this manner, highly potent, though not exactly equal to the prescribed, dose levels are delivered to the target volume in SBRT treatments.

The other characteristic of SBRT dose distributions that challenged the classic standard of dose distribution quality is the allowance of dose inhomogeneity inside the target. In particular, the enlargement of the volume of tissue exposed to relatively high doses may seem inevitable when simultaneously increasing dose within the target to above the dose prescribed to the target's outline. This is generally the case, however, when the goal is to equalize the dose in close shells near the target [22]. As long as an acceptable minimum dose is delivered to all parts of the target, higher doses may be desirable if they can be manipulated to occur within the central core of the tumor target where hypoxic, radio-resistant cells may be more prevalent.

14.5 SBRT – Delivery Techniques

The three conditions for dose distributions are achieved for SBRT treatments by combining multiple beams aimed at different sections of the target. This can be realized with multiple beams of varying weights or through appropriate intensity shaping of a limited number of non-coplanar beams. The angular directions of beams in SBRT radiation delivery is the main determinant of the dose shape distribution around the target [3, 4, 22–24]. Rapid and uniform fall-off of dose in all directions from the tumor is the consequence of the non-coplanar method of stereotactic radiation delivery employed originally at Karolinska Hospital and followed later by many other centers that utilize the SBRT technique [6, 14, 25]. Generally, the volume defined by the intersection of all beams is only slightly larger than PTV itself and it conforms closely to the shape of the target. The resulting isodose surface defined by full dose contributions from all beams matches closely, up to the margin defined by penumbra effects, the shape of the PTV. The speed of dose decrease away from the tumor is defined under these circumstances primarily by the rate of decline in the number of beams intersecting each other in subsequent shells that enfold the target. In turn, the rate of decline in the number of beams intersecting in subsequent shells is correlated with the level of uniformity in the used beams' directional distribution. The above heuristic analysis indicates that shaping optimal dose distributions for SBRT treatments favor the use of a large number of non-coplanar beams or arcs. Practical considerations often limit the number of beam directions, particularly for standard gantry-based (non-robotic) linear accelerator treatment units. With a limited number of directions it is even more important to choose them so that they create the smallest possible volume of intersection. This usually involves finding beam directions, from among the set of those that are admissible, that are maximally separated in three dimensions [4, 22, 23, 26]. The problem of separation of directions leads to counterintuitive results even in situations where there are no constraints imposed on the selection of angles [22].

In real cases of SBRT irradiation, additional limitations are added to avoid tissues of sensitive organs and to satisfy mechanical restrictions imposed by the equipment [4, 22]. Implementing SBRT treatments with traditional linear accelerators limits the number of beams per treatment to around 10. The factor that is influential in deciding the minimum number of beam directions, per treatment, is the dose at the skin for close-to-skin targets. It is prudent to keep the dose at the skin to below 30% of the prescribed dose. For PTV located less than 3 cm from the skin (e.g., tumors attached to chest wall of emaciated patients) this requirement is difficult to satisfy unless a truly large number of beam directions can be utilized for treatment performed with a traditional gantry-mounted accelerator or, more naturally, with a robot-mounted accelerator.

For SBRT treatments in which a large number of beam directions are seen as advantageous for dose shaping, the most isotropic distribution of directions in space may serve as the first order approximation for the choice of beam placement. Calculations have been performed in the past that can guide the choice of these isotropic beam directions. Nevertheless, for gantry-based accelerator treatments, the desirable highly isotropic placement of beam directions must respect mechanical restrictions imposed by the equipment (collisions of couch and gantry, etc.). These restrictions are considerably reduced for robotic-type accelerator treatments.

14.6 The Impact of Tissue Density Inhomogeneities in SBRT

One of the principal challenges for lung treatments is the proper quantitative evaluation of dose in the target and surrounding organs. The difficulty arises from dramatic changes in tissue density for chest treatments. For lung treatments, beams first have to pass through chest wall, then penetrate low density lung tissue, and finally deliver energy to the solid tumor cells. The chest wall is a combination of muscle and bone. Muscle has electron density close to the electron density of water, and bone has electron density higher than water. Overall, the disturbance of dose caused by the chest wall is negligible [17, 27, 28] but dose disturbance occurring at the tumor site, caused by the low density of lung tissue, can be significant [14, 27–31]. The dose decrease at the surface of the target caused by the loss of electronic equilibrium is compensated by dose increase, caused by enhanced penetration of photons through the lower density lung tissue. These two influences, one increasing and the other decreasing the dose at the surface of the target approximately cancel each other. As a result, a similar dose to the target outer edge in homogeneous and inhomogeneous media may be expected so long as the dose is calculated and prescribed to the periphery of the target [1–6, 11, 32, 33].Since delivery of the correct dose to the periphery of the tumor is more important in SBRT than any particular shape of dose inside the target, one may observe that the uncorrected heterogeneity calculations of dose are suitable for most SBRT treatments. This observation is attractive from a clinical point of view as it allows consistent dose prescription and dose delivery for all treated cases based on homogeneous media dose calculations. Calculations and phantom measurements (including RPC lung phantom) of a number of lung SBRT treatment cases have been performed at our institution that substantiated these conclusions. Investigations included two treatment-planning systems (Render Plan 3D and cmS XIO Plan) as well as Monte Carlo simulations. It should be noted, however, that assumptions justifying cancellation of dose heterogeneity effects at the surface of PTV are not universally valid. For example, they are not austerely applicable to treatments of small targets, requiring the smallest CyberKnife® (Accuray Incorporated, Sunnyvale, CA) collimators.

References

1. Blomgren H, Lax I, Naslund I. Radiosurgery for tumors in the body. *International Journal of Radiation Oncology Biology Physics* 1997; 39:328–328.
2. Blomgren H, Lax I, Naslund I, *et al.* Stereotactic high dose fraction radiation therapy of extracranial tumors using an accelerator. Clinical experience of the first thirty-one patients. *Acta Oncologica* 1995; 34:861–870.
3. Lax I, Blomgren H, Naslund I, *et al.* Stereotactic radiotherapy of malignancies in the abdomen. Methodological aspects. *Acta Oncologica* 1994; 33:677–683.
4. Papiez L, Timmerman R, DesRosiers C, *et al.* Extracranial Stereotactic Radioablation: Physical Principles. *Acta Oncologica* 2003; 42:882–894.
5. Timmerman R, Papiez L, Suntharalingam M. Extracranial stereotactic radiation delivery: Expansion of technology beyond the brain. *Technology in Cancer Research & Treatment* 2003; 2:153–160.
6. Timmerman R, Papiez L, McGarry R, *et al.* Extracranial stereotactic radioablation – Results of a phase I study in medically inoperable stage I non-small cell lung cancer. *Chest* 2003; 124:1946–1955.
7. vantRiet A, Mak ACA, Moerland MA, *et al.* A conformation number to quantify the degree of conformality in brachytherapy and external beam irradiation: Application to the prostate. *International Journal of Radiation Oncology Biology Physics* 1997; 37:731–736.
8. Knoos T, Kristensen I, Nilsson P. Volumetric and dosimetric evaluation of radiation treatment plans: radiation conformity index. *International Journal of Radiation Oncology, Biology, Physics* 1998; 42:1169–1176.
9. Graham MV, Matthews JW, Harms WB, *et al.* 3-Dimensional Radiation Treatment Planning Study for Patients with Carcinoma of the Lung. *International Journal of Radiation Oncology Biology Physics* 1994; 29:1105–1117.
10. Graham MV, Purdy JA, Emami B, *et al.* Clinical dose-volume histogram analysis for pneumonitis after 3D treatment for non-small cell lung cancer (NSCLC). *International Journal of Radiation Oncology Biology Physics* 1999; 45:323–329.
11. Lax I. Target Dose Versus Extratarget Dose in Stereotaxic Radiosurgery. *Acta Oncologica* 1993; 32:453–457.
12. Uematsu M, Shioda A, Tahara K, *et al.* Focal, high dose, and fractionated modified stereotactic radiation therapy for lung carcinoma patients – A preliminary experience. *Cancer* 1998; 82:1062–1070.
13. Wulf J, Hadinger U, Oppitz U, *et al.* Stereotactic radiotherapy of extracranial targets: CT-simulation and accuracy of treatment in the stereotactic body frame. *Radiotherapy & Oncology* 2000; 57:225–236.
14. Wulf J, Hadinger U, Oppitz U, *et al.* Stereotactic radiotherapy of targets in the lung and liver. *Strahlentherapie Und Onkologie* 2001; 177:645–655.
15. Nagata Y, Negoro Y, Aoki T, *et al.* Clinical outcomes of 3D conformal hypofractionated single high-dose radiotherapy for one or two lung tumors using a stereotactic body frame. *International Journal of Radiation Oncology Biology Physics* 2002; 52:1041–1046.
16. Papiez L, Moskvin V, Timmerman R. Dose allocation function as a tool for evaluation of SBRT treatments. *Medical Physics* 2004; 31:1824–1824.
17. Hadinger U, Thiele W, Wulf J. Extracranial stereotactic radiotherapy: evaluation of PTV coverage and dose conformity. *Zeitschrift fur Medizinische Physik* 2002; 12:221–229.
18. Giraud P, Antoine M, Larrouy A, *et al.* Evaluation of microscopic tumor extension in non-small-cell lung cancer for three-dimensional conformal radiotherapy planning.[comment]. *International Journal of Radiation Oncology, Biology, Physics* 2000; 48:1015–1024.
19. Herfarth KK, Debus J, Lohr F, *et al.* Extracranial stereotactic radiation therapy: Set-up accuracy of patients treated for liver metastases. *International Journal of Radiation Oncology Biology Physics* 2000; 46:329–335.
20. Harada T, Shirato H, Ogura S, *et al.* Real-time tumor-tracking radiation therapy for lung carcinoma by the aid of insertion of a gold marker using bronchofiberscopy. *Cancer* 2002; 95:1720–1727.
21. Papiez L, Langer M. On probabilistically defined margins in radiation therapy. *Physics in Medicine & Biology* 2006; 51:3921–3939.
22. Papiez L, Lu XY, Langer M. On the isotropic distribution of beam directions. *Mathematical Models & Methods in Applied Sciences* 2000; 10:991–1000.
23. Papiez L. On the equivalence of rotational and concentric therapy. *Physics in Medicine & Biology* 2000; 45:399–409.
24. Papiez L, Ringor M. Implications of a reconstruction formula for rotational therapy in treatment planning optimization. *Inverse Problems* 1997; 13:1519–1532.
25. Nagata Y, Negoro Y, Mizowaki T, *et al.* Clinical outcome of 3D conformal radiotherapy for solitary lung cancer using a stereotactic body frame. *Radiology* 2000; 217:140–141.
26. Webb S. The problem of isotropically orienting N converging vectors in space with application to radiotherapy planning. *Physics in Medicine & Biology* 1995; 40:945–954.
27. Butson MJ, Elferink R, Cheung T, *et al.* Verification of lung dose in an anthropomorphic phantom calculated by the collapsed cone convolution method. *Physics in Medicine and Biology* 2000; 45:N143–N149.
28. Engelsman M, Damen EMF, Koken PW, *et al.* Impact of simple tissue inhomogeneity correction algorithms on conformal radiotherapy of lung tumours. *Radiotherapy and Oncology* 2001; 60:299–309.
29. Saitoh H, Fujisaki T, Sakai R, *et al.* Dose distribution of narrow beam irradiation for small lung tumor. *International Journal of Radiation Oncology Biology Physics* 2002; 53:1380–1387.
30. Ahnesjo A. Collapsed Cone Convolution of Radiant Energy for Photon Dose Calculation in Heterogeneous Media. *Medical Physics* 1989; 16:577–592.
31. Butts JR, Foster AE. Comparison of commercially available three-dimensional treatment planning algorithms for monitor unit calculations in the presence of heterogeneities. *Journal of Applied Clinical Medical Physics* 2001; 2:32–41.
32. McGarry RC, Papiez L, Williams M, *et al.* Stereotactic body radiation therapy of early-stage non-small-cell lung carcinoma: Phase I study. *International Journal of Radiation Oncology Biology Physics* 2005; 63:1010–1015.
33. Timmerman R, McGarry R, Papiez L, *et al.* Initial report of a prospective phase II trial of stereotactic body radiation therapy for patients with medically inoperable stage I non-small cell lung cancer. *International Journal of Radiation Oncology Biology Physics* 2005; 63:S99–S99.

Selection of Treatment for Patients with Early Stage Non-Small Cell Lung Cancer in a Multidisciplinary Thoracic Oncology Program

15

Stuart M. Berman, Mark Huberman, Armin Ernst, David Feller-Kopman, David H. Roberts, Phillip Boiselle, J. Anthony Parker, Danielle McDonald, Nancy Rumplik, Elena A. Nedea, Simon Ashiku, Sidhu P. Gangadharan, Michael Goldstein, Susan Schumer, Sanjay R. Jain, Darren Brennan, Michael S. Buff, Robert L. Thurer, and Malcolm DeCamp

CONTENTS

15.1 Introduction

Lung cancer is the leading cause of cancer mortality in men and women. More than 170,000 Americans receive a diagnosis of lung cancer annually and the majority of them die of the disease [1]. There is considerable interest in improving treatment for lung cancer given its high impact on society. More than 80% of patients with lung cancer have non-small cell carcinoma (NSCLC). Surgery is the mainstay of treatment for patients with early stage NSCLC and results in cure of about 60-80% of patients with stage I (T1-2N0) disease [2–5].

Since the majority of patients with lung cancer are current or former smokers, a substantial number are not candidates for surgery because of significant respiratory or cardiac compromise. Patients may additionally have other medical problems such as advanced diabetes mellitus or severe peripheral vascular disease which may render them high risk candidates for resection of early stage lung cancer. A number of treatment options may be suitable for patients who are not candidates for lobectomy. These include more limited resections such as segmentectomy or wedge resection, but these operations are associated with higher rates of local recurrence and poorer survival than lobectomy [6]. Minimally invasive approaches such as thoracoscopic wedge resections may extend the option of surgery to some marginally fit patients, but many patients remain poor candidates even for these procedures.

Primary radiation therapy is a standard treatment for patients with early stage NSCLC who are medically inoperable. Conventional fractionated radiotherapy to doses of 60–66 Gy in 2-Gy fractions is associated with a high rate of local relapse in the range of 55–70% and survival at 5 years of 10–30% [7, 8]. Conventional radiation therapy may also result in irradiation of a clinically significant volume of normal lung in patients with little pulmonary reserve even when only the gross primary tumor is targeted. These poor results with conventional radiation therapy have led investigators to seek other approaches for management of medically inoperable patients with early stage NSCLC.

Stereotactic radiation therapy allows substantial reductions of treatment volumes, more precise targeting of tumors, hypofractionation, higher biologically effective doses, and marked reduction of overall treatment time compared to conventional radiation therapy. Timmerman et al. reported results from a Phase II study of stereotactic radiosurgery for T1-2N0M0 NSCLC in 70 medically inoperable patients [9]. Doses of 60–66 Gy were delivered in three fractions over 1–2 weeks. Kaplan-Meier local control at 2 years was 95% and 2-year overall survival was 54.7%. Patients with perihilar/central tumors had an 11-fold increased risk of severe toxicity compared with more peripheral tumor locations; 2-year freedom from severe toxicity was only 54% for the perihilar/central group. Onishi et al. reported a retrospective analysis of 245 patients from several Japanese institutions treated with hypofractionated high-dose stereotactic irradiation using varying dosing schemes [10]. Patients included those with both central and peripheral lung tumors. Local disease recurrence was 13.5% for the whole group with a rate of 8.1% for biologically effective dose (BED) ≥ 100 Gy compared with 26.4% for BED < 100 Gy ($p < 0.01$). Overall 5-year survival was 47% and cause-specific survival was 78%. Grade 2 radiation pneumonitis occurred in 4.1% of patients; Grade 3 or 4 radiation pneumonitis developed in 2.4%, and Grade 2 or 3 esophagitis was seen in 2.0% of patients. Chronic segmental bronchitis and wall thickening causing atelectasis of the peripheral lung was observed in only one patient (0.4%). Other investigators have reported good local control of early stage NSCLC in medically inoperable patients treated with stereotactic radiosurgery [11, 12].

The CyberKnife® (Accuray Incorporated, Sunnyvale, CA) is an image-guided robotic radiosurgery system which precisely tracks tumor motion during treatment. This allows reduction of the amount by which the gross tumor volume (GTV) is expanded to create the planning target volume (PTV) and therefore may further reduce the volume of normal lung irradiated compared to other radiosurgery systems. Treatment with CyberKnife currently requires implantation of radio-opaque fiducial markers into the lung tumor to serve as reference points for tumor tracking. The fiducial markers, which are tiny gold seeds, are usually implanted by percutaneous or bronchoscopic approaches. Preliminary reports suggest a good level of local control of early stage NSCLC with CyberKnife [13, 14]. These and other results are discussed in several chapters in this section.

Radiofrequency ablation (RFA) has been used successfully in the treatment of primary and metastatic hepatic tumors [15]. Goldberg et al. in 1995 demonstrated in experiments on rabbits that radiofrequency tissue ablation could be safely performed in lung parenchyma with a percutaneous approach and that tissue response to thermal injury was predictable and could be monitored with CT scanning [16]. Recently, several investigators have reported results using RFA for treatment of primary lung tumors. These series describe a heterogeneous patient population including patients with early stage lung cancer who are medically inoperable, operable patients who have refused surgery, patients with locally advanced lung cancer, and patients with metastases to lung. Fernando et al. reported a series of 18 patients with 21 peripheral non-small cell lung tumors treated with RFA; 9 patients had stage I disease and the remaining 9 had stage II through IV NSCLC [17]. Median tumor diameter was 2.8 cm. Generally a tissue ablation extending 0.5 to 1 cm beyond the tumor was sought. CT scans and PET scans were performed at 4 to 6 weeks following RFA and then at 3-month intervals; modified RECIST criteria were used to assess response [18]. At a median follow-up of 14 months, 15 patients were alive (83.3%); local progression occurred in 8 nodules (38.1%) in 6 patients (33%). Because of a case of fatal hemoptysis occurring 21 days after RFA of a central tumor metastasis included in a prior report, the authors advised against CT-guided RFA of central lesions close to the hilum.

Hiraki et al. reported a large series from Japan of 128 patients with 342 tumors treated with RFA (25 primary lung tumors in 24 patients and 317 metastatic lung neoplasms in 104 patients) [19]. All patients were unsuitable for surgical resection or refused to undergo resection. Mean tumor diameter was 1.7 cm (range, 0.3–9.4 cm). Median follow-up was 12 months (range, 6–47 months). Tumor responses were assessed with chest CT at 1, 3, 6, 9, and 12 months and thereafter at 6-month intervals whenever possible; no patients had PET scans. Local progression was observed in 94 (27%) tumors at a median of 7 months after the first ablation session. The overall primary technique effectiveness rates were 72% at 1 year, 60% at 2 years, and 58% at 3 years. Multivariate multilevel analysis for all 342 tumors showed that larger tumor size (hazard ratio = 1.97; 95% CI = 1.74-2.65; $p < 0.00001$) was an independent risk factor for local progression. Other investigators have reported that radiographic response rates after RFA are significantly lower for larger lung tumors. Akeboshi et al. described a significant difference in the rate of complete tumor necrosis, as determined by disappearance of FDG uptake on PET images and resolution of tumor enhancement on CT images, between tumors 3 cm or less and tumors larger than

3 cm (69% vs. 39%; p=0.04) [20]. Lee et al. reported complete necrosis, defined as complete resolution of enhancement on CT, in all six lung tumors (100%) smaller than 3 cm but in only 6 (23%) of 26 larger tumors ($p < 0.05$) treated with RFA [21]. Investigators at Rhode Island Hospital combined thermal ablation (37 RFA, 4 microwave ablation) with conventional fractionated radiation therapy (n=27) or brachytherapy (n=14) in the treatment of 41 patients with inoperable stage I or II NSCLC [22]. Median follow-up was 19.5 months. Local recurrence occurred in 11.8% of tumors smaller than 3 cm and in 33% of larger tumors (p=0.03). Overall survival rates were 70.4% at 2 years and 57.1% at 3 years. Outcomes in the brachytherapy and external beam radiation therapy groups did not differ significantly.

Having explored several options for treatment of lung cancer in medically inoperable patients, a question which arises is how does a physician choose which one is appropriate for a particular patient? Even assessing whether a patient is, in fact, medically inoperable depends on what type of operative intervention is planned. Certainly there are patients who are fit enough for a thoracoscopic wedge resection who are not fit enough for a pneumonectomy. The choice of treatment should depend little on which specialist the patient is initially referred to or which physician the patient sees first. For most patients with lung cancer, it is our experience that the best decisions are made when evaluation takes place in a multidisciplinary thoracic oncology program in which pulmonary physicians, thoracic surgeons, medical oncologists, radiation oncologists, radiologists, nuclear medicine physicians, interventional pulmonologists, and nurses collaborate to review the imaging, discuss the case, and arrive at a consensus recommendation to present to the patient.

15.2 The Multidisciplinary Clinic

The multidisciplinary clinic offers many advantages for both patient and physician. For patients, the alternative is to independently consult several physicians and try to comprehend and resolve recommendations which may differ in minor or significant ways [23]. Even the most well-informed patients rarely have the knowledge to make good decisions about subtle differences in treatment options and are usually eager for their physicians to talk to each other and to offer a consensus recommendation. Physician communication, which may otherwise be a time-consuming process, is greatly facilitated by the multidisciplinary clinic setting. Each oncologist knows best the specific benefits and risks of treatments within his own discipline for a particular patient. By discussing the case together, particularly in the setting of a thorough review of the imaging studies by radiologists and nuclear medicine physicians, a consensus regarding further diagnostic work-up and treatment can usually be achieved. This mutual decision-making process provides reassurance for physicians undertaking challenging cases that the treatment plan they will pursue has the backing of other experts. Physicians who participate in multidisciplinary clinics, where robust discussion and debate is encouraged, also learn a great deal from colleagues outside their specialty by understanding characteristics of a particular clinical scenario which other specialists consider important in recommending a specific approach. Other specialists are a resource rather than a limitation to an individual physician's freedom.

The impact of management by a multidisciplinary team on treatment and survival of patients with cancer has been studied. Forrest et al. examined survival of 243 patients with inoperable NSCLC (stages IIIA, IIIB, and IV) in 1997 and 2001, before and after the introduction of a multidisciplinary team in 1998 at the Royal Infirmary, Glasgow [24]. The multidisciplinary team consisted of two respiratory physicians, two surgeons, a medical oncologist, a clinical oncologist, a palliative care physician, a radiologist, and a specialist respiratory nurse. Median survival was significantly higher in patients treated in 2001 compared with 1997 (6.6 vs 3.2 months, $p < 0.001$). Introduction of the multidisciplinary team was associated with a change in the treatment of patients with inoperable NSCLC; in particular, more patients received chemotherapy and fewer patients received palliative care only. Junor et al. in 1994 reported a retrospective study of all 533 cases of ovarian cancer registered in Scotland in 1987 [25]. After adjustment for age, stage, pathology, degree of differentiation and presence of ascites, survival improved when patients were referred to a multidisciplinary team at a joint clinic ($p < 0.001$) and this was not solely due to the prescription of platinum chemotherapy.

Development of an effective multidisciplinary thoracic oncology program takes time and effort. Boyle et al. [26] and Hall and Weaver [27] note that multidisciplinary teams are not fully formed at their first meeting but develop through a series of stages; these include the *forming* stage where individuals assemble and the goals and objectives for the team are defined, the *storming* stage where the individuals and group react affectively to the requirements of the task and to interpersonal conflicts, the *norming* stage where effective cooperation between team members occurs through communication of ideas, opinions and information and where functional roles are defined, and finally, the *performing* stage where the team is effective and efficient in delivering outcomes [26, 27]. Leadership is important in a multidisciplinary cancer program and the leader must be able to identify different stages of the team's development and implement appropriate leadership approaches [27]. Kagan, however, cautions that clinics that are "run" by one dominating physician often stagnate because differing opinions are censored [23]. Practicing oncology as part of a well-functioning team may increase job satisfaction and reduce the risk of professional burnout, whereas poor team function increases stress [26].

15.2.1 Evaluation Procedure

At Beth Israel Deaconess Medical Center (BIDMC) in Boston, one of the major teaching hospitals of Harvard Medical School, we have a strong multidisciplinary thoracic oncology program with a very high level of physician collaboration and patient satisfaction. What follows is a description of how a patient with suspected clinical stage I NSCLC who is a marginal surgical candidate might be evaluated through our clinic and how treatment would be selected. The first step in the process is obtaining medical records and imaging studies from the referring physician and institution. A nurse practitioner with substantial experience in thoracic oncology coordinates intake of new patients, and with support from administrative staff, schedules initial appointments with a thoracic surgeon and radiation oncologist and sends imaging studies for review by our thoracic radiologists and nuclear medicine physicians. In patients who are referred from locations some distance away from the Boston area, specifically for consideration of CyberKnife treatment, one of the radiation oncologists trained in CyberKnife radiosurgery reviews the records and images to ascertain whether it is worthwhile for the patient to come to Boston to be evaluated at BIDMC. In cases where it is clear from this preliminary review that CyberKnife radiosurgery would not be appropriate, either the radiation oncologist or CyberKnife nurse coordinator calls the patient to tell them that CyberKnife would not be an option so that a trip to Boston is not made unnecessarily. We describe the process of multidisciplinary evaluation at BIDMC and some patients wish to be seen for a second opinion regarding management of their known or suspected lung cancer even if CyberKnife will not be an option.

The next step in the evaluation process is presentation of the case at our multidisciplinary thoracic oncology conference. This conference, which meets for 90 minutes every Thursday (except Thanksgiving) is an essential part of the program. Each case reviewed at the conference is briefly presented by a resident, fellow, or attending physician who has either seen the patient or reviewed the medical record. The relevant imaging studies are then presented on a large high-resolution display by one of the thoracic radiologists and/or nuclear medicine physicians. Scans from outside institutions are loaded into the PACS system, if possible, to facilitate presentation at the conference and future review. Generally about 40 staff physicians, residents, fellows, and nurses attend the conference; seats are reserved for the attending physicians who are the core members of the team because actual recommendations are formulated at this conference. Core team members presenting at the conference include thoracic surgeons, radiation oncologists, medical oncologists, pulmonary physicians, thoracic radiologists, nuclear medicine physicians, interventional pulmonologists, nurses, and physician assistants. After discussion of the imaging findings, there is a discussion about further diagnostic work-up and treatment. For the example of a patient with suspected clinical stage I non-small lung cancer, if CT scan and PET-CT support clinical stage I NSCLC, team members would discuss factors affecting fitness for the standard surgical procedure, lobectomy and mediastinal lymph node dissection. These factors include pulmonary function, cardiovascular fitness, weight loss, performance status, and nutritional status. If, based on the data presented at the conference, the patient appeared to be medically fit for lobectomy, a preliminary agreement among the team for lobectomy as the preferred treatment would be made at the conference. This recommendation would be finalized after the surgeon and radiation oncologist had seen and examined the patient that day and discussed the proposed treatment with the patient. Generally both surgeon and radiation oncologist try to spend at least some time simultaneously in the room with the patient so that the patient is reassured that there has been good communication between surgeon and radiation oncologist and that the recommendation represents the consensus of the whole multidisciplinary team.

In a patient with known or suspected stage I NSCLC seen in the multidisciplinary thoracic oncology clinic who on preliminary review is not clearly medically fit for lobectomy, further evaluation is recommended to assess fitness for surgery. This may include full pulmonary function tests, exercise oximetry, formal cardiopulmonary exercise testing measuring peak oxygen consumption, echocardiography, and other studies. Decision trees and format for this analysis are described elsewhere [28]. If this evaluative process suggests that the patient is not a good candidate for lobectomy, other surgical options such as segmentectomy or wedge resection, possibly done thoracoscopically, are discussed at the conference. If the patient is felt from both an anatomic and medical fitness perspective to be a satisfactory candidate for one of these compromise operations, often the consensus recommendation will be for limited resection. Although the local recurrence rate is higher and 5-year survival probably worse after limited resection than after lobectomy [6], local tumor control is 70–100% for T1N0 lesions treated with segmentectomy or wedge resection [6, 29, 30], and this probably compares favorably with data from series of patients treated with stereotactic radiosurgery or RFA. Given the relatively recent introduction of both stereotactic radiosurgery and RFA for treatment of lung tumors, there are no long-term outcome data available for these methods. For this reason, for patients who are satisfactory candidates for limited resection, our multidisciplinary team has generally favored this approach over stereotactic radiosurgery or RFA until there is longer term follow-up data from series employing these non-operative methods.

When cardiorespiratory evaluation suggests that a patient with clinical stage I NSCLC is not medically fit even for a limited resection, alternatives including conventional fractionated radiation therapy, radiosurgery (with CyberKnife in our institution), and RFA are discussed in the multidisciplinary thoracic oncology conference. As mentioned earlier in this chapter, conventional fractionated radiation therapy in this setting is associated with a high rate of local relapse, poor survival, and irradiation of a significant volume of normal lung; treatment with either radiosurgery or RFA probably produces better outcomes with less toxicity than conventional radiation therapy so one of these more focal methods is generally preferred for most patients. The choice between radiosurgery and RFA can be difficult since there are no reports of direct comparisons, randomized or otherwise, between the two methods. Series of patients with lung tumors larger than 3 cm in diameter treated with RFA have shown rates of complete necrosis as low as 23–39% for these lesions [20, 21]. This suggests that RFA may not be an ideal treatment for larger lesions and we have favored radiosurgery for lesions greater than 3 cm. For smaller lesions, RFA is a more viable treatment and it is considered a reasonable option particularly for patients that would have difficulty lying down for the 1.5–2 hour sessions required for CyberKnife radiosurgery or for those where an expeditious single-visit treatment is of paramount importance. Radiosurgery with CyberKnife is an appropriate option for peripheral tumors, and for these lesions we generally employ doses of 18–20 Gy per fraction for three fractions over 4–9 days. Although Timmerman [9] described an 11-fold increased risk of severe toxicity for central/perihilar lesions compared to peripheral lesions treated with 20–22 Gy×3, Japanese investigators [10, 11] reported acceptable toxicity treating central lesions with 12 Gy×4. We have used this latter fractionation scheme for central lesions with no severe toxicity so far, but our median follow-up is only 7 months. The method of fiducial marker placement is also discussed in the multidisciplinary conference. For peripheral lesions, we have generally preferred percutaneous placement, which is performed by the interventional radiologists; for more central lesions our interventional pulmonologists have successfully placed fiducial markers by bronchoscopic or superDimensional bronchoscopic approaches.

For all patients evaluated in the multidisciplinary thoracic oncology program, we strive to develop a consensus recommendation that we can present to the patient; we try to avoid mixed messages from different members of the team. We strongly believe that patients should not suffer because of physicians' failure to communicate with each other; the multidisciplinary clinic and weekly conference certainly provide a forum for efficient communication. Follow-up care is also coordinated through the program to avoid duplication and gaps. Our goal is to give the best care possible and for patients to feel reassured that all of our resources are coordinated to optimize their care.

Another important mission of the multidisciplinary thoracic oncology program is teaching residents and fellows. The opportunity for trainees to participate in discussions with all the members of the program during conference and in the clinic is invaluable and gives a broad perspective on the management of thoracic neoplasms; learning in the setting of collegial interdisciplinary collaboration serves as a model for trainees to emulate in future practice.

15.3 Case Reports

In order to illustrate further the process of evaluation and treatment selection in the multidisciplinary thoracic oncology program at BIDMC the following two cases are presented.

The first patient is a 72-year-old female with chronic obstructive pulmonary disease (COPD) and a 25 pack-year smoking history. Chest X-ray showed a left upper lung (LUL) lesion. Chest CT in May of 2005 showed a 2.1×3.5 cm mass in the LUL and a 1.7×1.2 cm mass in the superior segment of the left lower lung (LLL). PET-CT in June showed abnormal FDG-avidity of both lesions. Bronchoscopy and transbronchial needle aspiration (TBNA) of the LUL lesion was performed in September of 2005 and pathology showed chronic inflammatory changes; acid-fast bacilli culture (AFB) was positive compatible with mycobacterium avium complex (MAC), but there was no carcinoma. Bronchoscopy in October with transbronchial biopsy, brushings, and bronchoalveolar lavage (BAL) of the LLL lung mass showed atypical cells highly suspicious for NSCLC; stains for pneumocystis and fungi were negative and acid-fast and fungal cultures were negative. The patient was started on treatment for MAC but she did not tolerate this well. Repeat bronchoscopy in January of 2006 at an outside hospital showed chronic inflammation from a LUL specimen and TBNA from the LLL showed non-small cell carcinoma with squamous differentiation. The patient was seen in the thoracic oncology clinic at BIDMC in March 2006. The thoracic surgeon felt the patient was a possible candidate for surgery and recommended further evaluation and exercise testing. PET-CT showed intense uptake in a 4.4×3.0 cm mass (increased in size from 3.6×2.2 cm five months earlier) in the superior segment of the LLL with a standardized uptake value (SUV) of 25.6. The lesion was contiguous with the left hilum and extended laterally to the pleural surface; there was a 2.6×2.4 cm spiculated nodule in the LUL with SUV of 4.6 which had increased in size only slightly from 2.5×2.3 cm over the same 5-month time period (Fig. 15.1). Pulmonary function tests in March 2006 showed a forced expiratory volume in 1 sec (FEV1) of 0.68 L or 51% of predicted; exercise testing showed a maximal oxygen uptake of 6.5 ml/kg/min (the test was felt to be a good-quality exam with good effort). Because of the severe reduction in oxygen uptake on the exercise testing with the result well below 10 ml/kg/min, the thoracic surgeon did not feel the patient was a candidate for surgery even if only a bi-segmentectomy were performed. We, therefore, mutually agreed on a plan of CyberKnife radiosurgery as the treatment most likely to be effective and to have a tolerable risk of morbidity. RFA was felt a less satisfactory alternative because lesion size was greater than 3 cm [20, 21]. Because of the relatively central location of the lesion and proximity to the LLL bronchus, our interventional pulmonary group recommended fiducial marker placement by superDimensional bronchoscopy. These were successfully placed in April 2006 (Fig. 15.2). The patient underwent Cy-

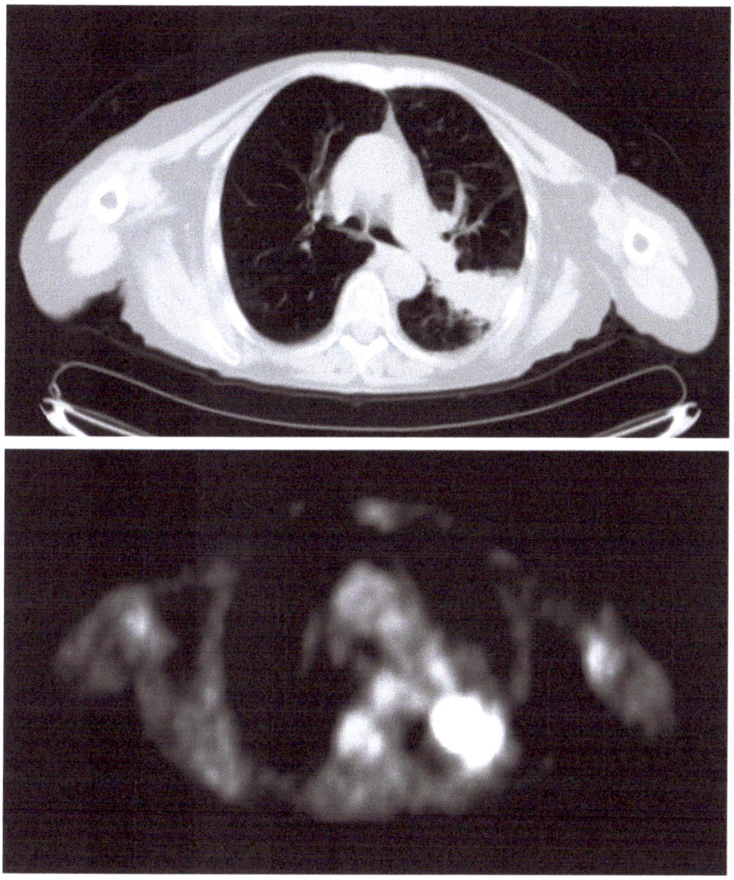

Fig. 15.1a,b. PET-CT prior to CyberKnife radiosurgery shows intense uptake in a 4.4×3.0 cm mass in the superior segment of the LLL with an SUV of 25.6; the lesion is contiguous with the L hilum and extends laterally to the pleural surface.

berKnife radiosurgery to the LLL NSCLC to a dose of 48 Gy in 4 fractions over 12 days (completed in mid-May 2006). This total dose and fractionation scheme was selected because of the proximity to the LLL bronchus and left hilum [9–11]. Further antibiotic therapy was recommended for the MAC infection under the care of an infectious disease specialist at an outside institution. The patient tolerated CyberKnife treatment well with only minimal increase in dyspnea and marked fatigue both of which were improving at follow-up one month after treatment. PET-CT performed three months after treatment (August 2006) showed that the known carcinoma in the LLL had significantly decreased in size with the solid component measuring only 1.3 cm and SUV decreased to 2.5 compared with 25.6 prior to treatment (Fig. 15.3). The LUL lesion, which was felt to represent infection, was stable in appearance with an SUV of 5.1. The only concerning finding on this PET-CT was a very anterior prevascular lymph node measuring 1.2×0.7 cm with SUV of 2.6; this was felt to be possibly related to the known infection in the LUL which had not been adequately treated. The patient was referred back to her infectious disease specialist to resume therapy for MAC. A follow-up PET-CT and visit in the thoracic clinic was scheduled for 6 months.

Our second illustrative case is a 68-year-old male with a history of tracheal stenosis who underwent a tracheal resection in 2003. He has multiple medical problems including coronary artery disease (CAD), diastolic congestive heart failure (CHF), COPD, pulmonary embolism, adrenal insufficiency, and a history of gastrointestinal bleeding. He has a pacemaker to treat complete heart block. He is wheelchair bound and has dyspnea with only minimal exertion. During CT angiography to rule out pulmonary embolism in May of 2005 he was

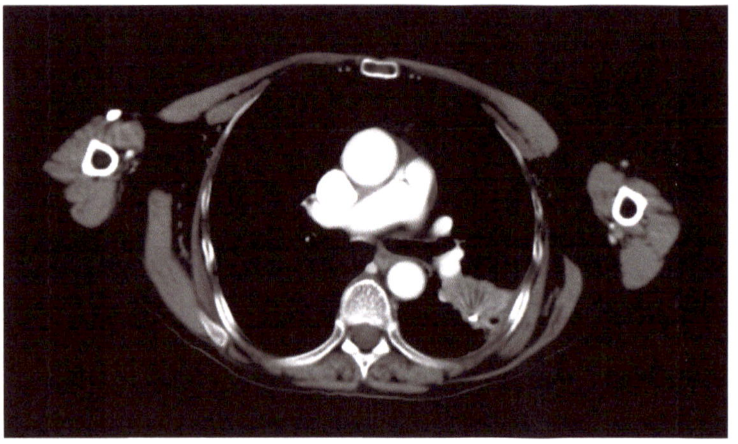

Fig. 15.2 CT scan at time of CyberKnife planning session shows one of the fiducial markers in the superior segment LLL mass; the fiducial markers were placed using superDimensional bronchoscopy.

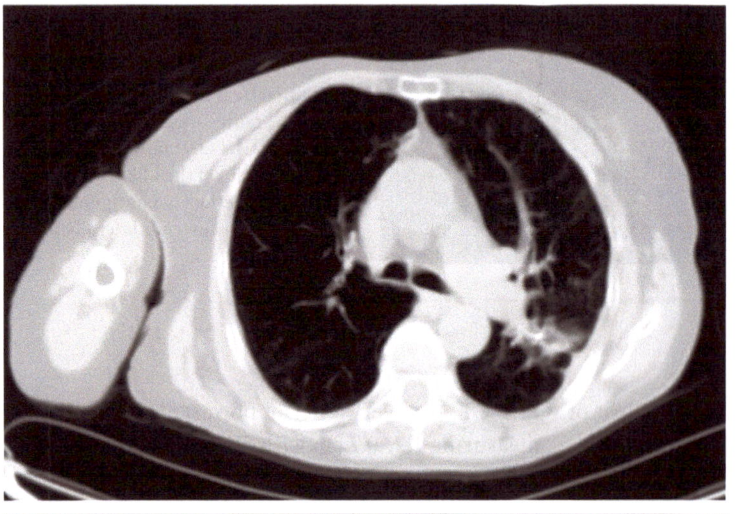

a

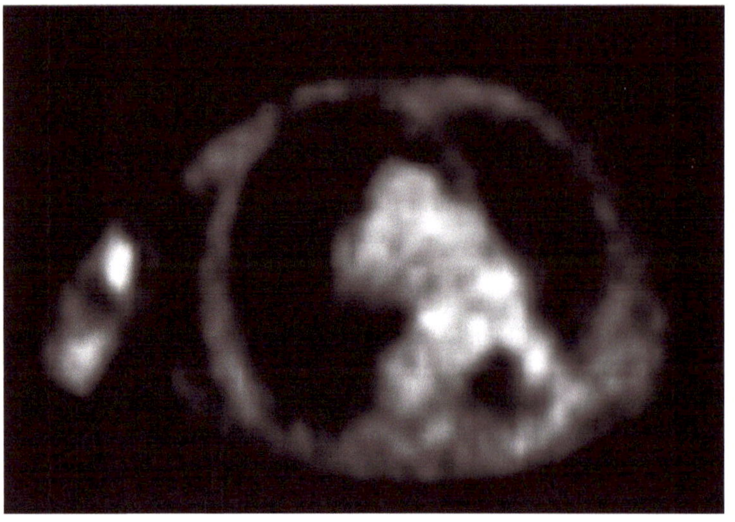

b

Fig. 15.3a,b. PET-CT three months after CyberKnife radiosurgery shows that the known carcinoma in the LLL has significantly decreased in size with the solid component measuring only 1.3 cm and SUV decreased to 2.5.

found to have a 1.3×1.0 cm nodular opacity in the superior segment of the LLL. PET-CT two months later showed the LLL lesion had an SUV of 3.6 and was felt to be suspicious for carcinoma vs. infection (Fig. 15.4); there was no other abnormal FDG-avidity. CT-guided percutaneous biopsy of the LLL lesion showed cytology positive for malignant cells compatible with poorly differentiated adenocarcinoma. The patient was evaluated in our multidisciplinary thoracic oncology clinic at BIDMC. Our thoracic surgeon did not think the patient was medically fit even for a thoracoscopic wedge resection because of his multiple medical problems and his poor general condition. The patient was strongly in favor of having treatment for his lung cancer. We discussed RFA and CyberKnife radiosurgery as possible treatments given the small size of the lesion. The consensus was that RFA would be a better treatment option than CyberKnife because (1) the patient did not think that he could lie down for the 1.5–2 hours required for the CyberKnife sessions, (2) RFA would require only a single treatment session, and (3) RFA would probably cause a smaller volume of lung injury than CyberKnife radiosurgery for this patient. In November of 2005 our interventional radiologist performed RFA to greater than 60°C using 1800 mA current for 12 minutes. A fiducial marker was placed at that time to allow CyberKnife radiosurgery if tumor persisted after RFA. The procedure was complicated by a small left pneumothorax treated with a chest tube. The patient also developed collapse of the right upper lung (RUL) adjacent to a large bleb which was

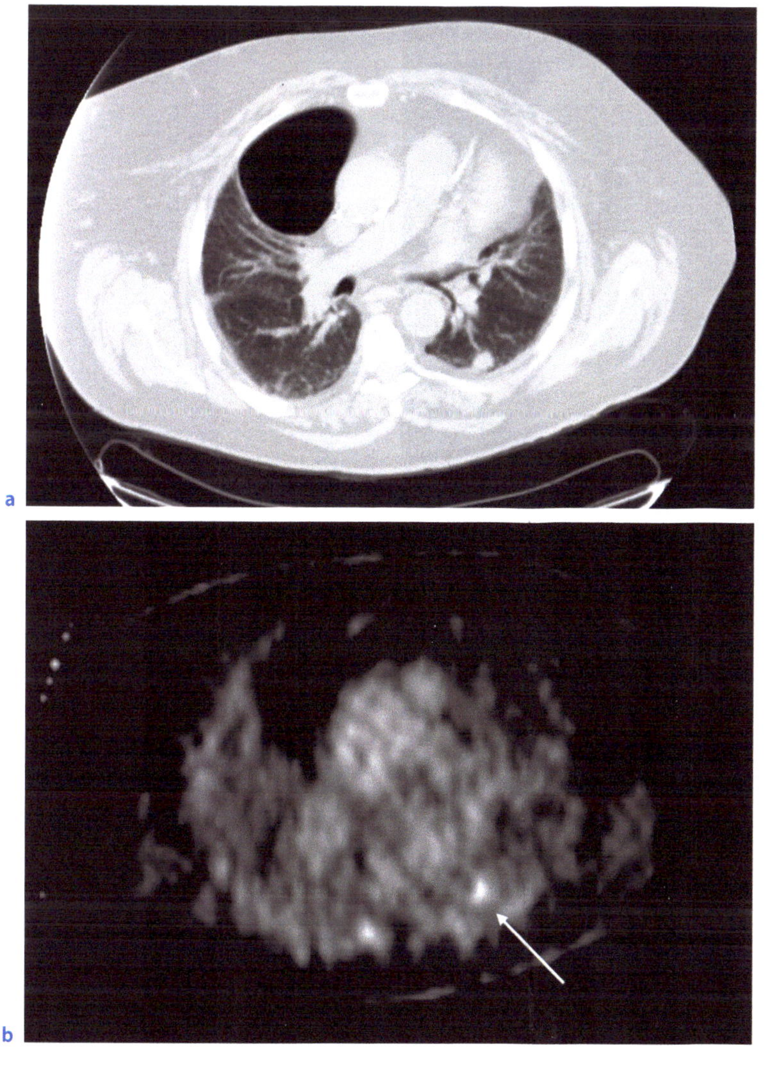

Fig. 15.4a,b. PET-CT prior to RFA shows the LLL nodule with SUV 3.6 which proved to be poorly differentiated adenocarcinoma. There is a large bleb anteriorly on the right.

felt to be secondary to aspiration or mucus plugging. He was admitted to intensive care (MICU) and intubated. He was hospitalized for 26 days and recovered further in a long-term care facility for three months. Follow-up PET-CT in late August of 2006 showed two FDG-avid foci in the LLL posteriorly; the more superior one was at the prior RFA site and had an SUV of 5.3 compared with SUV 4.3 prior to RFA (Fig. 15.5). Inferior and lateral to this was a new 2.0×1.4 cm lesion with an SUV of 8.3. Both lesions were felt to be indeterminate but were concerning for persistent or new carcinoma. There was no abnormal FDG-avidity in hilar or mediastinal lymph nodes or elsewhere. Unfortunately, the fiducial marker which had been placed at the time of RFA had migrated to the left lung base and could not be used for salvage treatment with CyberKnife. Neither biopsy nor implantation of new fiducial markers in the LLL lesions is considered a very safe option given the patient's protracted problems after the RFA. Conventional radiation therapy is not a good option because patient has no pulmonary reserve. We considered implanting fiducial markers into the chest wall adjacent to the lesion but these would not hold position in fat; there was profound chest-wall muscle wasting secondary to long-term steroid use for COPD. We will repeat a PET-CT after 4 months given that the lesions are indeterminate and may represent inflammation. Fiducial-less CyberKnife treatment with the new lung Xsight[TM] (Accuray Incorporated, Sunnyvale, CA) system may be an option if the follow-up PET-CT suggests persistent disease.

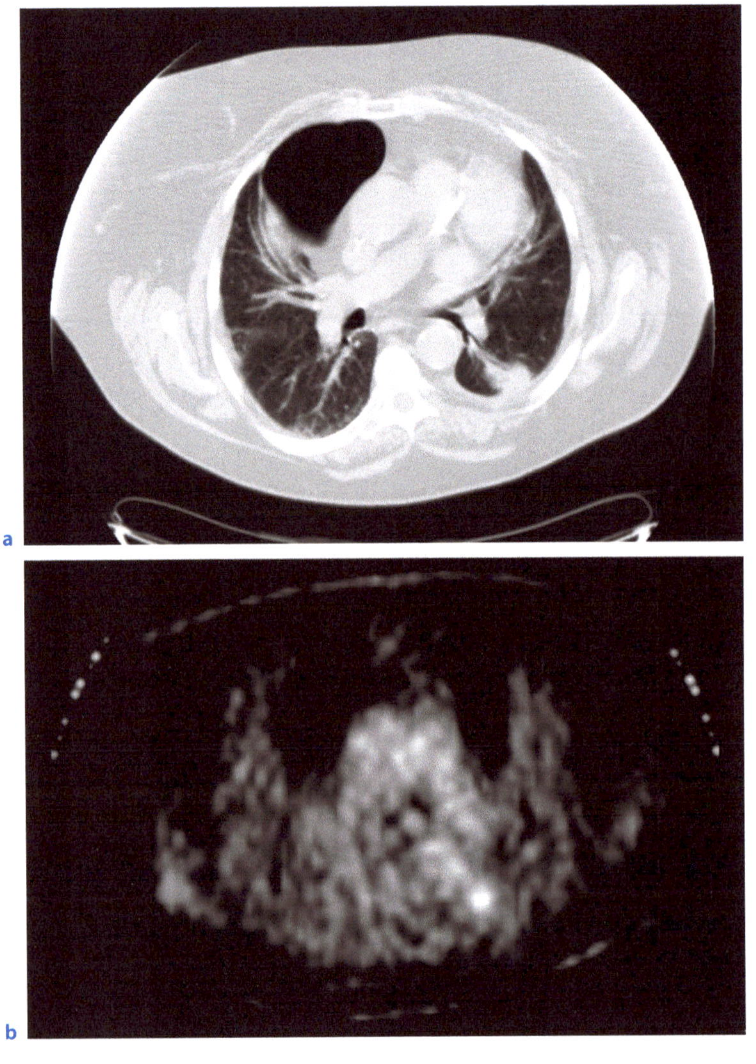

Fig. 15.5a,b. Follow-up PET-CT ten months after RFA shows an FDG-avid focus (SUV 5.3) in the LLL posteriorly at the prior RFA site.

15.4 Summary

We hope that these case presentations and the thoughts presented in this chapter illustrate the value to both patient and physician of a multidisciplinary thoracic oncology clinic and conference. We are confident that selection of treatment for patients with early stage NSCLC is best accomplished in this multidisciplinary setting where there is close collaboration and efficient communication among thoracic surgeons, radiation oncologists, medical oncologists, pulmonary physicians, thoracic radiologists, nuclear medicine physicians, interventional pulmonologists, nurses, and physician assistants. Each specialist is able to add his or her expertise to the development of a consensus plan. The multidisciplinary thoracic oncology program at Beth Israel Deaconess Medical Center also serves as the focal point for clinical and translational research protocol development and implementation. The program benefits patients through optimization of care and serves as an excellent educational resource for medical students, residents, fellows, and attending physicians. We are fortunate to enjoy the rewards of practicing medicine in a collaborative and supportive environment working alongside great colleagues from various disciplines.

References

1. Jemal A, Siegel R, Ward E, *et al.* Cancer statistics, 2006. *CA Cancer J Clin* 2006;56:106–130.
2. Dominioni L, Imperatori A, Rovera F, *et al.* Stage I nonsmall cell lung carcinoma: analysis of survival and implications for screening. *Cancer* 2000; 89:2334–2344.
3. Flehinger BJ, Kimmel M, Melamed MR. The effect of surgical treatment on survival from early lung cancer. Implications for screening. *Chest* 1992; 101:1013–1018.
4. Gharagozloo F, Tempesta B, Margolis M, *et al.* Video-assisted thoracic surgery lobectomy for stage I lung cancer. *Ann Thorac Surg* 2003; 76:1009–1014; discussion 1014–1005.
5. Mountain CF. Revisions in the International System for Staging Lung Cancer. *Chest* 1997; 111:1710–1717.
6. Ginsberg RJ, Rubinstein LV. Randomized trial of lobectomy versus limited resection for T1 N0 non-small cell lung cancer. Lung Cancer Study Group. *Ann Thorac Surg* 1995; 60:615–622; discussion 622–613.
7. Kaskowitz L, Graham MV, Emami B, *et al.* Radiation therapy alone for stage I non-small cell lung cancer. *Int J Radiat Oncol Biol Phys* 1993; 27:517–523.
8. Sibley GS, Jamieson TA, Marks LB, *et al.* Radiotherapy alone for medically inoperable stage I non-small-cell lung cancer: the Duke experience. *Int J Radiat Oncol Biol Phys* 1998; 40:149–154.
9. Timmerman R, McGarry R, Yiannoutsos C, *et al.* Excessive toxicity when treating central tumors in a phase II study of stereotactic body radiation therapy for medically inoperable early-stage lung cancer. *J Clin Oncol* 2006; 24:4833–4839.
10. Onishi H, Araki T, Shirato H, *et al.* Stereotactic hypofractionated high-dose irradiation for stage I nonsmall cell lung carcinoma: clinical outcomes in 245 subjects in a Japanese multiinstitutional study. *Cancer* 2004; 101:1623–1631.
11. Nagata Y, Takayama K, Matsuo Y, *et al.* Clinical outcomes of a phase I/II study of 48 Gy of stereotactic body radiotherapy in 4 fractions for primary lung cancer using a stereotactic body frame. *Int J Radiat Oncol Biol Phys* 2005; 63:1427–1431.
12. Wulf J, Baier K, Mueller G, *et al.* Dose-response in stereotactic irradiation of lung tumors. *Radiother Oncol* 2005; 77:83–87.
13. Le Q-T, Loo BW, Ho A, *et al.* Results of a phase I dose-escalation study using single-fraction stereotactic radiotherapy for lung tumors. *Journal of Thoracic Oncology • Volume , Number ,* 2006; 1:802–809.
14. Nuyttens JJ, Prevost JB, Praag J, *et al.* Lung tumor tracking during stereotactic radiotherapy treatment with the CyberKnife: Marker placement and early results. *Acta Oncol* 2006; 45:961–965.
15. Curley SA, Izzo F, Delrio P, *et al.* Radiofrequency ablation of unresectable primary and metastatic hepatic malignancies: results in 123 patients. *Ann Surg* 1999; 230:1–8.
16. Goldberg SN, Gazelle GS, Compton cc, *et al.* Radiofrequency tissue ablation in the rabbit lung: efficacy and complications. *Acad Radiol* 1995; 2:776 784.
17. Fernando HC, De Hoyos A, Landreneau RJ, *et al.* Radiofrequency ablation for the treatment of non-small cell lung cancer in marginal surgical candidates. *J Thorac Cardiovasc Surg* 2005; 129:639–644.
18. Herrera LJ, Fernando HC, Perry Y, *et al.* Radiofrequency ablation of pulmonary malignant tumors in nonsurgical candidates. *J Thorac Cardiovasc Surg* 2003; 125:929–937.
19. Hiraki T, Sakurai J, Tsuda T, *et al.* Risk factors for local progression after percutaneous radiofrequency ablation of lung tumors: evaluation based on a preliminary review of 342 tumors. *Cancer* 2006; 107:2873–2880.
20. Akeboshi M, Yamakado K, Nakatsuka A, *et al.* Percutaneous radiofrequency ablation of lung neoplasms: initial therapeutic response. *J Vasc Interv Radiol* 2004; 15:463–470.
21. Lee JM, Jin Gy, Goldberg SN, *et al.* Percutaneous radiofrequency ablation for inoperable non-small cell lung cancer and metastases: preliminary report. *Radiology* 2004; 230:125–134.
22. Grieco CA, Simon CJ, Mayo-Smith WW, *et al.* Percutaneous image-guided thermal ablation and radiation therapy: outcomes of combined treatment for 41 patients with inoperable stage I/II non-small-cell lung cancer. *J Vasc Interv Radiol* 2006; 17:1117–1124.

23. Kagan AR. The multidisciplinary clinic. *Int J Radiat Oncol Biol Phys* 2005; 61:967–968.
24. Forrest LM, McMillan DC, McArdle CS, *et al.* An evaluation of the impact of a multidisciplinary team, in a single centre, on treatment and survival in patients with inoperable non-small-cell lung cancer. *Br J Cancer* 2005; 93:977–978.
25. Junor EJ, Hole DJ, Gillis CR. Management of ovarian cancer: referral to a multidisciplinary team matters. *Br J Cancer* 1994; 70:363–370.
26. Boyle FM, Robinson E, Dunn SM, *et al.* Multidisciplinary care in cancer: the fellowship of the ring. *J Clin Oncol* 2005; 23:916–920.
27. Hall P, Weaver L. Interdisciplinary education and teamwork: a long and winding road. *Med Educ* 2001; 35:867–875.
28. Society BT, Britain SoCSoG, Party IW. BTS guidelines: guidelines on the selection of patients with lung cancer for surgery. *Thorax* 2001; 56:89–108.
29. Martini N, Bains MS, Burt ME, *et al.* Incidence of local recurrence and second primary tumors in resected stage I lung cancer. *J Thorac Cardiovasc Surg* 1995; 109:120–129.
30. Yoon H-E. Wide wedge resection for lung cancer ≤ 1 cm. American Society of Clinical Oncology; 2004. p. abstract #7247.

CyberKnife Frameless Image-Guided High-Dose Fractionated Stereotactic Radiosurgery with the Synchrony Motion Tracking Module in the Treatment of Single Small Peripheral Lung Tumors

16

Brian T. Collins, Kelly Erickson, Sean P. Collins, Gregory J. Gagnon, Sonja Dieterich, Donald A. McRae, Cristina Reichner, Thomas Chang, Carlos Jamis-Dow, Filip Banovac, Shakun Malik, and Eric D. Anderson

CONTENTS

16.1 Abstract

Curative surgery is not an option for many patients with clinical stage I non-small-cell lung carcinoma (NSCLC) because of associated comorbidities. Stereotactic radiosurgery with the CyberKnife® (Accuray Incorporated, Sunnyvale, CA) tumor tracking system may be an option for many of these medically inoperable patients. Here we provide a concise overview of the standard treatment options for stage I lung cancer and a practical summary of the Georgetown University Hospital stereotactic radiosurgery treatment protocol. We review preliminary outcomes for single small peripheral lung tumors uniformly treated with the CyberKnife using the Synchrony® (Accuray Incorporated, Sunnyvale, CA) motion tracking module.

16.2 Introduction

Approximately 170,000 new cases of lung cancer will be diagnosed in the United States in 2006 [1]. Only 15% of non-small-cell lung carcinoma (NSCLC) cases are clinical stage I [1]. The incidence of early stage NSCLC will rise in the future as spiral CT based lung cancer screening programs are implemented [2–5]. Peer-reviewed evidence has established lobectomy as the most efficacious treatment [6], but debate continues regarding the utility of less invasive options such as video-assisted lobectomy (VATS). Despite recent improvements in lobectomy, however, it remains a major operation that can cause significant morbidity [6, 7].

Sublobar resections (segmental or wedge resection) are less invasive procedures associated with suboptimal local control rates in prospective randomized studies [8]. Nonetheless, sublobar resection continues to be utilized in the treatment of marginally operable clinical stage I patients who are not candidates for radical surgery. Conventional radiation (50 Gy) and brachytherapy have been successful adjuvant therapies after surgery [9–15], both reliably sterilizing subclinical disease.

Treatment options for patients with clinical stage I NSCLC who are not surgical candidates are limited. Conventional radiotherapy is a suboptimal treatment with reported 5-year survival rates ranging from 6% to 30% [16]. Typically a small number of radiation fields are used and the clinical treatment volume is enlarged to compensate for respiratory tumor motion. Treatment is normally delivered over 6 to 7 weeks with conventional fraction sizes of 1.8–2.0

achieving total doses of 60–66 Gy to minimize toxicity in this typically frail patient population. The poor efficacy of conventional radiation for clinically localized NSCLC is largely attributed to poor local tumor control [16] secondary to inadequate total doses (< 100 Gy) [17, 18] and prolonged treatment times (> 3–4 weeks) [19, 20].

In the late 1980s two practical dose escalation treatment approaches were developed simultaneously to address the local control problem in clinically localized NSCLC: conformal conventionally fractionated radiation [21] and hypofractionated stereotactic radiosurgery [22]. Both approaches utilized newly developed treatment planning and delivery techniques to maintain tumor coverage while limiting normal tissue radiation doses, allowing for safe dose escalation.

The United States radiation oncology community championed the conventionally fractionated conformal delivery approach. This approach was embraced because it allowed for the treatment of the wide range of tumor volumes seen with localized NSCLC. Several similar single institution Phase I dose escalation trials and eventually the multi-institutional RTOG 9311 were initiated to assess the efficacy of this approach. In general, inoperable stage I-III NSCLC was treated with a 2-cm margin surrounding the gross tumor volume (primary tumor and involved lymph nodes). Accomplishing this routinely required treatment with multiple radiation fields utilizing CT based 3D computer treatment planning systems and custom patient immobilization devices. Recent results suggest that local control is improved with higher conventional doses in stage I-III and that this improvement has an appreciable affect on survival [23–25]. Limited mature results have been published with high-dose conformal radiotherapy (66 to 100 Gy) for lymph node negative inoperable NSCLC [26]. Unfortunately, despite delivering a median dose of 84 Gy over seven to ten weeks with three or more fields to these relatively small tumors, the local control rate and survival at five years was a disappointing 46% and 21%, respectively. Conventionally fractionated radiation alone, regardless of the total dose, is unable to eradicate gross disease reliably.

Thoracic stereotactic radiosurgery was developed by Blomgren et al. in the early 1990s at the Karolinska Hospital in Sweden [22]. The stereotactic approach developed for intracranial lesions was roughly applied to small pulmonary tumors. Utilizing a body frame along with abdominal compression to limit tumor motion and seven non-coplanar radiation fields, lesions could be treated reliably with small margins (10 mm) [27]. This enhanced accuracy facilitated the safe delivery of clinically effective doses of radiation quickly, making it an ideal non-invasive treatment modality for stage I NSCLC. Theoretically, by completing the treatment program in less than two weeks, the rapid repopulation of NSCLC cells seen with protracted conventional conformal treatment courses could be avoided [19]. Recently reported data from trials utilizing similar radiosurgery techniques suggest that extremely high doses (BED Gy_{10} = 180 Gy) can be delivered safely [28, 29] and data on the efficacy of moderate doses (BED Gy_{10} > 100 Gy) [30, 34], including some with extensive follow-up [35], suggest that the results will compare favorably to conventional radiation. Nyman et al. uniformly delivered 45 Gy in three fractions over a one-week period (BED Gy_{10} = 110) to inoperable stage I lung tumors. Local control and survival at five years was 80% and 30%, respectively. National Phase II studies are currently being completed in Japan, Scandinavia, and the United States at low (BED Gy_{10} = 106) and high (BED Gy_{10} = 180) stereotactic doses, respectively.

Thoracic stereotactic radiosurgery utilizing a body frame is a major step forward in the treatment of small peripheral lung cancers. Frame-based systems, however, have several limitations. Frames can restrict aiming of LINAC beams, thus limiting non-coplanar beam direction and hindering optimal dosimetry [36]. Furthermore, conformality is hindered by the limited number of beams utilized secondary to prolonged manual set-up times for each beam. Frame-based treatment with abdominal compression, breath holding or respiratory gating are uncomfortable for patients and exclude some patients from treatment (i.e., obese patients who cannot fit into the frame and patients with severe respiratory dysfunction who cannot tolerate breathing manipulation). Moreover, they only dampen the effect of respiration on tumor motion necessitating relatively large longitudinal margins (10 mm). Image verification during treatment is not typically utilized. For these reasons we were pleased that in mid-2004, after several years of experience with the conventional

CyberKnife system, the Synchrony motion tracking system was installed and quickly utilized to treat small peripheral lung tumors.

16.3 Georgetown University Hospital Stereotactic Radiosurgery Treatment Approach

Georgetown University Hospital obtained the CyberKnife radiosurgery system in early 2002 and experienced initial success treating intracranial lesions with bony landmarks utilized for tumor tracking. Further experience with the CyberKnife allowed for the effective treatment of spine tumors using implanted fiducials for tracking [37]. With the addition of the Synchrony Motion Tracking module in mid-2004, Georgetown University Hospital implemented a treatment policy for patients with small inoperable peripheral lung tumors.

Patients included in this analysis were reviewed by the Georgetown University Hospital multidisciplinary thoracic oncology team consisting of staff thoracic surgeons, pulmonologists, medical oncologists, radiologists, and radiation oncologists. Per institutional policy, when clinically feasible, all evaluated patients completed pretreatment CT imaging with IV contrast, PET imaging, and pulmonary function testing. Three to five gold fiducials measuring 1 mm in diameter by 5 mm in length were placed in close proximity to the tumors under CT guidance for tumor targeting and tumor motion tracking. Although this method has a known risk of pneumothorax [38], it was chosen in order to optimize the positioning and spacing of the fiducials. Tracking utilizing translational and rotational information requires that a minimum of three non-collinear fiducials be placed. Only conscious sedation and local anesthesia were routinely necessary.

Fine-cut (1.25 mm) treatment planning CT scans were obtained during an inhalation breath hold in the treatment position 7–10 days after fiducial placement to allow any hemorrhage to resolve which may obscure the target and to limit migration post-implantation. The gross tumor volumes (GTV) were contoured utilizing lung windows. A treatment plan using a 5 mm margin was generated using the On-Target TPS v. 5.2.1. This is a non-isocentric, inverse planning algorithm using the TPS default tissue heterogeneities for air, soft tissue and lung. For dosimetry, the effective-depth technique is used with air and soft tissue densities of 0 and 1, respectively, and the lung density of 0.3. Hundreds of small cylindrical beams ranging in collimator diameter from 20 to 30 mm were spread out over a large solid angle in each case to optimize the dose distribution. This treatment approach results in a relatively high central tumor dose with dose gradients conforming closely to the shape of the target [39], theoretically ideal for the treatment of the radiation-resistant tumor center and radiation-sensitive subclinical radial disease spread seen with NSCLC [40, 41]. No attempt was made to treat at risk but clinically negative lymph nodes (elective nodal irradiation). The radiation dose was prescribed to an isodose line that covered at least 95% of the planning treatment volume (GTV + 5 mm). All critical thoracic structures were contoured in order to ensure that incidental radiation delivered to these structures was tracked and limited. The percentage of the total lung volume receiving 15 Gy or more was also tracked and limited to 15%. Patients with tumors larger than 4 cm in maximum diameter or in close proximity to critical normal thoracic structures were excluded from this analysis and treated with alternative fraction schemes.

Radiation was delivered in three fractions of 15 to 20 Gy each to the prescription isodose line (BED Gy_{10} > 100 Gy). Fractions were 1–2 hours long and were delivered over 3 to 11 days. Lower doses were prescribed in the initial patients when concerns about adjacent structures arose, when patients had severe cardiopulmonary dysfunction and were felt to be too frail to tolerate any complication, when patients had prior conventional thoracic irradiation, and for some patients with single lung metastases for whom the benefit of treatment was considered uncertain. When considering patients for this study it was necessary to confirm that their tumors were in fact peripheral. Peripheral lesions in stereotactic radiosurgery, however, have only recently been conservatively defined by the Radiation Therapy Oncology Group (RTOG) as those that lay outside a region that extends 2 cm from the major central airways (i.e., carina, main bronchi, upper lobe bronchi, intermedius bronchus, middle lobe bronchus, lingular bronchus, and lower lobe bronchi). This conservative definition may prove to be clinically significant in the long run. Nevertheless, we chose to apply a more inclusive definition of peripheral lesions as those located far enough from sensitive central structures so that high doses of radiation (15–20 Gy/fraction) could be delivered to tumors while adhering to the dose limits for adjacent critical structures that are listed in Table 16.1.

Stereotactic radiosurgery treatment was delivered using the CyberKnife (G3) stereotactic radiosurgery system. Immediately prior to the first treatment each patient's chest was evaluated with fluoroscopy to estimate the amplitude of tumor motion, to assess tumor motion relative to critical normal structures, and to verify that the motion of the fiducials chosen for tracking correlated precisely with tumor motion. In preparation for treatment, patients were placed in the supine position on the treatment couch with their arms at their sides and allowed to breathe freely. Frames, abdominal compression, breath holding, and gating techniques were not utilized. Patient positioning was optimized utilizing digitally reconstructed radiographic (DDR) images of the patient derived from planning CT images and the CyberKnife remote automatic patient positioning system. Patients wore a form-fitting vest with 3 red light-emitting markers placed on its anterior surface in the region of maximum chest and upper abdominal excursion. These LEDs were visible to an adjustable camera array in the treatment room. The fiducials were located in multiple orthogonal X-ray images acquired with ceiling-mounted diagnostic X-ray sources and corresponding amorphous silicon image detectors secured to the floor on either side of the patient.

Prior to initiating treatment, an adaptive correlation model was produced between the fiducial positions, which were periodically imaged by the X-ray targeting system, and the red light-emitting markers, which were continuously imaged by the camera array. A valid model requires a minimum of three X-ray images and optimally the last ten fiducial images at varying points in the respiratory cycle. During treatment the fiducials were imaged prior to the delivery of every third beam for treatment verification and to update the correlation model, which may vary with time [42]. If fiducials were misidentified by the software or the correlation error exceeded 3 mm in more than one consecutive image, the treatment was stopped and the model was rebuilt with three images and optimized with additional images. Despite the complexity of human respiration, this was a rare event. During treatment delivery the tumor position was tracked using the marker signal and correlation model; the compact, lightweight 6-MV X-band LINAC was moved by the computer-controlled robotic arm to maintain alignment with the tumor throughout the complete respiratory cycle while hundreds of radiation beams were each delivered to the target.

After treatment, imaging, and pulmonary function testing were completed at 3, 6, 12, 18, and 24 months. Toxicity levels were scored according to the National Cancer Institute common terminology criteria version 3.

Table 16.1.

	Maximum Dose Limit (total for 3 fractions)
Spinal cord	18 Gy
Left ventricle	18 Gy
Esophagus	24 Gy
Main bronchus	30 Gy
Trachea	30 Gy
Aorta	30 Gy

16.4 Patients

Twenty-four consecutive patients were treated over a two-year period. No patient was excluded from treatment based on body habitus and all patients completed the planned treatment course without interruption. There were 10 men and 14 women ranging in age from 54 to 82 years, with a mean age of 70 years. Severe cardiopulmonary dysfunction was the leading reason patients were not eligible for surgical treatment of their stage I lung cancer. The maximum diameter of the tumors ranged from 0.9 cm to 3.5 cm (mean 2.0 cm). The mean GTV was 8 cc (range 1 to 14 cc). The majority (n=16, or 67%) of lesions involved the upper lobes, consistent with published surgical series. Two lesions (8%) were in the middle lobe and 6 (25%) in the lower lobes.

16.5 Treatment Parameters

Total doses ranged from 45 to 60 Gy (mean = 54 Gy) delivered in 3 equal fractions to a mean isodose line of 80% (range 75% to 90%). The mean biologically effective dose (BED Gy_{10}) delivered to the prescription isodose line was 150 Gy (range 110 to 180 Gy). The mean treatment time (beam-on to beam-off) was 82 min (range 53 to 120 min), which correlated with the mean number of MU delivered which was 11087 (range 7210–15268). The mean number of beams per treatment was 164 (range 87 to 270) and mean target verification X-ray images per treatment was 55 (29 to 90). The mean homogeneity index was 1.25 (range 1.11 to 1.33) and the mean conformality index was 1.69 (range 1.08 to 2.03). The mean percentage of lung receiving 15 Gy or greater was 7% (range 3% to 11%).

16.6 Tumor Response and Follow-Up

All evaluable lesions responded to CyberKnife treatment based on a reduction in their volume on CT imaging at 3-month follow-up. By 6 months, 6 lesions had resolved completely (25% complete response); see Figure 16.1. Nine lesions (38%), however, were completely obscured by radiation fibrosis at this time and were not evaluable; see Figure 16.2. Two single lung metastases failed locally. There have been no regional lymph node failures. Three patients with metastatic lesions have developed systemic metastatic progression. Four patients have died (17%), one of progressive metastatic disease and three of comorbid illnesses without recurrence.

16.7 Complications

Pneumothorax was seen in 30% of patients, and 17% of all patients required a chest tube for a clinically significant pneumothorax either during or immediately following the fiducial placement procedure. Acute toxicity consisting of mild fatigue was reported in the majority of patients. Grade III radiation pneumonitis was seen in 2 (8%) patients. Transient chest wall discomfort developed in 6 of 11 patients with tumors within 5 mm of the pleura.

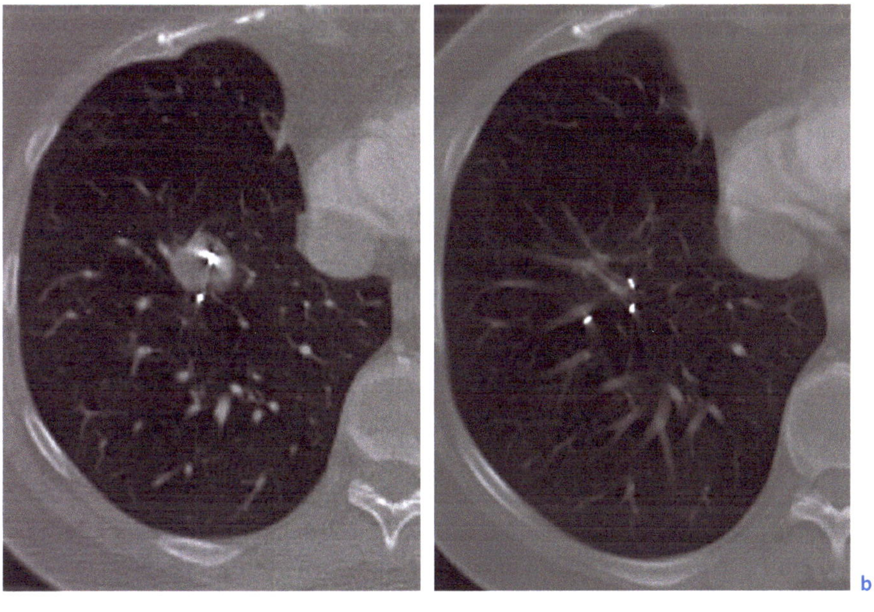

a b

Fig. 16.1a,b. Initial treatment planning CT (**a**) and CT at 12 months post-treatment (**b**).

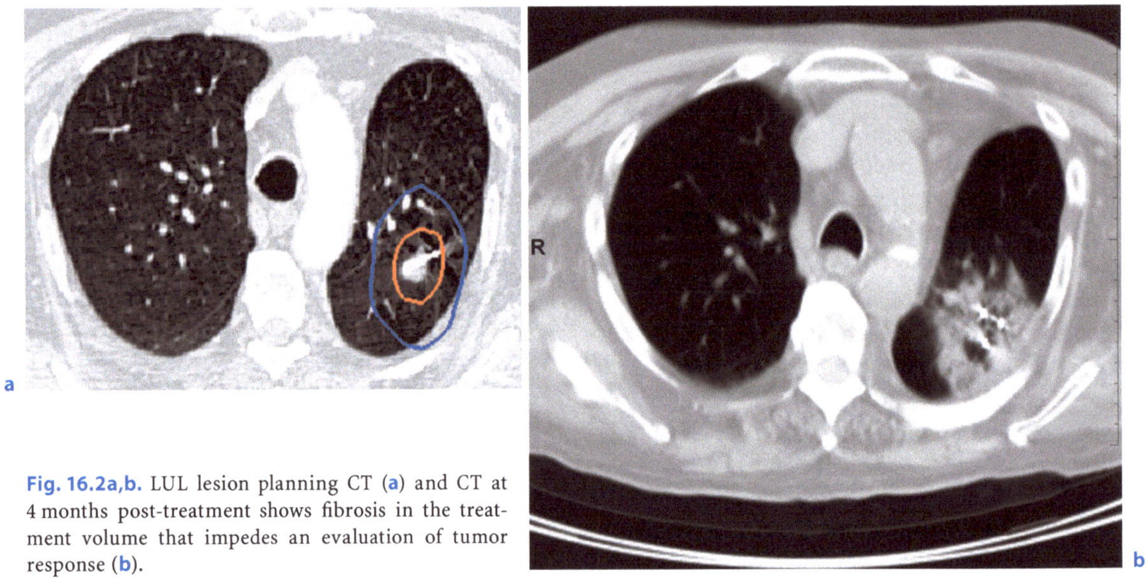

Fig. 16.2a,b. LUL lesion planning CT (**a**) and CT at 4 months post-treatment shows fibrosis in the treatment volume that impedes an evaluation of tumor response (**b**).

16.8 Discussion

Based on the early optimistic extracranial stereotactic radiosurgery findings published in late 2003 [29], we began in mid-2004 treating patients with inoperable peripheral stage I NSCLC or single small peripheral lung metastases with stereotactic radiosurgery using doses ranging from 45 to 60 Gy delivered in 3 fractions. Continuous tracking (using Synchrony) of tumor motion during treatment allowed us to deliver conformal dose distributions with tight margins (5 mm) around the tumors. Hundreds of beams were used to produce a relatively high central tumor dose and steep, uniform dose gradients that conformed closely to the shape of the tumors. We limited the maximum point doses allowed to critical normal structures and used this to define lesions as *peripheral*. We also limited the volume of lung receiving 15 Gy or greater to 15% or greater in order to minimize the risk of radiation pneumonitis. Twenty-four patients have been treated in 24 months. At a median follow-up of 12 months the local control rate is 92% and there have been no procedural mortalities. Thus, we conclude that radiosurgery with the CyberKnife for patients with small peripheral lung lesions is feasible, safe and, in the short term, effective.

Despite promising short-term results, critical issues concerning optimal treatment have yet to be fully addressed. High-dose radiation delivered precisely to small pulmonary nodules will cause lung damage that complicates interpretation of the tumor response in addition to causing manageable acute and chronic lung toxicity. At 3 months all evaluable tumors responded to treatment, as seen by a decrease in their volume. Six lesions responded completely to treatment at six months. However, with time, many of these lesions were obscured by radiation fibrosis clearly conforming to the high-dose radiation volume, making the assessment of tumor response impossible. PET activity within this region does not reliably indicate tumor recurrence because remodeling within dense fibrosis is itself PET avid. Thus, it is unclear how to classify these lesions, i.e., as responsive to treatment, stable, or even failures. Although biopsy could clarify this issue, patients in general have been unwilling to subject themselves to the risks of biopsy given their doctors perception that these changes merely represent a normal response to high-dose radiation and that such recurrences, even if diagnosed early, would not be curable.

Radiation fibrosis did not obscure the frank radiographic progression of two single lung metastases, occurring at 10 and 12 months. Local failure of optimally delivered radiosurgery can result from two causes: poor target definition and radiation resistance. Biopsy of the first lesion, a 3-cm bronchoalveolar lung cancer metastasis treated to 45 Gy (BED Gy_{10} = 110) in 3 fractions, revealed clear central radiation-induced fibrosis with active tumor cells repopulating the periphery. Local failure in this case likely resulted from poor target definition given the extensive subclinical lepidic spread typical of bronchoalveolar carcinoma [43]. The second local failure, a single cavitating squamous cell carcinoma lung metastasis treated to 60 Gy (BED Gy_{10} = 180) in three fractions, was not biopsied at the patient's request. We suspect that local progression of this tumor was secondary to hypoxia-induced radiation resistance [44].

High-dose thoracic radiotherapy, no matter how accurate, results in lung damage. The damage in this study may have been limited by the small lung volumes irradiated, and any changes may have been difficult to measure given the inherent complexity and normal variation of lung function. Nevertheless, as the percentage of total lung volume irradiated to a given dose increases, the risk of radiation pneumonitis and subsequent pulmonary fibrosis increases [45]. In the absence of an accepted parameter to gauge the risk of radiation pneumonitis for hypofractionated radiation treatment of small lung volumes, we simply chose to limit the percentage of lung receiving greater than 15 Gy. Although we were able to keep within these dose-volume constraints (3% to 11%), patients with severe pulmonary and cardiac dysfunction still experienced exacerbation of their chronic disease during early follow-up as seen in other trials [29, 46]. Classic Grade III radiation pneumonitis, however, occurred in only two patients with either prior extensive conventional thoracic radiation or concurrent Gefitinib treatment, a known cause of acute interstitial lung disease [47]. Deaths were due to metastatic disease or comorbid illnesses; no patients died secondary to pneumonitis, fibrosis, or local recurrence.

Finally, transient mild-to-moderate chest wall pain was seen in several patients with lesions within 5 mm of the pleura. It is hoped that by treating tumors throughout the respiratory cycle with tight margins that this injury, which in many cases may be unavoidable, will be mild. The use of hundreds of lightly weighted beams rather than a few heavily weighted ones has prevented the infrequent but potentially severe skin injury seen in prior stereotactic radiosurgery series [29].

The main contributor to CyberKnife treatment-related morbidity is the requirement of fiducial implantation prior to treatment. We used a percutaneous transthoracic placement method under CT guidance to optimize the positioning and spacing of the fiducials. This may result in pneumothorax, sometimes requiring a chest tube and a brief hospital stay. Our institution has developed a technique for placing fiducials in the lung parenchyma via bronchoscopy [48]. The side-effects profile is clearly superior to percutaneous transthoracic placement, and is our preferred approach for placing fiducials near central thoracic tumors. However, these bronchoscopic methods are currently ineffective in accurately placing fiducials for peripheral tumor CyberKnife radiosurgery. Small peripheral tumors will require advanced technology to place fiducials precisely via bronchoscopy. An electromagnetic navigation system for bronchoscopy is available which can localize peripheral lung tumors and reliably reach them with extended work channels for tumor biopsy and fiducial deployment [49, 50]. This system may enhance the safety of peripheral lung tumor fiducial placement. Finally, fiducial implantation and recovery necessitates a treatment delay of several weeks regardless of the method of placement. Ongoing research on fiducial-less lung tumor tracking may result in technology that obviates the need for fiducial placement [51].

16.9 Conclusion

Clinical Stage I NSCLC is curable [52]. The best results are seen when small peripheral tumors are radically resected [5, 52, 53]. Small peripheral tumors are also more likely to be cured with high-dose, primary radiation approaches, because there are fewer tumor cells to eradicate [17] and less subclinical lymphatic spread [5, 53]. Treating small moving targets reliably with tight margins (5 mm or less) requires an optimal radiation delivery system with robust image verification. The CyberKnife is such a system. The delivery of hundreds of beams coupled with the

Synchrony tumor tracking system results in ideal dose distributions within and reliable tight margins around the tumor. Early local control rates (>90%) and tolerability are encouraging. This stereotactic radiosurgery approach for small peripheral lung tumors will be optimized further in preparation for the anticipated increase in the number of such tumors diagnosed with the implementation of CT-based lung cancer screening programs.

References

1. Jemal A, Siegel R, Ward E, *et al.* Cancer statistics, 2006. *CA Cancer J Clin* 2006; 56:106–130.
2. Henschke CI, McCauley DI, Yankelevitz DF, *et al.* Early Lung Cancer Action Project: overall design and findings from baseline screening. *Lancet* 1999; 354:99–105.
3. Oken mm, Marcus PM, Hu P, *et al.* Baseline chest radiograph for lung cancer detection in the randomized Prostate, Lung, Colorectal and Ovarian Cancer Screening Trial. *J Natl Cancer Inst* 2005; 97:1832–1839.
4. Gohagan JK, Marcus PM, Fagerstrom RM, *et al.* Final results of the Lung Screening Study, a randomized feasibility study of spiral CT versus chest X-ray screening for lung cancer. *Lung Cancer* 2005; 47:9–15.
5. Henschke CI, Yankelevitz DF, Libby DM, *et al.* Survival of patients with stage I lung cancer detected on CT screening. *N Engl J Med* 2006; 355:1763–1771.
6. Allen MS, Darling GE, Pechet TT, *et al.* Morbidity and mortality of major pulmonary resections in patients with early-stage lung cancer: initial results of the randomized, prospective ACOSOG Z0030 trial. *Ann Thorac Surg* 2006; 81:1013-1019; discussion 1019–1020.
7. Handy JR, Jr., Asaph JW, Skokan L, *et al.* What happens to patients undergoing lung cancer surgery? Outcomes and quality of life before and after surgery. *Chest* 2002; 122:21–30.
8. Ginsberg RJ, Rubinstein LV. Randomized trial of lobectomy versus limited resection for T1 N0 non-small cell lung cancer. Lung Cancer Study Group. *Ann Thorac Surg* 1995; 60:615–622; discussion 622–613.
9. Birdas TJ, Koehler RP, Colonias A, *et al.* Sublobar resection with brachytherapy versus lobectomy for stage Ib nonsmall cell lung cancer. *Ann Thorac Surg* 2006; 81:434–438; discussion 438–439.
10. d'Amato TA, Galloway M, Szydlowski G, *et al.* Intraoperative brachytherapy following thoracoscopic wedge resection of stage I lung cancer. *Chest* 1998; 114:1112–1115.
11. Fernando HC, Santos RS, Benfield JR, *et al.* Lobar and sublobar resection with and without brachytherapy for small stage IA non-small cell lung cancer. *J Thorac Cardiovasc Surg* 2005; 129:261–267.
12. Group LCS. Effects of postoperative mediastinal radiation on completely resected stage II and stage III epidermoid cancer of the lung. . *N Engl J Med* 1986; 315:1377–1381.
13. Keller SM, Adak S, Wagner H, *et al.* A randomized trial of postoperative adjuvant therapy in patients with completely resected stage II or IIIA non-small-cell lung cancer. Eastern Cooperative Oncology Group. *N Engl J Med* 2000; 343:1217–1222.
14. Lee W, Daly BD, DiPetrillo TA, *et al.* Limited resection for non-small cell lung cancer: observed local control with implantation of I-125 brachytherapy seeds. *Ann Thorac Surg* 2003; 75:237-242; discussion 242–233.
15. Santos R, Colonias A, Parda D, *et al.* Comparison between sublobar resection and 125Iodine brachytherapy after sublobar resection in high-risk patients with Stage I non-small-cell lung cancer. *Surgery* 2003; 134:691–697; discussion 697.
16. Qiao X, Tullgren O, Lax I, *et al.* The role of radiotherapy in treatment of stage I non-small cell lung cancer. *Lung Cancer* 2003; 41:1–11.
17. Fletcher GH. Clinical dose-response curves of human malignant epithelial tumours. *Br J Radiol* 1973; 46:1–12.
18. Sibley GS, Jamieson TA, Marks LB, *et al.* Radiotherapy alone for medically inoperable stage I non-small-cell lung cancer: the Duke experience. *Int J Radiat Oncol Biol Phys* 1998; 40:149–154.
19. Saunders M, Dische S, Barrett A, *et al.* Continuous, hyperfractionated, accelerated radiotherapy (CHART) versus conventional radiotherapy in non-small cell lung cancer: mature data from the randomised multicentre trial. CHART Steering committee. *Radiother Oncol* 1999; 52:137–148.
20. Withers HR, Taylor JM, Maciejewski B. The hazard of accelerated tumor clonogen repopulation during radiotherapy. *Acta Oncol* 1988; 27:131–146.
21. Emami B, Purdy JA, Manolis J, *et al.* Three-dimensional treatment planning for lung cancer. *Int J Radiat Oncol Biol Phys* 1991; 21:217–227.
22. Blomgren H, Lax I, Naslund I, *et al.* Stereotactic high dose fraction radiation therapy of extracranial tumors using an accelerator. Clinical experience of the first thirty-one patients. *Acta Oncol* 1995; 34:861–870.
23. Bradley J, Graham MV, Winter K, *et al.* Toxicity and outcome results of RTOG 9311: a phase I–II dose-escalation study using three-dimensional conformal radiotherapy in patients with inoperable non-small-cell lung carcinoma. *Int J Radiat Oncol Biol Phys* 2005; 61:318–328.
24. Kong FM, Ten Haken RK, Schipper MJ, *et al.* High-dose radiation improved local tumor control and overall survival in patients with inoperable/unresectable non-small-cell lung cancer: long-term results of a radiation dose escalation study. *Int J Radiat Oncol Biol Phys* 2005; 63:324–333.
25. Rosenzweig KE, Fox JL, Yorke E, *et al.* Results of a phase I dose-escalation study using three-dimensional conformal radiotherapy in the treatment of inoperable nonsmall cell lung carcinoma. *Cancer* 2005; 103:2118–2127.
26. Chen M, Hayman JA, Ten Haken RK, *et al.* Long-term results of high-dose conformal radiotherapy for patients with medically inoperable T1-3N0 non-small-cell lung cancer: is low incidence of regional failure due to incidental nodal irradiation? *Int J Radiat Oncol Biol Phys* 2006; 64:120–126.
27. Lax I, Blomgren H, Naslund I, *et al.* Stereotactic radiotherapy of malignancies in the abdomen. Methodological aspects. *Acta Oncol* 1994; 33:677–683.

28. McGarry RC, Papiez L, Williams M, *et al.* Stereotactic body radiation therapy of early-stage non-small-cell lung carcinoma: phase I study. *Int J Radiat Oncol Biol Phys* 2005; 63:1010–1015.
29. Timmerman R, Papiez L, McGarry R, *et al.* Extracranial stereotactic radioablation: results of a phase I study in medically inoperable stage I non-small cell lung cancer. *Chest* 2003; 124:1946–1955.
30. Fukumoto S, Shirato H, Shimzu S, *et al.* Small-volume image-guided radiotherapy using hypofractionated, coplanar, and noncoplanar multiple fields for patients with inoperable Stage I nonsmall cell lung carcinomas. *Cancer* 2002; 95:1546–1553.
31. Nagata Y, Takayama K, Matsuo Y, *et al.* Clinical outcomes of a phase I/II study of 48 Gy of stereotactic body radiotherapy in 4 fractions for primary lung cancer using a stereotactic body frame. *Int J Radiat Oncol Biol Phys* 2005; 63:1427–1431.
32. Onishi H, Araki T, Shirato H, *et al.* Stereotactic hypofractionated high-dose irradiation for stage I nonsmall cell lung carcinoma: clinical outcomes in 245 subjects in a Japanese multiinstitutional study. *Cancer* 2004; 101:1623–1631.
33. Uematsu M, Shioda A, Suda A, *et al.* Computed tomography-guided frameless stereotactic radiotherapy for stage I non-small cell lung cancer: a 5-year experience. *Int J Radiat Oncol Biol Phys* 2001; 51:666–670.
34. Wulf J, Haedinger U, Oppitz U, *et al.* Stereotactic radiotherapy for primary lung cancer and pulmonary metastases: a noninvasive treatment approach in medically inoperable patients. *Int J Radiat Oncol Biol Phys* 2004; 60:186–196.
35. Nyman J, Johansson KA, Hulten U. Stereotactic hypofractionated radiotherapy for stage I non-small cell lung cancer-mature results for medically inoperable patients. *Lung Cancer* 2006; 51:97–103.
36. Takayama K, Nagata Y, Negoro Y, *et al.* Treatment planning of stereotactic radiotherapy for solitary lung tumor. *Int J Radiat Oncol Biol Phys* 2005; 61:1565–1571.
37. Degen JW, Gagnon GJ, Voyadzis JM, *et al.* CyberKnife stereotactic radiosurgical treatment of spinal tumors for pain control and quality of life. *J Neurosurg Spine* 2005; 2:540–549.
38. Whyte RI, Crownover R, Murphy MJ, *et al.* Stereotactic radiosurgery for lung tumors: preliminary report of a phase I trial. *Ann Thorac Surg* 2003; 75:1097–1101.
39. Papiez L, Timmerman R, DesRosiers C, *et al.* Extracranial stereotactic radioablation: physical principles. *Acta Oncol* 2003; 42:882–894.
40. Goldstein NS, Ferkowicz M, Kestin L, *et al.* Wedge resection margin distances and residual adenocarcinoma in lobectomy specimens. *Am J Clin Pathol* 2003; 120:720–724.
41. Giraud P, Antoine M, Larrouy A, *et al.* Evaluation of microscopic tumor extension in non-small-cell lung cancer for three-dimensional conformal radiotherapy planning. *Int J Radiat Oncol Biol Phys* 2000; 48:1015–1024.
42. Schweikard A, Shiomi H, Adler J. Respiration tracking in radiosurgery. *Med Phys* 2004; 31:2738–2741.
43. Travis WD, Colby TV, Corrin B. World Health Organization Pathology Panel: World Health Organization Histologic Typing of Lung and Pleural Tumors. In: International Histological Classification of Tumors. Vol 3rd. Berlin: Springer-Verlag; 1999.
44. Fowler JF, Tome WA, Fenwick JD, *et al.* A challenge to traditional radiation oncology. *Int J Radiat Oncol Biol Phys* 2004; 60:1241–1256.
45. Graham MV, Purdy JA, Emami B, *et al.* Clinical dose-volume histogram analysis for pneumonitis after 3D treatment for non-small cell lung cancer (NSCLC). *Int J Radiat Oncol Biol Phys* 1999; 45:323–329.
46. Paludan M, Traberg Hansen A, Petersen J, *et al.* Aggravation of dyspnea in stage I non-small cell lung cancer patients following stereotactic body radiotherapy: Is there a dose-volume dependency? *Acta Oncol* 2006; 45:818–822.
47. Inoue A, Saijo Y, Maemondo M, *et al.* Severe acute interstitial pneumonia and gefitinib. *Lancet* 2003; 361:137–139.
48. Reichner CA, Collins BT, Gagnon GJ, *et al.* The placement of gold fiducials for CyberKnife stereotactic radiosurgery using a modified transbronchial needle aspiration technique. *Journal of Bronchology* 2005; 12:193–195.
49. Schwarz Y, Greif J, Becker HD, *et al.* Real-time electromagnetic navigation bronchoscopy to peripheral lung lesions using overlaid CT images: the first human study. *Chest* 2006; 129:988–994.
50. Schwarz Y, Mehta AC, Ernst A, *et al.* Electromagnetic navigation during flexible bronchoscopy. *Respiration* 2003; 70:516–522.
51. Schweikard A, Shiomi H, Fisseler J. Fiducial-less Respiration tracking in radiosurgery. *Med Robotics & Computer Assisted Surgery* 2005; 1:19–27.
52. Wisnivesky JP, Yankelevitz D, Henschke CI. The effect of tumor size on curability of stage I non-small cell lung cancers. *Chest* 2004; 126:761–765.
53. Ketchedjian A, Daly BD, Fernando HC, *et al.* Location as an important predictor of lymph node involvement for pulmonary adenocarcinoma. *J Thorac Cardiovasc Surg* 2006; 132:544–548.

Initial Experience Treating Lung Tumors with the CyberKnife

17

William T. Brown, Xiaodong Wu, Beatriz E. Amendola, Mark Perman, Fahed Fayad, Silvio García, Hoke T. Han, Marco Amendola, Alberto de la Zerda, and James G. Schwade

CONTENTS

17.1 Abstract

Stereotactic body radiosurgery (SBRS) of lung tumors with the CyberKnife® (Accuray Incorporated, Sunnyvale, CA) achieves excellent rates of local disease control with limited toxicity to surrounding tissues. We retrospectively reviewed treatments and outcomes for 90 patients with 109 lung lesions treated at the CyberKnife Center of Miami between March 2004 and September 2006. This monotherapy review included 49 patients with 53 primary lung cancers, 27 patients with 42 pulmonary metastases, 6 patients with external beam failure and 8 patients treated by SBRS as a boost following or before conventionally fractionated radiotherapy (3DCRT or IMRT). In the primary tumor category, 43 patients remain alive. Thirty-two have been followed 1 to 25 months (median = 11.5 months). Forty-nine percent (21/43) of them have had a complete radiographic response and have been followed for a median of 18.5 months. Another 8 have evidence of at least a partial radiographic response. There have been 5 failures (5/43) within the PTV, for a local recurrence rate of 11%. Of the total 109 treated lesions, 97/109 (89%) showed radiographic evidence of at least a partial response to treatment. Six of the failures were in lesions <20 cc; 4 were in lesions 21–100 cc and 2 were in lesions >100 cc. All of the patients tolerated SBRS well with fatigue as the main toxicity. Two patients required hospitalization for Grade III radiation pneumonitis. We conclude that the delivery of precisely targeted, high dose, hypofractionated irradiation to lung tumors with the CyberKnife is well tolerated and has outcomes that are comparable with published results for other methods of SBRS.

17.2 Introduction

Lung cancer kills more Americans annually than any other malignancy. Non-small cell lung cancers account for 75% of 170,000 cases diagnosed annually. Approximately 80% of patients with non-small cell lung cancer present with locally advanced stages, rendering them practically incurable with standard treatment options. For the 15% to 20% with early stage disease, surgery has been, and remains, the treatment of choice, with 5-year survival rates of 60% to 80% [1]. Unfortunately, many potentially curable patients cannot tolerate or refuse surgical treatment. In general, older patients and those with a history of lung resection, serious cardiovascular disease, or severely compromised pulmonary function are at increased risk for surgical morbidity. As a result, approximately 50% of these will die within one year [2]. Currently accepted standards of care call for these patients to be treated with irradiation or chemotherapy, either alone or in combination. There is little consensus on appropriate drugs, doses, or treatment schedules for either modality. Five-year survival rates range from 10% to 30% for such patients treated with external beam irradiation [3, 4]. Dose escalation studies suggest a dose-response relationship for both survival and local control [5], thereby implying a potential for improvement if prescription doses can be increased safely.

In 1995, Blomgren et al. [6] first reported on the successful application of the principles of intracranial stereotactic radiosurgery (SRS) to extracranial sites, including the lung. SRS, as first established by Professor Leksell with the Gamma Knife, was characterized by both geometric accuracy and a steep dose fall-off. Unfortunately, the Gamma Knife was incapable of treating lesions below the second cervical vertebral body. Blomgren et al. described the use of a stereotactic body frame with a fixation device for stereotactic radiation of extracranial lesions. Like the Gamma Knife, using an external frame enabled precise localization and positioning. Because the lesions could not be directly observed during treatment, relatively large margins were necessary to compensate for tumor motion and set-up uncertainties. Using a similar approach, Timmerman et al. [7] published their preliminary results on the use of a modified linear accelerator for medically inoperable-Stage I non-small cell lung cancers. They found that in the 37 patients enrolled in the study, 87% responded, with a complete response noted in 27%. Wulf et al. [8] similarly found that 92% of their patients with primary lung cancers had local control at a median follow-up of 11 months (range, 2–61 months). The overall survival rate after 1 year was 52%, with 60% having no evidence of systemic disease progression.

Based upon the reported success of others using linac-based, stereotactic, high-dose, hypofractionated radiotherapy [7–11], we became interested in treating patients with lung cancer with the CyberKnife. By utilizing real-time tracking, the CyberKnife with Synchrony® (Accuray Incorporated, Sunnyvale, CA). Respiratory Tracking allows a reduction in the tumor margin required to prevent a geographic miss. Between March 2004 and September 2006, we have treated 90 patients with lung tumors. All of the diagnoses were histologically confirmed. These patients included patients with early-stage NSCLC, recurrent lung cancer, metastasis, and patients treated with SBRS following or preceding conventional radiotherapy.

17.3 Treatment Methodology

17.3.1 Patient Selection and Pretreatment Evaluation

Prior to acceptance for treatment, both a thoracic surgeon and a radiation oncologist evaluated each patient. Pretreatment evaluation included a computed tomography (CT) scan of the chest, abdomen, and pelvis, a co-registered PET/CT scan, pulmonary function tests (PFTs), complete blood count (CBC) and chemistry, and tumor markers, such as carcinoembryonic antigen (CEA), if applicable.

All patients with primary tumors were determined to have technically resectable, but medically inoperable tumors or refused surgery. Patients were deemed ineligible for surgery if they lacked adequate respiratory reserve, had cardiac dysfunction or chronic heart disease, pulmonary hypertension, diabetes mellitus with severe end organ damage, vascular disease, general frailty, or severe cerebral disease. Informed consent was obtained before proceeding with treatment. We did not exclude patients receiving other forms of antineoplastic therapy such as chemotherapy, biological therapy, and vaccine therapy.

SBRS of lesions in the hilar regions has been reportedly linked with bronchial stenosis leading to atelectasis of large portions of the lung distal to the lesion. We therefore chose to follow the RTOG [12] recommendations to exclude central tumors within 2 cm of the proximal bronchial tree. Patients with evidence of mediastinal disease (mediastinal lymph nodes >1 cm, abnormal hilar or mediastinal FDG uptake observed on PET scan), pleural or pericardial effusions (malignant or benign), pneumothorax, or inability to lay down on the CT couch and CyberKnife table in a reproducible manner were also excluded. Tumors had to be well defined radiographically and number less than four lesions per lung.

17.3.2 Treatment Preparation and Planning

Descriptions of CyberKnife treatment planning for pulmonary malignancies have been published [13, 14]. Briefly, our patients undergo CT-guided transthoracic placement of gold fiducial markers into the tumor using pre-loaded needles. Following an interval of 5-7 days to allow for stabilization of the fiducials, the treatment planning CT was obtained. Continuous axial slices (1.5 mm thick) with contrast (125 cc Omnipaque 350) were obtained while patients held their breath in expiration. Treatment plans were subsequently developed and evaluated by the team. The planning target volume (PTV) routinely included a 3-5 mm margin beyond the gross tumor volume (GTV) to include microscopic extension and to accommodate targeting uncertainty. The dose was prescribed to the PTV, usually at the 60-85% isodose line. The inverse planning module was used to maximize dose conformality. To ensure an acceptably high dose gradient, 50 to 150 or more non-coplanar, non-isocentric beams were often required. A typical treatment plan for a 13 cc lesion is shown in Figure 17.1.

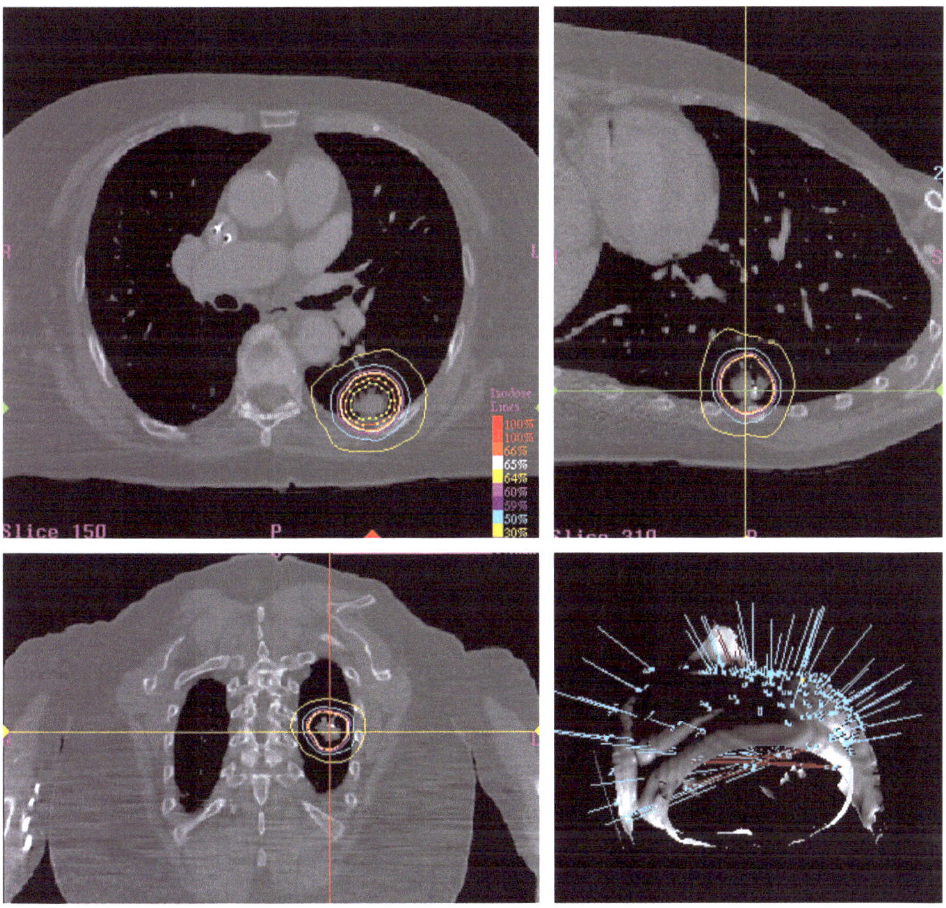

Fig. 17.1 A typical treatment plan for a 13 cc NSCLC lesion, delivering 60 Gy in 3 fractions to the 65% isodose line using 60 beams and results in a V(15)=4.6%.

Treatment followed shortly afterwards and was completed in three to eight days. Follow-up included CT scan one month post-treatment, and PET/CT scans every three months thereafter.

17.4 Results

Between March 2004 and September 2006, 90 patients with 109 lung lesions were treated. There were 53 previously unirradiated primary tumors, 42 unirradiated metastases, 6 tumors that failed prior to external beam, and 8 lesions treated in planned combination with external beam radiotherapy. Two patients received treatment for synchronous primary lesions.

We consider local recurrence at follow-up to be an enlarging lesion on CT or increasing FDG avidity on PET not explainable by the appearance of radiation pneumonitis or fibrosis. Whenever possible, biopsy was attempted to establish histological proof of recurrence/persistence. A complete response was defined as a disappearance of the lesion in CT scans and a partial response was defined as the presence of a residual (stable or smaller) abnormality on a scan. In our series of 49 Stage I patients with 53 primary NSCLC, 5 patients died from intercurrent disease, and 49% (21/43) of these patients have had a complete response. There have been 5 failures (5/43) within the PTV, for a local recurrence rate of 11%.

17.4.1 Primary Tumors

We treated 53 primary lesions in 49 patients (5 patients received treatment for synchronous primary lesions, Table 17.1). Twenty-four patients had Stage IA and 25 had Stage IB tumors. Average treatment volumes were 8 cc for the Stage IA and 34 cc for the Stage IB lesions. We employed a wide range of dosing schemes in this patient population. Total dose was dependent upon tumor size, its proximity to vital structures, co-morbid conditions and physician preference. Twenty lesions received 60 Gy/3 fractions. Another 10 received 55.5 Gy/3 fractions. The other 23 received less. The prescribed dose was typically to the 70% isodose line and delivered a heterogeneous distribution with higher doses at the center of the tumor.

In the Stage IA patients, 3 patients died from comorbidities. All 24 patients showed a response except for a woman with a 5 cc tumor that was treated with 55.5 Gy/3 fractions. This patient had an increase in the size of her lesion from 1.5 cm to 3 cm with increasing FDG uptake at 3 months. Fine-needle aspiration (FNA) established that she suffered from

Table 17.1 Characteristics of patients, tumors, and treatment parameters for patients with primary, metastatic, and recurrent tumors, and those treated with CyberKnife as a boost to external beam radiotherapy. All data are presented as means (and ranges).

	Primary	Metastases	Recurrences	Boost
# Patients	49	27	6	8
Gender				
Male	17	15	2	4
Female	32	12	4	4
Age	84 yrs (51–93)	66 yrs (33–91)	61 yrs (41–84)	76 yrs (55–92)
PTV Volume	21 ml (2–71)	40 ml (1–152)	125 ml (5–338)	42 ml (6.1–182)
BED	114 Gy (42–180)	76 Gy (20–180)	46 Gy (23–79)	44 Gy (23–79)
Total Dose	51 Gy (16–60)	30 Gy (5–60)	24 Gy (15–36)	20 Gy (10–36)
# Fractions	2.94 (1–3)	2.5 (1–3)	3.2 (1–5)	2.5 (1–3)
Isodose Line	70% (60–85)	71% (58–84)	69% (60–80)	66% (60–85)

persistent disease and went on to undergo successful lobectomy in spite of her poor pulmonary function. Two patients experienced recurrences in the PTV at 9 and 12 months, and died of progressive disease at 20 and 22 months. Both received 60 Gy/3 fractions to PTVs measuring 11 and 13 cc. Three patients went on to develop hilar/mediastinal metastases. Two patients were lost to follow-up. Eleven patients have complete radiographic responses, while 4 more show partial responses.

Twenty-five Stage IB patients were treated. Three have expired from intercurrent disease. Two Stage IB patients have recurred in the PTV. The first received 60 Gy/3 fractions to a 39 cc tumor. The second recurred 6 months post-treatment. This second patient had a 35 cc PTV, treated to 60 Gy. Four developed metastases. Two patients have been lost to follow-up. Ten others enjoy a complete response, while 4 show a partial response.

17.4.2 Stage I Complete Responses

Twenty-one Stage I patients have had a complete response thus far (Table 17.2). With a median follow-up of 18.5 months, our complete response rate is 49% (21/43). Another 8 patients exhibited a partial radiographic response. Only 11% (5/43) failed within the PTV

Table 17.2 Number of patients experiencing a complete response after CyberKnife treatment and their TMN stage, dose, and follow-up duration.

# Patients	21 (out of 49)
Stage	
IA	11 patients
IB	10 patients
Total Dose (mean)	
< 50 Gy	5 patients
50–59 Gy	6 patients
> 60 Gy	10 patients
Follow-up (months)	
0–12	6 patients
13–24	14 patients
>24	1 patients

17.4.3 Lung Metastases

Twenty-seven patients with 42 lung metastases (11 patients with multiple metastases) have been treated with the CyberKnife (Table 17.1). Of these patients, 3 exhibited radiographic evidence of persistent or recurrent disease. One occurred in a breast cancer patient with 3 lung lesions. This patient received 10 Gy in a single fraction, 25 Gy in a single fraction, and 22 Gy in a single fraction, respectively, to each lesion. Recurrent disease was detected 9 months later in one of the lesions and all were retreated successfully with 51 Gy/3, 51 Gy/3, and 48 Gy/3 fractions with radiographic complete responses. The second occurred in a lesion treated with 15 Gy/1 fraction. This patient was also successfully salvaged 6 months later with 15 Gy/3 fractions. The third was a lesion treated to 48 Gy/3 fractions, which was retreated 3 months later with another 48 Gy/3 fractions. Thirty-seven of 42 tumors (88%) were stable-to-improved on follow-up scans.

17.4.4 External Beam Salvage

In our series, 6 patients were retreated with the CyberKnife for PET-proven recurrence after external beam radiotherapy (Table 17.1). One patient subsequently failed due to disease persistence after undergoing 25 Gy/5 fractions (BED = 48) to a PTV measuring 338 cc. Two patients have been lost to follow-up, 2 had complete responses and 1 has stable disease 2 years post-salvage.

17.4.5 External Beam Boost

In this subset we treated 8 patients with CyberKnife as a boost combined with irradiation to definitive doses (Table 17.1). Two patients died from co-morbid disease. Two patients have been lost to follow-up. One patient (external beam + 10 Gy/3 fractions, CK) experienced an in-field recurrence, salvaged with another course of CyberKnife (12 Gy/1 fraction). Three patients remain without radiographic evidence of disease.

17.4.6
CyberKnife Recurrences / Persistent Disease

As summarized in Table 17.3, 12 of 109 treated lesions failed in the PTV (11%). Six of the failures occurred in lesions receiving at least 48 Gy. Six of the failures were in lesions < 20 cc; 4 were in lesions 21–100 cc and 2 were in lesions > 100 cc.

Table 17.3 Summary of patients with lesions that failed treatment in the PTV.

Patient	Category	PTV (cc)	Dose (Gy)/fractions
1	IA	5	55.5/3
2	IA	11	60/3
3	IA	13	60/3
4	IB	39	60/3
5	IB	35	60/3
6	Metastasis	31	10/1
6	Metastasis	2	25/1
6	Metastasis	24	22/1
7	Metastasis	12	15/1
8	Metastasis	4	48/3
9	Salvage	338	25/5
10	Boost	182	10/3

17.4.7
Complications

The most commonly encountered acute toxicity was mild, self-limiting fatigue. Typically it occurs soon after treatment and lasts on average 2 to 4 weeks. Four patients developed symptomatic radiation pneumonitis. Three of these patients were in the Stage 1A group. Two of them responded to outpatient care while a third required hospitalization. The breast cancer patient with 3 lung metastases that failed CyberKnife developed Grade III radiation pneumonitis that required oxygen, steroids, and hospitalization. Following retreatment 9 months later, she again developed Grade II radiation pneumonitis (RP) that resolved without hospitalization. Grade I–II esophagitis developed in 3 patients.

17.5
Discussion

17.5.1
Primary Lung Cancer

Patients with early stage disease for whom surgery is deemed inappropriate may be considered for SBRS with curative intent. Typically this population includes the elderly and those with compromised cardiovascular status, pulmonary disease, or other co-morbidities. Results using SBRS with linac-based systems have been encouraging. The use of hypofractionated, large dose irradiation with a biologically equivalent dose (BED) of ≥ 100 Gy produces improved local control, recurrence-free and overall survival compared to conventional irradiation. In the Phase I study of Timmerman et al. [7], a dose escalation trial for T1 and T2 lesions, a local failure rate of 10/47 (21%) was obtained, and in all but one case the fractionated dose was ≤ 16 Gy (46 Gy, total). A follow-up Phase II study evaluated 70 patients with T1 and T2 lesions. Sixty to 66 Gy/3 fractions were delivered. With a median follow-up of 18 months, the local control was 95%. The median overall survival was 32 months with an overall survival of 55% at 2 years. Median time to toxicity was 10 months and occurred in 17% of peripheral lesions [15].

The initial report of the CyberKnife for the treatment of lung cancer detailed the results of a Phase I trial [14]. Fifteen patients with primary lung cancer and 8 with metastases were included. A dose of 15 Gy was delivered in a single fraction. Nine patients were treated with breath-hold, the others with Synchrony. At a mean of 7 months follow-up, 2 patients had a complete radiographic response, 15 showed a partial response, 4 were stable, and 2 showed disease progression. A second study of CyberKnife treatment of lung cancer followed [16]. In this dose escalation trial, 9 patients received 15 Gy and 10 received 25 Gy. Once again, treatment was delivered in a single fraction. At a mean follow-up period of 8 months, 4 patients failed locally, all in the lower dose group.

Forty-eight patients with primary lung tumors were treated with the CyberKnife in Miami between March 2004 and September 2006. As of October 2006, 32 patients remain alive and have been followed 1 to 25 months (median = 11.5 months). Sixteen complete responses have been followed for a median of 18.5 months. Five of these patients were treated in 2004, 10 in 2005, and one this year. Employing a range of dosing schemes patterned as much as possible on RTOG 0236, we achieved an overall local control rate (complete plus partial responses) of 67% (29/43). When Grade I pneumonitis or radiation fibrosis is present radiographically, it becomes difficult to differentiate treatment sequelae from recurrent disease. In each case of radiation pneumonitis, the first PET/CT at 3 months demonstrated a complete response. At 6 to 9 Months, FDG avidity and ground glass appearance of the PTV were noted. In each case, the radiologist reported the likelihood of a local recurrence. Three of the 4 patients were biopsied. One was found to have inflammatory disease and the other 2 had biopsy-proven recurrence.

17.5.2 Lung Metastases

For oligometastatic disease, the International Registry of Lung Metastases reported that after complete metastatectomy survival was 36% at 5 years and 26% at 10 years. As with primary and recurrent lesions, not all patients are candidates for surgery. Wulf et al. reported on 41 patients with 51 pulmonary metastases not amenable to surgery that were treated with stereotactic radiotherapy [8]. Patients received either 3 × 10 Gy, 3 × 12–12.5 Gy or 1 × 26 Gy. The median follow-up was 9 months (range, 2–37 months). The actuarial local control rate was 80% ≥1 year after treatment and was significantly improved by increasing the dose from 3 × 10 Gy to 3 × 12–12.5 Gy or 1 × 26 Gy (p = 0.038). The overall survival rate after 1 and 2 years was 85% and 33%, respectively. After 12 months, 35% of patients with pulmonary metastases were without systemic progression. No severe acute or late toxicity was observed. We found that in our patients, the control rate for metastatic lesions was 88%.

17.5.3 Recurrent NSCLC

Local recurrence of lung cancer following conventional irradiation poses considerable challenges when considering reirradiation. Okamoto et al. [17] reported on retreatment with external beam irradiation with the aim of prolonging survival or relieving symptoms. A response was noted in 14 of 18 patients treated definitively, while 12 of 16 patients experienced significant palliation. The overall survival rate after reirradiation was 43% at one year and 27% at two years. Toxicity was high, with 19 patients experiencing symptomatic radiation pneumonitis and symptomatic radiation esophagitis in 6 patients. Wu et al. [18] also found that Grade I–II radiation pneumonitis developed in 22% of their patients undergoing reirradiation. The local progression-free rate at one and two years was 51% and 42%, respectively. Kramer et al. [19] published their results utilizing hypofractionated irradiation for symptomatic, recurrent NSCLC. Patients received 2 fractions of 8 Gy separated by a week. The performance score improved in 45% of these patients.

We have treated 6 patients, in our limited experience, using the CyberKnife for external beam salvage. Three were stable or improved radiographically while 2 were lost to follow-up. One of the 6 had persistent disease after having undergone 25 Gy/5 fractions for a paratracheal recurrence. Due to our concern for the development of tracheal stenosis, a relatively large PTV (338 cc) received a relatively low dose per fraction, resulting in a BED = 48 Gy.

17.5.4 Toxicity

The volume of irradiated lung correlates with the risk of radiation-induced pneumonitis. If the volume of the total lung receiving more than 20 Gy (V_{20}) exceeds approximately 38%, the risk of radiation pneumonitis increases dramatically [20]. With this dose level, it is thought that the diffusion capacity for carbon monoxide decreases. It has been shown that the loss may occur with doses as low as 13 Gy [21] while others report pneumonitis not occurring until doses reach 30 Gy [22]. The effects of chemotherapy may amplify the risk of developing radiation pneumonitis [23].

We found that our patients tolerated the CyberKnife treatments well. The most common acute side effect encountered was fatigue, typically occurring at or closely following treatment, lasting several days. Only 4 of 89 (4.5%) patients developed clinical pneumonitis that required treatment. Two of them required hospitalization, and all responded to medical care. Symptomatic, but mild esophagitis developed in 3 patients (3.4%), once again all responding to conservative measures.

17.5.5 Outlook

Radiosurgery for lung lesions is in its infancy, and it is difficult to make firm recommendations regarding optimal treatment parameters. For early stage lung tumors, despite the demonstrated effectiveness of a dose scheme of 60 Gy in 3 fractions, local failures are certainly not absent. Although the reasons for these local failures are not fully known, we are led to speculate on how we should alter treatment parameters to increase the likelihood of cure based on available data and our experience. It seems appropriate to consider alteration along at least three lines: shortened treatment times, enlarged GTV-to-PTV margins, and increased biologically effective doses.

Treatment times should be shortened to prevent loss of radiobiological effectiveness. Fowler et al. have reported loss of biological effectiveness in 1 hour to be 10–15% [24]. Thus, a much shortened treatment duration (compared with the 90 minutes or longer treatment time we currently obtain) would be advantageous. When the demand of high conformity necessitates longer treatment times, a higher dose might be needed to compensate for the loss of biological effect. In addition, although the CyberKnife is capable of irradiating with a minimal geometric miss (~1.5 mm), an adequate margin should be carefully considered due to the presence of microscopic disease. We are altering the margins arranged around the GTV, from 3–5 mm to 5–8 mm in order to cover microscopic extension of tumor cells [25].

Doses may be escalated further for adequate local control. Higher doses may be needed to kill hypoxic or radioresistant cells in the tumor and microscopic cells outside the high-dose PTV region. We are treating T1 tumors and smaller T2 tumors with additional fractions to reduce the possibility of early and late toxicity. An increase in fractionation may also enhance tumor cell kill by allowing cell re-oxygenation. Two examples of such alternated fractionation schemes might be 16 Gy by 4 fractions (105 Gy NTD) or 13.5 Gy by 5 fractions (103 Gy NTD in 2 Gy fractions, assuming an alpha/beta ratio of 20). Systematic research is necessary to clearly establish optimal treatment parameters.

17.6 Conclusions

While long-term survival rates and late effects cannot be established at this time, early results from this population suggest that the CyberKnife is a safe and effective alternative for the treatment of certain pulmonary malignancies. Volumes of irradiated lung are well below recommended limits, and little severe acute toxicity was encountered. As follow-up of these patients continues, long-term outcomes and toxicity data will be acquired. These data will be reported, and when combined with results from on-going multi-institutional studies, should provide further insight into the role of the CyberKnife in treating lung cancer.

References

1. Flehinger BJ, Kimmel M, Melamed MR. The effect of surgical treatment on survival from early lung cancer. Implications for screening. *Chest* 1992; 101:1013–1018.
2. McGarry RC, Song G, des Rosiers P, *et al.* Observation-only management of early stage, medically inoperable lung cancer: poor outcome. *Chest* 2002; 121:1155–1158.
3. Qiao X, Tullgren O, Lax I, *et al.* The role of radiotherapy in treatment of stage I non-small cell lung cancer. *Lung Cancer* 2003; 41:1–11.
4. Sibley GS, Jamieson TA, Marks LB, *et al.* Radiotherapy alone for medically inoperable stage I non-small-cell lung cancer: the Duke experience. *Int J Radiat Oncol Biol Phys* 1998; 40:149–154.
5. Narayan S, Henning GT, Ten Haken RK, *et al.* Results following treatment to doses of 92.4 or 102.9 Gy on a phase I dose escalation study for non-small cell lung cancer. *Lung Cancer* 2004; 44:79–88.
6. Blomgren H, Lax I, Naslund I, *et al.* Stereotactic high dose fraction radiation therapy of extracranial tumors using an accelerator. Clinical experience of the first thirty-one patients. *Acta Oncol* 1995; 34:861–870.
7. Timmerman R, Papiez L, McGarry R, *et al.* Extracranial stereotactic radioablation: results of a phase I study in

medically inoperable stage I non-small cell lung cancer. *Chest* 2003; 124:1946–1955.
8. Wulf J, Haedinger U, Oppitz U, *et al.* Stereotactic radiotherapy for primary lung cancer and pulmonary metastases: a noninvasive treatment approach in medically inoperable patients. *Int J Radiat Oncol Biol Phys* 2004; 60:186–196.
9. McGarry RC, Papiez L, Williams M, *et al.* Stereotactic body radiation therapy of early-stage non-small-cell lung carcinoma: phase I study. *Int J Radiat Oncol Biol Phys* 2005; 63:1010–1015.
10. Nyman J, Johansson KA, Hulten U. Stereotactic hypofractionated radiotherapy for stage I non-small cell lung cancer-mature results for medically inoperable patients. *Lung Cancer* 2006; 51:97–103.
11. Onishi H, Araki T, Shirato H, *et al.* Stereotactic hypofractionated high-dose irradiation for stage I nonsmall cell lung carcinoma: clinical outcomes in 245 subjects in a Japanese multiinstitutional study. *Cancer* 2004; 101:1623–1631.
12. Radiation Therapy Oncology Group. RTOG 0236: A Phase II Trial of Stereotactic Body Radiation Therapy (SBRT) in the Treatment of Patients with Medically Inoperable Stage I/II Non-Small Cell Lung Cancer. 2004.
13. Le QT, Loo BW, Ho A, *et al.* Results of a phase I dose-escalation study using single-fraction stereotactic radiotherapy for lung tumors. *Journal of Thoracic Oncology* 2006; 1:802–809.
14. Whyte RI, Crownover R, Murphy MJ, *et al.* Stereotactic radiosurgery for lung tumors: preliminary report of a phase I trial. *Ann Thorac Surg* 2003; 75:1097–1101.
15. Timmerman R, McGarry R, Yiannoutsos C, *et al.* Excessive toxicity when treating central tumors in a phase II study of stereotactic body radiation therapy for medically inoperable early-stage lung cancer. *J Clin Oncol* 2006; 24:4833–4839.
16. Le QT, Ho A, Cotrutz C, *et al.* Single fraction stereotactic radiosurgery (SFSR) for lung tumors-a phase I dose escalation trial. *J Clin Oncol* 2004; 22:7231.
17. Okamoto Y, Murakami M, Yoden E, *et al.* Reirradiation for locally recurrent lung cancer previously treated with radiation therapy. *Int J Radiat Oncol Biol Phys* 2002; 52:390–396.
18. Wu KL, Jiang GL, Qian H, *et al.* Three-dimensional conformal radiotherapy for locoregionally recurrent lung carcinoma after external beam irradiation: a prospective phase I-II clinical trial. *Int J Radiat Oncol Biol Phys* 2003; 57:1345–1350.
19. Kramer GW, Gans S, Ullmann E, *et al.* Hypofractionated external beam radiotherapy as retreatment for symptomatic non-small-cell lung carcinoma: an effective treatment? *Int J Radiat Oncol Biol Phys* 2004; 58:1388–1393.
20. Marks LB. Dosimetric predictors of radiation-induced lung injury. *Int J Radiat Oncol Biol Phys* 2002; 54:313–316.
21. Gopal R, Tucker SL, Komaki R, *et al.* The relationship between local dose and loss of function for irradiated lung. *Int J Radiat Oncol Biol Phys* 2003; 56:106–113.
22. Fay M, Tan A, Fisher R, *et al.* Dose-volume histogram analysis as predictor of radiation pneumonitis in primary lung cancer patients treated with radiotherapy. *Int J Radiat Oncol Biol Phys* 2005; 61:1355–1363.
23. Schild SE, McGinnis WL, Graham D, *et al.* Results of a Phase I trial of concurrent chemotherapy and escalating doses of radiation for unresectable non-small-cell lung cancer. *Int J Radiat Oncol Biol Phys* 2006; 65:1106–1111.
24. Fowler JF, Welsh JS, Howard SP. Loss of biological effect in prolonged fraction delivery. *Int J Radiat Oncol Biol Phys* 2004; 59:242–249.
25. Giraud P, Antoine M, Larrouy A, *et al.* Evaluation of microscopic tumor extension in non-small-cell lung cancer for three-dimensional conformal radiotherapy planning. *Int J Radiat Oncol Biol Phys* 2000; 48:1015–1024.

Stereotactic Radiosurgery for Early Stage Non-Small Cell Lung Cancer: Rationale, Patient Selection, Results and Complications

18

Neil A. Christie, Steven Burton, Arjun Pennathur, and James D. Luketich

CONTENTS

18.1 Introduction

Lung cancer is the most common cause of cancer death [1]. While advanced disease portends a very poor prognosis, the subset of patients who present with localized disease are potentially curable. Anatomic lobectomy is the current standard of care for early stage lung cancer [2]. Sub-lobar resection is generally reserved for patients with inadequate pulmonary reserve. For patients with small tumors (2 cm or less) new data suggests that sub-lobar resection may be adequate [3, 4]. It has also become apparent that patients with very marginal pulmonary reserve may safely tolerate lung resection or even have improved pulmonary function afterwards [5, 6].

For non-operative candidates historically standard radiation was offered but generally with little hope for cure and poor survival rates of 20-40% at 5 years [7, 11]. Survival rates and cure are more likely in smaller (T1) tumors as opposed to larger tumors and more advanced stages of disease [9, 12, 13]. Advanced radiation techniques with three-dimensional planning, multiple beam paths, and respiratory motion compensation have resulted in the ability to administer high dose hypofractionated radiation or stereotactic radiosurgery to lung tumors. While such focused high dose radiation alone will never be more effective than sub-lobar resection, there is evidence that in appropriately selected patients such therapy may be curative [14, 19]. Early results are encouraging but results are not mature and recently late toxicity has been reported [20, 21]. In this chapter we review some of the findings that guide decisions about the proper therapy for patients with non-small cell lung cancer (NSCLC), and early outcomes and complications observed after stereotactic radiosurgery and radiotherapy for lung cancer.

18.2 Rationale and Patient Selection

Therapy for localized lung cancer may involve surgery, non-surgical therapy, or in some cases no therapy at all. Determination of the appropriate therapy for a given patient considers tumor characteristics such as size and location combined with patient characteristics such as cardiopulmonary reserve, medical co-morbidities, and functional status. Surgical excision may consist of a lobectomy, an anatomic segmentectomy, or a non-anatomic wedge resection. Post-operative adjuvant radiation may be administered in some cases.

A lobectomy is a larger resection with a presumed improved local control rate and improved survival [2]. It is often, but not always [5], associated with a loss of lung function and decreased quality of life [22]. The sub-lobar resection removes less lung and results in a local control rate that is usually presumed to be worse but with better preservation of lung function [2, 23, 24].

The Lung Cancer Study Group performed a prospective multi-institutional randomized trial comparing limited resection with lobectomy for patients with peripheral T1 N0 non-small cell lung cancer [2]. One hundred and 22 patients underwent limited resection and 125 underwent lobectomy. Of the limited resections, 40 were wedge resections and 82 were segmental resections. The local recurrence rate of 17% in the limited resection group was significantly higher than the 6% recurrence rate seen with lobectomy.

Landreneau and colleagues performed a retrospective analysis of patients who had undergone either lobectomy or sub-lobar resection. Local recurrence in 139 patients who underwent sub-lobar resection was 16% which was significantly higher than the 5% local recurrence seen in patients undergoing lobectomy [25].

El-Sherif found in a retrospective review of sub-lobar resection patients that pathologic resection margins less then 1 cm were associated with an increased rate of local recurrence [26].

Adjuvant radiation of resection margins after sub-lobar resection has been explored in an attempt to improve local control. Application of conventional radiation techniques to resection margins is limited by respiratory motion and sensitivity of the normal lung to radiation damage in patients with limited pulmonary reserve. Birdas and colleagues developed a technique of lung brachytherapy where radioactive ^{125}I seeds are imbedded in vicryl sutures and sewn to a polyglyconate mesh [27]. The mesh is secured to the suture line and delivers 100 to 125 Gy of radiation to a depth of 0.5 cm. This has been shown to decrease local recurrence rates substantially, down to levels seen with lobectomy. Fernando et al. found a similar beneficial effect in brachytherapy mesh in reducing local recurrences for patients with sub-lobar resection [28].

Not all studies, however, have shown an increased local recurrence rate with sub-lobar resection alone. Okada et al. reported 70 patients with tumors less than 2 cm in diameter who all underwent surgical resection with intra-operative margin analysis and noted no local recurrences [29]. Similar findings were seen in another Japanese study where 53 patients with tumors 2 cm or less underwent segmental resection with the development of only one local recurrence [30]. Intuitively, one would expect to see better survival rates with a more extensive resection. In the prospective Lung Cancer Study Group trial an incremental improvement in survival was found for patients who underwent lobectomy as compared to sub-lobar resection [2]. This has not been seen in all studies, however. Yoshikawa et al. reported an excellent 5-year survival of 92% in patients with tumors 2 cm or less who underwent segmentectomy [4]. Okada et al. performed a retrospective analysis of over 1200 patients who had undergone resection for Stage I NSCLC to determine factors influencing survival [3]. For tumors 2 cm or less in diameter, no significant difference in survival was seen with lobectomy or sub-lobar resection. For tumors 3 cm or greater in diameter, survival after lobectomy was significantly better than after segmental resection (81% 5-year survival rate for lobectomy versus 63% for segmentectomy). No patients survived for 5 years after wedge resection. Inoue demonstrated that increased tumor size correlates with the rate of lymph node metastases [31].

One of the main reasons to perform sub-lobar lung resections is to preserve lung function. The Lung Cancer Study Group demonstrated preserved lung function at 6 and 12 months follow-up with preservation of FEV1 after sub-lobar resection [2]. Harada also demonstrated better preservation of lung function after a sub-lobar resection [24]. In some patients, however, lung resection can result in a post-operative improvement of lung function. The National Emphysema Treatment Trial Study group demonstrated that, in appropriately selected patients with severe emphysema, lung resection can result in an improvement in subjective and objective measures of lung function. This was achieved with an overall operative mortality of only 2.9% [32]. In the subset of patients with upper-lobe predominant emphysema and low exercise capacity, lung resection was associated with a significant decrease in mortality rate compared to patients who were medically treated [33].

Choong et al. reported on 21 patients with severe emphysema and compromised lung function who underwent combined lung cancer resection and lung volume reduction surgery [6]. There was no perioperative mortality and all patients demonstrated an improvement in lung function postoperatively. Survival was 100% at one year and 67% at 5 years. Korst et al. reported an improvement in postoperative FEV1 by 4% in a subset of patients who underwent lobectomy for cancer and had a pre-operative FEV1 less than 60% of predicted [5].

In spite of all this research, it remains true that not all patients are candidates for general anesthesia and lung resection. For example, a subset of patients in the National Emphysema Treatment Trial who had extremely poor lung function, as defined by an FEV1 less than 20% and a DLCO less than 20%, had a perioperative mortality of 29% [33]. Comorbidity and functional status also determine the outcome of patient's therapy for Stage I NSCLC. The Charlson Score is a weighted index of comorbidity for 19 medical conditions, and was developed on the basis of the relative risk of death that reflects both the number and severity of comorbidities. The Cumulative Illness Rating Scale likewise measures patients' extent and severity of medical illness. The Karnofsky Performance Status scale measures patients' mobility and functional status. Comorbidity and Karnofsky scores have been shown to be independent predictors of outcome for early stage lung cancer; independent of other factors such as stage or treatment modality [34]. Although overall survival was greater with surgery than for radiation therapy, patients with severe co-morbidity and poor functional status may experience equivalent outcomes.

Although age was not a predictor in the Firat et al. study [34], the age at which surgical resection of lung tumors should still be considered is a controversial issue. Port et al. reported on outcomes of surgical resection in octogenarians. They observed acceptably low operative mortality and morbidity, relatively short hospital stays, and an 82% 5-year survival rate in those patients with Stage I NSCLC [35]. Mery et al. reported no difference in survival between lobectomies and limited resection in terms of survival time for elderly patients after a review of 14,555 patients with Stage I or II NSCLC in the Surveillance, Epidemiology, and End Results (SEERS) database [36]. Curative surgery was performed in only 70% of patients older than 75 years of age as compared to 86–92% of younger patients. The median survival rate of 28 months, seen in patients greater than 75 years of age, was significantly less than that seen in younger patients (47 to 71 months). For younger patients lobectomies conferred better survival times than limited resection after two years. There was no difference in survival between lobectomies and limited resection in terms of survival, however, for the elderly patients. The statistical difference in long-term survival between those patients undergoing lobectomy and those undergoing limited resections disappeared at an age of 71 years. It seems, therefore, although excellent results can be seen in elderly patients with above average functional status, caution should be undertaken when considering surgical resection over non-surgical therapy in very elderly patients.

The development of minimally invasive surgical techniques has facilitated the safe conduct of operations in patients who are elderly or have limited lung function and co-morbidities. Onaitis et al. recently reported the Duke University experience with minimally invasive lobectomy in 500 consecutive patients [37]. Conversion to open thoracotomy was required in only 1.6% of cases. Operative mortality was only 1% and perioperative morbidity was also very low. The median length of hospitalization was only three days. It is generally acknowledged that the survival rate with thoracoscopic resections is at least equivalent to that seen with open surgery [23].

Some would argue that life expectancy in patients with severe impairment of cardiopulmonary function or other severe co-morbidities is such that the cancer will not likely impact on their survival and that no cancer therapy is appropriate. In the NETT trial, for a group of patients with severe COPD, the majority of patients were still alive 5 years after entering the trial [32]. McGarry et al. evaluated the outcome of observation only management for Stage I lung cancer in a debilitated group of patients who were offered no therapy [38]. In the majority of these patients cancer was the cause of death.

18.3 Radiation Therapy

In spite of advances in surgical techniques, perioperative surgical care, and appreciation of the lung volume reduction effect, there is still a subset of patients who have poor functional status, medical comorbidities, compromised pulmonary reserve, and advanced age for whom any surgical intervention is contravened and yet who are still likely to die prematurely of their lung cancer if untreated. Heretofore, conventional radiation had generally been the standard therapy. Conventional radiation offers a relatively poor chance of cure for primary lung tumors and is generally considered a palliative rather than curative treatment [9–11]. It also leads to significant damage to surrounding non-neoplastic lung tissue. Radical radiotherapy with conventional techniques shows long-term survival rates of 6% to 30% [7, 8, 39–41]. Smaller tumors, however, seem to have a more favorable prognosis with primary radiotherapy. T staging appears to be an independent predictor of survival after radiotherapy. In subset analysis, patients with small tumors (T1) were found to have 5-year survival rates of 30–50% [9, 13].

Higher radiation doses enhance local tumor control. In one study, patients receiving greater than 70 Gy had better local control and cancer specific survival than those treated with lower doses [39]. Several studies have reported a benefit to dose escalation with a dose response relationship to both local control and survival. When dose escalation has been attempted using conventional treatment techniques and values, however, there is dose limiting toxicity. Dose escalation is limited primarily by the risk of radiation pneumonitis [42, 43]. Dose volume histogram data shows that the risk of toxicity increases as the area of adjacent normal lung irradiated increases [44, 45]. Radiation fibrosis seems to depend on the volume of the lung radiated above a threshold of 20 to 30 Gy.

Poor local control with conventional radiation likely results from inaccurate tumor targeting, failure to satisfactorily conform the dose distribution to the target volume, and an associated failure to deliver an adequate dose of radiation. Three-dimensional conformal treatment planning provides more accurate dose targeting by incorporating three-dimensional anatomic information from diagnostic scans into the planning process. It allows higher doses to be delivered to the target by improved shaping of the radiation portals so as to avoid normal structures. Three-dimensional conformal radiotherapy has allowed safe dose escalation to total doses as high as 90 or 100 Gy. Local failure is still a significant problem at 30–40% local control and 3-year survival rates of only 30% [46].

Improvements in local control can be achieved if the biologically effective dose (BED) to the tumor can be escalated further without damaging nearby tissue. To do this in the lung requires some form of compensation for respiratory tumor motion. Respiratory motion is significant and results in an increased field requiring radiation, thereby limiting the overall maximum dose to the tumor. One study evaluated the amplitude of maximum intra-thoracic organ motion during breathing [47]. CT scan images were taken during different phases of respiration. Greatest lung displacements were observed just above the diaphragm and smallest displacements were observed at the lung apices and adjacent to the carina. Lung tumor movement during breathing is considerable and an appreciation of organ motion is necessary for the definition of appropriate treatment margins. Studies taking into account motion of lung tumors show improvement in local control [48].

Therapy with breath-hold to facilitate tumor location has also shown improved local control [49, 50]. Similarly, clinicians have evaluated gating techniques to allow administration of radiation only during a certain phase of respiration in free-breathing patients [51, 53]. Another approach which has been explored is body immobilization [54] and dynamic radiation tracking of the tumor position during radiation therapy. It too has been shown to be an effective technique to deliver accurately high dose radiation while minimizing risk of pneumonitis [55].

18.4 Stereotactic Radiosurgery

The evolution of more precise, conformal radiotherapy techniques has culminated in stereotactic radiosurgery, which holds the most promise for curative radiotherapy in patients with limited lung tumors. Stereotactic targeting utilizes a variety of systems to decrease the effects of lung motion that would translate into target motion as well as improved localization techniques. These systems allow dramatic reduction of treatment values facilitating hypofractionation with markedly increased daily doses and a significantly reduced overall treatment time. Initial systems were frame-based and used abdominal compression to limit lung tumor motion and thereby facilitate treatment with small margins [56]. No dynamic tracking systems are commercially available in frame-based stereotactic radiosurgery systems.

The CyberKnife® (Accuray Incorporated, Sunnyvale, CA) is a frameless stereotactic radiosurgery system consisting of a linear accelerator mounted on a robotic arm. Multiple highly focused beams of radiation (typically 100–200) converge at the tumor site to produce a very conformal treatment volume. Orthogonal X-ray cameras frequently image gold fiducial markers implanted near or within the tumor. By registering these intraoperative X-rays with digitally reconstructed images from the planning CT, the CyberKnife system precisely localizes the tumor in space. A camera array in the treatment room continuously images externally applied light-emitting diodes (LEDs) on the patient. The location of the LEDs is correlated with the location of the internal fiducials; the correlation model allows for continuous dynamic respiratory tracking and treatment throughout the respiratory cycle.

Percutaneous fiducial placement is done under local anesthesia. Pneumothorax is a complication in 13–26% of cases. Clinically insignificant pulmonary contusion is also seen in some cases. Bronchoscopic deployment is also possible by deploying the fiducials with a 19–21 gauge Wang transbronchial needle. It is associated with a much lower risk of pneumothorax but requires conscious sedation [57].

CyberKnife treatment planning results in radiation treatment plans with excellent conformity of the radiation to the tumor and a very rapid fall-off of radiation in the adjacent normal lung. In fact, the CyberKnife system is so accurate and the conformality so high that the radiation can be precisely contoured to the tumor with a very rapid fall-off of radiation in the surrounding lung parenchyma. While this may facilitate the safe administration of extremely high doses of radiation to the tumor, treatment plans may not include an adequate margin of radiation treatment to the tumor margin where radiologically occult microscopic disease may be present. Historically radiotherapy fields for early stage NSCLC encompassed the primary tumor and regional lymphatics in the ipsilateral hilum and mediastinum. This prophylactic treatment was based on identified risk of occult nodal involvement from surgical series ranging up to 20% and surgical data indicating better control with more extensive resections. Nonetheless, large radiotherapy fields are potentially poorly tolerated in the population of patients with limited pulmonary reserve. More recent retrospective experience shows similar survival rates with fields limited to the tumor alone, as compared to fields including prophylactic treatment to lymph node chains [58–61]. With stereotactic radiosurgery a dose margin surrounding the GTV that is high enough to kill microscopic disease may have to be programmed by the clinician. There is published evidence that suggests that this margin should be 8 mm for NSCLC [62]. As long-term efficacy data for CyberKnife treatment (and other stereotactic, hypofractionated approaches) accumulates, analyses of failure patterns may result in guidelines for margin parameters that will maximize tumor control without damaging healthy lung tissue.

18.5 Results and Complications

The gradual shift in the last two decades from conventional radiotherapy to 3D conformal radiotherapy to precisely delivered high-dose, hypofractionated radiation culminated in stereotactic body radiotherapy (SBRT) and CyberKnife radiosurgery. An extensive review of SBRT can be found in Chapter 13 by Dr. McGarry of this volume, but a brief review here reveals the basis for treatment choices made by CyberKnife users.

Timmerman et al. [17] conducted a landmark dose escalation study of stereotactic radiosurgery in 37 patients with medically inoperable lung cancer. Patients were treated in 3 fractions administered over a 2-week period using a frame-based radiosurgery system. The dose started at 8 Gy per fraction and was escalated to 20 Gy. Patients tolerated the treatment well, with generally minor acute side effects and no late toxicity at a median follow-up of 15.2 months. A complete response was seen in 27% of patients and partial response was seen in 60%.

Cancer progressed in 13/37 patients; only 6/13 patients progressed locally. Local recurrences were more common at lower doses and occurred at a median of 13 months after treatment. No patients treated with over 18 Gy per fraction showed progression. This study showed the feasibility of dose escalation with acute and subacute toxicity that was limited and manageable.

A multi-institutional study from Japan also showed good tolerability to high-dose, hypofractionated radiotherapy in 245 heterogeneous subjects [16]. Higher doses were again associated with better response rates. With a median follow-up of 24 months, significant pulmonary toxicity was seen in only 2.4% of patients. Local progression was seen in only 14.5% of patients overall and in only 8.3% of patients with high dose therapy. Similarly promising outcomes were reported in a handful of other studies on SBRT for lung tumors [14, 15, 18, 56, 63-65].

Excellent tumor control was accompanied by significant toxicity in a later study by Timmerman et al. [21]. They reported on the outcome of SBRT for 70 patients with T1 or T2 lung tumors (< 7 cm) N0M0. T1 tumors were treated with 60 Gy in 3 fractions and T2 tumors were treated with 66 Gy in 3 fractions. Overall local control was excellent with a local control rate of 95% at two years. Median overall survival was 32 months with a 2-year survival rate of 54.7%. Major response rate was 60% and the remaining patients had stable disease. Only 10 patients had recurrence of tumor; three locally at 9–33 months post-treatment and the rest outside of the radiation field. Fifty-eight had Grade 1–2 toxicity consisting mostly of fatigue, musculoskeletal discomfort, and radiation pneumonitis. These toxicities occurred within 1–2 months of treatment and resolved by 3–4 months. Eight patients had Grade 3–4 toxicity evidenced by decreases in pulmonary function tests, pneumonia, pleural effusions, apnea, and skin reactions. These more severe toxicities occurred 7.6 months after treatment (range 1.1 to 25.1 months). Six patients had treatment-related deaths (Grade 5 toxicity). Deaths were seen at 0.6, 3.9, 12.1, 12.8, 13.8, and 19.5 months after treatment. Four deaths were secondary to bacterial pneumonia, one from complications of a pericardial effusion, and one from massive fatal hemoptysis after treatment of a central lesion. Four of the patients who died were treated for central or peri-hilar tumors. Patients with peripheral tumors had a 2-year freedom from toxicity of 83% compared to patients with central tumors who had a 2-year freedom from toxicity of only 54%. Gross tumor volumes greater than 10 ml had an 8-times greater risk of high-grade toxicity than smaller tumors. Survival was worse in patients with T2 tumors, worse in patients with cardiac co-morbidities as compared to pulmonary co-morbidities, and worse in patients with an FEV1 < 40 versus those with FEV1 > 40%. The authors concluded that treatment with 60 Gy in 3 fractions to central tumors may be too toxic. Japanese researchers have shown good tumor control with less toxicity to central tumors by delivering 48 Gy in 4 fractions [15].

The first report of CyberKnife therapy for isolated lung tumors was by Whyte et al. [66]. Nine of their 23 patients were treated with a breath-hold technique and the rest with the CyberKnife's respiratory tracking system. All tumors were deemed unresectable by radiologic criteria (T4), because patients were medically inoperable, or because patients refused surgery. Radiation was delivered in a single fraction of 15 Gy. Procedural complications consisted of pneumothorax in 7 patients, only one of whom required a chest tube. There was no radiation esophagitis or clinically apparent radiation pneumonitis. At radiologic follow up of 1–3 months, response was complete in 2 patients, partial in 15 patients, stable in 4 patients, and progressive in 2 patients.

A handful of studies on CyberKnife treatment for lung cancer have been presented at professional meetings. Berger et al. reported multi-fraction treatment of 12 patients with Stage I NSCLC using 24–60 Gy in 3 fractions [67]. At a median follow-up of 93 days, a complete response was seen in 23%, partial response in 54%, and local progression in 8% of patients. Distant failure was seen in 15% of patients. No symptomatic radiation pneumonitis was seen and asymptomatic radiation pneumonitis occurred in 15% of patients. Collins et al. reported on 28 patients with small peripheral lung tumors treated with 45–60 Gy in 3 fractions over a 1–2 week period [68]. Mild fatigue and temporary soft tissue discomfort were the only appreciable acute side affects. Limited subacute radiation pneumonitis was seen in 6% of patients and was uniformly associated with prior extensive conventional radiotherapy. With limited short-term follow-up, no local progression was seen in any case. Finally, Bellairs et al. reported on 61 patients with solitary lung tumors who underwent stereotactic radiosurgery [69]. Mean tumor volume was 34.8 cc and mean dose was 36 Gy in 3 fractions. The most common side effect was mild fatigue. Nine patients developed shortness of breath which resolved a few weeks after treatment. With a median 7 month follow-up (range 1–21 months) local control was noted in 96% of the cases.

At the University of Pittsburgh, we initially treated 32 patients with CyberKnife stereotactic radiosurgery. All patients were assessed by a thoracic surgeon, medical oncologist, and radiation oncologist to determine the appropriateness of stereotactic radiosurgery. One-to-four fiducials were percutaneously placed under CT guidance. A median dose of 60 Gy, delivered in 3 fractions of 20 Gy to the 80% isodose line, was administered. The initial response rate was assessed by CT scan and PET scan at three months. Patients were subsequently monitored for time to progression and survival with clinical evaluations, CT scans, and PET scans every three months. Of the treated group, 25/32 had NSCLC and 7/32 had solitary pulmonary metastases. All patients were treated over a two-year period and had a median age of 68 years. Percutaneous fiducial placement resulted in pneumothorax requiring a pleural catheter in 9/32 patients. An initial complete response was seen in 6/32 patients, a partial response in 11/32 patients, and stable disease in 6/32 patients. Early progression of disease occurred in 9/32 patients. At a median follow-up of five months, 12/32 patients remained progression free. The time to progression in the remaining 20 patients was six months. There were no procedure related mortalities with this treatment and no clinically significant radiation pneumonitis.

We subsequently analyzed another group of 21 patients with Stage I NSCLC who were medically inoperable and were treated with 3 fractions of 20 Gy. In this group with uniformly poor lung function, CT-guided fiducial placement resulted in pneumothorax requiring a pigtail catheter in 10 (47%) patients. An initial complete response was observed in 7 (33%) patients (see Figs. 18.1–18.3), partial response in 5 (24%) patients, stable disease in 5 (24%) patients, and 1 patient was not evaluated (5%). Early progression occurred in 3 (14%) patients. During follow-up, local progression occurred in 9 patients (43%) and the median time to progression was 12 months. The remaining patients were locally progression-free at a median follow-up of 8.5 months (range 2–18 months). There were no procedure-related deaths. The estimated probability of survival at 1 year was 90% and median survival was 26.4 months.

Le at al. recently reported the results of a single institution Phase I study of dose escalation using single fraction stereotactic radiosurgery in non-operative candidates with the CyberKnife radiosurgery system [20]. Thirty-two patients with T1 and T2 N0 NSCLC or solitary metastases were treated with doses of 15, 25, and 30 Gy in a single fraction. Significant toxicity was observed after 25 Gy in patients who had received prior thoracic radiotherapy. Large tumors (volume > 50 cc) and centrally located tumors were associated with increased toxicity. Dose escalation to 30 Gy was only applied to previously un-irradiated patients and treatment volumes less than 50 cc. Complications occurred at doses greater than 25 Gy and included four cases of Grade 2 to 3 pneumonitis, one pleural effusion and three possible treatment-related deaths. All 3 treatment-related deaths occurred in patients that received chemotherapy either before or after radiosurgery, and in two cases toxicity was possibly triggered by gemcitabine-based chemotherapy delivered up to six months after completion of radiosurgery. Most pulmonary toxicities occurred late in follow-up,

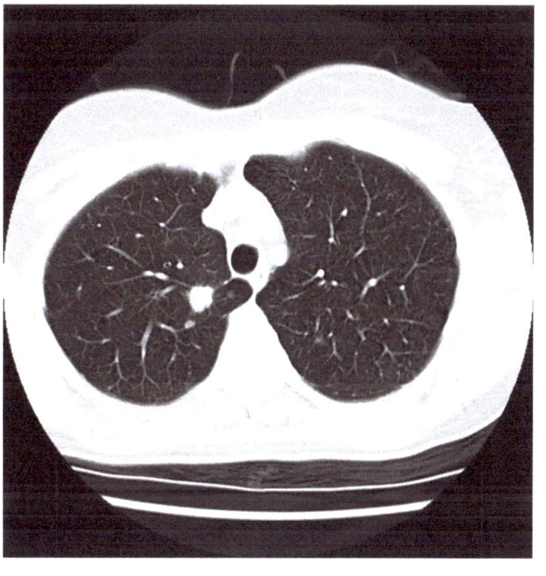

Fig. 18.1 Pre-treatment axial CT image of a right upper lobe lesion.

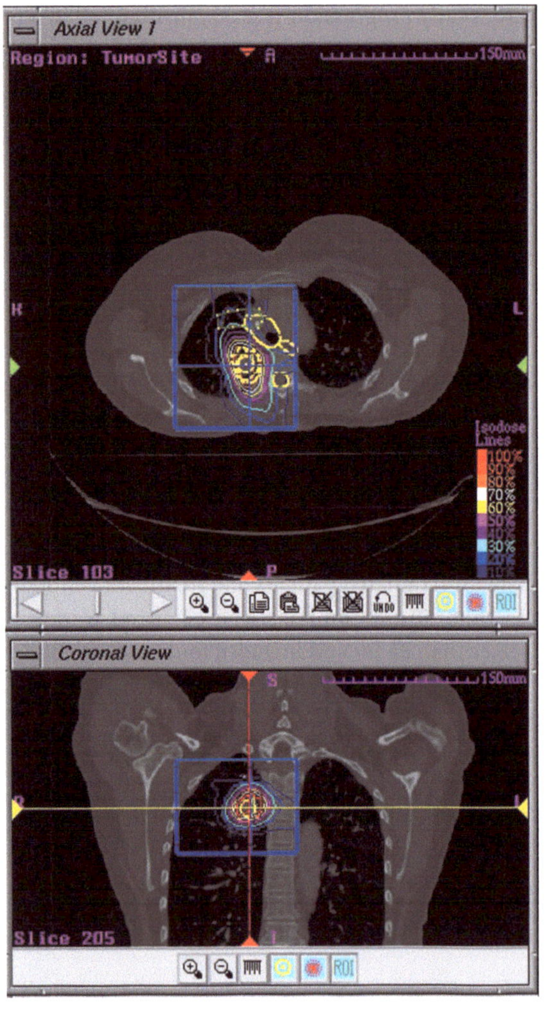

Fig. 18.2 Axial and coronal treatment plan images demonstrating right upper lobe target and isodose lines.

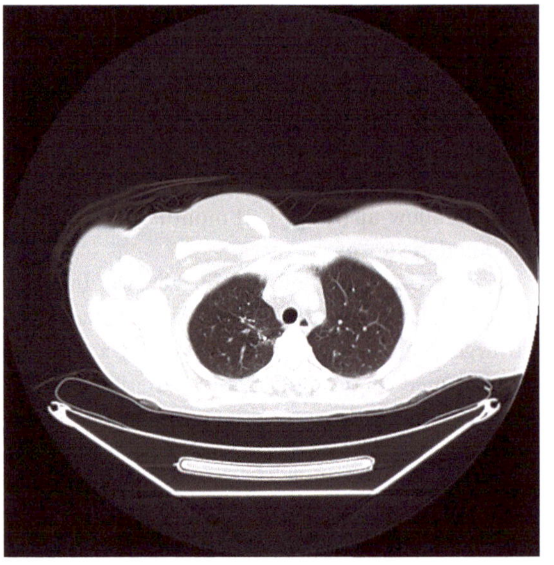

Fig. 18.3 Post-treatment axial CT image. Residual scar is visible, but no evidence of the right upper lobe tumor. Note fiducials in treatment field.

usually 5 to 6 months after radiosurgery. The 1-year freedom from local progression was 91% for doses greater than 20 Gy and 54% for doses less than 20 Gy. The response rates (complete and partial) were the same at all dose levels, but the complete response rate was higher at doses over 25 Gy. Local relapse rates were increased at doses less than 20 Gy. Only 1 of 7 patients treated with over 20 Gy showed local relapse. The 1-year freedom from local relapse was 100% for T1 tumors and 83% for T2 tumors treated with doses greater than 20 Gy.

These early studies demonstrated the feasibility and safety of stereotactic radiosurgery in a single high fraction for patients with sever co-morbidities, poor lung function, and solitary lung tumors. While these short term results are very encouraging, more recent reports with longer term follow-up of patients treated with multi-fractionated high dose radiotherapy demonstrate significant delayed morbidity and even mortality in a subset of patients. The rate of initial non-responders, as well as initial responders who subsequently progressed was also significant. These findings suggest there is significant room for improvement.

18.6 Conclusions

In this chapter we have reviewed research on surgical resection, radiotherapy, and stereotactic radiosurgery in the treatment of lung cancer. While anatomic lobectomy is still the optimal therapy for early stage lung cancer, a subset of patients with tumors 2 cm or less who are properly staged and have no evidence of nodal metastases may be successfully treated with sub-lobar resection, provided adequate resection margins can be achieved. The appreciation that there exists a subset of patients who can be cured with sub-lobar resection encourages us on the prospect of stereotactic radiosurgery as a curative therapy in that group of patients. The dose escalation which has been facilitated by stereotactic radiosurgery techniques has resulted in better local control rates than have been previously seen with conventional radiation. In patients with small (< 3 cm) peripheral tumors, excellent local tumor control with follow-up extending out to 2 years has been reported. Larger tumors (> 3 cm) are less likely to be controlled by radiosurgery, a finding which parallels outcomes of sub-lobar resection for tumors greater than 2–3 cm. In addition, larger tumors and central tumors may not be safely treated by 60 Gy in 3 fractions or 25 Gy or more in a single fraction. The nature of intrathoracic radiation toxicity with high dose stereotactic radiosurgery may be different from the radiation pneumonitis seen with conventional radiation. Significant interaction of radiation lung toxicity with chemotherapeutic agents and previous thoracic irradiation has also been identified.

Further studies and long-term follow-up are needed to identify optimal dose and fractionation schedules. Attention to radiation fall-off at the margins of the gross tumor volume will also likely prove to be important in achieving optimal local control. These studies should ultimately allow optimization of stereotactic radiosurgery for delivery of effective and safe therapy of lung tumors and also facilitate the identification of tumor and patient factors which will predict where therapy will most likely be curative.

References

1. Jemal A, Siegel R, Ward E, *et al.* Cancer statistics, 2006. *CA Cancer J Clin* 2006; 56:106–130.
2. Ginsberg RJ, Rubinstein LV. Randomized trial of lobectomy versus limited resection for T1 N0 non-small cell lung cancer. Lung Cancer Study Group. *Ann Thorac Surg* 1995; 60:615-622; discussion 622-613.
3. Okada M, Nishio W, Sakamoto T, *et al.* Effect of tumor size on prognosis in patients with non-small cell lung cancer: the role of segmentectomy as a type of lesser resection. *J Thorac Cardiovasc Surg* 2005; 129:87–93.
4. Yoshikawa K, Tsubota N, Kodama K, *et al.* Prospective study of extended segmentectomy for small lung tumors: the final report. *Ann Thorac Surg* 2002; 73:1055–1058; discussion 1058–1059.
5. Korst RJ, Ginsberg RJ, Ailawadi M, *et al.* Lobectomy improves ventilatory function in selected patients with severe COPD. *Ann Thorac Surg* 1998; 66:898–902.
6. Choong CK, Meyers BF, Battafarano RJ, *et al.* Lung cancer resection combined with lung volume reduction in patients with severe emphysema. *J Thorac Cardiovasc Surg* 2004; 127:1323–1331.
7. Chen M, Hayman JA, Ten Haken RK, *et al.* Long-term results of high-dose conformal radiotherapy for patients with medically inoperable T1-3N0 non-small-cell lung cancer: is low incidence of regional failure due to incidental nodal irradiation? *Int J Radiat Oncol Biol Phys* 2006; 64:120–126.
8. Qiao X, Tullgren O, Lax I, *et al.* The role of radiotherapy in treatment of stage I non-small cell lung cancer. *Lung Cancer* 2003; 41:1–11.
9. Graham MV, Pajak TE, Herskovic AM, *et al.* Phase I/II study of treatment of locally advanced (T3/T4) non-oat cell lung cancer with concomitant boost radiotherapy by the Radiation Therapy Oncology Group (RTOG 83–12): long-term results. *Int J Radiat Oncol Biol Phys* 1995; 31:819–825.
10. Kaskowitz L, Graham MV, Emami B, *et al.* Radiation therapy alone for stage I non-small cell lung cancer. *Int J Radiat Oncol Biol Phys* 1993; 27:517–523.
11. Wigren T, Kellokumpu-Lehtinen P, Ojala A. Radical radiotherapy of inoperable non-small cell lung cancer. Irradiation techniques and tumor characteristics in relation to local control and survival. *Acta Oncol* 1992; 31:555–561.
12. Dosoretz DE, Galmarini D, Rubenstein JH, *et al.* Local control in medically inoperable lung cancer: an analysis of its importance in outcome and factors determining the probability of tumor eradication. *Int J Radiat Oncol Biol Phys* 1993; 27:507–516.
13. Jeremic B, Classen J, Bamberg M. Radiotherapy alone in technically operable, medically inoperable, early-stage (I/II) non-small-cell lung cancer. *Int J Radiat Oncol Biol Phys* 2002; 54:119–130.
14. McGarry RC, Papiez L, Williams M, *et al.* Stereotactic body radiation therapy of early-stage non-small-cell lung carcinoma: phase I study. *Int J Radiat Oncol Biol Phys* 2005; 63:1010–1015.
15. Nagata Y, Takayama K, Matsuo Y, *et al.* Clinical outcomes of a phase I/II study of 48 Gy of stereotactic body radiotherapy in 4 fractions for primary lung cancer using a stereotactic body frame. *Int J Radiat Oncol Biol Phys* 2005; 63:1427–1431.
16. Onishi H, Araki T, Shirato H, *et al.* Stereotactic hypofractionated high-dose irradiation for stage I nonsmall cell lung carcinoma: clinical outcomes in 245 subjects in a Japanese multiinstitutional study. *Cancer* 2004; 101:1623–1631.
17. Timmerman R, Papiez L, McGarry R, *et al.* Extracranial stereotactic radioablation: results of a phase I study in medically inoperable stage I non-small cell lung cancer. *Chest* 2003; 124:1946–1955.
18. Wulf J, Baier K, Mueller G, *et al.* Dose-response in stereotactic irradiation of lung tumors. *Radiother Oncol* 2005; 77:83–87.
19. Zimmermann FB, Geinitz H, Schill S, *et al.* Stereotactic hypofractionated radiation therapy for stage I non-small cell lung cancer. *Lung Cancer* 2005; 48:107–114.
20. Le Q-T, Loo BW, Ho A, *et al.* Results of a phase I dose-escalation study using single-fraction stereotactic radiotherapy for lung tumors. *Journal of Thoracic Oncology • Volume , Number ,* 2006; 1:802–809.

21. Timmerman R, McGarry R, Yiannoutsos C, *et al.* Excessive toxicity when treating central tumors in a phase II study of stereotactic body radiation therapy for medically inoperable early-stage lung cancer. *J Clin Oncol* 2006; 24:4833–4839.
22. Handy JR, Jr., Asaph JW, Skokan L, *et al.* What happens to patients undergoing lung cancer surgery? Outcomes and quality of life before and after surgery. *Chest* 2002; 122:21–30.
23. McKenna RJ, Jr., Houck W, Fuller CB. Video-assisted thoracic surgery lobectomy: experience with 1,100 cases. *Ann Thorac Surg* 2006; 81:421–425; discussion 425–426.
24. Harada H, Okada M, Sakamoto T, *et al.* Functional advantage after radical segmentectomy versus lobectomy for lung cancer. *Ann Thorac Surg* 2005; 80:2041–2045.
25. Landreneau RJ, Sugarbaker DJ, Mack MJ, *et al.* Wedge resection versus lobectomy for stage I (T1 N0 M0) non-small-cell lung cancer. *J Thorac Cardiovasc Surg* 1997; 113:691–698; discussion 698–700.
26. Elsherif A, Fernando H, Santos R, *et al.* Margin and local recurrence after sublobar resection of non-small cell lung cancer. Society of Surgical Oncology; in press.
27. Birdas TJ, Koehler RP, Colonias A, *et al.* Sublobar resection with brachytherapy versus lobectomy for stage Ib nonsmall cell lung cancer. *Ann Thorac Surg* 2006; 81:434–438; discussion 438–439.
28. Fernando HC, Santos RS, Benfield JR, *et al.* Lobar and sublobar resection with and without brachytherapy for small stage IA non-small cell lung cancer. *J Thorac Cardiovasc Surg* 2005; 129:261–267.
29. Okada M, Yoshikawa K, Hatta T, *et al.* Is segmentectomy with lymph node assessment an alternative to lobectomy for non-small cell lung cancer of 2 cm or smaller? *Ann Thorac Surg* 2001; 71:956–960; discussion 961.
30. Bando T, Yamagihara K, Ohtake Y, *et al.* A new method of segmental resection for primary lung cancer: intermediate results. *Eur J Cardiothorac Surg* 2002; 21:894–899; discussion 900.
31. Inoue M, Minami M, Shiono H, *et al.* Clinicopathologic study of resected, peripheral, small-sized, non-small cell lung cancer tumors of 2 cm or less in diameter: pleural invasion and increase of serum carcinoembryonic antigen level as predictors of nodal involvement. *J Thorac Cardiovasc Surg* 2006; 131:988–993.
32. Fishman A, Martinez F, Naunheim K, *et al.* A randomized trial comparing lung-volume-reduction surgery with medical therapy for severe emphysema. *N Engl J Med* 2003; 348:2059–2073.
33. Patients at high risk of death after lung-volume-reduction surgery. *N Engl J Med* 2001; 345:1075–1083.
34. Firat S, Bousamra M, Gore E, *et al.* Comorbidity and KPS are independent prognostic factors in stage I non-small-cell lung cancer. *Int J Radiat Oncol Biol Phys* 2002; 52:1047–1057.
35. Port JL, Kent M, Korst RJ, *et al.* Surgical resection for lung cancer in the octogenarian. *Chest* 2004; 126:733–738.
36. Mery CM, Pappas AN, Bueno R, *et al.* Similar long-term survival of elderly patients with non-small cell lung cancer treated with lobectomy or wedge resection within the surveillance, epidemiology, and end results database. *Chest* 2005; 128:237–245.
37. Onaitis MW, Petersen RP, Balderson SS, *et al.* Thoracoscopic lobectomy is a safe and versatile procedure: experience with 500 consecutive patients. *Ann Surg* 2006; 244:420–425.
38. McGarry RC, Song G, des Rosiers P, *et al.* Observation-only management of early stage, medically inoperable lung cancer: poor outcome. *Chest* 2002; 121:1155–1158.
39. Bradley JD, Ieumwananonthachai N, Purdy JA, *et al.* Gross tumor volume, critical prognostic factor in patients treated with three-dimensional conformal radiation therapy for non-small-cell lung carcinoma. *Int J Radiat Oncol Biol Phys* 2002; 52:49–57.
40. Alberts WM. The future and lung cancer: room for optimism? *Cancer Control* 2000; 7:13–14.
41. Baumann M, Appold S, Petersen C, *et al.* Dose and fractionation concepts in the primary radiotherapy of non-small cell lung cancer. *Lung Cancer* 2001; 33 Suppl 1: S35–45.
42. Abratt RP, Morgan GW. Lung toxicity following chest irradiation in patients with lung cancer. *Lung Cancer* 2002; 35:103–109.
43. Byhardt RW, Martin L, Pajak TF, *et al.* The influence of field size and other treatment factors on pulmonary toxicity following hyperfractionated irradiation for inoperable non-small cell lung cancer (NSCLC) – analysis of a Radiation Therapy Oncology Group (RTOG) protocol. *Int J Radiat Oncol Biol Phys* 1993; 27:537–544.
44. Bradley J, Graham MV, Winter K, *et al.* Toxicity and outcome results of RTOG 9311: a phase I–II dose-escalation study using three-dimensional conformal radiotherapy in patients with inoperable non-small-cell lung carcinoma. *Int J Radiat Oncol Biol Phys* 2005; 61:318–328.
45. Kong FM, Hayman JA, Griffith KA, *et al.* Final toxicity results of a radiation-dose escalation study in patients with non-small-cell lung cancer (NSCLC): predictors for radiation pneumonitis and fibrosis. *Int J Radiat Oncol Biol Phys* 2006; 65:1075–1086.
46. Lagerwaard FJ, Senan S, van Meerbeeck JP, *et al.* Has 3-D conformal radiotherapy (3D CRT) improved the local tumour control for stage I non-small cell lung cancer? *Radiother Oncol* 2002; 63:151–157.
47. Giraud P, De Rycke Y, Dubray B, *et al.* Conformal radiotherapy (CRT) planning for lung cancer: analysis of intrathoracic organ motion during extreme phases of breathing. *Int J Radiat Oncol Biol Phys* 2001; 51:1081–1092.
48. Shimizu S, Shirato H, Ogura S, *et al.* Detection of lung tumor movement in real-time tumor-tracking radiotherapy. *Int J Radiat Oncol Biol Phys* 2001; 51:304–310.
49. Engelsman M, Sharp GC, Bortfeld T, *et al.* How much margin reduction is possible through gating or breath hold? *Phys Med Biol* 2005; 50:477–490.
50. Mah D, Hanley J, Rosenzweig KE, *et al.* Technical aspects of the deep inspiration breath-hold technique in the treatment of thoracic cancer. *Int J Radiat Oncol Biol Phys* 2000; 48:1175–1185.
51. Keall PJ, Kini VR, Vedam SS, *et al.* Potential radiotherapy improvements with respiratory gating. *Australas Phys Eng Sci Med* 2002; 25:1–6.
52. Tada T, Minakuchi K, Fujioka T, *et al.* Lung cancer: intermittent irradiation synchronized with respiratory motion--results of a pilot study. *Radiology* 1998; 207:779–783.
53. Willoughby TR, Forbes AR, Buchholz D, *et al.* Evaluation of an infrared camera and X-ray system using implanted fiducials in patients with lung tumors for gated radiation therapy. *Int J Radiat Oncol Biol Phys* 2006; 66:568–575.

54. Nagata Y, Negoro Y, Aoki T, *et al.* Clinical outcomes of 3D conformal hypofractionated single high-dose radiotherapy for one or two lung tumors using a stereotactic body frame. *Int J Radiat Oncol Biol Phys* 2002; 52:1041–1046.
55. Harada T, Shirato H, Ogura S, *et al.* Real-time tumor-tracking radiation therapy for lung carcinoma by the aid of insertion of a gold marker using bronchofiberscopy. *Cancer* 2002; 95:1720–1727.
56. Blomgren H, Lax I, Göranson H, *et al.* Radiosurgery for tumors in the body: clinical experience using a new method *Journal of Radiosurgery* 1998; 1:63–74.
57. Reichner CA, Collins BT, Gagnon GJ, *et al.* The placement of gold fiducials for CyberKnife stereotactic radiosurgery using a modified transbronchial needle aspiration technique. *Journal of Bronchology* 2005; 12:193–195.
58. Krol AD, Aussems P, Noordijk EM, *et al.* Local irradiation alone for peripheral stage I lung cancer: could we omit the elective regional nodal irradiation? *Int J Radiat Oncol Biol Phys* 1996; 34:297–302.
59. Williams TE, Thomas CR, Jr., Turrisi AT, 3rd. Counterpoint: better radiation treatment of non-small cell lung cancer using new techniques without elective nodal irradiation. *Semin Radiat Oncol* 2000; 10:315–323.
60. Robertson JM, Ten Haken RK, Hazuka MB, *et al.* Dose escalation for non-small cell lung cancer using conformal radiation therapy. *Int J Radiat Oncol Biol Phys* 1997; 37:1079–1085.
61. Slotman BJ, Antonisse IE, Njo KH. Limited field irradiation in early stage (T1-2N0) non-small cell lung cancer. *Radiother Oncol* 1996; 41:41–44.
62. Giraud P, Antoine M, Larrouy A, *et al.* Evaluation of microscopic tumor extension in non-small-cell lung cancer for three-dimensional conformal radiotherapy planning. *Int J Radiat Oncol Biol Phys* 2000; 48:1015–1024.
63. Hara R, Itami J, Kondo T, *et al.* Stereotactic single high dose irradiation of lung tumors under respiratory gating. *Radiother Oncol* 2002; 63:159–163.
64. Hiraoka M, Nagata Y. Stereotactic body radiation therapy for early-stage non-small-cell lung cancer: the Japanese experience. *Int J Clin Oncol* 2004; 9:352–355.
65. Nyman J, Johansson KA, Hulten U. Stereotactic hypofractionated radiotherapy for stage I non-small cell lung cancer--mature results for medically inoperable patients. *Lung Cancer* 2006; 51:97–103.
66. Whyte RI, Crownover R, Murphy MJ, *et al.* Stereotactic radiosurgery for lung tumors: preliminary report of a phase I trial. *Ann Thorac Surg* 2003; 75:1097–1101.
67. Berger B, Wood R, Fiedler J, *et al.* Initial Report of CyberKnife Radiosurgery for Non-Surgical Lung Tumors. CyberKnife Society Users' Meeting. Carlsbad, CA; 2006.
68. Collins BT, Collins SP, Erickson K, *et al.* CyberKnife® Frameless Image-guided Fractionated High-dose Stereotactic Radiosurgery with the Synchrony™ Motion Tracking Module in the Treatment of Small Peripheral Lung Tumors. CyberKnife Society Users' Meeting. Carlsbad, CA; 2006.
69. Bellairs E, Fink A, Palmer M, *et al.* Extracranial Radiosurgery for Lung Tumors. CyberKnife Society Users' Meeting. Carlsbad, CA; 2006.

Toxicity and Efficacy of Treating Mediastinal and Hilar Lesions

19

Andrew S. Fink, June A. Kim, D. Ross McBride, Micheal J. Hervey II, Penny J. Sinner, Jodi L. Mammenga, Guy R. Sherwood, and Lori Copsey

CONTENTS

19.1 Introduction

When pondering the choice of therapy for a cancer, clinicians must weigh the risks and benefits of each modality individually for each patient, taking into account age, coexisting medical problems, general fitness, therapeutic toxicity, and therapeutic efficacy. This chapter will focus on lung cancer and other tumors in an area of the chest fraught with potential complications: the mediastinum and hilum. We will briefly review the literature on toxicity and efficacy of radiotherapy to the central chest, and then present some of our early experience with stereotactic body radiotherapy (SBRT) to this difficult area.

19.2 The Mediastinum

The mediastinum (Fig. 19.1) is the central chest bounded by the sternum anteriorly, spine posteriorly, the thoracic inlet superiorly, and the diaphragm inferiorly. It is sandwiched by lung parenchyma and pleura laterally. The hilar vessels and bronchi are an extension of the mediastinum into the parenchyma. The parenchyma has "parallel functionality"; structural redundancy makes it resistant to complete dysfunction caused by a single small lesion. In contrast, the hilum and mediastinum are packed with radiation-sensitive structures that have "serial functionality", where injury to a small portion (usually of a tubular structure) may render the whole organ non-functional. This includes the esophagus, tracheobronchial tree, great vessels, and neurologic structures such as the spinal cord posteriorly, brachial plexus superiorly, phrenic and vagus nerves, and thoracic duct. The thymus gland, lymph nodes, and heart make up the remainder of the mediastinum. These are considered the "organs at risk".

Most cancer treated in the mediastinum is metastatic, from either the lung (most frequently) or elsewhere in the body. Inherently, therefore, most cancer treated in this area is already advanced, usually Stage III or IV. As such, therapeutic approaches here are often palliative, though for isolated spread they may be intended to cure. Furthermore, surgery is seldom an option for treatment. This leaves chemotherapy, which is usually not considered curative, and radiation. The use of radiation to this area is limited by its toxicity, which is explored in the next section.

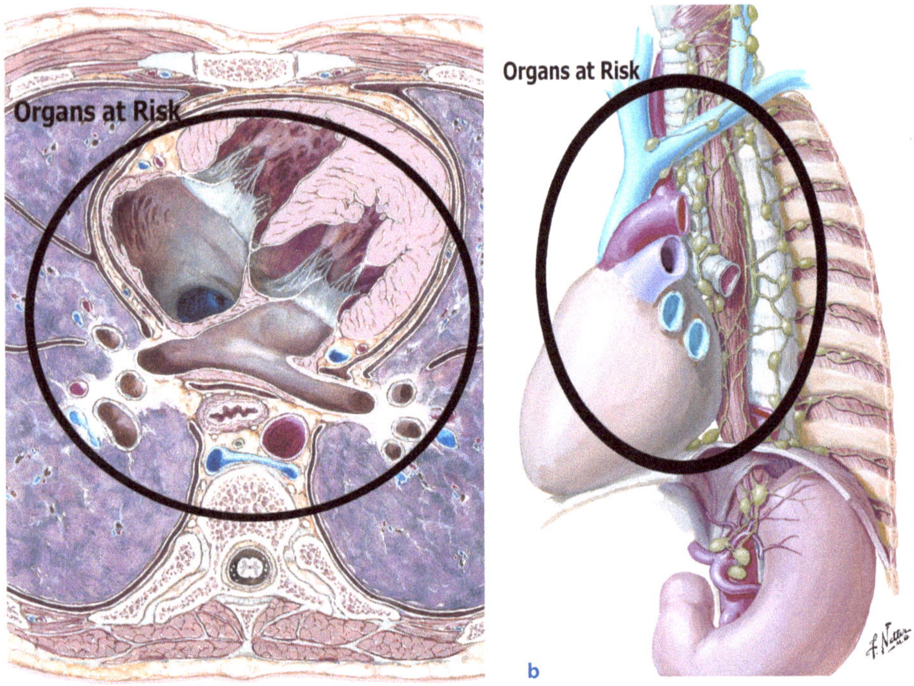

Fig. 19.1a,b. Axial and sagittal views of the mediastinum.

19.3 Lung Parenchyma

It has long been recognized that if too much lung tissue receives too much radiation, patients die. At lower doses and smaller volumes patients may experience dyspnea, or perhaps cough or hemoptysis. These are the early symptoms of pneumonitis, along with occasional chest pain, fever, and chest fullness or congestion. Symptom severity and rapidity of onset are dependent on the amount of lung involved in treatment.

"Early" radiation-induced lung toxicity, radiation pneumonitis occurring within weeks of treatment, and "late" toxicity, fibrosis occurring within months to years, are probably the most well studied of radiation injuries in the chest. Pneumonitis is felt to result primarily from alveolar and end bronchiolar injury, with reversible build-up of non-cellular inflammatory fluid in the alveoli and interstitial spaces of the alveoli and capillaries [1, 2]. Fibrosis may be the outcome of a more serious episode of pneumonitis [3], with cellular injury great enough that recovery is not possible. Clinical pneumonitis occurs in about 5–15% of radiated patients with lung cancer [1, 4, 5] or other cancers involving the chest such as the breast [6], lymphoma [7], etc.

Toxicity scales that focus primarily on clinical symptoms are probably the most valuable for clinical practice. Grade 1 and 2 are usually considered "acceptable" in terms of dyspnea and activities of daily living, whereas Grade 3 or higher are more serious and dose-limiting in escalation studies. A CT scan is usually the radiographic test of choice. A focal area of fibrosis is expected in the area of treatment and can be difficult to distinguish radiographically from other causes of infiltrates. Ventilation perfusion scans (V/Q scans), arterial blood gases (ABGs), and pulmonary function tests (PFTs) may be helpful in assessment of pneumonitis, with the greatest changes noted in the perfusion portion of the V/Q scan and diffusion capacity (DLCO) in ABGs. These changes presumably reflect the interstitial alveolar edema and endothelial injury.

Most early studies on tolerance were based on whole lung radiation, which is perhaps helpful but is becoming less relevant with newer techniques using image guidance and 3D-CT localization that decrease target volumes. Emami et al. [8] updated previous historical normal tissue tolerances by analyzing the tolerances of normal tissues of concern in an era of 3D treatment planning. TD 5/5 represents the dose at which there is a probability of 5% risk of complications within 5 years. Similarly, a TD 50/5 represents the dose at which there is a probability of 50% risk of complications within 5 years. These doses were "calculated" relative to the volume of the organ at risk. For lung, the TD 5/5 for 1/3, 2/3, or 3/3 of lung volume treated was 4500, 3000, and 1750 cGy, respectively. For a 50% 5-year chance of a complication of pneumonitis, the dose was 6500, 4000, and 2450 cGy, respectively. Predictably, the larger the volume of normal lung included in a treatment field, the smaller the dose necessary to create symptoms of pneumonitis or fibrosis.

Graham et al. found that data derived from dose volume histograms and, in particular, the total lung volume receiving more than 20 Gy (V_{20}), could be correlated to the incidence of severe pneumonitis [9]. In their analysis of 99 medically inoperable non-small cell lung cancer (NSCLC) patients treated with 3D-conformal radiation therapy (3D-CRT), patients with a $V_{20} < 22\%$ were free of pneumonitis, and those with $V_{20} > 40\%$ had a 23% incidence of Grade 3–5 pneumonitis. Three died of pneumonitis.

Similarly, Kong et al. [10] recently published toxicity results of a 3D-CRT dose escalation study involving 109 patients with NSCLC followed for 5 years, or until death. Though no Grade 4 or 5 toxicity occurred, 14.6% experienced Grade 2 or 3 pneumonitis (requiring steroid course or oxygen respectively), and 13.8% experienced Grade 2 to 3 fibrosis. They found that V_{20}, V_{13}, lung effective volume (Veff), the mean lung dose (MLD), and normal tissue complication probability (NTCP) all correlated with pneumonitis, while, interestingly, PFTs, tumor location in parenchyma, tumor size by gross tumor volume (GTV), and total tumor dose were not correlated. Although 3D-CRT toxicity rates may not be directly applicable to SBRT, they do establish the validity of dosimetric measures such as the V_{20} and MLD for comparison. It appears that with increasing conformality, toxicity risk is more dependent on volume treated than dose.

In 2001 Uematsu et al. [11] reported on the use of SBRT for 50 patients with early stage NSCLC. This was followed with an update at the ASTRO meeting in 2003 [12]. Doses ranged from 50 to 60 Gy in 5–10 fractions, and 18 patients had prior conventional radiotherapy. Other than the fibrosis expected in the ablated treatment area of the tumor, there were no clinical pneumonitis or fibrosis reported after a median follow-up of 36 months in the 33 patents alive at that time. Of the 17 who died, none died of pulmonary complications. No Grade 3 or higher toxicity was reported by either Hof et al. or Nagata et al. [13, 14]. The latter authors reported on 45 patients with Stage 1A and B NSCLC treated with 48 Gy in 4 fractions (biologically effective dose, or BED = 105 assuming an α/β ratio of 10). These authors obtained a mean V_{20} of 4.5% (range 1–11.6%).

In 2005 McGarry et al. reported on a Phase I dose escalation trial of SBRT that involved 47 patients with Stage T1A and B N0M0 NSCLC [15]. Doses ranged from 24 Gy in 3 fractions to 72 Gy in 3 fractions. The median follow-up was 15.2 months. Three patients had Grade 3–5 pneumonitis, with 2 at the highest dose of 72 Gy. In 2006 Bauman et al. reported on SBRT in medically inoperable Stage I lung cancer in the Nordic countries. Doses ranged from 30–48 Gy 2–4 fractions. Grade 3–4 toxicity included 1 pneumonitis, 2 atelectases, 2 "decreased lung function", and 1 pneumonia (by RTOG criteria) [16]. The rate of serious parenchymal complications was 4% in this series. Finally, in 2006, Le et al. reported on a Phase I dose-escalation trial using single fraction SBRT; for lung tumors, mimicking the single fraction approach to brain tumors [17]. Of 32 patients, 4 were identified as having pneumonitis after "central" SBRT, the remaining 28 were undefined. All received 25 Gy in a single fraction. Two were Grade 2, but one died of a "PE and recall pneumonitis" and another died after developing an "effusion and pneumonitic changes".

Despite the complications mentioned above, these current and increasingly mature studies indicate that SBRT is well tolerated and appears to have less pneumonitis and pulmonary fibrosis than standard radiotherapy. Nearly all follow-up scans indicate radiographic changes in the treated area (presumably fibrosis), but the volumes are apparently small enough not to cause significant clinical symptoms. Most of these studies, however, have short enough follow-up that further pulmonary fibrosis is not yet completely ruled out and they emphasized earlier stage lung cancers that tended to be more peripheral. Therapy focused on the mediastinum and hilum may include less parenchyma but more of other organs at risk such as the esophagus.

19.4 Esophagus

Early esophageal toxicity at standard doses of radiation is primarily a mucosal effect with esophagitis and possible ulceration. It usually manifests within 2 weeks of starting conventional radiation therapy with odynophagia or dysphagia. While it may persist for several weeks after stopping therapy, it is usually self limited and easily treated [18]. Mild symptoms are frequent, but serious symptoms are uncommon with standard doses of radiation for lung cancer, on the order of 1–2% [19–21]. Severe esophagitis may lead to inanition and can be dose-limiting or lead to breaks in therapy with a subsequent drop in treatment efficacy.

Late toxicity may be observed at higher doses [18, 22] and is primarily muscular in origin with fibrosis leading to strictures or dysmotility. Chronic mucosal ulceration and, uncommonly, perforation, can occur as well. Emami et al. analyzed the late toxic effects of stricture and perforation [8]. They found the TD 5/5 to be 6000, 5800, and 5500 cGy for 1/3, 2/3, and 3/3 of the esophageal volume, and the TD 50/5 to be 7200, 7000, and 6800 cGy for 1/3, 2/3, and 3/3 of the volume, respectively. This is a mild volume effect that may reflect the serial nature of the esophagus (see also Bradley et al. [23]). However, several studies suggest that these measures are not as critical as the dose [24, 25]. This might be relevant for SBRT in particular, where potentially small esophageal volumes could get large doses. Being a serial organ, a focal lesion (e.g., a stricture) could still render the esophagus non-functional.

Adding sensitizers, especially concurrent chemotherapy, can dramatically increase esophagitis rates to 20–46% in conventional radiation therapy [18, 26]. In 2003, Singh et al. [27] reported their experience with 207 consecutive NSCLC patients treated with 3D-CRT at 70 Gy in 2-Gy fractions. Grade 3–5 esophageal toxicity (RTOG criteria) was found in 16 patients (8%). Seven of these had both acute and chronic toxicity. One died of esophageal perforation from an ulcer. The most relevant risk factors on univariate analysis were the use of concurrent (but not sequential) chemotherapy and the maximal esophageal point dose. Fourteen of the 16 who had severe esophagitis received concurrent chemotherapy and radiation with a maximal radiation point dose > 58 Gy. For the two who did not get chemotherapy, the point doses were > 69 Gy. This rate of esophagitis with escalated doses and concurrent chemotherapy has been confirmed by others, though the benefits in lung cancer control with chemoradiation efficacy are likely worth it [28].

The radioprotectant Amifostine, an organic thiophosphate that preferentially scavenges oxygen free radicals in normal tissue, has been studied extensively in esophagitis. Some studies have found it to be helpful; others have not, including the RTOG 9801 [29].

Using SBRT, contrary to conventional radiation, no esophagitis greater than Grade 2 was reported in a series of studies with a combined, heterogeneous group of 280 lung cancer patients [11, 13–16]. Onimaru et al. [30], however, described a patient bleeding to death from a "radiation-induced ulcer" 5 months after receiving 48 Gy in 8 fractions to a 3.5 cm tumor "posterior to the right main bronchus". They attributed the injury to a hot spot observed upon recontouring. These authors treated 58 lesions in 46 patients, and 39 tumors were either "large or centrally located". They observed no other adverse reactions except a Grade 2 chest wall pain that resolved after 8 months. Wulf et al. [31] also reported a single Grade 3 esophageal ulceration after treating "a tumor close to the mediastinum" with 30 Gy in 3 fractions. Nyman et al. [32] reported 4 cases of Grade 1 esophagitis out of 45 patients treated with 45 Gy in 3 fractions. There was no Grade 2 or greater esophagitis.

These studies appear to substantiate that the decreased volumes and escalated doses found in SBRT may be superior to conventional radiation in regard to acute esophageal toxicity. In terms of late toxicity, however, these conclusions may be premature as stricture formation can take years to develop, and few patients in these reports were treated for central lesions. Also, many SBRT studies have a paucity of information regarding adjuvant chemotherapy, obviously an important factor in the incidence of esophageal toxicity. Nevertheless, these results are encouraging regarding use of SBRT in the mediastinum.

19.5 Airways

The airways include the trachea, carina, mainstem bronchi, lobar bronchi, and tertiary and subsegemental bronchi out to the terminal bronchioles. The "major airways" extend to the hila, are lined by a mucosa containing goblet cells and ciliated columnar epithelium, and have walls of cartilage, smooth muscle, glands, and an elastic stroma. All branches are in close proximity to lymphatics and pulmonary vessels. The trachea has the esophagus posteriorly and great vessels arching around it. These anatomical relationships are significant because radiation injury can be quickly fatal if a fistula from an airway develops into a blood vessel or the esophagus.

Injury to the major airways causes dyspnea, chronic cough, and possibly hemoptysis. Early injury is more likely a mucosal problem with bronchitis, edema, and exudate. Late injury manifests as stricture with downstream atelectasis or potentially fistulization. Grading criteria are based on the clinical manifestations of atelectasis, dyspnea, cough, and hemoptysis, and do not include the anatomical findings of bronchitis, ulceration, stenosis, or fistula noted on bronchoscopy.

There is surprisingly little literature on stenosis. Neither Emami et al. [8] nor McDonald et al. [4] specifically mention stenosis. RTOG, SWOG have no specific criteria for it. Nevertheless, reports of bronchial stenosis after 3D-CRT, in some cases leading to death, have been published [33, 34]. Endobronchial high-dose-rate (HDR) brachytherapy, which may also be relevant to the mediastinal use of SBRT, has been associated with an 11% risk of radiation bronchitis and stenosis in early studies [35]. More recent studies have obtained similar rates [36].

SBRT is not exempt. Uno et al. [37] reported in their series of 20 lesions that one patient treated with 50 Gy in 5 fractions developed chronic cough and bronchial stenosis in a segmental bronchus that, it was discovered in later review, received 90% of the dose. Onishi et al. [38] reported on SBRT performed on Stage I NSCLC patients compiled from 11 Japanese institutions. It included 245 patients, 90 of whom had T2 lesions, treated with a range of doses and fractions. Although a heterogeneous group, it is noteworthy that only 6.9% had greater than Grade I toxicity, and only a few had Grade 3 or greater pneumonitis, esophagitis, or dermatitis. The majority were treated with BEDs greater than 100 Gy. A single patient out of 245 was noted to have "chronic segmental bronchitis and wall thickening causing atelectasis." No further details were supplied.

McGarry et al. reported on one Grade 2 bronchitis and one Grade 3 tracheal necrosis in 47 patients treated with SBRT in a dose escalation trial [15]. These individuals received 60 Gy in 3 fractions and 72 Gy in 3 fractions respectively (>180 Gy BED in both cases). Song et al. [39] reported one bronchoscopically confirmed stenosis and a second likely bronchial stenosis (patient refused bronchoscopy but had lobar collapse downstream from the lesion). These two (out of 17) patients were treated for pulmonary metastases to the hilar region with 35 and 41 Gy over 3 fractions. Of interest, the former had only "scar" at the tumor site at 11 months follow-up (with no tumor on bronchoscopy), and the latter had a "partial response" at 9 months.

Several interesting aspects may be ascertained from these SBRT studies. First, where specified, most cases occurred when the target was "centrally located". Second, only "symptomatic" stenoses were confirmed by bronchoscopy; it seems probable that there were other sub-clinical cases of stenosis. Perhaps cases characterized as "parenchymal fibrosis" in other studies are in fact permanent collapse from a bronchial injury "upstream", making this entity more common than realized. Third, stenosis seems to occur as a late effect, much like the fibrotic effects in other organs at risk (e.g. esophagus). Stenosis was identified more amongst those who were surviving longer, and in some cases in those without other evidence of disease. Finally, of the studies that involved central radiation at high doses, no dose volume information or incidence of concurrent or other chemotherapy was given.

Unfortunately, when considering the airways, one must also consider the unnerving complication of hemoptysis. Hemoptysis is a communication between the airway and a blood vessel. It may be as simple as mucosal inflammation with minor streaking, or it may be massive, causing rapidly fatal hemorrhage when it involves the pulmonary artery or other major vessels. Fatal hemorrhage is not uncommon in HDR brachytherapy [40]. But this population is predisposed to hemoptysis, usually with endobronchial tumor and is often treated palliatively, perhaps with additional CRT. Using "single shot" SBRT, Wulf et al. [31] reported one episode of fatal hemorrhage from the pulmonary artery, but they were unsure if this was due to the treatment or tumor infiltration. Le et al. [17] reported a death from a 25-Gy single-fraction treatment of a "central" tumor leading to a tracheoesophageal fistula and eventually fatal hemoptysis 6 months after therapy. Interestingly, an autopsy "confirmed the fistula without evidence of cancer". Timmerman et al. [41] also described an episode of fatal hemorrhage 19.5 months after an apparent hemoptysis. These early reports suggest that single fraction treatment of the mediastinum or doses with high BEDs may be at greater risk for this frightening complication. Still, hemoptysis has otherwise not been reported, and it is unlikely to be a common entity.

19.6 Heart

There is much less to say about the heart. The main radiation-induced injury is a result of pericardial reactions with effusions or, generally later, restrictive pericarditis. Most studies come from mantle radiation of Hodgkin's disease, which is associated with a 9.5% cumulative incidence of pericarditis at 10 years after doses higher than 41 Gy covering 60% of the cardiac silhouette [42]. Emami et al.'s selected endpoint for heart TD 5/5 and 50/5 was "pericarditis" (6000 cGy and 7000 cGy, respectively, for one third of the heart volume in the field) [8]. In addition, cardiomyopathy that is presumed to be mediated by injury to the myocardial capillaries, mild fibrosis, and radiation-induced atherosclerosis have been reported [43]. Cardiac valves appear to be clinically unaffected at standard doses.

Takayama et al. used dose-volume histograms to analyze dose distributions to the heart and other organs at risk due to treatment of peripheral lung lesions with 48 Gy in 4 fractions [44]. Only 5 of 48 patients received a cardiac dose per fraction higher than 5 Gy and no cardiac toxicity was noted in any of the patients. McGarry et al. [15] described two episodes of Grade 2 and 3 pericardial effusion in patients receiving 54 Gy or 66 Gy in 3 fractions, respectively. No other information was given. Timmerman et al. [41] reported on a death from "complications of a pericardial effusion", in a single patient (out of 70), that was in a Phase II trial of SBRT treating patients with 60–66 Gy in 3 fractions. He noted that this tumor was "adjacent to the mediastinum superior to the hilum". In no other studies are cardiac toxicities mentioned, however, and the relatively high doses noted in the protocols used by McGarry et al. and Timmerman et al. had significantly higher BEDs and may be indicative of a dose limitation to the mediastinum and hilar regions of the central chest. These reports suggest that the most serious heart-related concern for clinicians treating with SBRT may be compensating for the movement of the tumor caused by the beating heart.

19.7 Our Early Experience

This section will discuss our early experience at Healtheast St. Joseph's Hospital in St. Paul, Minnesota, with the safety and efficacy of robotic SBRT in the mediastinum and hilum.

19.8 Patients and Methods

A total of 30 patients were treated for tumors originating in the lung and oligometastases from a variety of other organs. This study was undertaken to assess the early toxicity and crude local control for the treatment of tumors located in close proximity to the hilum and mediastinum, specifically those that fit within the primary bronchial "exclusion zone" described in RTOG 0236 (see Fig. 19.2), including those within 2 cm of the tracheobronchial tree or being T3 on the basis of direct contact with the mediastinal pleura, parietal pericardium, or mediastinal fat [45].

Sixteen of these tumors were previously treated with radiation and 21 were previously treated with chemotherapy (though none were receiving chemo-

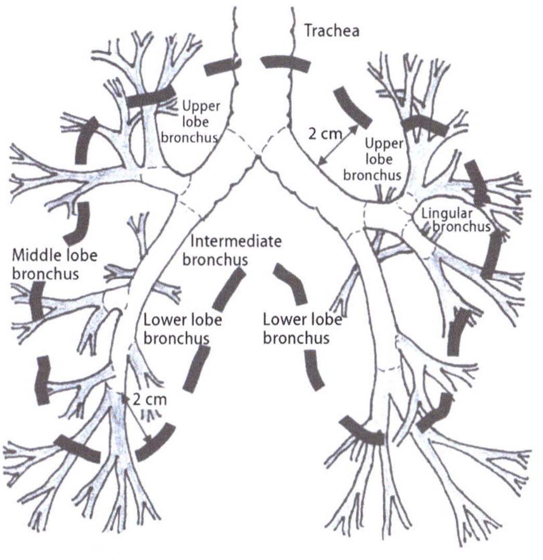

Fig. 19.2 Two-centimeter exclusion zone of the proximal bronchial tree, from RTOG protocol 0236.

therapy during a 2-month window around the time of radiation). All potential surgical candidates were previously seen by an outside thoracic surgeon and were considered unresectable, usually based on the higher stage of their disease but occasionally for medical reasons (e.g., poor pulmonary function). In most cases these patients had exhausted every other avenue of treatment.

Current CT scans or PET/CT scans of the tumors were evaluated. The tumor was then classified as a primary lung cancer, a lung cancer recurrence, or a metastasis from a distant site. Some were treated definitively, while in others, SBRT was performed for palliation of pain or, more commonly, for impending obstruction. The latter criteria were based on bronchoscopic findings or CT evidence, often with clinical signs such as focal wheezing or intermittent atelectasis. Patients were informed of the treatment risks, including possible fatality, and signed an Institutional Review Board approved consent to participate in this research.

Patients were excluded if there were more than two lesions within the region. A patient was also not considered a candidate if gross invasion of the main pulmonary artery was noted on CT. If gross involvement of the esophagus or a main airway branch was a possibility based upon imaging or clinical symptoms, an endoscopic exam of the bronchus or esophagus was performed for direct visualization of the airways. Proximity of the tumor to these structures was not considered a specific contraindication, and many tumors were in direct contact with the airways, esophagus, or smaller braches of the pulmonary artery.

Potential candidates were screened for presenting symptoms including dyspnea, supplemental oxygen use, pain, cough, hemoptysis, hoarseness, swallowing problems, and fatigue. KPS scores were calculated. The maximum size (longest diameter) of the tumor was identified; the largest tumor was 13.5 cm.

Treatment was performed with the CyberKnife® (Accuray Incorporated, Sunnyvale, CA) which utilizes fiducials placed within the tumor to allow tumor motion tracking during treatment. The fiducials were placed percutaneously or bronchoscopically. One patient experienced a pneumothorax secondary to percutaneous placement that required a chest catheter overnight but was otherwise uneventful. A simulation CT was performed 7–12 days following placement, utilizing 1.25 mm thick slices with the patient immobilized in a Vac Loc bag in the treatment position.

Contouring was performed utilizing lung windows for tumors within the lung parenchyma, or mediastinal windows for tumors in or abutting the mediastinum and hilum. Critical structures included the esophagus, heart, tracheobronchial tree, spinal cord, pulmonary artery, and lungs. The clinical tumor volume (CTV) was often equivalent to the gross tumor volume (GTV), though occasionally a 2 to 5 mm margin was used to account for microscopic extension. However, the planning tumor volume (PTV) equaled the CTV as movement of the tumor was accounted for with the CyberKnife's ability to follow targets during treatment using the Synchrony® (Accuray Incorporated, Sunnyvale, CA) Respiratory Motion Tracking System.

Dose was determined by the size of the lesion, its relationship to nearby critical structures, and prior radiotherapy. Varying doses and fractionation schemes were therefore used. In general, a more conservative dose was used when there was prior radiation or proximity of the tumor to a critical structure. Dosing of critical structures was based on known tolerances. Higher doses were administered when there was no prior radiation, and the lesion was (relatively) further away from critical structures in the central chest. Because of the varied regimens, prescription doses were converted into biologic equivalent doses (BEDs) assuming an α/β ratio of 10 Gy, and then were somewhat arbitrarily segregated into BED ranges. The lowest range (<72 Gy), corresponded to high-end conventional radiation doses. A BED of 100 was also chosen as a minimum acceptable dose cut-off because some SBRT literature suggests that doses greater than 100 Gy are necessary for high rates of local control in lung lesions.

Treatments usually occurred on consecutive days or every other day depending on patient convenience but were completed within 10 days regardless. The patients were all treated as outpatients, and received a follow-up phone call the day after, a week after, and a month after treatment. Patients were then followed with a physical exam and CT scan or PET/CT scan every three months thereafter. Follow-up included assessment for acute toxicity (less than 3 months) and late toxicity (greater than 3 months) using NCI CTCAE v3 criteria. Local control was defined as a radiologically stable lesion, a partial response (decrease in longest diameter by 30%), or a complete response (disappearance of all target lesions).

19.9 Results

Table 19.1 gives the characteristics of the patients and tumors treated. Thirty patients with 36 tumors located in or abutting the mediastinum and hilum as geographically defined above (in or abutting the "bronchial exclusion zone" and mediastinum) were treated from March 2004 through July 2006.

All patients had a KPS of 60 or higher. Sixteen of the 30 patients had previous radiation to their lesion (and thus to some portion of the "exclusion zone"). Doses ranged from 2000 to 6000 cGy in 1–5 fractions. Corresponding BEDs ranged from 42–180 Gy but only a single patient had a BED >125 Gy. Max dose for 5% of the GTV ranged from 2603 cGy to 7462 cGy (median 4618 cGy). The doses to surrounding critical structures were also kept as low as possible and within tolerances for standard radiation schedules. Lesion size, using longest diameter, ranged from 0.8–

Table 19.1. Patient, target, and treatment characteristics

	30 patients, 36 lesions
Gender	
Male	16 (53%)
Female	14 (47%)
Age	
Range	33–79
Median	65.5
Mean	62.6
Pre-treatment KPS	
Range	60–100
Median	90
Mode	100
Type of Cancer and Stage	
Primary Lung	16 (53%)
Stage of Primary Lung	
Stage I	5
Stage II	2
Stage III	5
Stage IV	4
Histological Type of Primary Lung	
NSCLC (unspecified)	13
Adenocarcinoma	2
SCLC	1
Recurrent Lung	5 (17%)
NSCLC (unspecified)	3
SCLC	2
Metastatic	9 (30%)
RCCa	6
Breast	1
Colon	1
Gastroesophageal	1
Tumor Size	
Range	0.8–1.35 cm
Median	3.8 cm
Mean	4.3 cm
GTV	
Range	1.1–708.6 cc
Median	33.5 cc
Mean	78.3 cc
Prescription Dose	
Range	2000–6000 cGy
Median	3600 cGy
Mean	3644 cGy
Fraction range	1-5
Max dose (5% of GTV)	
Range	2603–7462 cGy
Median	4618 cGy
Mean	4735 cGy
BED dose (α/β = 10)	
Range	43.2–180 Gy
Median	75.6 Gy
Mean	80.1 Gy
Prior Radiation	16 (53%)
Prior Chemotherapy	21 (70%)

13.5 cm (median 3.8 cm). The GTVs ranged from 1.1 to 708 cc (median 33.5 cc). The median follow-up for the survival analysis was 9.6 months (range 0.9–26.2 months). The overall Kaplan-Meier survival rate of patients receiving CyberKnife treatment for mediastinal and hilar lesions was 34% at 26 months (Fig. 19.3).

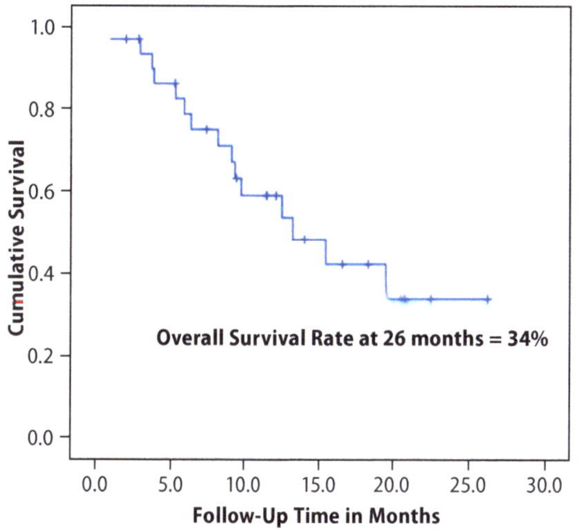

Fig. 19.3 Overall Kaplan-Meier Survival Curve for patients receiving CyberKnife treatment for mediastinal and hilar lesions (n=30).

19.9.1 Efficacy

Information regarding post-treatment tumor size was available for 32 of 36 lesions treated. No information on post-treatment lesion size was available on 4 lesions (one patient died one month after treatment, one patient left the country, one patient was recently treated, and one was lost to follow-up); therefore they were excluded from the local control analysis. For the 32 lesions with information regarding post-treatment lesion size, the median follow-up time to assess local control was 8.2 months (range 2.0–26.2 months).

Twenty-six of the 32 lesions had local control (stability, partial response, or complete response) with a crude local control rate of 81% (Table 19.2). To determine if there was a correlation between local control and dose, local control was compared amongst the different BED ranges using the Pearson's Chi-Square test. There was no association between BED and local control in this population ($p = 0.77$).

Table 19.2 Crude local control rates and biological equivalent doses (BED)

BED	Lesions controlled / lesions treated (local control) (n = 32)
< 72	10/13 (77%)
72–100	8/9 (89%)
> 100	8/10 (80%)
All lesions	26/32 (81%)

19.9.2 Toxicity

Cancer patients are often fatigued, for reasons that include their tumor burden, grueling therapies, depression, and so forth. Out of 30 patients treated, 11 fit criteria for Grade 1 fatigue and 10 for Grade 2 fatigue before CyberKnife treatment. Post-treatment, 13 patients experienced no change in fatigue, 10 patients experienced less fatigue, and 6 patients experienced greater fatigue. Three of the six patients that experienced greater fatigue post-CyberKnife treatment experienced Grade 3 fatigue, but all three started at Grade 2. In only one case was Grade 3 fatigue present at 3 months (and the patient was receiving further chemotherapy at that time).

Cancer often causes pain. Before treatment, 19 patients had no pain, 4 people complained of Grade 1 pain, 4 had Grade 2 pain, and 3 had the equivalent of Grade 3 pain. This latter group had involvement of the posterior mediastinum and/or vertebral bodies. Much of the lower Grades of pain did not necessarily appear to be caused by the tumor in the area of treatment (i.e., it was caused by other tumors or other non-cancer causes outside of the treatment field). Pain was unchanged in 19 patients post-treatment, improved for 6 patients, and worsened for 4 patients. In the four patients that experienced increased pain post-treatment one patient who experienced no pain pre-treatment experienced Grade 1 pain acutely. This patient's pain resolved within 3 months. The other 3 patients experienced Grade 3 pain post-treatment (one went from Grade 1 to 3 and two patients went

Table 19.3 Selected Symptom and Side Effect Grades (NCI CTCAE v3)

Symptom/ Side Effect	Presenting Symptom (n=30)	Acute Side Effect (n=29)	Late Side Effect (n=25)
Cough			
Grade 0	17	23	20
Grade 1	12	4	3
Grade 2	1	2	2
Radiation Dermatitis			
Grade 0	NA	27	24
Grade 1	NA	2	0
Grade 2	NA	0	1
Dyspnea			
Grade 0	18	19	19
Grade 1	5	0	1
Grade 2	3	5	0
Grade 3	4	5	5
O_2 use			
0 Liters	26	data not available*	20
1 Liter	0	data not available*	0
2 Liters	2	data not available*	3
3 Liters	1	data not available*	0
4 Liters	1	data not available*	2
Bleeding			
Grade 0	NA	27	25
Grade 1	NA	1	0
Grade 5	NA	1	0

*All four patients requiring O_2 long-term began using it within 3 months post-treatment. Data was not collected on patients that required short-term O_2 use.

from Grade 2 to 3). Two of these patients continued to experience Grade 3 pain three months post-treatment. Amongst the 3 patients with Grade 3 pain pre-treatment, one patient's pain resolved completely, one decreased to Grade 2, and one patient was lost to follow-up.

Table 19.3 describes the change in Grades from pre-treatment to post-treatment (both acute and late effects) for selected symptoms and side effects. Thirteen patients presented with a Grade 1 or 2 cough. Acutely, after CyberKnife treatment, 17 patients experienced no change in their cough, 9 patients improved, and 3 got worse. For the 3 patients who got worse the highest Grade of cough reported was Grade 2.

Dermatitis was experienced by three patients, two with Grade 1 and one with Grade 2. The two Grade-1 patients experienced dermatitis acutely whereas the Grade-2 patient experienced dermatitis 15 months post-treatment after starting Erbitux.

Inoperable chest cancer patients often have dyspnea of varying degrees. COPD is common in these patients as many are smokers. But potentially impending bronchial obstruction and atelectasis can also contribute to dyspnea. Four patients started with dyspnea during activities of daily living; all of these were on oxygen (Grade 3 by CTCAE criteria). One patient got worse after treatment and required 4 liters of oxygen during the day and night (increased from a baseline requirement of 3 liters of oxygen at night). Three patients maintained their pretreatment levels of dyspnea and oxygen requirements (see Table 19.3). One patient experienced Grade 1 dyspnea pre-treatment and Grade 3 dyspnea post-treatment, and now requires 2 liters of oxygen.

Bleeding was, fortunately, not common after treatment despite the frequent proximity of these tumors to large vascular structures, often including smaller branches of vessels within the target. Only 2 patients had post-treatment bleeding, one Grade 1 and the other Grade 5. Both were on Coumadin for atrial fibrillation. The former had his coumadin regulated (INR was 4), and was still able to continue with therapeutic anticoagulation with tight control and no further bleeding. Unfortunately, the latter individual had an INR >6 and refused to remain at his local ER for therapeutic intervention after what was, in retrospect, a "sentinel bleed". His tumor was large, 7.5 cm, located in the AP window, and he had had conventional radiation to 6100 cGy 15 months before receiving 30 Gy in 3 fractions from the CyberKnife (BED = 60 Gy). This is the only death that was related to the CyberKnife treatment in these 30 patients over this relatively short follow-up period.

19.10 Discussion

To date there has been little published on SBRT treatment to the mediastinum and hilum. Most studies have focused on early-stage lung cancers, which favor more peripheral lesions, or have given little or no "geographic" tumor location information when

Table 19.4 Toxicity for selected patients with a Grade 3 or higher acute or late side effect

Side Effect	Location	Tumor Size (cm)	Acute/ Late Grade	Prior Radiation (dose cGy)	Regimen (cGy)	Prescription Dose (cGy)	BED Dose (Gy)	D5% for organ at risk	V_{20} %	Representation of pre-treatment CT
Dyspnea	Left lung	2.4	Acute & Late 3	No	4×800	3200	57.6	1520	1.3	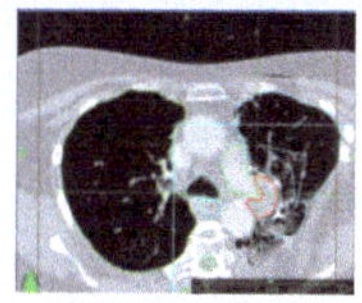
Dyspnea	Left lower lung	3.0	Acute & Late 3	No	3×1600	4800	124.8	2760	3.4	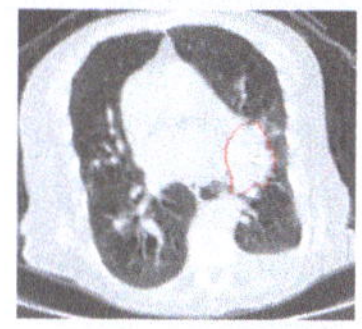
Dyspnea	Right post lung	5.8	Acute & Late 3	No	4×1200	4800	105.6	3727	5.5	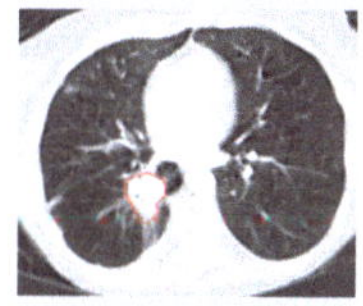
Dyspnea	Right lower lobe	3.9	Acute & Late 3	No	4×1200	4800	105.6	2560	3.5	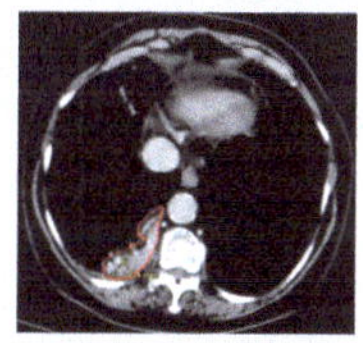
Dyspnea	Right lower lobe	7.2	Acute & Late 3	Yes (5600)	4×800	3200	57.6	2880	5.0	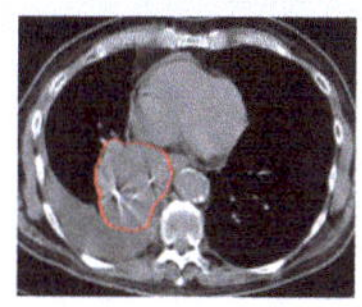
Bleeding	Left upper lobe	7.5	Acute 5	Yes (6120)	3×1000	3000	60.0	2006	NA	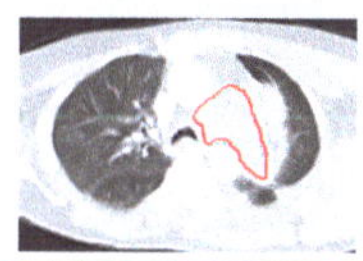

pointing out that lesions were in or near the central chest. The RTOG 0236 prospective trial has helped solve this dilemma by creating a geographic zone around the mediastinum with definitive measurable borders around the airways and by excluding T3 tumors "involving the central chest and structures of the mediastinum". Of course, the intention of the study is to clarify the role of SBRT in peripheral Stage I/II lung cancers and will thus exclude central tumors. There is a reference to trials relating to the central chest that are "under development" by the same study group [41]. Several SBRT studies that report toxicity implied that the treated tumor was "in the central chest" or "near the hilum or mediastinum"; otherwise information specific to the mediastinum and hilum is sparse.

There are other limitations of published studies as well. Timmerman et al. [41] recently illuminated the issue of longevity in onset of late toxicity. Some of their Grade 3 and 4 toxicities occurred up to 25.1 months after treatment (median 7.6 months). Their fatal complications, in 6 of 70 patients, occurred from weeks to a sobering 19.5 months after SBRT treatment; four were from pneumonias, 1 from complications of pericardial effusion, and 1 from a fatal hemorrhage at an area of recurrence in the carina (the latter two mentioned above). Le et al.'s series included 3 patient deaths from "central" single fraction SBRT, all occurring 5–6 months after therapy [17]. Uematsu, however, in a follow-up on his first 50 patients treated with SBRT for Stage I NSCLC, indicated that their complication rate changed little with a median follow-up of >60 months (range 45–90 months) [12]. They found no definite adverse effects related to treatment except for 2 minor bone fractures and 6 patients with temporary pleural pain. Yet the majority of SBRT studies have shorter follow-up and could conceivably miss serious late toxicities. Unfortunately, our early results suffer the same shortcoming.

We presented here information regarding local control. Local control of 82% in this early follow-up is somewhat less than that reported in other SBRT series ranging from 85–95%. The lower local control rate may be accounted for by the fact that our study included tumors that were larger, closer to dose limiting critical structures, and often associated with dose limiting past radiation. As with other studies, these patients have usually exhausted other first-line therapeutic options. Overall, our early results are still consistent with studies of SBRT treatment of tumors in more peripheral locations within the chest.

By stratifying the BEDs for control we hoped to find a correlation between dose and local control. We found none, although with the small numbers and disparate tumor variables (size, type, location, etc.), we did not expect to find a definitive trend. A dose escalation trial specific to this "exclusion zone" would be helpful.

For the studies that mention the mediastinum and hilum (or "central chest") as being a site of target treatment, few other specifics are given except perhaps the prescribed dose to the tumor. But in determining safety, "the devil is in the details". These details are rarely supplied with enough information to make definitive conclusions regarding actual SBRT risks. For example, details regarding prior radiation to the treatment area or use of chemotherapy regimens with known "recall" reactions are often not mentioned. As with the non-SBRT literature presented above, dose-volume metrics of tumors and critical structures need to be considered as well. Our inclusion of the data on the "percent of the total lung volume exceeding 20 Gy" (V_{20}), a common parameter in the 3D-CRT and conventional radiation literature for predicting pneumonitis, showed values too low to be meaningful. This perhaps illustrates that adequate metrics, for the most part, have not yet been clearly identified in SBRT. No doubt including V_{20} (or another "V" value) for the "organs at risk" in the central chest would be helpful for analyzing organ toxicity in the future. In addition, more basic science research needs to be done to understand radiobiological differences between SBRT and other forms of radiation to help identify the predictive factors for toxicity, not just in the mediastinum and hilum, but in SBRT in general [46].

At least in the short run, acute toxicity of the central chest appears to be short lived and tolerable. Though several individuals experienced Grade 3 toxicities, many had a baseline of dyspnea, fatigue, and pain that was only slightly less than what they experienced after treatment and it usually returned to baseline (or better) within a time frame that was acceptable to the patients.

There was one death (Grade 5 toxicity) from massive hemoptysis. This gentleman had minor occasional hemoptysis pretreatment and was on warfarin. His primary care was elsewhere (and the family declined autopsy). SBRT treatment of patients on Coumadin with tumors in this region may ultimately prove to be contraindicated.

Dose analysis for Grade 3–5 toxicity was not useful with the few patients noted here, and BEDs range considerably. However, the doses we chose were, for the most part, significantly less than the BEDs that Timmerman et al. [41] published recently (doses of 60 or 66 Gy in three fractions with BEDs >180 Gy). It may be that the information here will be helpful in ultimately reaching a safe dose range for the mediastinum as it appears the central chest is indeed more prone to severe complications at the higher dose/low volumes found to be safe in more peripheral lung lesions using SBRT.

Nevertheless, the majority of our patients did not have major toxicity, despite high doses and large tumors. Two examples are presented here:

Case 1 is a 67-year-old man with metastatic renal cell cancer. Figures 19.4 and 19.5 show a large (9.5 cm) mediastinal metastasis splaying the carina and extending along the esophagus into the right hilum. The patient received 3000 cGy in 5 fractions. Figure 19.6 shows a significant response at the level of carina with return to more normal anatomy. The patient had no acute or early signs of late toxicity, and is alive and active at 18-months follow-up.

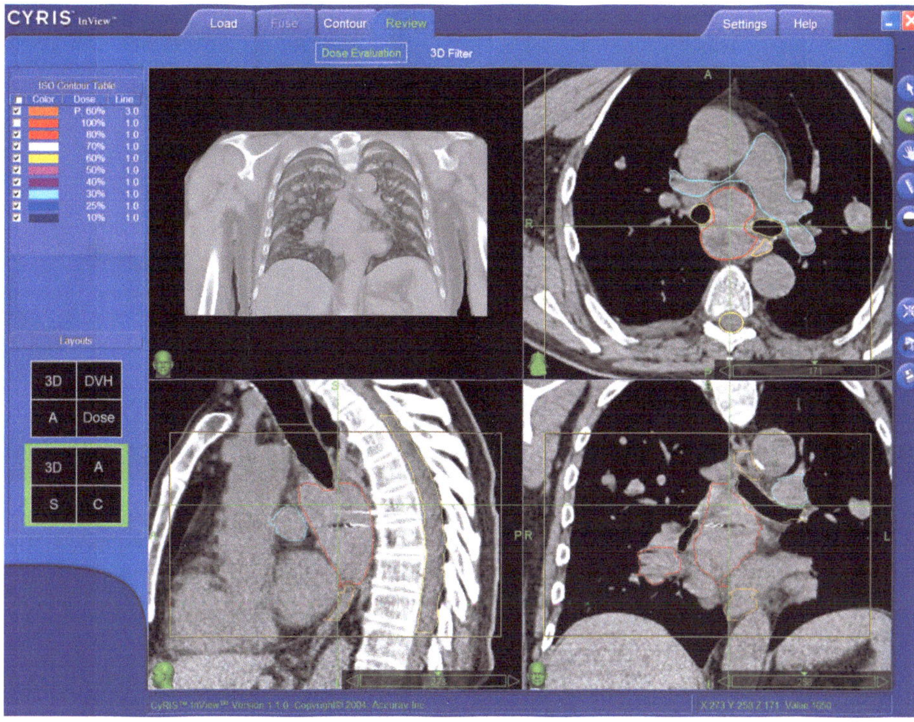

Fig. 19.4 Simulation CT after contouring in illustrative Case 1.

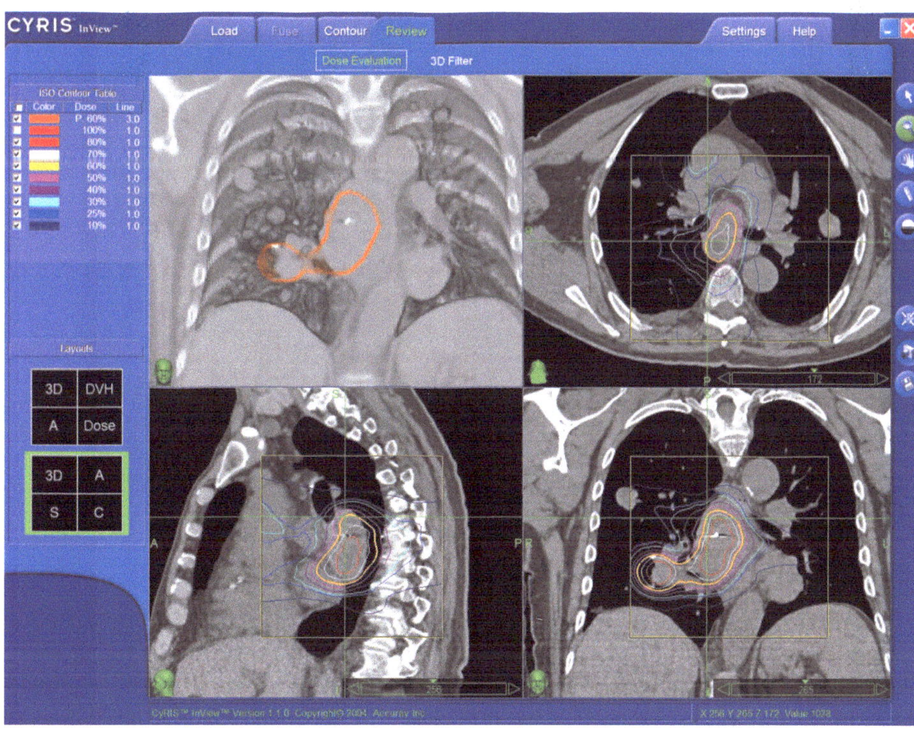

Fig. 19.5 Representation of plan in illustrative Case 1. The patient received 3000 cGy in 5 fractions (no previous RT).

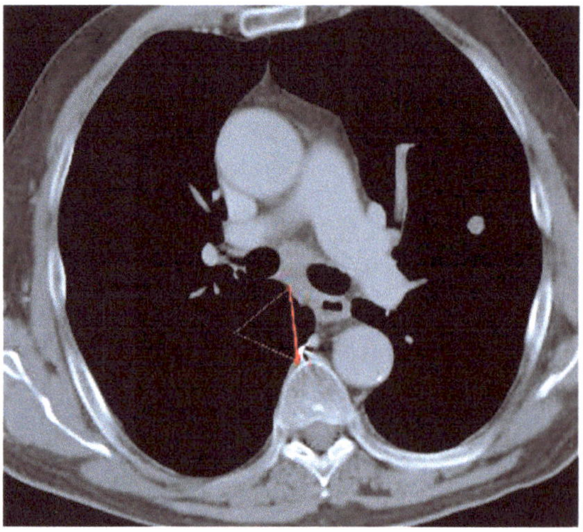

Fig. 19.6 Six months after CyberKnife in illustrative Case 1. Note the significant response at the level of carina with return to more normal anatomy. No acute or early signs of late toxicity. Patient is alive and active. The lesion has not grown at 18 months follow-up.

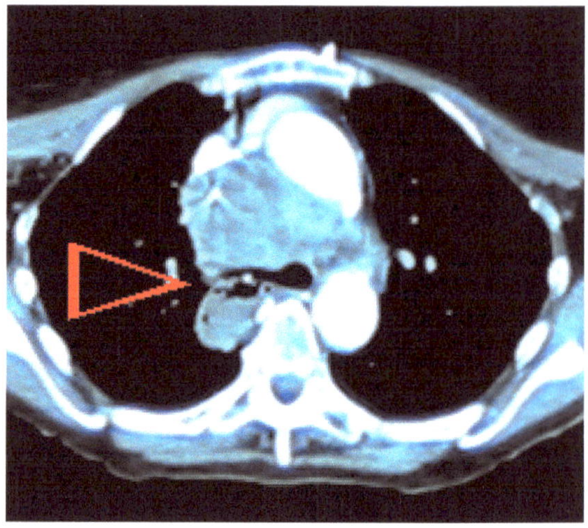

Fig. 19.7 Recurrence near gastroesophageal anastomosis in mid chest in illustrative Case 2. Arrow shows narrowed bronchus with large tumor anteriorly. Wheezing and rhonchi present.

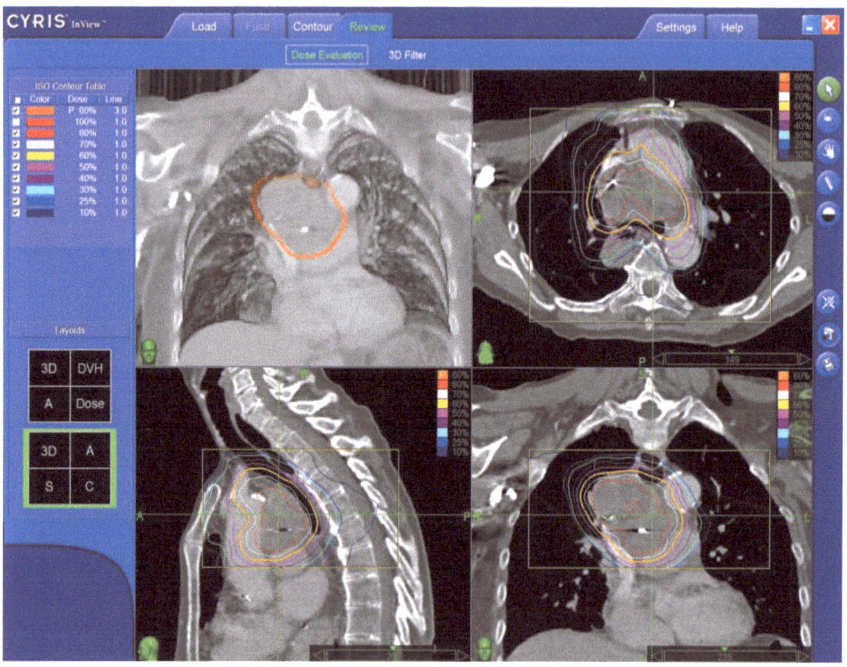

Fig. 19.8 Representation of plan in illustrative Case 2. The patient received 4000 cGy in 5 fractions.

Case 2 is a 72-year-old woman 2 years after esophageal resection for adenocarcinoma of the GE junction (see Figs. 19.7 and 19.8). A recurrence was noted near the gastroesophageal anastomosis in her mid-chest. She presented with rhonchi and wheezing developing from bronchial compression. She was treated with 4000 cGy in 5 fractions. Her abdominal metastasis was also treated. The tumor shows nearly complete regression (see Fig. 19.9). Figure 19.9 shows that her right mainstem bronchus is open and the SVC is back in normal position. This site (and an additional site treated in the abdomen) remains PET negative.

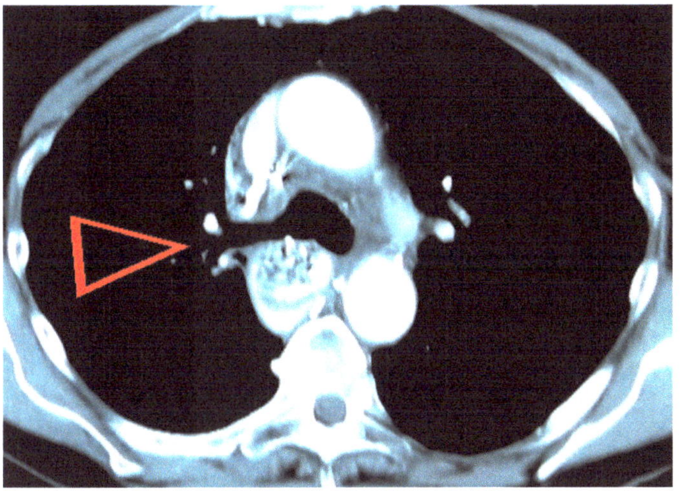

Fig. 19.9 CT 6 months later in illustrative Case 2 shows nearly complete regression of tumor. Note: Right mainstem bronchus is open and SVC is back in normal position. This site (and an additional site treated in the abdomen) remains PET negative and unchanged at 24-months follow-up. She had Grade 2 fatigue but no other acute or late toxicity. Wheezing and rhonchi are now absent.

References

1. Gross NJ. Pulmonary effects of radiation therapy. *Ann Intern Med* 1977; 86:81–92.
2. Rubin P, Siemann DW, Shapiro DL, *et al.* Surfactant release as an early measure of radiation pneumonitis. *Int J Radiat Oncol Biol Phys* 1983; 9:1669–1673.
3. Monson JM, Stark P, Reilly JJ, *et al.* Clinical radiation pneumonitis and radiographic changes after thoracic radiation therapy for lung carcinoma. *Cancer* 1998; 82:842–850.
4. McDonald S, Rubin P, Phillips TL, *et al.* Injury to the lung from cancer therapy: clinical syndromes, measurable endpoints, and potential scoring systems. *Int J Radiat Oncol Biol Phys* 1995; 31:1187–1203.
5. Thomas C. Thoracic Tumors: Overview. In: Gunderson LL, Tepper JE, editors. Clinical Radiation Oncology. London: Churchill Livingston; 2000. p. 580.
6. Polansky SM, Ravin CE, Prosnitz LR. Pulmonary changes after primary irradiationfor early breast carcinoma. *AJR Am J Roentgenol* 1980; 134:101–105.
7. Shapiro SJ, Shapiro SD, Mill WB, *et al.* Prospective study of long-term pulmonary manifestations of mantle irradiation. *Int J Radiat Oncol Biol Phys* 1990; 19:707–714.
8. Emami B, Lyman J, Brown A, *et al.* Tolerance of normal tissue to therapeutic irradiation. *Int J Radiat Oncol Biol Phys* 1991; 21:109–122.
9. Graham MV, Purdy JA, Emami B, *et al.* Clinical dose-volume histogram analysis for pneumonitis after 3D treatment for non-small cell lung cancer (NSCLC). *Int J Radiat Oncol Biol Phys* 1999; 45:323–329.
10. Kong FM, Hayman JA, Griffith KA, *et al.* Final toxicity results of a radiation-dose escalation study in patients with non-small-cell lung cancer (NSCLC): predictors for radiation pneumonitis and fibrosis. *Int J Radiat Oncol Biol Phys* 2006; 65:1075–1086.
11. Uematsu M, Shioda A, Suda A, *et al.* Computed tomography-guided frameless stereotactic radiotherapy for stage I non-small cell lung cancer: a 5-year experience. *Int J Radiat Oncol Biol Phys* 2001; 51:666–670.
12. Uematsu M. CT-guided stereotactic radiation therapy (SRT) for stage I NSCLC: 8 year results of 50 initial patients IJROBP ; 57:S281. American Society for Therapeutic Radiation Oncology. Salt Lake City, Utah; 2003.
13. Hof H, Herfarth KK, Munter M, *et al.* Stereotactic single-dose radiotherapy of stage I non-small-cell lung cancer (NSCLC). *Int J Radiat Oncol Biol Phys* 2003; 56:335–341.
14. Nagata Y, Takayama K, Matsuo Y, *et al.* Clinical outcomes of a phase I/II study of 48 Gy of stereotactic body radiotherapy in 4 fractions for primary lung cancer using a stereotactic body frame. *Int J Radiat Oncol Biol Phys* 2005; 63:1427–1431.
15. McGarry RC, Papiez L, Williams M, *et al.* Stereotactic body radiation therapy of early-stage non-small-cell lung carcinoma: phase I study. *Int J Radiat Oncol Biol Phys* 2005; 63:1010–1015.
16. Baumann P, Nyman J, Lax I, *et al.* Factors important for efficacy of stereotactic body radiotherapy of medically inoperable stage I lung cancer. A retrospective analysis of patients treated in the Nordic countries. *Acta Oncol* 2006; 45:787–795.
17. Le QT, Loo BW, Ho A, *et al.* Results of a phase I dose-escalation study using single-fraction stereotactic radiotherapy for lung tumors. *Journal of Thoracic Oncology* 2006; 1:802–809.
18. Bradley J, Movsas B. Radiation esophagitis: Predictive factors and preventive strategies. *Semin Radiat Oncol* 2004; 14:280–286.
19. Dillman RO, Seagren SL, Propert KJ, *et al.* A randomized trial of induction chemotherapy plus high-dose radiation versus radiation alone in stage III non-small-cell lung cancer. *N Engl J Med* 1990; 323:940–945.
20. Sause W, Kolesar P, Taylor SI, *et al.* Final results of phase III trial in regionally advanced unresectable non-small cell lung cancer: Radiation Therapy Oncology Group, Eastern Cooperative Oncology Group, and Southwest Oncology Group. *Chest* 2000; 117:358–364.
21. Werner-Wasik M. Treatment-related esophagitis. *Semin Oncol* 2005; 32:S60–66.

22. Coia LR, Myerson RJ, Tepper JE. Late effects of radiation therapy on the gastrointestinal tract. *Int J Radiat Oncol Biol Phys* 1995; 31:1213–1236.
23. Bradley J, Deasy JO, Bentzen S, *et al.* Dosimetric correlates for acute esophagitis in patients treated with radiotherapy for lung carcinoma. *Int J Radiat Oncol Biol Phys* 2004; 58:1106–1113.
24. Choy H, LaPorte K, Knill-Selby E, *et al.* Esophagitis in combined modality therapy for locally advanced non-small cell lung cancer. *Semin Radiat Oncol* 1999; 9:90–96.
25. Werner-Wasik M, Pequignot E, Leeper D, *et al.* Predictors of severe esophagitis include use of concurrent chemotherapy, but not the length of irradiated esophagus: a multivariate analysis of patients with lung cancer treated with nonoperative therapy. *Int J Radiat Oncol Biol Phys* 2000; 48:689–696.
26. Choy H, Akerley W, Safran H, *et al.* Multiinstitutional phase II trial of paclitaxel, carboplatin, and concurrent radiation therapy for locally advanced non-small-cell lung cancer. *J Clin Oncol* 1998; 16:3316–3322.
27. Singh AK, Lockett MA, Bradley JD. Predictors of radiation-induced esophageal toxicity in patients with non-small-cell lung cancer treated with three-dimensional conformal radiotherapy. *Int J Radiat Oncol Biol Phys* 2003; 55:337–341.
28. Curran W, Scott C, Langer C. Phase III comparisons of sequential vs. concurrent chemoradiation for patients with unresected stage III non-small cell lung cancer (NSCLC): Report of Radiation Therapy Oncology Group (RTOG) 9410. 9th World Conference on Lung Cancer. Tokyo; 2000. p. 303.
29. Movsas B, Scott C, Langer C, *et al.* Randomized trial of amifostine in locally advanced non-small-cell lung cancer patients receiving chemotherapy and hyperfractionated radiation: radiation therapy oncology group trial 98-01. *J Clin Oncol* 2005; 23:2145–2154.
30. Onimaru R, Shirato H, Shimizu S, *et al.* Tolerance of organs at risk in small-volume, hypofractionated, image-guided radiotherapy for primary and metastatic lung cancers. *Int J Radiat Oncol Biol Phys* 2003; 56:126–135.
31. Wulf J, Hadinger U, Oppitz U, *et al.* Stereotactic radiotherapy of targets in the lung and liver. *Strahlenther Onkol* 2001; 177:645–655.
32. Nyman J, Johansson KA, Hulten U. Stereotactic hypofractionated radiotherapy for stage I non-small cell lung cancer–mature results for medically inoperable patients. *Lung Cancer* 2006; 51:97–103.
33. Hayakawa K, Mitsuhashi N, Saito Y, *et al.* Short communication: adverse chronic effects of high-dose irradiation on proximal bronchus in patients treated for bronchogenic carcinoma. *Br J Radiol* 1993; 66:477–479.
34. Miller KL, Shafman TD, Anscher MS, *et al.* Bronchial stenosis: an underreported complication of high-dose external beam radiotherapy for lung cancer? *Int J Radiat Oncol Biol Phys* 2005; 61:64–69.
35. Speiser BL, Spratling L. Radiation bronchitis and stenosis secondary to high dose rate endobronchial irradiation. *Int J Radiat Oncol Biol Phys* 1993; 25:589–597.
36. Hennequin C, Tredaniel J, Chevret S, *et al.* Predictive factors for late toxicity after endobronchial brachytherapy: a multivariate analysis. *Int J Radiat Oncol Biol Phys* 1998; 42:21–27.
37. Uno T, Aruga T, Isobe K, *et al.* Radiation bronchitis in lung cancer patient treated with stereotactic radiation therapy. *Radiat Med* 2003; 21:228–231.
38. Onishi H, Araki T, Shirato H, *et al.* Stereotactic hypofractionated high-dose irradiation for stage I nonsmall cell lung carcinoma: clinical outcomes in 245 subjects in a Japanese multiinstitutional study. *Cancer* 2004; 101:1623–1631.
39. Song DY, Benedict SH, Cardinale RM, *et al.* Stereotactic body radiation therapy of lung tumors: preliminary experience using normal tissue complication probability-based dose limits. *Am J Clin Oncol* 2005; 28:591–596.
40. Hara R, Itami J, Aruga T, *et al.* Risk factors for massive hemoptysis after endobronchial brachytherapy in patients with tracheobronchial malignancies. *Cancer* 2001; 92:2623–2627.
41. Timmerman R, McGarry R, Yiannoutsos C, *et al.* Excessive toxicity when treating central tumors in a phase II study of stereotactic body radiation therapy for medically inoperable early-stage lung cancer. *J Clin Oncol* 2006; 24:4833–4839.
42. Cosset JM, Henry-Amar M, Girinski T, *et al.* Late toxicity of radiotherapy in Hodgkin's disease. The role of fraction size. *Acta Oncol* 1988; 27:123–129.
43. Stewart JR, Fajardo LF, Gillette SM, *et al.* Radiation injury to the heart. *Int J Radiat Oncol Biol Phys* 1995; 31:1205–1211.
44. Takayama K, Nagata Y, Negoro Y, *et al.* Treatment planning of stereotactic radiotherapy for solitary lung tumor. *Int J Radiat Oncol Biol Phys* 2005; 61:1565–1571.
45. Timmerman R, Papiez L, McGarry R, *et al.* Extracranial stereotactic radioablation: results of a phase I study in medically inoperable stage I non-small cell lung cancer. *Chest* 2003; 124:1946–1955.
46. Timmerman RD, Story M. Stereotactic body radiation therapy: a treatment in need of basic biological research. *Cancer J* 2006; 12:19–20.

Section 5

Liver and Pancreas Treatment – Techniques and Experiment

A Review of Stereotactic Body Radiotherapy for Liver and Pancreas

20

Klaus K. Herfarth and Marc W. Münter

CONTENTS

20.1 Abstract

The first published experience with stereotactic body radiotherapy was reported on the treatment of liver malignancies at the Karolinska Institute in Stockholm, Sweden in the early 1990s. Since then, two different treatment schedules have evolved: hypofractionated radiation therapy, mostly using 3 fractions, and the radiosurgical approach, using single-dose therapy. As opposed to the stereotactic treatment of liver tumors, little has been published about stereotactic treatments of pancreatic tumors. The following chapter reviews the history of stereotactic radiation therapy of liver and pancreas tumors including recent updates of ongoing clinical trials.

20.2 Introduction

In 1954, Philipps et al. published the first successful palliative radiation treatment of liver metastases. Symptoms were reduced in more than 50% of the cases [1]. The palliative effect of whole liver radiation therapy has been established [2–4]. Dose escalation of whole liver radiation treatment, however, resulted in a significant increase in liver toxicity [5]. Radiation-induced liver damage (RILD) as a dose-limiting toxicity appears 4–8 weeks after radiation therapy. Clinical symptoms include weight gain, increased abdominal girth, ascites and a substantial rise in alkaline phosphatase.

The occurrence of RILD is not only dependent on the dose, it is also dependent on the irradiated volume [6, 7]. Emami et al. estimated a TD 5/5 (tolerance dose that results in a normal tissue complication probability of 5% within 5 years after treatment) of 30 Gy for whole liver radiation. If 1/3 or 2/3 of the liver could be spared the TD 5/5 was estimated to increase to 35 Gy and 50 Gy, respectively [8]. Based on more recent data from the University of Michigan using three-dimensional treatment planning and conformal therapy, Dawson et al. estimated an even more pronounced volume effect with a TD 5/5 of 31 Gy, 47 Gy and 90 Gy for whole liver, 2/3 liver, or 1/3 liver irradiation, respectively [9].

The most extensive published experience of partial liver irradiation together with or without whole liver irradiation and intra-arterial chemotherapy has come from the University of Michigan [10–12]. A total of 203 inoperable patients with normal liver function had been irradiated for hepatocellular cancer (HCC; n=58), cholangiocellular cancer (CCC; n=47), and liver metastases (n=98) from 1987 to 1999. Forty-one patients were treated with whole liver irradiation (24–36 Gy), 20 patients were treated with whole liver irradiation followed by a boost to a partial liver volume (to 45–66 Gy), and 142 patients were treated with partial liver irradiation alone (48–90 Gy) in doses of 1.5–1.65 Gy delivered twice daily (bid). The median dose was 52.5 Gy (range 24–90 Gy). Simultaneously, intra-arterial chemotherapy with 5-FU (n=169) or bromodeoxyuridine (BUdR, n=34) was administered. Treatment plans and total dose were adjusted to an expected normal liver toxicity level of 10% using a modified Lyman-NTCP model [10, 12]. As predicted, in 19 patients (9%) RILD of RTOG-Grade 3 or greater (treatment required) was observed. Six patients had received whole liver irradiation, 6 whole liver irradiation plus local radiation boost, and 7 were treated by partial liver irradiation alone. The strongest parameter predicting liver toxicity was the mean liver dose. In patients with hepatic toxicity the mean liver dose was 37 Gy (NTCP 0.17) compared to 31 Gy in patients without RILD (NTCP 0.04), which is in accordance with findings of Emami et al. [8] who published volume-related tolerance doses for the whole liver of TD5/5 = 30 Gy and TD50/5 = 40 Gy. Analyzing non-dosimetric prognostic factors by logistic regression Dawson et al. found a significantly increased risk of RILD for hepatobiliary carcinoma (compared to metastases), use of BUdR, and in the male gender. In the group receiving 5-FU, male patients with hepatobiliary cancer had the highest risk for RILD. Additionally, the results were found to be consistent with the "threshold hypothesis" of Jackson et al. [13], who assumed that the risk for RILD could be kept near zero if the partial liver irradiation volume could be kept below a threshold volume regardless of the dose. Dawson et al. suggested that doses as high as 100 Gy might be safely administered for small volumes of normal liver tissue (approximately 1/3 of whole liver). This dose escalation might be beneficial, because the clinical results of the Michigan patients revealed an improved local tumor control with increased dose [11]. In a newer trial, Dawson et al. used the Michigan data for dose escalation in stereotactic liver tumor treatment. As of June 2006, they had treated 79 patients (33 HCC, 12 CCC and 34 metastases) in 6 fractions up to a total dose of 57 Gy, and had not yet observed dose-limiting toxicity [14].

Care must be taken if cirrhotic liver is irradiated. All University of Michigan data were collected on patients with normal liver function. If liver function is impaired, however, the risk of developing RILD increases and more liver tissue has to be spared than in healthy liver patients. Seong et al. [15] combined focal liver irradiation in 50 patients with HCC (Child class A n=38, Child class B n=12) and transarterial-chemo-embolization (TACE). The total dose (30–60 Gy) was determined by the fraction of the non-tumor liver volume receiving more than 50% of the prescribed dose given in 1.8 Gy daily fractions. Six patients were observed with RILD, unfortunately dose-volume relations for these patients were not analyzed. The same group published an analysis of dose-response relation in local radiotherapy for HCC in 158 patients [16]. Ninety percent of patients had liver cirrhosis (Child A 74%, Child B 26%); patients with advanced liver cirrhosis (Child class C) were excluded. The tumor size ranged from < 5 cm (11%), 5–10 cm (54%), and > 10 cm (35%). The average 3D-conformal planned dose was 48.2 +/- 7.9 Gy (25.2–59.4 Gy). While statistical evaluation revealed that the total radiation dose was the only significant factor determining tumor response, hepatic toxicity also increased with dose. Eleven patients showed RILD: 4.2% (n=1) of all patients in the category of < 40 Gy, 5.9% (n=3) from 40–50 Gy, and 8.4% (n=7) with doses > 50 Gy. Liver cirrhosis of Child class B seemed to be a risk factor in development of RILD, but the number of cases was small: 0/16 patients < 40 Gy, 2/13 patients 40–50 Gy (15.4%), and 2/20 patients > 50 Gy (10%). Nevertheless, the evaluation demonstrates that partial liver irradiation can be performed on large volumes even in patients with impaired liver function. Liver function, however, should be evaluated first.

A stereotactic treatment approach for liver malignancies should achieve even better normal tissue sparing than conventional or conformal planning and delivery techniques. The first steps in the development of stereotactic radiation therapy of liver malignancies were taken at the Karolinska Institute in Stockholm, Sweden [17].

20.3 Hypofractionation

Blomgren and Lax published their data on stereotactic irradiation of liver tumors first in 1995 followed by an update in 1998 [18, 19]. After having negative experiences with single-dose therapy (discussed below), they mainly used hypofractionated radiotherapy. The fractionation and the overall time of treatment varied widely. Doses ranged from 2 × 8 Gy to 3 × 15 Gy or 4 × 10 Gy. The dose was prescribed to the planning target volume (PTV) encompassing the 65% isodose, which resulted in maximal total doses of 20–82 Gy. Treatment times varied from 3 to 44 days [18]. This group treated 20 primary intrahepatic cancers in 11 patients. The clinical target volume (CTV) was a median of 22 cc (range of 3–622 cc). They observed no local failure at a mean follow-up of 12 months; however, two fatal cases of RILD were reported. Both patients had liver cirrhosis. The first patient presented with a 57 cc HCC nodule due to hepatitis C and liver cirrhosis. The tumor was treated with 3 × 15 Gy to the periphery of the PTV. The patient developed ascites 20 days after completion of the treatment and died the next month. The other patient had a large (293 cc) HCC treated with 3 × 10 Gy to the periphery of the PTV. This patient also developed ascites in the first 6 weeks after treatment and died shortly thereafter. Unfortunately, there is no detailed information about the size of the liver, the degree of pretherapeutic liver impairment, or the mean liver dose. Therefore, no definite conclusions about the risk assessment can be drawn from this published data. Apart from these fatal side effects, patients experienced nausea, fever, or chills for a few hours after radiosurgery.

Ten patients with 20 metastases were also treated at the Karolinska institute using the hypofractionated stereotactic regimen. The median CTV was 24 cc with a range of 2–263 cc. Tumor response was evaluated after a mean follow-up of 9.6 months. All tumors showed response to the therapy. One local recurrence was observed 6 months after therapy. Again, patients experienced nausea, fever, and chills a few hours after the procedure. Patients received a prophylactic treatment of paracetamol and antiemetics. One patient suffered from hemorrhagic gastritis a few weeks after treatment. In this case, one third of the stomach wall had been exposed to 7 Gy for two treatment sessions. Parts of the duodenum were exposed to 4 × 5 Gy in another patient who developed a duodenal ulcer, which was treated conservatively.

These early data from Stockholm indicate the feasibility and the possible success of a hypofractionated stereotactic treatment of liver tumors. Unfortunately, no clear relation between treatment success and dose-volume parameters can be drawn from these data due to the wide range of doses and fractionation schemes. The Stockholm group has continued to treat patients with hepatic cancer with the stereotactic approach, although new data have not been published.

Wulf et al. from the University of Würzburg in Germany adopted parts of the Stockholm treatment approach [20]. They treated 24 patients with liver tumors (one CCC and 23 metastases). The median CTV was 50 cc (range 9–512 cc). All but one patient were treated with 3 × 10 Gy to the 65% isodose at the periphery of the PTV. One patient was treated by 4 × 7 Gy, also normalized to the periphery of the PTV. The reason for this other fractionation was a close proximity of the target to the esophagus. The crude local control was 83% with a mean follow-up of 9 months. The actuarial local control after 12 months was reported to be 76%. Four recurrences occurred 3, 8, 9 and 17 months after treatment. All were treated with 3 × 10 Gy. Three of the failures occurred marginally. The median survival of these patients was calculated to 20 months. The morbidity of the treatment was low: only 7/24 patients reported side effects of Grade 1 or 2 according to the WHO classification. The side effects were mostly related to one fraction and included fever, chills, or pain with a typical onset a few hours after irradiation, sometimes accompanied by nausea and/or vomiting. The symptoms ceased spontaneously or were successfully treated with paracetamol or prednisolon. Only one patient showed longer lasting fatigue, weakness, and loss of appetite. In 2006, an update of the Würzburg data was published after treatment of 39 patients with 51 metastases and 5 patients with HCC [21]. The applied dose was either 3 × 10 Gy to the 65% isodose line (low-dose group) or 3 × 12.5 Gy to the 65% isodose or 1 × 26 Gy to the 80% isodose (the last two regimens constituting the high-dose group). Actuarial local control was 92% and 66% after 12 and 24 months, respectively, with a median follow-up of 15 months. There was a trend to higher local control in the high dose group (100% and 82% at 1 and 2 years vs. 86% and 56%; p=0.077). In addition, dose was the only significant factor predicting local control in a multivariate analysis (p=0.0089) [21].

In the US, a multi-center University of Colorado dose escalation trial for the hypofractionated SBRT of liver metastases was initiated in 2004. The initial dose level was 12 Gy delivered three times for a total minimal target dose of 36 Gy in 5 to 10 days. Dose escalation is performed in steps of 2 Gy per fraction (6 Gy total) up to a total dose of 60 Gy, or upon determination of a maximum tolerated dose (MTD). The first results of this trial have recently been published [22]. Eligible patients had one to three liver metastases, tumor diameter < 6 cm, and adequate liver function. At least 700 mL of normal liver received a total dose < 15 Gy. Dose-limiting toxicity (DLT) included acute Grade 3 liver or intestinal toxicity or any acute Grade 4 toxicity. The MTD was exceeded if 2/6 patients in a cohort experienced DLT. Eighteen patients were enrolled and the median aggregate gross tumor volume was 18 cc. No patient experienced a DLT, and dose was escalated to 60 Gy in 3 fractions without reaching MTD. The authors concluded that biologically potent doses of SBRT are well tolerated in patients with limited liver metastases. A Phase II trial is currently active and first results have recently been published [23]. As of June 2006, 36 patients had been enrolled (18 in Phase I and 18 in Phase II). Of these, 21 patients were evaluable with at least 6 months follow-up (median follow-up 19 months). The estimated actuarial local control at 18 months was 93%. Only one local recurrence occurred 15 months after 3 × 18 Gy. No Grade 4 or Grade 5 toxicity had been observed and only 1 Grade 3 toxicity was observed (skin ulceration) [23]. It is not known how many of these tumors were centrally located in the liver. Initial reports of a similar trial with dose escalation up to 3 × 22 Gy in lung tumors had no major toxicity when first published [24]. A recent update including Phase II data, however, showed major toxicity with a longer follow-up (median 17.5 months) [25]. Eight of the 70 patients (11%) developed Grade 3 or 4 toxicity and 6 patients (9%) died as a result of the treatment. Centrally located tumors had a high risk (about 46%) of developing severe toxicity within 2 years of receiving 3 × 20 Gy to the periphery of the PTV [25].

20.4 Radiosurgery

Blomgren and Lax also started with single-dose therapy for liver tumors [19]. Six tumors in 5 patients were radiosurgically treated. The prescribed dose to the periphery of the PTV was a median 15.5 Gy (range 7.7–30 Gy). No recurrences were observed during a median follow-up of 5 months; however, one patient died 2 days after treatment. This patient had a 229-cc HCC and a cirrhotic liver. The tumor was treated with 30 Gy applied to the periphery of the PTV. The isocenter dose was 48 Gy. The patient was already icteric and showed signs of ascites at the time of treatment. The other four patients showed marginal recurrences during follow-up, as indicated in a later paper [18]. These two circumstances forced Blomgren and Lax to abandon the single-fraction approach for large liver tumors.

In 1997, a Phase I/II trial was initiated at the German Cancer Research Center in Heidelberg (Germany) to assess the feasibility and the clinical outcome of single-dose treatment of liver tumors [26]. Non-resectable tumors in the liver were included. As many as 3 tumors were included, or 4 if two of them were less than 3 cm from each other. Single lesions could not exceed 6 cm in diameter, and none of the tumors could be immediately adjacent to the gastrointestinal tract (distance >6 mm). Patients were excluded if they had insufficient liver function. Thirty-seven patients were treated for a total of 60 tumors during 40 treatment sessions. The targets included 4 primary hepatic tumors and 56 metastases (mainly colorectal cancer or breast cancer). The median target volume was 10 cc (1–132 cc). The dose was prescribed to the isocenter with the 80% isodose encompassing the PTV and was escalated from 14 Gy to 26 Gy based on the liver dose in the dose-volume histogram. That is, the dose to 30% of the liver was increased from 6 Gy to 12 Gy while the maximum dose to 50% of the liver was escalated from 4 Gy to 7 Gy. Dose, however, was reduced during dose escalation if normal tissue constraint of nearby gastrointestinal organs (esophagus 14 Gy, stomach 12 Gy, small bowel 12 Gy) did not allow higher doses. After initial dose escalation, an actuarial local tumor control of 81% at 18 months was achieved with a mean follow-up of 9.5 months. All patients received prophylactic dexamethasone before and after radiation therapy. The actuarial 2-year survival was 59%. Patients treated with curative intent showed a significantly longer survival (actuarial 87% at 2 years) than patients with additional extrahepatic tumor manifestations at the time of treatment (median survival 12 months) [27]. An update of these study patients with a mean follow-up of 17 months was published in 2003 [28]. Two patients developed late local recurrences 4 years after therapy. The actuarial local control remained unchanged at 81% after 18 months. An example of tumor response after single dose irradiation is shown in Figure 20.1.

A follow-up trial was initiated after these promising initial results. More patients were radiosurgically treated according to the initial Phase II protocol until recruitment of the follow-up trial could be started. A combined total of 78 patients were treated until spring 2003. The mean follow-up has been 12 months and the actuarial local tumor control dropped to 72% at 12 months. Analysis of the increased failure rate revealed that patients with colorectal metastases showed significantly worse local tumor control than patients with other histologies [28]. In particular, all 11 patients who had already received chemotherapy using CPT-11 or oxaliplatine showed local (infield and marginal) recurrences during the first 15 months after therapy. Therefore, higher doses and/or larger safety margins should be used especially if colorectal cancer metastases are treated.

Side effects of the treatment were minimal [26], including mild nausea or loss of appetite for 1–2 weeks in about one third of the patients. Hiccups occurred

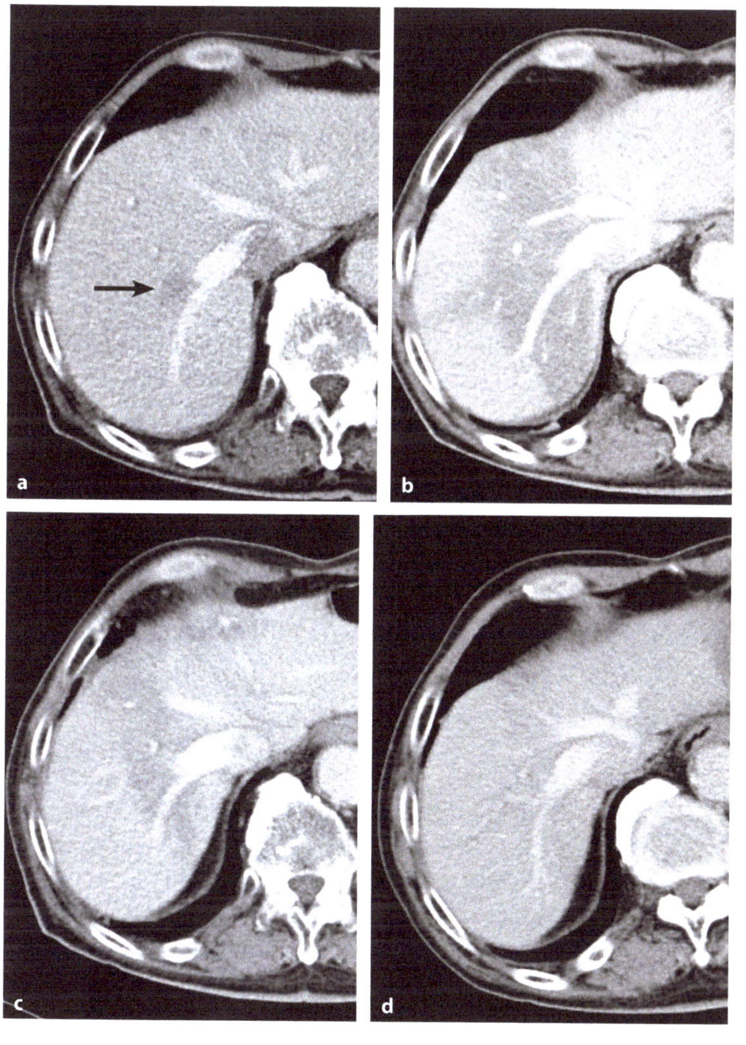

Fig. 20.1a–d. Follow-up of a solitary liver metastasis for a primary colorectal cancer after radiosurgery. The metastasis is located adjacent to the right anterior to the intermedial liver vein: **a** before treatment, **b** 6 weeks after irradiation (perifocal radiation reaction is visible), **c** 12 months (volume decrease of radiation reaction) and **d** 18 months after treatment.

in 2 patients and one patient developed fever. There were no clinical signs of RILD. On the other hand, a focal liver reaction occurred after radiation (described below).

The hypofractionated and the single-dose approach in the stereotactic radiation treatment of liver metastases should be evaluated in a Phase III study. The StRaL-trial (Stereotactic Radiation Therapy of Liver Metastases) was a prospective randomized multi-center trial that started patient recruitment in March 2003 with a planned enrollment of 276 patients over 5 years. Inclusion criteria were a maximum of 3 surgically inoperable liver metastases. The maximum size of the tumors depended on the number of targets: 5 cm for 1 target, 4 cm for 2 targets, and 3 cm for 3 targets. The primary study goal was the comparison of local tumor control after single- or multi-fraction treatment. Secondary endpoints were survival, morbidity, and quality of life. The study was designed to prove the equivalence of the treatment arms. Patients in arm A received a single-dose radiation therapy of 28 Gy normalized to the isocenter with the 80% isodose (22.4 Gy) encompassing the PTV. Patients in arm B received hypofractionated therapy with 3 × 12.5 Gy normalized to the 65% isodose (encompassing the PTV). Normal tissue constraints are listed in Table 20.1.

Unfortunately, the trial closed early in 2006 after recruiting only 18 patients in 3 German centers. This failure may be attributed to a dramatic drop in the number of patients in Germany treated with stereotactic radiotherapy and instead treated with radiofrequency ablation (RFA) or laser induced thermo therapy (LITT). No difference could be observed between the two groups.

Table 20.1 Target doses and normal tissue constraints for the German prospective randomized multi-center trial StRaL.

	Arm A 1×28 Gy/isocenter		Arm B 3×12.5 Gy/65% isodose	
	relative dose/fx	absolute dose/fx	relative dose/fx	absolute dose/fx
Isocenter	100%	28 Gy	100%	19.2 Gy
Minimum PTV	80%	22.4 Gy	65%	12.5 Gy
Liver (30% Vol.)	43%	12 Gy	36%	7 Gy
Liver (50% Vol.)	25%	7 Gy	26%	5 Gy
Esophagus (max.)	43%	12 Gy	36%	7 Gy
Stomach (max.)	43%	12 Gy	36%	7 Gy
Duodenum (max.)	43%	12 Gy	36%	7 Gy
Colon (max.)	43%	12 Gy	36%	7 Gy
Myelon (max.)	43%	12 Gy	36%	7 Gy

20.5 Focal Liver Reaction

The focal liver reaction after single dose radiation therapy has been examined by the Heidelberg group. All patients who were followed using multiphasic CT scanning showed a sharply demarcated focal radiation reaction (see Fig. 20.1). Tumor and radiation reaction could be well differentiated in the portal-venous contrast-enhanced CT scans. Liver vessels run through the liver reaction and were not displaced, as they are by expanding tumors. A detailed evaluation and characterization of this focal radiation reaction in 36 of the Heidelberg patients was published in 2003 [30].

The area of radiation reaction was hypodense in the majority of the non-enhanced CT scans. Three different types of reaction appearance could be defined based on the liver density in the portal-venous and the late phase after contrast agent administration:

- **Type 1 reaction:** Hypodensity in portal-venous contrast phase, isodensity in the late contrast phase
- **Type 2 reaction:** Hypodensity in portal-venous contrast phase, hyperdensity in the late contrast phase
- **Type 3 reaction:** Isodensity/hyperdensity in portal-venous contrast phase, hyperdensity in the late contrast phase.

The median onset of the reaction was 1.8 months. Types 1 and 2 reactions usually showed up earlier and type 3 reactions appeared later, resulting in a shift during follow-up towards type 3 appearances. In addition, the volume of the radiation reaction decreased with time (Fig. 20.1). The most dramatic shrinkage was observed during the first months after appearance leading to speculation that the complete liver reaction proceeded through different radiological stages (type 1, 2 and 3 appearances). The histological basis of these stages was not determined since no biopsies were taken. Others had reported a type 2 appearance after single dose radiation therapy, however, and a veno-occlusive disease was histologically confirmed [31].

Based on reconstruction of the dose-volume histograms, the mean threshold dose was 13.7 Gy with a wide range (8.9 to 19.2 Gy) given in a single fraction. One reason for this large variance might be that the volume decreased greatly between the initial detection and the later follow-up examinations. The examination might not have detected larger reaction volumes and, therefore, the calculated threshold doses might have been overestimated. This was supported by the significant correlation between the threshold dose and the time of detection (correlation coefficient r=0.709). In addition to the time factor, other factors which might influence individual radiation sensitivity (e.g., additional toxic liver agents such as alcohol) might have also contributed to the variance. More data is needed to strengthen our confidence in these threshold doses.

Recently, we have published data using a rabbit model for evaluation of the radiation reaction. Small parts of the liver of New Zealand white rabbits were treated with up to a single dose of 45 Gy (isocenter). Nevertheless, no radiation reaction comparable to that seen in humans could be detected [32].

In summary, SBRT has been shown to be a feasible and completely non-invasive treatment modality complementing the present armamentarium in the fight against potentially curable localized primary and secondary liver tumors. The acceptance and the ultimate success of this new treatment modality will, to a large extent, depend upon our willingness to prove its capabilities in the framework of multimodality treatment approaches.

20.6 Radiosurgery for Pancreatic Cancer

Radiosurgery for pancreatic cancer is a new treatment option, not commonly in clinical use for this disease. The poor prognosis and high incidence of this tumor should encourage establishing new treatment concepts in order to improve the outcome. The 5-year survival is still less than 5%, and pancreatic cancer is the fourth leading cancer-related cause of death in the United States. For inoperable or locally advanced pancreatic cancer, radiochemotherapy is the therapy of choice as shown by Moertel et al. [33] and additionally by a study conducted by the Gastrointestinal Tumor Study Group [34]. Normally patients receive a CT based 3D planned radiotherapy in a four field technique. The dose limiting organ at risk is the small intestine. Therefore, normally a dose of approximately 45 Gy to 50.4 Gy in a single fraction of 1.8 Gy is applied. Recent studies are combining modern fractionated techniques like stereotactic guided inverse planned intensity modulated radiotherapy (IMRT) with chemotherapy and "targeted therapies" in order to improve outcome and achieve a higher number of resectable patients after initial therapy [35]. A second new and promising approach is the use of radiosurgery for locally advanced pancreatic cancer a single modality or as a boost after fractionated radiochemotherapy. In a Phase I study, Koong et al. [36] performed a dose escalation for locally advanced pancreatic cancer using the CyberKnife® (Accuray Incorporated, Sunnyvale, CA). The authors treated 15 patients at three dose levels starting with 15 Gy (3 patients) and escalating the dose from 20 Gy (5 patients) to 25 Gy (7 patients). No Grade 3 or higher acute gastrointestinal toxicity was seen at any dose. The authors stopped the study before a dose-limiting toxicity was reached. For all patients treated with 25 Gy the clinical objective of this study (local tumor control) was achieved and all patients progressed systemically. The median overall survival in the group treated with 25 Gy was 8 months with a median follow-up of 4.5 months. Only five of the 15 patients developed a Grade I or II acute gastrointestinal toxicity. For the patients receiving 25 Gy the mean dose to 50% of the duodenum was 14.5 Gy and the mean dose to 5% of the duodenum 22.5 Gy.

Koong et al. [37] followed-up this study by combining 45 Gy by IMRT with concurrent 5-FU followed by 25 Gy stereotactic radiosurgery for locally advanced pancreatic cancer. A total of 16 patients finished the study protocol as planned. No patient experienced a Grade IV acute gastrointestinal toxicity. Grade III acute toxicity was seen in two patients and 11 patients suffered from a Grade I toxicity. This small study showed excellent local control rates of 94% (15 of 16 patients). Unfortunately, the time to progression was a disappointing 17.5 weeks post-therapy, and the site of first progression was in all cases distant. According to Kaplan-Meier functions, the estimated 6-month survival was 80% and the estimated 1-year survival was 15%. Robotic-based radiosurgery was applied in both studies mainly by using a breath hold technique. Patients were trained to hold their breath for 15–20 s. By using a breath-holding technique, the position of the pancreas was reproducible on average within 2.5 mm in all three dimensions [38]. A disadvantage of radiosurgery with breath-holding is the long treatment time. Patients were treated in a range from 3 to 6 hours [36]. An option to reduce the treatment time is the respiratory tracking system of the CyberKnife (Synchrony®, Accuray Incorporated). This technique retargets the robotic linac in real-time by continuous tracking of infrared detectors attached to the chest wall. Tumor motion is tracked by correlating the position of the internal fiducials on intraoperative X-rays with the infrared detectors. In the study of Koong et al., the radiographic markers were mainly implanted using CT guidance [36]. In two patients, the fiducials were placed during an open laparotomy. A different approach to implant fiducials for real-time image guidance was published by Pishvaian et al. [39] which describes an endoscopic ultrasound-guided technique. This approach was successful in 11 of 13 patients in different locations of the mediastinum and the abdomen including in patients with pancreatic cancer. Only one patient developed an infectious complication within 30 days.

Radiosurgery for pancreatic cancer is, in conclusion, a new and promising treatment modality, which could improve the outcome of pancreatic cancer in combination with fractionated radiotherapy and new systemic approaches.

References

1. Phillips R, Karnofsky DA, Hamilton LD, *et al.* Roentgen therapy of hepatic metastases. *Am J Roentgenol Rad Ther Nucl Med* 1954; 71:826–834.
2. Borgelt BB, Gelber R, Brady LW, *et al.* The palliation of hepatic metastases: results of the radiation therapy oncology pilot study. *Int J Radiat Oncol Biol Phys* 1981; 7:587–591.
3. Leibel SA, Pajak TF, Massullo V, *et al.* A comparison of misonidazole sensitized radiation therapy to radiation therapy alone for the palliation of hepatic metastases: results of a radiation oncology group randomized prospective trial. *Int J Radiat Oncol Biol Phys* 1987; 13:1057–1064.
4. Sherman DM, Weichselbaum R, Order SE, *et al.* Palliation of hepatic metastasis. *Cancer* 1978; 41:2013–2017.
5. Russell AH, Clyde C, Wasserman TH, *et al.* Accelerated hyperfractionated hepatic irradiation in the management of patients with liver metastases: results of the RTOG dose escalating protocol. *Int J Radiat Oncol Biol Phys* 1993; 27:117–123.
6. Haddad E, Le Bourgeois JP, Kuentz M, *et al.* Liver complications in lymphomas treated with a combination of chemotherapy and radiotherapy: preliminary results. *Int J Radiat Oncol Biol Phys* 1983; 9:1313–1319.
7. Poussin-Rosillo H, Nisce LZ, D'Angio GH. Hepatic radiation tolerance in Hodgkin's Disease patients. *Radiology* 1976; 121:461–464.
8. Emami B, Lyman J, Brown A, *et al.* Tolerance of normal tissue to therapeutic irradiation. *Int J Radiat Oncol Biol Phys* 1991; 21:109–122.
9. Dawson LA, Ten Haken RK, Lawrence TS. Partial irradiation of the liver. *Semin Radiat Oncol* 2001; 11:240–246.
10. Lawrence TS, Ten Haken RK, Kessler ML, *et al.* The use of 3-D dose volume analysis to predict radiation hepatitis. *Int J Radiat Oncol Biol Phys* 1992; 23:781–788.
11. Dawson LA, McGinn CJ, Normolle D, *et al.* Escalated focal liver radiation and concurrent hepatic artery fluorodeoxyuridine for unresectable intrahepatic malignancies. *J Clin Oncol* 2000; 18:2210–2218.
12. Dawson LA, Normolle D, Balter JM, *et al.* Analysis of radiation-induced liver disease using the lyman NTCP model. *Int J Radiat oncol Biol Phys* 2002; 53:810–821.
13. Jackson A, Ten Haken RK, Robertson JM, *et al.* Analysis of clinical complication data for radiation hepatitis using a parallel architecture model. *Int J Radiat Oncol Biol Phys* 1995; 31:883–891.
14. Dawson LA, Eccles C, Craig T. Individualized image guided iso-NTCP based liver cancer SBRT. *Acta Oncol* 2006; 45:856–864.
15. Seong J, Park HC, Han KH, *et al.* Clinical results and prognostic factors in radiotherapy for unresectable hepatocellular carcinoma: a retrospective study of 158 patients. *Int J Radiat Oncol Biol Phys* 2003; 55:329–336.
16. Park HC, Seong J, Han KH, *et al.* Dose-response relationship in local radiotherapy for hepatocellular carcinoma. *Int J Radiat Oncol Biol Phys* 2002; 54:150–155.
17. Lax I, Blomgren H, Näslund I, *et al.* Stereotactic radiotherapy of malignancies in the abdomen. *Acta Oncol* 1994; 33:677–683.
18. Blomgren H, Lax I, Göranson H, *et al.* Radiosurgery for tumors in the body: clinical experience using a new method. *J Radiosurgery* 1998; 1:63–74.
19. Blomgren H, Lax I, Näslund I, *et al.* Stereotactic high dose fraction radiation therapy of extracranial tumors using an accelerator. *Acta Oncol* 1995; 34:861–870.
20. Wulf J, Hädinger U, Oppitz U, *et al.* Stereotactic radiotherapy of targets in the lung and liver. *Strahlenther Onkol* 2001; 177:645–655.
21. Wulf J, Guckenberger M, Haedinger U, *et al.* Stereotactic radiotherapy of primary liver cancer and hepatic metastases. *Acta Oncol* 2006; 45:838–847.

22. Schefter TE, Kavanagh BD, Timmerman RD, *et al.* A Phase I trial of stereotactic body radiation therapy (SBRT) for liver metastases. *Int J Radiat Oncol Biol Phys* 2005; 62:1371–1378.
23. Kavanagh BD, Schefter TE, Cardenes HR, *et al.* Interim analysis of a prospective phase I/II trial of SBRT for liver metastases. *Acta Oncol* 2006; 45:848–855.
24. Timmerman R, Papiez L, McGarry R, *et al.* Extracranial stereotactic radioablation: results of a phase I study in medically inoperable stage I non-small cell lung cancer. *Chest* 2003; 124:1946–1955.
25. Timmerman R, McGarry R, Yiannoutsos C, *et al.* Excessive toxicity when treating central tumors in a phase II study of stereotactic body radiation therapy for medically inoperable early-stage lung cancer. *J Clin Oncol* 2006; 24:4833–4839.
26. Herfarth KK, Debus J, Lohr F, *et al.* Stereotactic single dose radiation therapy of liver tumors: results of a phase I/II trial. *J Clin Oncol* 2001; 19:164–170.
27. Herfarth KK, Debus J, Lohr F, *et al.* Stereotaktische Bestrahlung von Lebermetastasen. *Radiologe* 2001; 41:64–68.
28. Herfarth KK, Münter MW, Debus J. Strahlentherapeutische Behandlung von Lebermetastasen. *Chir Gastroenterol* 2003; 19:359–363.
29. Herfarth KK, Debus J. Stereotactic radiation therapy of liver tumors. *Radiother Oncol* 2003; 68 (Suppl.1):S45.
30. Herfarth KK, Hof H, Bahner ML, *et al.* Assessment of focal liver reaction by multiphasic CT after stereotactic single-dose radiotherapy of liver tumors. *Int J Radiat Oncol Biol Phys* 2003; 57:444–451.
31. Willemart S, Nicaise N, Struyven J, *et al.* Acute radiation-induced hepatic injury: evaluation by triphasic contrast enhanced helical CT. *Br J Radiol* 2000; 73:544–546.
32. Herfarth KK, Munter MW, Groene HJ, *et al.* Absence of tissue reaction after focal high-dose irradiation of rabbit liver. *Acta Oncol* 2006; 45:865–869.
33. Moertel CG, Frytak S, Hahn RG, *et al.* Therapy of locally unresectable pancreatic carcinoma: a randomized comparison of high dose (6000 rads) radiation alone, moderate dose radiation (4000 rads + 5-fluorouracil), and high dose radiation + 5-fluorouracil: The Gastrointestinal Tumor Study Group. *Cancer* 1981; 48:1705–1710.
34. Treatment of locally unresectable carcinoma of the pancreas: comparison of combined-modality therapy (chemotherapy plus radiotherapy) to chemotherapy alone. Gastrointestinal Tumor Study Group. *J Natl Cancer Inst* 1988; 80:751–755.
35. Krempien R, Muenter MW, Huber PE, *et al.* Randomized phase II–study evaluating EGFR targeting therapy with cetuximab in combination with radiotherapy and chemotherapy for patients with locally advanced pancreatic cancer–PARC: study protocol [ISRCTN56652283]. *BMC Cancer* 2005; 5:131.
36. Koong AC, Le QT, Ho A, *et al.* Phase I study of stereotactic radiosurgery in patients with locally advanced pancreatic cancer. *Int J Radiat Oncol Biol Phys* 2004; 58:1017–1021.
37. Koong AC, Christofferson E, Le QT, *et al.* Phase II study to assess the efficacy of conventionally fractionated radiotherapy followed by a stereotactic radiosurgery boost in patients with locally advanced pancreatic cancer. *Int J Radiat Oncol Biol Phys* 2005; 63:320–323.
38. Murphy MJ, Martin D, Whyte R, *et al.* The effectiveness of breath-holding to stabilize lung and pancreas tumors during radiosurgery. *Int J Radiat Oncol Biol Phys* 2002; 53:475–482.
39. Pishvaian AC, Collins B, Gagnon G, *et al.* EUS-guided fiducial placement for CyberKnife radiotherapy of mediastinal and abdominal malignancies. *Gastrointest Endosc* 2006; 64:412–417.

Multidisciplinary Overview of Local-Regional Therapies for Liver Malignancies

21

Robert M. Goldstein, Brian D. Berger, and John K. O'Connor

CONTENTS

21.1 Introduction

This chapter reviews, from the surgeon's perspective, the various local-regional treatments for hepatic malignancies and how the use of stereotactic radiosurgery fits into current general surgical practice. Understanding these modalities is important in surgical practice and enables a rational approach to both surgical and non-surgical therapies for hepatic malignancies. Some of these treatments are clearly in the surgical domain as these therapies may be best given via laparoscopic or open surgical approach. Additionally, these local-regional modalities are increasingly being used as neoadjuvant or adjuvant therapy, particularly in Hepatocellular Carcinoma (HCC). Surgeons are faced with the challenge of adopting these alternative techniques into their practice. Familiarity with these therapies allows the surgeon, with their unique expertise and skill, to participate actively in the nonsurgical management of these lesions. Several of the more prominent non-surgical local-regional therapies are reviewed here.

21.2 Hepatocellular Cancer

Hepatocellular carcinoma (HCC) is the fifth most common solid-organ tumor. Worldwide, it is responsible for nearly a million deaths annually, and its incidence is on the rise in Western countries [1–3]. Surgical resection is clearly the treatment of choice for HCC. Recent advances in surgical management have reduced the mortality rate associated with surgery and at most major medical centers operative mortality is generally less than 5% [3]. Additionally, long-term survival after resection of HCC has improved, with an overall 5-year survival rate of about 50% achieved in recent years [4]. However, the majority of patients with HCC are not candidates for surgery and resection rates remain low, in the range of 10–37% [5–7]. Liver transplantation is a clearly established alternative curative treatment for HCC. Liver transplant offers excellent survival in patients with solitary HCCs smaller than 5 cm or those with up to three nodules each smaller than 3 cm without extrahepatic or vascular spread. Five-year survival rates are greater than 70% for these patients [8, 9]. Indeed, some of these local-regional modalities are increasingly being used as a "bridge" to transplantation for controlling cancer growth while patients with transplantable HCC wait for suitable donor organs. The use of liver transplantation is, of course, limited by the shortage of donor organs. Nonsurgical local-regional alternatives are especially important considering the dismal results of systemic chemotherapy, with a response rate less than 20% and no significant survival benefit when compared with symptomatic management [10, 11]. Because of the limited eligibility for surgical treatment and poor results of chemotherapy, the past decade has seen great efforts directed toward the development of effective nonsurgical therapeutic modalities for HCC [12]. There is a growing list of local-regional therapeutic options for HCC, and several of the more prominent techniques are reviewed here including stereotactic radiosurgery (SRS), transarterial chemoembolization (TACE), transarterial radiotherapy, and radiofrequency ablation (RFA).

21.2.1 Stereotactic Radiosurgery

Conventional external beam radiation therapy has been used for decades in the treatment of HCC, but now has a more limited role due to the potential of damage to the normal liver at cytotoxic radiotherapy doses. Modern three-dimensional conformal radiotherapy (3D-CRT) can minimize radiation dose to surrounding normal tissues, deliver radiation to the tumor with more precision, and has resulted in improved tolerance. In unresectable HCC, a pilot study of 3D-CRT resulted in a tumor response rate of 58% and good liver tolerance [13].

However, the risk of radiation induced liver disease (RILD), a clinical syndrome of anicteric hepatomegaly, ascites, and elevated liver enzymes occurring after radiotherapy, has led to the development of even more conformal radiation treatment. Stereotactic radiosurgery has the potential to maximize dose to liver tumors, thereby increasing tumor cell kill, while minimizing risk of radiation injury by reducing radiation dose to normal liver tissue. Image-guided radiation therapy (IGRT) with pretreatment CT scanning and/or peritumoral fiducials allows more accurate patient set-up and treatment delivery. Implanted fiducials and robotic radiosurgery allows real-time tumor tracking with respiration during treatment, may further increase the precision of stereotactic radiosurgery, and may potentially enable the delivery of higher radiation doses, thus delivering better response rates.

Early experience with radiosurgery for liver tumors was reported by Blomgren et al. [14]. Eleven patients with 20 tumors, including nine patients with HCC, one patient with intrahepatic duct cancer, and one patient with an embryonic cancer, were treated with stereotactic radiosurgery. The total dose within the target ranged from 14 to 45 Gy given in one to four fractions. The results were promising, with stable disease in 25% of tumors, a 60% partial response rate, and a 10% complete response rate. All patients experienced fever and nausea following treatment. One patient died two days after a single dose of 30 Gy to a large HCC in the left liver lobe, and two patients developed ascites within 3–6 weeks of treatment and died. Herfarth et al. performed a Phase I/II trial of radiosurgery for primary and metastatic liver tumors. [15]. Sixty neoplastic liver lesions were treated with single fraction stereotactic radiosurgery using escalating doses from 14 to 26 Gy. All treatments were given without major side effects and the local control rate for all lesions was 75%, 71%, and 67% at 6, 12, and 18 months, respectively. While most patients had metastatic disease, this study did demonstrate the feasibility of hepatic stereotactic radiosurgery with high local tumor control rate and low morbidity.

21.2.2 Radiofrequency Ablation

Radiofrequency ablation (RFA) is the use of a high frequency alternating current (100 to 500 KHz) for hyperthermic ablation of liver tumors. The heat created around the probe is conducted into the surrounding tissue, causing coagulative necrosis at temperatures between 50 and 100 degrees Celsius [16]. Various advances in radiofrequency electrodes and techniques, such as cooled-tip and expandable electrodes and pulsed delivery, allow complete necrosis of spherical lesions 3 to 5 cm in diameter and also reduce the number of treatment sessions. RFA can be performed percutaneously or approached via laparoscopy or laparotomy [17]. Percutaneous RFA is performed under ultrasound or CT guidance with local anesthesia and can be an outpatient procedure. However, in patients with large HCCs (> 5 cm), patients with high risk of bleeding, those with tumors adjacent to critical normal tissues at risk of thermal injury, or with deeply positioned lesions not reachable by percutaneous puncture, a laparoscopic or open approach for RFA may be necessary.

RFA is an effective means of treating HCC, particularly in smaller lesions, well-differentiated tumors and noninfiltrative HCCs. After a single session of RFA for HCCs smaller than 3 to 5 cm, several studies have shown 80 to 90% complete tumor necrosis [18–20]. In another study of 126 HCCs larger than 3 cm treated with RFA, complete necrosis was observed in 61% of tumors smaller than 5 cm and in only 24% of tumors larger than 5 cm. In the same study, for tumors smaller than 5 cm, complete response rates were 51% for noninfiltrative HCCs compared with only 22% for infiltrative forms [21]. Long-term survival of patients undergoing RFA in HCC has been reported in two studies. Tateishi et al. noted a 5-year overall survival rate of 54% in 319 patients treated with RFA for HCC and Lencioni et al. noted a 5-year survival rate of 41% [22, 23].

Two randomized studies comparing RFA to ethanol injection have shown that RFA offers more effective local tumor control in HCC. Lin et al. found improved rates of complete tumor necrosis in fewer treatment sessions with RFA compared to ethanol injection. The overall survival and cancer-free survival rate were highest in the RFA group [24]. Lencioni et al. noted 2-year survival rates of 98% in patients undergoing RFA for small HCC with cirrhosis versus 88% in patients undergoing percutaneous ethanol injection (PEI). Two-year local recurrence-free survival rates were also improved in the RFA group compared to those receiving PEI, 96% versus 62%, respectively [25].

RFA is generally well tolerated, even in patients with cirrhosis. Published studies indicate severe complication rates ranging from 0 to 12% and mortality rates ranging from 0.1% to 0.5% [26, 27]. Major acute complications include bleeding, liver failure, pneumothorax, liver abscess, intestinal perforation, hematoma, intraperitoneal hemorrhage, and bile duct stenosis. Needle tract seeding has also been seen with an incidence around 0.5% [28]. Relative contraindications for RFA include lesions bordering the gastrointestinal tract, particularly the colon and hepatic hilum, due to higher rates of complications seen when treating such lesions [29]. One major limitation to RFA is tumor vascularity and proximity of lesions to large vessels. Tumor vascularity or nearby major vessels may act as a heat sink, effectively cooling the area around the RFA probe by drawing heat away, thus increasing the risk of treatment failure. The Pringle maneuver (occlusion of tumor blood supply) during RFA can be an effective means to reduce the cooling effect of blood flow [30].

21.2.3 Transarterial Chemoembolization

Transarterial chemoembolization (TACE) is a regional therapy that has been used for unresectable HCC for several decades. During TACE, iodized poppy seed oil (Lipiodol) and chemotherapy such as doxorubicin, mitomycin C, or cisplatin are administered through the feeding artery of the tumor, followed by embolization with gelatin sponge particles. The concurrent use of transarterial chemotherapy and embolization has several advantages. First, transarterial embolization alone can induce tumor necrosis in HCC, with complete necrosis rates of 64% at 5 years [31]. Second, Lipiodol is selectively retained in malignant tissue and helps to concentrate the cytotoxic agents into the tumor, thereby increasing their effectiveness [32]. A randomized trial by Yoshikawa et al. showed that infusion of a Lipiodol–cytotoxic drug mixture produced a significantly better response rate and survival than the cytotoxic drug alone [33]. Last, the retention of Lipiodol results in intense tumor staining, which may assist in monitoring the tumor's response to treatment.

TACE is an effective therapy for inoperable HCC with several large retrospective studies revealing tumor response rates (reduced size or complete resolution of radiographic abnormality) of 29–62% and 1-year survival rates of 50-76% [34–40]. While some retrospective studies have shown improved survival with TACE when compared to symptomatic treatment, three randomized controlled studies have not found differences in survival. The potential benefit of TACE may be counteracted by its deleterious effect on liver function. In two of the randomized trials, more than 50% of patients had liver failure following TACE [41, 42].

Otherwise, TACE is generally well tolerated. Poon et al. recently analyzed their data of 484 patients with HCC undergoing TACE for inoperable or recurrent HCC. The overall treatment complication and mortality rates were 23% and 4.3%, respectively [43]. Complications included liver failure, liver abscess, tumor rupture, peptic ulcer, acute cholecystitis, acute pancreatitis, and renal failure. Additionally, the authors noted a self-limiting TACE syndrome consisting of nausea, fever, and abdominal pain. Although TACE has been used since the 1980s, there is no formal consensus regarding ideal patient selection. Several studies have shown that poorer outcomes can be expected in patients with large tumors, portal vein invasion, and poor liver function [12]. Other contraindications to TACE include extra-hepatic metastasis and severe arteriovenous shunting. With careful patient selection to reduce the risk of liver failure, TACE should continue to be an important palliative treatment for patients with unresectable and nontransplantable HCC, especially those with tumors larger than 5 cm or multiple tumors that are not favorable for local ablative therapy [12].

Currently, TACE is more commonly performed than either transarterial chemotherapy or transarterial embolization alone. However, transarterial embolization does play an essential role in controlling hemorrhage from ruptured HCC and is considered the first-line emergency treatment for patients with tumor rupture. Additionally, patients with effective hemostasis by transarterial embolization may then undergo elective resection of the tumor [44].

One additional benefit of TACE, shared with some other local-regional therapies, is that it can be repeated. Nevertheless, caution must be exercised and it may be preferable to repeat treatment based upon tumor progression rather than to proceed with a planned sequence of treatments. A recent trial showed fewer complications and better survival with repetition based on tumor response and patient tolerance than with planned repetition of TACE for three cycles [45].

TACE has also been used preoperatively on HCC for hepatic resection and may reduce post-operative recurrence [46, 47]. Two randomized trials, however, showed no difference in recurrence rate with preoperative TACE compared with hepatic resection alone [48, 49]. The benefit of preoperative TACE may be offset by an adverse effect on liver function. In a prospective, but nonrandomized study, the 5-year disease-free survival rate of patients who received preoperative TACE was significantly better than that for patients with resection alone (51% vs. 21%), but the overall 5-year survival rate was not significantly different (43% vs. 38%) because of a higher incidence of postoperative liver failure in the former group [12, 50]. While current evidence does not validate its routine use preoperatively, TACE may downstage tumors for patients with HCC of borderline resectability and increase the chance of curative resection [51, 52].

21.2.4 Transarterial Radiotherapy

Similar to TACE, the principle of transarterial radiotherapy is a targeted cytotoxic agent administered arterially and carried in a medium that is selectively retained in the tumor. In transarterial radiotherapy, the cytotoxic agent is a radioactive isotope such as iodine-131 or yttrium-90. Transarterial iodine-131 is injected with Lipiodol, which, as previously noted, is selectively retained in malignant tissue. The treatment produces a wide range of tumor responses ranging from 17% to 92% [53–56]. In addition to being an effective cytotoxic treatment, transarterial iodine-131 appears to be well tolerated, perhaps even better than TACE. A randomized trial comparing transarterial iodine-131 and TACE showed no significant difference in tumor response rate (24% vs. 25%, respectively) and identical 2-year survival rates, 22% in each arm [57]. Tolerance to iodine-131 was significantly better than with TACE. There were significantly fewer severe side effects in the transarterial iodine-131 group (3 vs. 29 in the TACE group) and lower incidence of arterial thrombosis (1 vs. 10 in the TACE group). Although well tolerated, not all patients are candidates for transarterial radiotherapy. For instance, portal vein thrombosis may be a contraindication to transarterial iodine-131 therapy. In a study of 24 patients with HCC and portal vein thrombosis, transarterial iodine-131 treatment produced only a 12% partial response rate and a 6% one-year survival with 42% of patients having liver failure [58]. Thus, transarterial iodine-131 may produce lower response rates and higher rates of complications in patients with HCC and portal vein thrombosis than alternative treatments.

Transarterial iodine-131 may also be of value in the adjuvant setting, decreasing local recurrence of HCC. In a randomized trial, patients with HCC who underwent curative resection were randomly assigned transarterial iodine-131 or no further treatment. The administration of iodine-131 significantly decreased the incidence of local recurrence from 59% to 28.5% and resulted in significantly improved disease-free and overall survival [59].

Yttrium-90 is a beta-emitting radioisotope that penetrates tissue with an average depth of 2.5 mm and a maximum depth of 11 mm. Yttrium-90 is integrated into glass microspheres and selectively injected into the hepatic artery. The yttrium-90 microspheres are preferentially deposited in the hypervascular tumor due to their size. The microspheres are about 20–30 μm in diameter, and become entrenched in the pre-capillary end-arterioles of the tumor and act as point sources of radiation, delivering intense local radiation to tumor [60].

In a study by Carr et al., 65 patients with unresectable HCC were treated with transarterial yttrium-90 microspheres. The median dose delivered was 134 Gy; 38.4% of the patients had a partial response as measured by computed tomography scan and 64.6% had a substantial decrease in tumor vascularity. The treatment was generally well tolerated with 9 episodes of abdominal pain and 2 episodes of acute cholecystitis, requiring cholecystectomy. Another prominent finding was prolonged and profound lymphopenia in more than 75% of the patients, which was without clinical significance [61]. Lau et al. treated 71 patients with unresectable HCC using transarterial yttrium-90 and all patients showed a response either by serum AFP or ferritin levels, 26.7% of patients had a partial response by CT scan, and overall median survival was 9.4 months [62].

21.2.5 Local-Regional Therapies as a Bridge to Transplantation

Because the waiting time for liver transplantation can be tragically long, the need to control tumor growth in patients with HCC who are waiting for this operation is critical. Increasingly, local-regional therapies are being used as "bridge" to liver transplantation with the goal of controlling local tumor growth while transplant patients are awaiting a graft.

There are no randomized trials in the use of pretransplant TACE. However, a study comparing patients with pretransplant TACE to historical controls without TACE found that the TACE induced marked tumor necrosis but no improvement in survival [63]. Furthermore, those receiving pretransplant TACE appeared to have an increased risk of acute posttransplant infections. A multimodality approach with TACE before transplant followed by postoperative chemotherapy is used by some centers for patients regarded as unsuitable candidates for transplantation, those with HCCs larger than 5 cm. Schwartz et al. from Mount Sinai Medical Center have treated 11 patients with recognized HCC 5 cm or greater with pretransplant TACE and intraoperative and postoperative chemotherapy. They noted no deaths and only one recurrence with a short mean follow-up of 14 months [64].

We have recently begun using CyberKnife® (Accuray Incorporated, Sunnyvale, CA) radiosurgery as a bridge to transplantation. We treated four patients with HCC who ultimately had a liver transplant using doses ranging from 33 to 42 Gy delivered in 3 fractions. There were no complications associated with the CyberKnife treatment. The average time from CyberKnife treatment to liver transplant was 54 days. Explant pathology revealed no viable tumor at the treatment site in two patients. There was viable tumor in the remaining two patients but they both had short intervals (less than six weeks) between radiosurgery and transplantation. This preliminary review suggests that patients with end stage liver disease and HCC tolerate tumor-directed CyberKnife radiosurgery without progressive decompensation. The CyberKnife appears to offer promise as part of a multimodality treatment regimen for patients with end-stage liver disease and HCC. Of course, this is preliminary work, and considerably more research is required to determine whether CyberKnife radiosurgery will play a role in the treatment of these patients.

21.3 Liver Metastasis

Liver metastasis is a persistent oncologic problem. The primary source of liver metastasis is colorectal cancer, followed by lung cancer. Approximately 50% of patients with stages II and III colorectal cancer will develop liver metastases, either at the initial presentation or as recurrent disease. The prognosis of untreated liver metastasis is poor, with 5-year survivals of less than 3% [65]. Surgical resection of liver metastasis has led to improved outcome for these patients. Criteria for resectability include a limited number of lesions, lack of major vascular involvement or other location precluding resection with negative surgical margins, absent or very limited extrahepatic disease, and adequate hepatic functional reserve. In patients that can undergo resection, the surgery can be performed with less than a 5% mortality rate at major centers and a 5-year survival rate of 25-40% [66–72]. The largest series in the literature is from Fong et al. who reported clinical, pathologic, and outcome data for 1001 consecutive patients undergoing liver resection for metastatic colorectal cancer [72]. The overall 5-year survival rate was 37% and the 10-year survival rate was 22%. Seven factors were found to be significant predictors of decreased survival: positive surgical margins, extrahepatic disease, node-positive primary, disease-free interval from primary to metastases of <12 months, number of hepatic tumors >1, largest hepatic tumor >5 cm, and carcinoembryonic antigen level >200 ng/ml. Clearly, the long-term survival data are promising in a select subgroup of patients. Unfortunately, only a small proportion of patients with liver metastases are candidates for surgical resection, and in these patients, alternative local-regional therapies are available.

21.3.1 Stereotactic Radiosurgery

Stereotactic radiosurgery has emerged as a popular modality for treating liver metastasis. In addition to the experiences by Herfarth et al. and Blomgren et al. noted above, Wulf et al. reported their experience with stereotactic radiosurgery in 23 patients with liver metastases. Radiation was given in three fractions of 10 Gy each and local control was defined as complete or partial remission [73]. For all liver metastases the actuarial local control at 1 and 2 years was 76% and 61%, respectively, and the overall survival at 1 and 2 years was 71% and 43%, respectively. The treatment was well tolerated and no acute Grade 3–5 side effects were observed in patients treated with radiosurgery for liver metastasis.

The interim results of a multi-institutional Phase I/II trial of stereotactic radiosurgery for liver metastasis were recently reported [74]. All patients had a maximum tumor diameter of <6 cm, 3 or less discrete lesions, and treatment planning confirming that at least 700 cm^3 of normal liver receives no more than 15 Gy. The most common primary sites were lung, colorectal, and breast. Thus far, 36 patients have received stereotactic radiosurgery; 18 in the Phase I portion and 18 in the Phase II portion. In the Phase I portion of the trial the radiation dose was escalated from 36 Gy in three fractions to 60 Gy in three fractions. All patients in the Phase II portion of the trial received 60 Gy in three fractions. Among 21 patients with ≥6 months follow-up (median 19 months), only one episode of Grade 3 toxicity occurred, in subcutaneous tissue superficial to the liver. No Grade 4 toxicity occurred. For 28 discrete lesions treated, the 18-month actuarial local control estimate was 93%. This interim analysis has shown that in a multi-institutional setting, stereotactic radiosurgery can be given safely with excellent in-field tumor control.

21.3.2 Radiofrequency Ablation

Radiofrequency ablation (RFA) has also been used for patients with unresectable liver metastases. The technique is similar to that used to treat hepatocellular cancer. Solbiati et al. reported results for 117 patients with 179 metachronous liver metastases from colorectal cancer treated with RFA [75]. Recurrent tumors were retreated in this study population. The median survival was estimated to be 36 months and the 3-year survival rate was estimated to be 46%. Interestingly, survival was not significantly related to the number of metastases treated. In 77 patients (66%), new metastases were observed at follow-up.

Overall, there was local recurrence in 39% of liver metastases treated with RFA. The percentage of patients with no new metastases after initial treatment at 2 years was 35%. The treatment was well tolerated. Only two of 117 patients developed severe complications, one patient with a perforated right colon adjacent to an exophytic metastasis and another patient with abdominal pain and signs of a small intraperitoneal hemorrhage. Both patients recovered well.

Another study compared the outcome of patients with solitary colorectal liver metastases treated by RFA or by surgery [76]. Twenty-five patients with solitary colorectal liver metastases were treated by RFA. The indications for RFA were extrahepatic disease in seven patients, vessel involvement in nine patients, and co-morbidity in nine patients. Their outcome was compared with 20 patients who were treated by liver resection for solitary metastases and who had no evidence of extrahepatic disease. Median survival after liver resection was 41 months with a 3-year survival rate of 55%. There was one postoperative death and morbidity was minimal. Median survival after RFA was 37 months with a 3-year survival rate of 53%. The authors conclude that survival after resection and RFA of a solitary colorectal liver metastasis is comparable, that RFA is less invasive, and that it can be performed with an overnight stay or even as an outpatient procedure.

21.3.3 Transarterial Chemoembolization

The results of TACE in liver metastasis are variable, largely due to the disparity in patient selection criteria and regimens used between the different studies [77]. Because of this a wide range of results are reported, with median survivals ranging from 9 to 62 months and the radiographic response rates ranging from 14% to 76%. When compared to systemic chemotherapy, however, TACE has produced reasonable response rates but no consistent improvement in survival for patients with liver metastases [78–83].

TACE has been used in the adjuvant setting for patients undergoing liver resection. In a trial of 109 patients with 1 to 3 potentially resectable liver metastases, patients were randomized to no adjuvant therapy or postoperative TACE combined with systemic 5-FU [84]. TACE combined with systemic 5-FU significantly increased the 4-year recurrence-free rate from 25% in the control group to 46%. The 4-year liver recurrence-free rate was also improved with TACE combined with systemic 5-FU, from 43% in the control group to 67% in the chemotherapy group. Despite the improved recurrence rates, however, TACE plus 5-FU did not result in significantly improved median and overall survival.

Like other local-regional therapies, TACE is useful when other systemic treatments have failed. Tellez et al. reported 30 patients who underwent chemoembolizations for liver metastasis in patients who had experienced failure with one or more systemic treatments [85]. The chemoembolization regimen consisted of a bovine collagen medium with cisplatin, mitomycin C, and doxorubicin. Radiologically, a 25% decrease in the size of the lesion occurred in 63% of the tumors and a decrease of at least 25% of the baseline carcinoembryonic antigen level occurred in 95% of patients. Median survival for all 30 patients was 8.6 months after the initiation of chemoembolization and 29 months after the initial diagnosis of metastasis to the liver. Toxicity included a "postembolization syndrome", consisting of fever > 101 degrees Fahrenheit, pain in the right upper quadrant, nausea, and vomiting. Lethargy was also a common occurrence (> 60%) and lasted up to 6 weeks. Overall, chemoembolization was felt to be a feasible treatment modality for patients with colorectal carcinoma metastasis to the liver who have experienced failure with other systemic treatments.

21.3.4 Transarterial Radiotherapy

Gray et al. reported the results of a Phase III trial of 74 patients with non-resectable liver metastases from primary colorectal adenocarcinoma. Patients were randomized to yttrium-90 and TACE or TACE alone [86]. The partial and complete response rate was significantly greater for patients receiving yttrium-90 and TACE when measured by tumor area, tumor volume, and CEA. The median time to disease progression in the liver was significantly longer for patients receiving yttrium-90 and TACE in comparison to patients receiving TACE alone. The one-, two-, three-, and five-year survival for patients receiving yttrium-90 and TACE was 72%, 39%, 17% and 3.5%, respectively, compared to 68%, 29%, 6.5% and 0% for TACE alone. Also, there was no increase in Grade 3–4 treatment related toxicity and no reduction in quality of life for patients receiving yttrium-90 and TACE in comparison to patients receiving TACE alone. The combination of a single injection of yttrium-90 and TACE is substantially more effective in increasing tumor responses and progression free survival than the same regimen of TACE alone.

21.3.5 Baylor Radiosurgery Center Experience

The Baylor Radiosurgery Center at Baylor University Medical Center in Dallas, TX is dedicated solely to intra- and extra-cranial radiosurgery in a multidisciplinary manner. It is the policy of the Baylor Radiosurgery Center for all patients to undergo evaluation by the radiation oncologist and a surgeon prior to proceeding with radiosurgery. The surgeon is intimately involved in the entire process of radiosurgical treatment at Baylor from patient selection and joint review of diagnostic imaging to radiosurgery planning and treatment. The close collaboration between the two disciplines is essential to ensuring the proper selection and quality management of patients for radiosurgery.

From April 2005 until November 2006, the Baylor Radiosurgery Center has treated 59 patients with abdominal radiosurgery using the CyberKnife. A cohort of 49 liver patients with a median follow-up of 12 months has been analyzed to review the early toxicity and outcome of radiosurgery for hepatic tumors. Of the 49 patients, 29 had metastatic disease (83% colorectal primary, 7% breast primary, and 10% other). Sixteen patients had primary hepatocellular carcinoma. The median size of liver metastasis was 36 cm^3 (range 6–128 cm^3) and for hepatocellular cancer the median size was 66 cm^3 (range 8–793 cm^3). All patients had peritumoral fiducials (3–6) placed prior to treatment planning for tumor tracking. Treatment planning in all patients consisted of dual-phase CT scans. MRI scans were utilized in some patients to help define tumor volumes and some patients were planned with a PET/CT. All patients were treated using the Synchrony™ respiratory tracking system (Accuray Incorporated, Sunyvale, CA). Most patients received radiosurgery in three fractions, over three consecutive days. For metastatic tumors the median dose was 14 Gy per fraction and the median total dose was 39 Gy. For primary hepatocellular carcinomas, the median dose was 13 Gy per fraction and the median total dose was 37 Gy.

Normal tissue dose constraints for all intraabdominal radiosurgery at Baylor are as follows: the maximum dose to the spinal cord is not to exceed 18 Gy in 3 fractions, the maximum dose of the duodenum, stomach, small, and large bowel is not to exceed 24 Gy in 3 fractions, 35% of the liver or a 700 cm^3 volume is not to exceed 15 Gy in 3 fractions, 2/3 of right kidney is kept to below 15 Gy in three fractions.

All patients tolerated their treatments well and no patients were unable to complete the prescribed course of radiosurgery. After radiosurgery, patients are followed up by both the radiation oncologist and surgeon. Contrast-enhanced MRI and/or CT's are obtained at 3-month post-treatment intervals to assess treatment response. By CTC v2.0 guidelines, only 4% of all liver patients developed Grade 3 or higher toxicity including gastric ulcers (2%) and hepatoduodenal fistula (2%). Of 22 evaluable patients, only a single patient experienced local progression at 9 months follow-up.

Of these patients, a subset of 24 patients with 27 liver metastasis from a colorectal primary was treated with radiosurgery. The median follow-up is 12 months. The median tumor size was 39 cm^3 and the median prescription isodose volume was 51 cm^3. Most patients received radiosurgery in 3 fractions to a median dose of 39 Gy (range 24–54 Gy). The dose was prescribed to a median isodose line of 62% (range 55–80%). In this group, 33% of patients developed acute toxicity and all were less than Grade 3. Nausea occurred in 4 (17%) patients and fatigue occurred in 4 (17%) patients. The only other side effect was skin erythema in 2 (8%) patients and both of these were Grade 1. As above, early overall local control is excellent. Additionally, the median pretreatment CEA was 7 μg/L (range 2–290) and decreased to a median value of 2 μg/L (range 1.3–102) following radiosurgery. Overall, 79% of patients had decreased CEA levels following radiosurgery.

Overall, we have found robotic radiosurgery to be a safe and well tolerated procedure with minimal acute toxicity and with promising early results in the management of patients with liver metastases and hepatocellular carcinoma.

21.4 Conclusion

Surgery remains the treatment of choice for HCC. Nevertheless, because most patients will not be eligible for surgical resection or transplantation, a familiarity with various local-regional therapies is important in surgical practice. Indeed, the use of these treatments is expanding as more outcome data become available and the technology matures. It is likely that these nonsurgical local-regional modalities will play an increasingly important role in the management of HCC. The precise role of these various local-regional therapies is best defined by Phase III trials. Despite a relative lack of data directly comparing these modalities, several treatment algorithms have been proposed [12, 87]. Surgical resection for select patients with liver metastases may prolong survival. However, if the criteria for resectability are not met, including a limited number of lesions, location precluding resection with negative margins, limited extrahepatic disease, and adequate hepatic functional reserve other local-regional modalities are emerging as viable alternatives.

References

1. Taylor-Robinson SD, Foster GR, Arora S, *et al.* Increase in primary liver cancer in the UK, 1979-94. *Lancet* 1997; 350:1142–1143.
2. El-Serag HB, Mason AC. Rising incidence of hepatocellular carcinoma in the United States. *N Engl J Med* 1999; 340:745–750.
3. Devita V. Cancer: Principles and Practice of Oncology. Sixth ed: Lippincott Williams and Wilkins; 2001.
4. Poon RT, Fan ST, Lo cm, *et al.* Improving survival results after resection of hepatocellular carcinoma: a prospective study of 377 patients over 10 years. *Ann Surg* 2001; 234:63–70.
5. Colella G, Bottelli R, De Carlis L, *et al.* Hepatocellular carcinoma: comparison between liver transplantation, resective surgery, ethanol injection, and chemoembolization. *Transpl Int* 1998; 11 Suppl 1:S193–196.
6. Fan ST, Lo cm, Liu CL, *et al.* Hepatectomy for hepatocellular carcinoma: toward zero hospital deaths. *Ann Surg* 1999; 229:322–330.
7. Fong Y, Sun RL, Jarnagin W, *et al.* An analysis of 412 cases of hepatocellular carcinoma at a Western center. *Ann Surg* 1999; 229:790–799; discussion 799–800.
8. Mor E, Kaspa RT, Sheiner P, *et al.* Treatment of hepatocellular carcinoma associated with cirrhosis in the era of liver transplantation. *Ann Intern Med* 1998; 129:643–653.
9. Bismuth H, Majno PE, Adam R. Liver transplantation for hepatocellular carcinoma. Semin Liver Dis 1999; 19:311–322.
10. Nerenstone SR, Ihde DC, Friedman MA. Clinical trials in primary hepatocellular carcinoma: current status and future directions. *Cancer Treat Rev* 1988; 15:1–31.
11. Mathurin P, Rixe O, Carbonell N, *et al.* Review article: Overview of medical treatments in unresectable hepatocellular carcinoma–an impossible meta-analysis? *Aliment Pharmacol Ther* 1998; 12:111–126.
12. Poon RT, Fan ST, Tsang FH, *et al.* Locoregional therapies for hepatocellular carcinoma: a critical review from the surgeon's perspective. *Ann Surg* 2002; 235:466–486.
13. Cheng SH, Lin YM, Chuang VP, *et al.* A pilot study of three-dimensional conformal radiotherapy in unresectable hepatocellular carcinoma. *J Gastroenterol Hepatol* 1999; 14:1025–1033.
14. Blomgren H, Lax I, Naslund I, *et al.* Stereotactic high dose fraction radiation therapy of extracranial tumors using an accelerator. Clinical experience of the first thirty-one patients. *Acta Oncol* 1995; 34:861–870.
15. Herfarth KK, Debus J, Lohr F, *et al.* Stereotactic single-dose radiation therapy of liver tumors: results of a phase I/II trial. *J Clin Oncol* 2001; 19:164–170.
16. Buscarini L, Rossi S. Technology for Radiofrequency Thermal Ablation of Liver Tumors. *Semin Laparosc Surg* 1997; 4:96–101.
17. Wood TF, Rose DM, Chung M, *et al.* Radiofrequency ablation of 231 unresectable hepatic tumors: indications, limitations, and complications. *Ann Surg Oncol* 2000; 7:593–600.
18. Goletti O, Lencioni R, Armillotta N, *et al.* Laparoscopic radiofrequency thermal ablation of hepatocarcinoma: preliminary experience. *Surg Laparosc Endosc Percutan Tech* 2000; 10:284–290.
19. Montorsi M, Santambrogio R, Bianchi P, *et al.* Radiofrequency interstitial thermal ablation of hepatocellular carcinoma in liver cirrhosis. Role of the laparoscopic approach. *Surg Endosc* 2001; 15:141–145.
20. Livraghi T, Goldberg SN, Lazzaroni S, *et al.* Small hepatocellular carcinoma: treatment with radio-frequency ablation versus ethanol injection. *Radiology* 1999; 210:655–661.
21. Livraghi T, Goldberg SN, Lazzaroni S, *et al.* Hepatocellular carcinoma: radio-frequency ablation of medium and large lesions. *Radiology* 2000; 214:761–768.
22. Tateishi R, Shiina S, Teratani T, *et al.* Percutaneous radiofrequency ablation for hepatocellular carcinoma. An analysis of 1000 cases. *Cancer* 2005; 103:1201–1209.
23. Lencioni R, Della Pina C, Bartolozzi C. Percutaneous image-guided radiofrequency ablation in the therapeutic management of hepatocellular carcinoma. *Abdom Imaging* 2005; 30:401–408.

24. Lin SM, Lin CJ, Lin cc, *et al.* Radiofrequency ablation improves prognosis compared with ethanol injection for hepatocellular carcinoma < or =4 cm. *Gastroenterology* 2004; 127:1714–1723.
25. Lencioni RA, Allgaier HP, Cioni D, *et al.* Small hepatocellular carcinoma in cirrhosis: randomized comparison of radio-frequency thermal ablation versus percutaneous ethanol injection. *Radiology* 2003; 228:235–240.
26. Giorgio A, Tarantino L, de Stefano G, *et al.* Percutaneous sonographically guided saline-enhanced radiofrequency ablation of hepatocellular carcinoma. *AJR Am J Roentgenol* 2003; 181:479–484.
27. Rhim H. Percutaneous radiofrequency ablation therapy for patients with hepatocellular carcinoma during occlusion of hepatic blood flow: comparison with standard percutaneous radiofrequency ablation therapy. *Cancer* 2003; 98:433–434; author reply 434–435.
28. Livraghi T, Meloni F. Treatment of hepatocellular carcinoma by percutaneous interventional methods. *Hepatogastroenterology* 2002; 49:62–71.
29. Meloni MF, Goldberg SN, Moser V, *et al.* Colonic perforation and abscess following radiofrequency ablation treatment of hepatoma. *Eur J Ultrasound* 2002; 15:73–76.
30. Patterson EJ, Scudamore CH, Owen DA, *et al.* Radiofrequency ablation of porcine liver in vivo: effects of blood flow and treatment time on lesion size. *Ann Surg* 1998; 227:559–565.
31. Matsui O, Kadoya M, Yoshikawa J, *et al.* Subsegmental transcatheter arterial embolization for small hepatocellular carcinomas: local therapeutic effect and 5-year survival rate. *Cancer Chemother Pharmacol* 1994; 33 Suppl: S84–88.
32. Bhattacharya S, Novell JR, Winslet MC, *et al.* Iodized oil in the treatment of hepatocellular carcinoma. *Br J Surg* 1994; 81:1563–1571.
33. Yoshikawa M, Saisho H, Ebara M, *et al.* A randomized trial of intrahepatic arterial infusion of 4'-epidoxorubicin with Lipiodol versus 4'-epidoxorubicin alone in the treatment of hepatocellular carcinoma. *Cancer Chemother Pharmacol* 1994; 33 Suppl:S149–152.
34. Ikeda K, Kumada H, Saitoh S, *et al.* Effect of repeated transcatheter arterial embolization on the survival time in patients with hepatocellular carcinoma. An analysis by the Cox proportional hazard model. *Cancer* 1991; 68:2150–2154.
35. Hsieh MY, Chang WY, Wang LY, *et al.* Treatment of hepatocellular carcinoma by transcatheter arterial chemoembolization and analysis of prognostic factors. *Cancer Chemother Pharmacol* 1992; 31 Suppl:S82–85.
36. Yang CF, Ho YJ. Transcatheter arterial chemoembolization for hepatocellular carcinoma. *Cancer Chemother Pharmacol* 1992; 31 Suppl:S86–88.
37. Yamamoto K, Masuzawa M, Kato M, *et al.* Analysis of prognostic factors in patients with hepatocellular carcinoma treated by transcatheter arterial embolization. *Cancer Chemother Pharmacol* 1992; 31 Suppl:S77–81.
38. Bismuth H, Morino M, Sherlock D, *et al.* Primary treatment of hepatocellular carcinoma by arterial chemoembolization. *Am J Surg* 1992; 163:387–394.
39. Ngan H, Lai CL, Fan ST, *et al.* Treatment of inoperable hepatocellular carcinoma by transcatheter arterial chemoembolization using an emulsion of cisplatin in iodized oil and gelfoam. *Clin Radiol* 1993; 47:315–320.
40. Mondazzi L, Bottelli R, Brambilla G, *et al.* Transarterial oily chemoembolization for the treatment of hepatocellular carcinoma: a multivariate analysis of prognostic factors. *Hepatology* 1994; 19:1115–1123.
41. Pelletier G, Ducreux M, Gay F, *et al.* Treatment of unresectable hepatocellular carcinoma with lipiodol chemoembolization: a multicenter randomized trial. Groupe CHC. *J Hepatol* 1998; 29:129–134.
42. Okada S. Transcatheter arterial embolization for advanced hepatocellular carcinoma: the controversy continues. *Hepatology* 1998; 27:1743–1744.
43. Poon RT, Ngan H, Lo cm, *et al.* Transarterial chemoembolization for inoperable hepatocellular carcinoma and postresection intrahepatic recurrence. *J Surg Oncol* 2000; 73:109–114.
44. Zhu LX, Wang GS, Fan ST. Spontaneous rupture of hepatocellular carcinoma. *Br J Surg* 1996; 83:602–607.
45. Ernst O, Sergent G, Mizrahi D, *et al.* Treatment of hepatocellular carcinoma by transcatheter arterial chemoembolization: comparison of planned periodic chemoembolization and chemoembolization based on tumor response. *AJR Am J Roentgenol* 1999; 172:59–64.
46. Shimada M, Takenaka K, Gion T, *et al.* Prognosis of recurrent hepatocellular carcinoma: a 10-year surgical experience in Japan. *Gastroenterology* 1996; 111:720–726.
47. Imaoka S, Sasaki Y, Shibata T, *et al.* A pre-operative chemoembolization therapy using lipiodol, cisplatin and gelatin sponge for hepatocellular carcinoma. *Cancer Chemother Pharmacol* 1989; 23 Suppl:S126–128.
48. Wu cc, Ho YZ, Ho WL, *et al.* Preoperative transcatheter arterial chemoembolization for resectable large hepatocellular carcinoma: a reappraisal. *Br J Surg* 1995; 82:122–126.
49. Yamasaki S, Hasegawa H, Kinoshita H, *et al.* A prospective randomized trial of the preventive effect of pre-operative transcatheter arterial embolization against recurrence of hepatocellular carcinoma. *Jpn J Cancer Res* 1996; 87:206–211.
50. Gerunda GE, Neri D, Merenda R, *et al.* Role of transarterial chemoembolization before liver resection for hepatocarcinoma. *Liver Transpl* 2000; 6:619–626.
51. Harada T, Matsuo K, Inoue T, *et al.* Is preoperative hepatic arterial chemoembolization safe and effective for hepatocellular carcinoma? *Ann Surg* 1996;224:4–9.
52. Majno PE, Adam R, Bismuth H, *et al.* Influence of preoperative transarterial lipiodol chemoembolization on resection and transplantation for hepatocellular carcinoma in patients with cirrhosis. *Ann Surg* 1997; 226:688–701; discussion 701–683.
53. Raoul JI, Bretagne JF, Caucanas JP, *et al.* Internal radiation therapy for hepatocellular carcinoma. Results of a French multicenter phase II trial of transarterial injection of iodine 131-labeled Lipiodol. *Cancer* 1992; 69:346–352.
54. Kajiya Y, Kobayashi H, Nakajo M. Transarterial internal radiation therapy with I-131 Lipiodol for multifocal hepatocellular carcinoma: immediate and long-term results. *Cardiovasc Intervent Radiol* 1993; 16:150–157.
55. Leung WT, Lau WY, Ho S, *et al.* Selective internal radiation therapy with intra-arterial iodine-131-Lipiodol in inoperable hepatocellular carcinoma. *J Nucl Med* 1994; 35:1313–1318.
56. Maini CL, Scelsa MG, Fiumara C, *et al.* Superselective intra-arterial radiometabolic therapy with I-131 lipiodol

in hepatocellular carcinoma. *Clin Nucl Med* 1996; 21:221-226.
57. Raoul JL, Guyader D, Bretagne JF, *et al.* Prospective randomized trial of chemoembolization versus intra-arterial injection of 131I-labeled-iodized oil in the treatment of hepatocellular carcinoma. *Hepatology* 1997; 26:1156-1161.
58. de Baere T, Taourel P, Tubiana JM, *et al.* Hepatic intra-arterial 131I iodized oil for treatment of hepatocellular carcinoma in patients with impeded portal venous flow. *Radiology* 1999; 212:665-668.
59. Lau WY, Leung TW, Ho SK, *et al.* Adjuvant intra-arterial iodine-131-labelled lipiodol for resectable hepatocellular carcinoma: a prospective randomised trial. *Lancet* 1999;353:797-801.
60. Garrean S, Joseph Espat N. Yttrium-90 internal radiation therapy for hepatic malignancy. *Surg Oncol* 2005; 14:179-193.
61. Carr BI. Hepatic arterial 90Yttrium glass microspheres (Therasphere) for unresectable hepatocellular carcinoma: interim safety and survival data on 65 patients. *Liver Transpl* 2004; 10:S107-110.
62. Furuse J, Satake M, Iwasaki M, *et al.* Percutaneous ethanol injection under interventional radiographic computed tomography-fluoroscopic guidance for the treatment of small hepatocellular carcinomas. *Int J Clin Onco* 1998; 3:102-107.
63. Oldhafer KJ, Chavan A, Fruhauf NR, *et al.* Arterial chemoembolization before liver transplantation in patients with hepatocellular carcinoma: marked tumor necrosis, but no survival benefit? *J Hepatol* 1998; 29:953-959.
64. Schwartz ME, Sung M, Mor E, *et al.* A multidisciplinary approach to hepatocellular carcinoma in patients with cirrhosis. *J Am Coll Surg* 1995; 180:596-603.
65. Wagner JS, Adson MA, Van Heerden JA, *et al.* The natural history of hepatic metastases from colorectal cancer. A comparison with resective treatment. *Ann Surg* 1984; 199:502-508.
66. Fong Y, Salo J. Surgical therapy of hepatic colorectal metastasis. *Semin Oncol* 1999; 26:514-523.
67. Adson MA, van Heerden JA, Adson MH, *et al.* Resection of hepatic metastases from colorectal cancer. *Arch Surg* 1984; 119:647-651.
68. Gayowski TJ, Iwatsuki S, Madariaga JR, *et al.* Experience in hepatic resection for metastatic colorectal cancer: analysis of clinical and pathologic risk factors. *Surgery* 1994; 116:703-710; discussion 710-701.
69. Hughes KS, Simon R, Songhorabodi S, *et al.* Resection of the liver for colorectal carcinoma metastases: a multi-institutional study of patterns of recurrence. *Surgery* 1986; 100:278-284.
70. Schlag P, Hohenberger P, Herfarth C. Resection of liver metastases in colorectal cancer-competitive analysis of treatment results in synchronous versus metachronous metastases. *Eur J Surg Oncol* 1990; 16:360-365.
71. Rosen CB, Nagorney DM, Taswell HF, *et al.* Perioperative blood transfusion and determinants of survival after liver resection for metastatic colorectal carcinoma. *Ann Surg* 1992; 216:493-504; discussion 504-495.
72. Fong Y, Fortner J, Sun RL, *et al.* Clinical score for predicting recurrence after hepatic resection for metastatic colorectal cancer: analysis of 1001 consecutive cases. *Ann Surg* 1999; 230:309-318; discussion 318-321.
73. Wulf J, Hadinger U, Oppitz U, *et al.* Stereotactic radiotherapy of targets in the lung and liver. *Strahlenther Onkol* 2001; 177:645-655.
74. Kavanagh BD, Schefter TE, Cardenes HR, *et al.* Interim analysis of a prospective phase I/II trial of SBRT for liver metastases. *Acta Oncol* 2006; 45:848-855.
75. Solbiati L, Livraghi T, Goldberg SN, *et al.* Percutaneous radio-frequency ablation of hepatic metastases from colorectal cancer: long-term results in 117 patients. *Radiology* 2001; 221:159-166.
76. Oshowo A, Gillams A, Harrison E, *et al.* Comparison of resection and radiofrequency ablation for treatment of solitary colorectal liver metastases. *Br J Surg* 2003; 90:1240-1243.
77. Vogl TJ, Zangos S, Eichler K, *et al.* Colorectal liver metastases: regional chemotherapy via transarterial chemoembolization (TACE) and hepatic chemoperfusion: an update. *Eur Radiol* 2006.
78. Wagman LD, Kemeny mm, Leong L, *et al.* A prospective, randomized evaluation of the treatment of colorectal cancer metastatic to the liver. *J Clin Oncol* 1990; 8:1885-1893.
79. Kemeny N, Daly J, Reichman B, *et al.* Intrahepatic or systemic infusion of fluorodeoxyuridine in patients with liver metastases from colorectal carcinoma. A randomized trial. *Ann Intern Med* 1987; 107:459-465.
80. Chang AE, Schneider PD, Sugarbaker PH, *et al.* A prospective randomized trial of regional versus systemic continuous 5-fluorodeoxyuridine chemotherapy in the treatment of colorectal liver metastases. *Ann Surg* 1987; 206:685-693.
81. Rougier P, Laplanche A, Huguier M, *et al.* Hepatic arterial infusion of floxuridine in patients with liver metastases from colorectal carcinoma: long-term results of a prospective randomized trial. *J Clin Oncol* 1992; 10:1112-1118.
82. Kemeny N, Cohen A, Seiter K, *et al.* Randomized trial of hepatic arterial floxuridine, mitomycin, and carmustine versus floxuridine alone in previously treated patients with liver metastases from colorectal cancer. *J Clin Oncol* 1993; 11:330-335.
83. Reappraisal of hepatic arterial infusion in the treatment of nonresectable liver metastases from colorectal cancer. Meta-Analysis Group in Cancer. *J Natl Cancer Inst* 1996; 88:252-258.
84. Kemeny mm, Adak S, Gray B, *et al.* Combined-modality treatment for resectable metastatic colorectal carcinoma to the liver: surgical resection of hepatic metastases in combination with continuous infusion of chemotherapy-an intergroup study. *J Clin Oncol* 2002; 20:1499-1505.
85. Tellez C, Benson AB, 3rd, Lyster MT, *et al.* Phase II trial of chemoembolization for the treatment of metastatic colorectal carcinoma to the liver and review of the literature. *Cancer* 1998; 82:1250-1259.
86. Gray B, Van Hazel G, Hope M, *et al.* Randomised trial of SIR-Spheres plus chemotherapy vs. chemotherapy alone for treating patients with liver metastases from primary large bowel cancer. *Ann Oncol* 2001; 12:1711-1720.
87. Llovet JM. Updated treatment approach to hepatocellular carcinoma. *J Gastroenterol* 2005; 40:225-235.

Stereotactic Radiosurgery for Liver Malignancies Using the CyberKnife

22

Karyn A. Goodman

CONTENTS

22.1 Abstract

Radiotherapy for liver tumors has been limited by the low tolerance of the whole liver to radiation. Improvements in imaging and localizing techniques allow more accurate targeting of radiotherapy. Our Phase I dose escalation study evaluated the safety and feasibility of single-fraction radiosurgery with the CyberKnife® (Accuray Incorporated, Sunnyvale, CA) for patients with primary or metastatic liver malignancies. Patients underwent abdominal CT scans with dual phase contrast to identify lesions and CT-guided percutaneous placement of gold spherical fiducials for targeting purposes. Treatments were delivered with motion tracking using X-ray detection of fiducial and surface light emitting diode positions. The dose was escalated from 18 Gy to 30 Gy increasing in increments of 4 Gy, prescribed to the minimal isodose line that completely encompassed the planning target volume (PTV). Acute gastrointestinal toxicity was scored according to the common toxicity criteria adverse event (CTCAE), version 3. Response to treatment was determined by serial high-resolution CT and PET scanning and/or MRI. Twenty patients were treated to 26 sites at four dose levels. Sixteen patients had metastatic disease involving the liver, 2 patients had hepatocellular carcinoma (HCC), and 2 patients had intrahepatic cholangiocarcinoma (IHCC). The mean treatment volume was 36.0 cm^3 (range 0.8–146.6 cm^3). With a median follow-up of 7 months (range: 2–30 months), Grade 1 toxicity was reported by 7 patients, consisting of nausea, abdominal pain, fatigue, and fever. Follow-up scans for 19 patients treated to 25 sites have demonstrated interval decrease in size in 17 sites, stable disease in 3 sites, and local progression in 5 sites. Seven lesions exhibited a complete response. Stereotactic radiosurgery to primary or metastatic malignancies of the liver is feasible and safe. We did not reach a liver tolerance dose.

22.2 Introduction

Unresectable hepatocellular carcinoma (HCC) and intrahepatic cholangiocarcinomas (IHCC), as well as liver metastases from other primary cancers are fatal conditions. In the United States, there were an estimated 17,550 new cases of primary liver cancer diagnosed and 15,420 deaths in 2005 [1]. The worldwide incidence of primary liver cancer is much more significant with an estimated 1 million patients diagnosed annually [2]. HCC represents approximately 70–85% of the cases, with IHCC representing the remainder of the cases. The incidence of HCC has been steadily increasing in the U.S. with a 75% increase over the past decade [3]. The most common cause of HCC worldwide is hepatitis B, with other major causes including hepatitis C, alcoholism, and primary biliary cirrhosis. There is an estimated 0.5% annual risk of developing HCC in chronic hepatitis B patients and 5% in chronic hepatitis C patients. A survival benefit exists if total excision is possible via partial hepatectomy or transplantation with 5-year survival rates ranging from 12% to 55% [2]. Unfortunately, only 15–35% are resectable at the time of presentation. Median survival in unresectable patients is less than 4 months [2].

Metastatic lesions to the liver from other primary sites are also a common cause of morbidity and mortality [4]. The most common metastatic lesion in the liver is from colorectal adenocarcinoma. Approximately 50% of patients with colorectal adenocarcinoma will develop metastatic disease to the liver. This is approximately 50,000 patients per year, half of whom have the liver as the only site of known metastatic disease [5]. If untreated, 3-year survival is 3% with no survivors at the 5-year time point. With surgery, there are reported 5-year survival rates of 25–37% [6]. Similar to primary liver disease, the majority of liver metastases are unresectable.

Alternatives to surgery, such as radiofrequency ablation (RFA), cryoablation, and alcohol injection have been attempted and have shown some benefit. These techniques, however, suffer from many limitations due to lesion location, underlying liver dysfunction and high recurrence rates [7]. Historically, radiotherapy has been limited due to the low tolerance of the whole liver to irradiation. Conformal techniques and image guidance have allowed for the delivery of focal ablative doses of radiation. Dose escalation studies with fractionated regimens have shown a clear dose response for HCC, IHCC, and liver metastases [8–10].

Stereotactic radiosurgery (SRS) techniques have been applied to the treatment of liver tumors, further refining our ability to target the tumor and minimize normal tissue toxicity [11–14]. Because of the spatial precision of SRS, it is feasible to administer a tumoricidal radiation dose in only a few outpatient treatments. By minimizing the irradiation of surrounding healthy tissue, it may be possible to decrease the rate of complications and thereby allow for dose escalation and ultimately better local tumor control. This approach also exploits the potentially beneficial effect of hypofractionation whereby higher doses of radiation in fewer fractions can overcome the effects of sub-lethal damage repair.

The use of SRS requires highly accurate localization and patient immobilization techniques to reduce uncertainties in the positioning of the delivered dose. With this type of highly targeted radiotherapy, there is an increased risk of geographical miss, especially with a moving target such as the liver [15]. Historically, SRS in the abdomen has been limited by the movement and deformation of intra-abdominal organs with respiration. Several methods to account for this movement have been developed in order to ensure that the high-dose region coincides with the tumor and does not inadvertently treat normal tissue [16, 17]. Some of these techniques include respiratory gating, active breathing control (ABC), deep inspiration breath hold (DIBH), abdominal compression, or some combination.

22.3 Robotic Radiosurgery for Liver Lesions

There have been several reports in the literature with encouraging results of SRS for liver lesions. Herfarth and colleagues from Heidelberg, Germany performed a Phase I/II dose-escalation trial of stereotactic single-dose radiation therapy in 37 patients with 60 liver lesions, the majority of which were metastases [11]. The dose was escalated from 14 Gy to 26 Gy (reference point), with the 80% isodose surrounding the planning target volume. Median tumor size was 10 cm^3 (range, 1 to 132 cm^3). All patients tolerated the treatment well without any major side effects. Eleven patients experienced intermittent loss of appetite or mild nausea for 1–3 weeks after treatment. None of the treated patients developed clinically detectable radiation-induced liver disease (weight gain, ascites, and newly developed increase of alkaline phosphatase concentration). The overall actuarial local tumor control rates were 75%, 71%, and 67% at 6 months, 12 months, and 18 months of follow-up, respectively. There was a statistically significant difference in Kaplan-Meier estimates of local tumor control between tumors treated with 14–20 Gy vs. 22–26 Gy: 0% of tumors were controlled in the early patients treated with low doses at 18 months, but 81% were controlled when treated with 22–26 Gy. Stratification by size did not reveal a significant difference in the local control rate for larger ($\geq 15\ cm^3$) compared with smaller targets ($< 15\ cm^3$) in the 22–26 Gy range. No major side effects have been observed [11]. Investigators at the Karolinska Institute in Stockholm, Sweden have reported their experience with SRS for primary and metastatic liver tumors, treated with 15–45 Gy, delivered in one to five fractions. Seventy-five evaluable tumors in 50 patients were treated with volumes ranging from 2 cm^3 to 732 cm^3 (mean, 73 cm^3). With a mean follow-up of 12 months (range 1.5–38 months) approximately 30% of tumors demonstrated growth arrest, nearly 40% were reduced in size, and 32% disappeared by imaging studies, which were usually performed at 2- to 3-month intervals after radiosurgery. Four of the tumors were classified as local failures (5.3%) [18]. Unfortunately, the mean survival time was only 13.4 months (range 1.5–39 months), with the predominant cause of death related to progressive liver cirrhosis and progression of extrahepatic disease [19]. Wulf et al. reported the experience from the University of Würzburg treating 24 hepatic lesions with 10 Gy × 3 fractions using stereotactic radiosurgery. Their actuarial local control was 76% at 1 year, and 61% at two years with no Grade 3 or higher toxicities [20].

More recently, Schefter et al. reported the results of their Phase I study of stereotactic body radiation therapy (SBRT) for liver metastases. Eighteen patients with 24 lesions were treated with doses escalated from 36 Gy to 60 Gy in 3 fractions. They were able to dose escalate to 60 Gy (20 Gy × 3) without experiencing any dose limiting toxicity (DLT) [13]. Twelve of the 18 patients were alive at the time of their analysis, a median of 7.1 months after enrollment on the protocol. A Phase II study is underway to evaluate the local control rates for 60 Gy in 3 fractions.

Investigators at the University of Rochester conducted a study of 72 patients treated to 182 sites of metastasis with stereotactic radiosurgery. Patients had a median of 2 lesions (range 1–6) with a median diameter of 2.7 cm (0.5–12.2 cm). The median total dose was 45 Gy (17.5–56 Gy) delivered in 2–10 fractions. With a median follow-up time of 12 months, they reported an 88% in-field local control rate and no Grade 3 or higher toxicity [21].

22.4 Robotic Radiosurgery for Primary and Metastatic Liver Lesions

At Stanford University, we are investigating the use of single fraction SRS with the CyberKnife robotic radiosurgery system. Recent technological advances in the CyberKnife made it possible to image and treat tumors within the abdominal cavity [22]. We conducted a Phase I dose escalation trial of CyberKnife radiosurgery to deliver a single fraction to either primary liver tumors or hepatic metastases. With the data from Heidelberg establishing the relative safety of large single-fraction stereotactic radiation therapy to the liver tumor [15–17], a starting dose level of 18 Gy was selected with escalation in 4 Gy increments to a maximum dose of 30 Gy. The 18 Gy starting dose level was chosen since it was prescribed to the isodose line encompassing the tumor volume and the maximum dose could be as much as 30% higher, resulting in a point dose equivalent to the 22 Gy dose level prescribed to a reference point in the German study. Acute and late toxicities were scored based on the CTCAE version 3 and response to treatment was determined by serial CT and/or PET scanning. Between February 2004 and September 2006, 20 patients have been treated on our study. A total of 26 sites were treated at the 4 specified dose levels: 18 Gy (n=3), 22 Gy (n=8), 26 Gy (n=12), 30 Gy (n=2), and one patient received 24 Gy to one site (Table 22.1). Sixteen patients had hepatic metastases, 2 patients had IHCC, and 2 had recurrent HCC. With a median follow-up of 7 months (range 2–30 months), acute Grade 1 toxicity (nausea, abdominal pain, fatigue, and fever) was reported by 7 patients. These symptoms were self-limited, with the duration of symptoms ranging for 3 hours for some patients with fever and nausea to two weeks for fatigue. One patient was diagnosed

Table 22.1 Patient, Treatment and Response Characteristics of 20 Patients Treated on Phase I Protocol

	Age	Diagnosis	# of Lesions	Dose (Gy)	Dmax (Gy)	GTV (cm^3)	Site	Acute Toxicity Grade	Late Toxicity Grade	Follow-up Time (mos)	Response	Vital status
1	72	Colorectal ca	1	18	25.7	21.7	Porta hepatis LNs	0	2 (duodenal ulcer, post EBRT)	30	PD	AWD
2	72	Cholangiocarcinoma	1	18	25.7	33.0	Left lobe	0	0	29	CR	AWD
3	51	Pancreatic ca	1	18	22.5	11.4	Right lobe	0	0	8	PR	DOD
4	43	Small cell carcinoma of the bile duct	1	22	29.3	30.6	Porta hepatis mass	1 (nausea)	2 (duodenal ulcer)	9	PD	DOD
5	75	Colorectal ca	1	22	26.5	24.0	Right lobe	0	0	16	PD	AWD
6	64	Pancreatic ca	1	22	28.9	27.0	Right + left lobes	0		2	No f/u scans	DOD
7	23	Neuroblastoma	2	22, 22	30.1, 29.3	42.6, 32.2	Porta hepatis lymph node, right lobe	1 (abdominal pain)	0	14	PR, CR	AWD
8	49	Neuroendocrine ca	2	24, 22	33.8, 28.9	7.5, 24.6	Right lobe	1 (fatigue)	0	18	PR, PR	AWD
9	84	Hepatocellular ca	1	22	34.9	74.5	Inf right lobe	0	0	7	CR	AWD
10	34	Sinonasal undifferentiated ca	1	26	41.3	76.6	Right lobe (dome)	0	0	5	CR	DOD
11	57	Pancreatic ca	1	26	32.5	8.0	Right lobe	0	0	6	CR	DOD
12	75	Gastric ca	2	26, 26	33.3, 36.1	26.9, 13.2	Right lobe	0	0	5	PR, PR	DOD
13	64	Ovarian ca	1	26, 26	30.6, 36.6	6.8, 36.1	Ant right lobe	0	0	9	SD, SD	AWD
14	62	Colorectal ca	2	26, 26	39.4, 32.9	94.7, 0.8	Ant right lobe	1 (nausea)	0	8	PR, PR	AWD
15	45	Colorectal ca	1	26	39.4	41.6	Peripheral right lobe	1 (nausea)	0	6	CR	AWD
16	78	Ovarian ca	1	26	34.7	53.2	Porta hepatis lymph node	2 (duodenal ulcer)	0	4	PR	AWD
17	72	Hepatocellular ca	1	26	38.8	12.5	Right lobe	1 (nausea)	0	6	PR	AWD
18	62	Colorectal ca	1	22, 26	31.0, 37.1	146.6, 21.4	Dome, caudate lobe	1 (low grade fever)	0	4	PD, SD	AWD
19	61	Breast Ca	1	30	45.5	29.8	Right Lobe	0	0	6	CR	AWD
20	45	Cholangiocarcinoma	1	30	46.6	39	Left lobe	0	0	6	PD	AWD

Complete Response (CR), Partial Response (PR), Progressive Disease (PD), Stable Disease (SD), Alive with Disease (AWD), Dead of Disease (DOD), Gross Tumor Volume (GTV), Lymph Nodes (LNs), External Beam Radiotherapy (EBRT)

with a Grade 2 duodenal ulcer within 2 months of treatment. Two patients developed late Grade 2 duodenal ulcers at 8 months and 23 months after CyberKnife treatment. The ulcers were symptomatic for several months for all of these patients. Follow-up scans were available for 19 patients treated to 25 sites and have demonstrated a complete response at 7 sites, partial response at 10 sites, stable disease at 3 sites, and local progression at 5 sites. For patients with progressive disease the median time to progression was 5.5 months (range 2–15 months). Six patients have died of distant disease with a median of 5 months between treatment and death.

PET-CT scans have been particularly important in assessing response as there are many changes in the liver parenchyma after high dose irradiation of the liver that may be related to radiation changes and are difficult to interpret [23]. Figure 22.1a–c demonstrates PET-CT responses seen for a patient treated on this protocol, with an initial response by PET seen at the early follow-up time but recurrence of FDG-avid disease at 21 weeks for the lower dose level. PET responses have been more durable at the higher dose levels.

The preliminary results of this Phase I dose escalation trial are very encouraging with minimal toxicity related to high single-fraction SRS to the liver. The dose-response data cannot be adequately assessed as the number of patients at the highest dose level is small; however, there appears to be some improvement in response even at the 26 Gy level. There are several long-term survivors on this study suggesting that SRS may have impacted on overall survival in a select group of patients.

22.5 Technical Considerations for Robotic Radiosurgery for Liver Lesions

22.5.1 Set-up Procedure

All patients treated with liver SBRT undergo implantation of 3–5 fiducial seeds into liver tumors either through CT-guidance by an interventional ra-

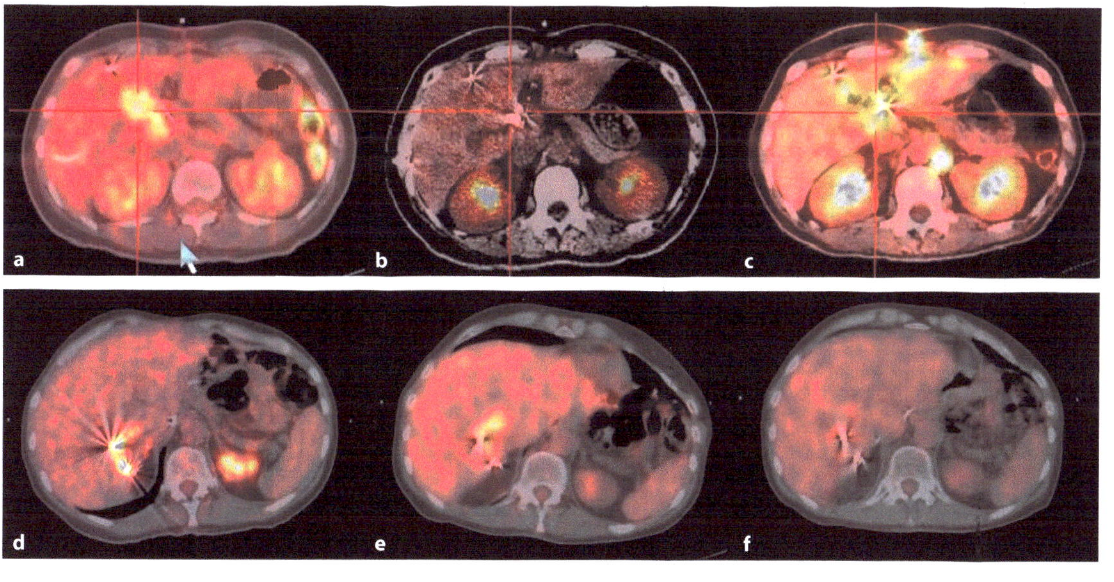

Fig. 22.1. a–c Example of FDG-PET response for patient with liver metastasis from biliary small cell carcinoma before a, 10 weeks after b, and 21 weeks after c CyberKnife SRS to 22 Gy in one fraction. At this lower dose level, the response was not durable. d–f 52 year old patient with metastatic gastric cancer treated with CyberKnife SRS to 30 Gy in one fraction. Fused F-18 FDG PET/CT images prior to d, five weeks e, and ten weeks f post-treatment are displayed. There is progressive interval decrease in size and focality of pathologic FDG uptake, with activity near background levels on the final image. The standardized uptake value (SUV) decreased from 5.9 to 3.5 to 2.9. Continued improvement is noted between five e and ten weeks f post-treatment although no additional radiation was administered during this interval.

diologist or during an open bypass procedure. These seeds are made of 100% gold and are radiographically visualized by kilovoltage X-rays. The seeds measure 5 mm in length and 1 mm in diameter. Because we observed that these seeds can migrate several mm in the first few days after implantation, we typically wait at least 5 days before performing subsequent treatment planning scans.

For each patient, a custom Alpha Cradle® (Smithers Medical Products, Inc., North Canton, OH) immobilization device is constructed to hold the body in a reproducible position. Patients are positioned supine with their arms above their head, resting on the support of the Alpha Cradle. With patients in the treatment position, a biphasic liver protocol CT scan (1.25 mm cuts) is performed for high resolution delineation of the tumor and surrounding structures. Intravenous contrast is administered in a rapid bolus such that the arterial phase is obtained when the patient is being coached to remain in the expiratory phase. In addition, we perform an FDG-PET scan at the treatment planning session so that there is accurate co-registration of the tumor with the areas of hypermetabolic activity. Furthermore, we also perform 4D-CT scans, in which CT data (2.5 mm cuts) are acquired synchronously with a respiratory signal, to evaluate temporal changes of the anatomy as a function of the respiratory phase during the imaging, in order to correct for respiratory related liver tumor movement. Patients are given an audio coaching CD to review prior to the set-up procedure to determine a comfortable breathing rhythm.

22.5.2 Target Definition

The tumor volume is delineated on cross-sectional images from the planning biphasic CT scan using the Varian Eclipse (Varian Medical Systems, Inc., Palo Alto, CA) treatment planning system. 4D-CT scans are reconstructed and tumor motion is evaluated on an Advantage Workstation® (GE Healthcare). The primary tumor is contoured on both the end-expiratory

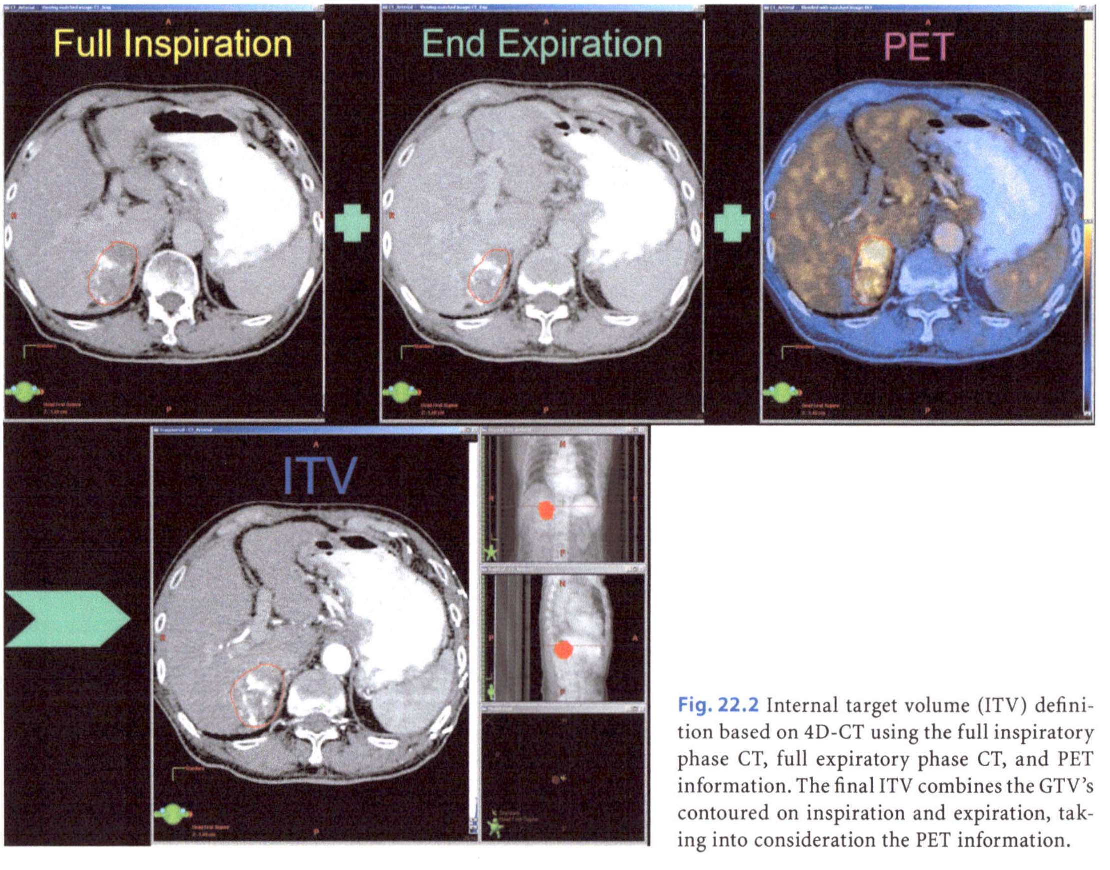

Fig. 22.2 Internal target volume (ITV) definition based on 4D-CT using the full inspiratory phase CT, full expiratory phase CT, and PET information. The final ITV combines the GTV's contoured on inspiration and expiration, taking into consideration the PET information.

(50–60%) and end-inspiratory (0%) phases (Fig. 22.2). These scans are registered with the biphasic CT scan based on fiducial matching, such that the implanted fiducials are superimposed between them, rather than based on DICOM coordinates. Thus, the different respiratory phases of the 4D-CT are registered by a fiducial translation. This allows for evaluation of the target volume deformation and rotation, rather than just superior-inferior translation, which is tracked by the Synchrony system® (Accuray Incorporated, Sunnyvale, CA). The liver lesion(s) is/are outlined on the different respiratory phases as the GTVarterial phase, GTVinspiratory phase, GTVexpiratory phase, and GTV-PET. The internal target volume (ITV) is the union of the GTVs defined on the arterial phase scan as well as the end-expiratory scan, and the end-inspiratory scan from the 4D-CT [24]. The ITV is needed as liver rotation and deformation are not accounted for during the treatment delivery. The free-breathing PET-CT information is also fused with the exhalation arterial phase CT scan and is used for confirmation of tumor position and delineation of normal liver and adjacent tissues. No GTV-CTV margin is added for subclinical extension. An additional margin is added for patient set-up error based on the size of the tumor. Typically, the margin for the PTV is not more than 2 mm, unless it is a small tumor where it can be up to 3–5 mm.

22.5.3 Treatment Planning

Images from the planning CT and all contours are transferred to the CyberKnife planning system. The SRS treatment plan is generated to cover the PTV. For CyberKnife plans, the SBRT dose (*Dmin*) is prescribed to the isodose line that completely surrounds the tumor (Fig. 22.3). Tuning structures are used to reduce the dose to the adjacent stomach, kidney, and/or small bowel. The *Dmax* is the maximum dose delivered and is always within the PTV.

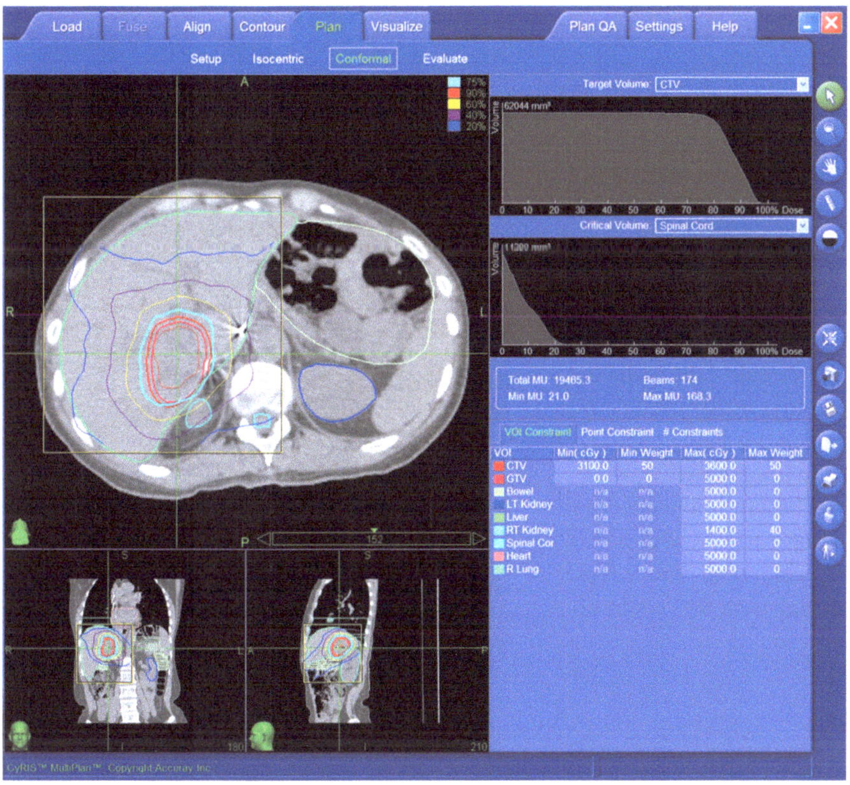

Fig. 22.3 Treatment plan with isodose lines showing sharp dose gradient for patient with 2 separate lesions treated by CyberKnife SRS. The red line represents the tumor volume. The green line is the 76% isodose line and the purple line is the 50% isodose.

22.5.4 Image Guidance for Treatment Delivery

For patients treated on the CyberKnife, the Synchrony system is used for real-time targeting of tumors. A series of light-emitting diodes (LEDs) are placed upon the chestwall and movement is detected by wall mounted cameras in the treatment room. Using motion tracking software, the Synchrony system identifies, updates, and then correlates external body surface movement with movement of the internal tumor fiducials. Throughout the procedure, the Synchrony system monitors the target and modifies the correlation model as needed to follow changes in tumor motion. However, only translational respiratory-related movement of liver tumors is accounted for by this system. The liver motion is complex due to organ deformation and rotation [25]. As described above, a margin is built into the ITV during the target definition process to account for some of these motions.

22.6 Conclusions

CyberKnife radiosurgery for liver lesions is becoming accepted as a safe, convenient, and effective option for patients with primary or metastatic tumors of the liver. Given the particularly grim prognosis faced by patients with unresectable tumors in the liver, a minimally invasive option with minimal toxicity is particularly appealing in this patient population since it is not expected to impair quality of life. Moreover, as systemic agents become more effective in prolonging survival among patients with metastatic disease, particularly colorectal and breast cancer, aggressive local therapy with SRS could ultimately impact overall survival.

References

1. Jemal A, Murray T, Ward E, *et al.* Cancer statistics, 2005. *CA Cancer J Clin* 2005; 55:10–30.
2. Cha C DR, DeMatteo RP, Blumgart LH. Surgery and ablative therapy for hepatocellular carcinoma. *J Clin Gastroenterol* 2002; 35:S130–137.
3. El-Serag HB, Mason AC. Rising incidence of hepatocellular carcinoma in the United States. *N Engl J Med* 1999; 340:745–750.
4. Kemeny NE, Fong Y. Treatment of Liver Metastases. In: Frei H, editor. Cancer Medicine 6ed. Hamilton, Ontario: BC Decker; 2003.
5. Blumgart LH, Fong Y. Surgical options in the treatment of hepatic metastasis from colorectal cancer. *Curr Probl Surg* 1995; 32:333–421.
6. Choti MA, Sitzmann JV, Tiburi MF, *et al.* Trends in long-term survival following liver resection for hepatic colorectal metastases. *Ann Surg* 2002; 235:759–766.
7. De Carlis L, Giacomoni A, Pirotta V, *et al.* Surgical treatment of hepatocellular cancer in the era of hepatic transplantation. *J Am Coll Surg* 2003; 196:887–897.
8. Crane CH, Abbruzzese JL, Evans DB, *et al.* Is the therapeutic index better with gemcitabine-based chemoradiation than with 5-fluorouracil-based chemoradiation in locally advanced pancreatic cancer? *Int J Radiat Oncol Biol Phys* 2002; 52:1293–1302.
9. Park HC, Seong J, Han KH, *et al.* Dose-response relationship in local radiotherapy for hepatocellular carcinoma. *Int J Radiat Oncol Biol Phys* 2002; 54:150–155.
10. Dawson LA, McGinn CJ, Normolle D, *et al.* Escalated focal liver radiation and concurrent hepatic artery fluorodeoxyuridine for unresectable intrahepatic malignancies. *J Clin Oncol* 2000; 18:2210–2218.
11. Herfarth KK, Debus J, Lohr F, *et al.* Stereotactic single-dose radiation therapy of liver tumors: results of a phase I/II trial. *J Clin Oncol* 2001; 19:164–170.
12. Sampson MC, Huggins C, Schell M, *et al.* Hypofractionated Stereotactic Body Radiation Therapy for Liver Metastases. *Int J Radiat Oncol Biol Phys* 2005; 63:s15.
13. Schefter TE, Kavanagh BD, Timmerman RD, *et al.* A phase I trial of stereotactic body radiation therapy (SBRT) for liver metastases. *Int J Radiat Oncol Biol Phys* 2005; 62:1371–1378.
14. Wulf J, Hadinger U, Oppitz U, *et al.* Stereotactic boost irradiation for targets in the abdomen and pelvis. *Radiother Oncol* 2004; 70:31–36.
15. Shirato H, Seppenwoolde Y, Kitamura K, *et al.* Intrafractional tumor motion: lung and liver. *Semin Radiat Oncol* 2004; 14:10–18.
16. Rietzel E, Rosenthal SJ, Gierga DP, *et al.* Moving targets: detection and tracking of internal organ motion for treatment planning and patient set-up. *Radiother Oncol* 2004; 73 Suppl 2:S68–72.

17. Rosu M, Dawson LA, Balter JM, *et al.* Alterations in normal liver doses due to organ motion. *Int J Radiat Oncol Biol Phys* 2003; 57:1472–1479.
18. Blomgren H., Lax I., Goranson H., *et al.* Radiosurgery for tumors in the body: clinical experience using a new method. *J Radiosurg* 1998; 1:63–74.
19. Fuss M, Thomas CR, Jr. Stereotactic body radiation therapy: an ablative treatment option for primary and secondary liver tumors. *Ann Surg Oncol* 2004; 11:130–138.
20. Wulf J, Hadinger U, Oppitz U, *et al.* Stereotactic radiotherapy of targets in the lung and liver. *Strahlenther Onkol* 2001; 177:645–655.
21. Sampson MC, Huggins C, Schell M, *et al.* Hypofractionated Stereotactic Body Radiation Therapy for Liver Metastases. *Int J Radiat Oncol Biol Phys* 2005; 63:S15.
22. Schweikard A, Shiomi H, Adler JR. Respiration tracking in radiosurgery. *Medical Physics* 2004; 31:2738–2741.
23. Herfarth KK, Hof H, Bahner ML, *et al.* Assessment of focal liver reaction by multiphasic CT after stereotactic single-dose radiotherapy of liver tumors. *Int J Radiat Oncol Biol Phys* 2003; 57:444–451.
24. ICRU62. Prescribing, Recording and Reporting Photon Beam Therapy (supplement to ICRU Report 50). Bethesda, MD ICRU; 1999.
25. Balter JM, Ten Haken RK, Lawrence TS, *et al.* Uncertainties in CT-based radiation therapy treatment planning associated with patient breathing. *Int J Radiat Oncol Biol Phys* 1996; 36:167–174.

CyberKnife Radiosurgery for Pancreatic Cancer

23

Scott G. Soltys, Karyn A. Goodman, and Albert C. Koong

CONTENTS

23.1 Abstract

Pancreatic cancer remains one of the most lethal cancer diagnoses. The high mortality rate is primarily related to its advanced stage at diagnosis and the high rate of distant metastases. The treatment results of resected and locally advanced pancreatic cancer via conventional chemoradiotherapy have been disappointing, with a high rate of local disease progression and metastases. Past investigation has suggested improved outcomes with radiotherapy dose escalation. Extracranial stereotactic radiosurgery using CyberKnife® (Accuray Incorporated, Sunnyvale, CA) technology is a novel approach to the administration of radiation in a single outpatient treatment, while minimizing the irradiation of surrounding normal structures. Phase I/II studies have shown it to be safe and effective in providing local tumor control and palliation of local symptoms. Furthermore, the decreased treatment time with radiosurgery minimizes the delay in initiation of systemic therapy. The role of pre-operative radiosurgery in combination with gemcitabine chemotherapy has many theoretical advantages over conventional neo-adjuvant or post-operative fractionated radiotherapy in combination with sensitizing chemotherapy. The movement of the pancreas due to respiration poses a challenge in the planning of stereotactic radiosurgery. In this chapter, the use of Image Guided Radiotherapy techniques (IGRT), such as the Synchrony® (Accuray Incorporated, Sunnyvale, CA) Respiratory Motion Tracking system and respiratory gated 4D PET-CT scanning for defining the tumor volume is reviewed.

23.2 Introduction

Pancreatic cancer remains one of the most lethal cancer diagnoses. In 2006, it is estimated that 33,730 new cases will be diagnosed and there will be 32,300 deaths [1]. The five-year survival rates for all stages combined are less than 5% [1]. The high mortality rate associated with this disease is primarily related to its advanced stage at diagnosis.

Currently, complete surgical resection offers the only potential for cure. Among patients who undergo surgery with curative intent, 30–50% are found to be unresectable intra-operatively [2]. Ultimately, only 20% of patients are resectable at diagnosis. Over 40% of patients have locally advanced, non-metastatic disease and the remainder have evidence of distant metastases at diagnosis [3]. Thus, the majority of patients require non-surgical therapy, including chemotherapy and radiation therapy.

Unfortunately, to date, studies have demonstrated unsatisfactory results for long-term survival among locally advanced and metastatic pancreatic cancer patients [4, 5] for whom the median survival is 6–10 months and 3–6 months, respectively [6]. Even with aggressive chemoradiotherapy for locally advanced disease, median survival is improved by only 2–4 months [5, 7]. In addition, local control remains poor with conventional chemoradiotherapy. Patients with local progression can experience significant morbidity, including weight loss, weakness, fatigue, pain, nausea, vomiting, and anorexia. Many pancreatic cancer patients die with local tumor progression, and improved methods of controlling the primary cancer are needed.

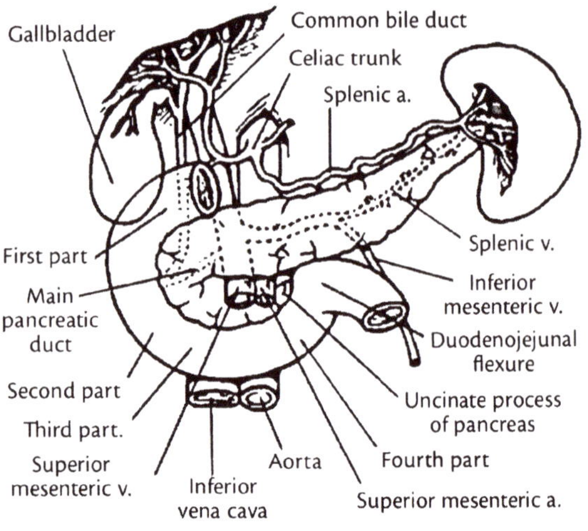

Fig. 23.1 Anatomy of the pancreas and surrounding structures. Source: http://ect.downstate.edu/courseware/haonline/figimgs/39_3.gif

23.2.1 Anatomy and Staging

The pancreas is an exocrine and endocrine organ that lies in the retroperitoneum at the level of the first and second lumbar vertebral bodies. Over 90% of tumors arise from the exocrine portion of the gland as adenocarcinomas. The pancreas is bounded by the duodenum, jejunum, stomach, splenic hilum, porta hepatis, and kidneys (Fig. 23.1). It is subdivided into anatomic regions: the head (which includes the uncinate process), the neck, the body, and the tail. It surrounds the superior mesenteric vein (SMV), where the uncinate process wraps behind the vein. Tumors arising to the right of the SMV are considered part of the head; those arising to the left of the SMV, up to the left border of the aorta are classified as part of the body. Those arising between the left border of aorta and the spleen are considered pancreatic tail lesions. Over 70% of pancreatic cancers arise from the head of the pancreas. Due to the close approximation to the biliary tree, they may obstruct sections of the common bile duct. The classic radiographic finding for a head of pancreas tumor is the 'double duct sign', in which there is dilatation of both the pancreatic duct and the common bile duct.

There is a rich lymphatic supply to the pancreas. The primary drainage is to the celiac, superior and inferior pancreaticoduodenal, porta hepatis, superior mesenteric, pancreaticosplenic, and para-aortic nodes [8]. The venous drainage is through the portal system; thus the liver is at highest risk for distant metastases, followed by the peritoneum and lung.

Although there is a TNM staging system for pancreatic cancer, most often patients are classified into those with resectable, locally advanced, or metastatic disease. Unresectability may be established by extrapancreatic involvement, including extensive peripancreatic lymphatic involvement, and/or distant metastases. In addition, tumors are deemed unresectable if there is encasement or occlusion of the SMV or the SMV-portal vein confluence, or if there is direct involvement of the superior mesenteric artery (SMA), inferior vena cava, aorta, or celiac axis as defined by the absence of a fat plane between the low density tumor and these structures on CT scan (Fig. 23.2). Some centers are revisiting this criterion and are demonstrating the feasibility of SMV reconstruction [9].

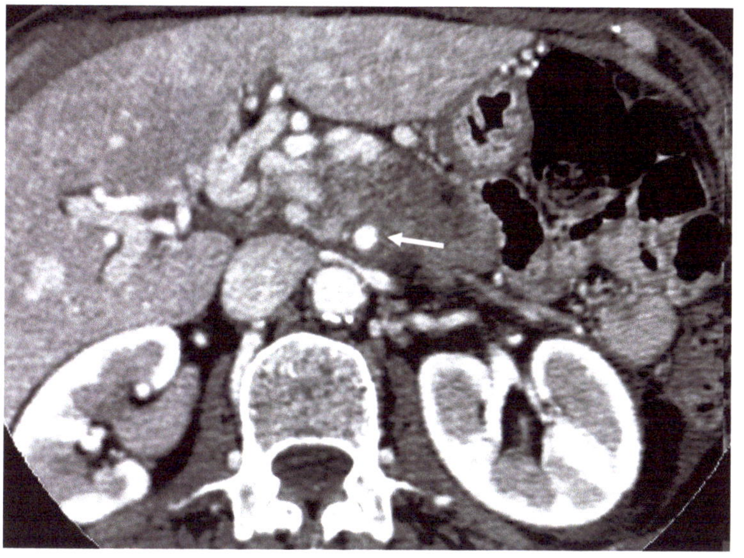

Fig. 23.2 Radiographic evidence of unresectable pancreatic cancer due to encasement of the Superior Mesenteric Artery (SMA) (*arrow*).

23.2.2 Locally Advanced Pancreatic Cancer

Randomized studies of chemoradiation have demonstrated a significant improvement in median survival for the combination of external beam radiotherapy and bolus 5-FU chemotherapy over surgical bypass, chemotherapy, or radiotherapy alone [5, 7]. A landmark study from the Gastrointestinal Tumor Study Group (GITSG) compared standard radiotherapy of 60 Gy split-course with two week breaks after 20 Gy and 40 Gy, versus concurrent bolus 5-FU and radiotherapy (either 40 Gy or 60 Gy split course) [5]. Median survival was only 6 months in the radiation alone arm and 10 months in the combined modality arms. There was no significant difference between the 40 Gy and 60 Gy arms, and both chemoradiation arms demonstrated good palliation of pain with no severe toxicity. Up to 27% of patients had local progression as a component of failure. Although this study is criticized for using split-course radiotherapy and for having small numbers of patients, it nevertheless has established a role for combined modality therapy in the management of locally advanced pancreatic cancer. A subsequent GITSG study demonstrated enhanced survival of patients with locally unresectable pancreatic cancer treated with chemoradiation over those treated with chemotherapy alone (streptozocin, mitomycin, and 5-FU) [7]. Again, however, 47% of patients failed at their primary site.

In most medical centers in the United States, chemoradiotherapy with 50.4 Gy (non-split regimen) concurrent with 5-FU-based chemotherapy became the standard regimen. Unfortunately, 5-year survivors are rare and local progression remains common. The disappointing results with conventional treatment of pancreatic cancer have led to a variety of clinical investigations using newer systemic cytotoxic agents and increasingly more aggressive radiation therapy.

Despite countless studies investigating the use of new chemotherapeutic agents and combinations, only gemcitabine has demonstrated clinical benefit over single agent 5-FU. Burris et al. [4] randomized 160 patients with locally advanced or metastatic pancreatic cancer to receive either gemcitabine or 5-FU. There was a small but statistically significant advantage favoring gemcitabine for median survival (5.7 months versus 4.4 months), 1-year overall survival (18% versus 2%), and median time to progressive disease (9 weeks versus 4 weeks).

In addition, this study only enrolled symptomatic patients. The concept of clinical benefit response (CBR) was developed to assess more accurately nontraditional measures of treatment effects, such as pain reduction and performance status. CBR was significantly better in the gemcitabine group, with 24% of patients experiencing some durable pain reduction versus only 5% in the 5-FU arm [4].

Multiple Phase I/II studies [10–14] have investigated the use of combinations of cytotoxic drugs or targeted agents with gemcitabine, none of which have shown any significant improvement in tumor response rates, time to tumor progression, or improvement in median survival as compared with single-agent gemcitabine [15].

While the early manifestation of metastatic disease in pancreatic cancer has fueled the multitudes of studies on systemic agents, local progression remains a significant source of morbidity and mortality. Modern radiation therapy has increasingly used conformal fields and dose escalation to enhance tumor control. Further dose escalation with conventional radiation therapy techniques risks injury to adjacent abdominal organs. Efforts to increase radiation dose to the pancreatic tumor without risking normal tissue injury have generally required relatively invasive techniques, such as the use of intra-operative radiation therapy (IORT) to deliver a single fraction of radiation to the primary tumor.

The Radiation Therapy Oncology Group (RTOG) investigated the use 20 Gy of IORT, followed by concurrent bolus 5-FU and 50.4 Gy external beam radiotherapy for locally unresectable, non-metastatic pancreatic cancer. For the 51 analyzable patients, the median survival time was 9 months and the 18 month actuarial overall survival was 9%. Local control was not reported. While the feasibility of IORT was demonstrated, there did not appear to be any survival advantage over historical results with conventional therapy [16].

Mohiuddin et al. and Nishimura et al. [17, 18] published their results and complications following IORT with or without subsequent chemoradiotherapy in 49 and 71 patients, respectively, with locally advanced pancreatic cancer. The prescribed dose ranged from 12 Gy (when large volumes of bowel were included in the field) to 33 Gy. Acute Grade 3/4 complications within 6 months of IORT occurred in 14% of patients in the Mohiuddin series. Nineteen percent of those surviving longer than 6 months experienced late complications and bowel obstruction [17]. All of these patients received IORT (15–20 Gy) followed by chemotherapy and chemoradiotherapy with an additional 40–55 Gy. In the Nishimura et al. series, Grade 3/4 complications included gastrointestinal ulcer (10%), perforation (3%), abscess (1%), and intestinal fibrosis (4%) [18]. Even with the dose escalation seen in these IORT series, local progression still occurred in 31–33% of patients.

Using a different approach to escalate radiotherapy dose, investigators from the University of Michigan conducted a Phase I dose escalation trial among 37 patients with unresectable or incompletely resected pancreatic cancer [19]. Radiation dose escalation was achieved by increasing the fraction size and keeping the duration of RT at 3 weeks. The radiation fields were planned using three-dimensional radiotherapy planning and covered only the gross target volume (GTV) with a 1.0 cm margin (i.e., no elective nodal RT and no respiratory movement compensation). Doses ranged from 24–42 Gy using 1.6–2.8 Gy fractions. Standard dose gemcitabine was administered concurrently with the radiation. At the final planned dose level (42 Gy in 2.8 Gy fractions), dose-limiting gastrointestinal toxicity was noted in 2/6 assessable patients. With a median follow-up of 22 months, local progression was noted in 7/37, regional progression in 3/37, and distant progression in 25/37 patients. The median survival was 11.6 months. The presence of metastatic disease, at study entry, in 14/37 patients did not have a significant impact on survival. Based on this study, the reduction in the field size did not appear to result in excess locoregional failure [20].

23.2.3 CyberKnife Radiosurgery for Locally Advanced Pancreatic Cancer

As reviewed above, in the United States, most patients with locally advanced pancreatic cancer are treated with 50.4 Gy fractionated radiotherapy and concurrent 5-FU based chemotherapy. Local progression of disease, however, is common and remains a significant source of morbidity and impairment in quality of life. As seen in Table 23.1, local failure with conventional chemoradiotherapy occurs in over one quarter of patients. With the apparent radiation dose escalation afforded by radiosurgery, improved local control may be realized with this treatment.

The rationale for continued study of radiation dose escalation of the GTV is underscored by the importance of preventing local tumor progression. The locoregional effects of metastatic spread to adjacent organs and to regional lymphatics often produce much of the morbidity and mortality related to this disease.

While palliative surgical bypass can alleviate the obstructive symptoms and jaundice related to biliary ductal involvement, local progression of tumor in the retroperitoneum can cause severe pain and deteriorate quality of life. A less invasive means of controlling tumor progression has been studied for several years at Stanford University by Koong and colleagues, who have treated the pancreas using single-fraction, stereotactic CyberKnife radiosurgery.

Recent technological advances in the CyberKnife made it possible to image and treat tumors within the abdominal cavity. As a result, these investigators were able to deliver high dose radiation with precise tumor targeting while minimizing the irradiation of surrounding healthy tissue, using radiosurgical techniques. Because of the spatial precision with which the CyberKnife can administer radiation, it is feasible to administer a tumoricidal radiation dose in a single outpatient treatment. This technique is particularly suited for patients with unresectable pancreatic cancer, where any improvement in local control contributes to palliation of local symptoms.

In a Phase I dose escalation trial, 15 patients with locally advanced pancreatic cancer were treated with a single fraction of stereotactic radiosurgery; 3/15 received 15 Gy, 5/15 received 20 Gy and 7/15 received 25 Gy [21]. The GTV treated ranged from 19.2 to 71.9 cc (mean: 32.9 cc, median: 29.0 cc). Prior to the development of the Synchrony tumor tracking technology, a breath-holding technique was used for approximately 50% of the patients [22]. Treatment times with the breath holding technique were longer than with Synchrony and ranged from 3 to 6 hours with the majority of patients treated in less than 4 hours. A representative treatment plan is seen in Figure 23.3.

Of the 15 patients, three had received prior therapy, two received concurrent 5-FU and 50 Gy radiotherapy and one received chemotherapy alone. Within the 12-week follow-up period after radiosurgery, there were no Grade 3 or higher gastrointestinal acute toxicities observed. Grade 1 nausea lasting less than 24 hours was reported in 2 patients and Grade 2 diarrhea in another patient. Two patients experienced moderate abdominal pain requiring analgesics. In both of these patients, the symptoms resolved within 24 hours, and no further workup was indicated. There were no significant changes in

Table 23.1 Treatment Characteristics and Local Failure Rates of Selected Trials

Study	Treatment Intent	Treatment	Number of Patients	Local Failure as a Component of Progression
GITSG [5]	Definitive in Unresectable	60 Gy 40 Gy + 5-FU 60 Gy + 5-FU	25 78 78	24% 26% 27%
GITSG [7]	Definitive in Unresectable	54 Gy + 5-FU + SMF SMF alone	22 21	46% 48%
EORTC [29]	Post-Operative Adjuvant	Observation 40 Gy + 5-FU	103 104	36% 33%
ESPAC-1 [34]	Post-Operative Adjuvant	2×2 design (see text)	237	42% overall*
Mohiuddin [17]	Definitive in Unresectable	IORT 10-20 Gy→5-FU + EBRT 40–50 Gy	49	31%
Nishimura [18]	Definitive in Unresectable	IORT 20-33 Gy→EBRT 40–60 Gy	42 analyzed	33%
Michigan [19]	Definitive in Unresectable	Gemcitabine + Hypo-fractionated EBRT 24–42 Gy	37	19%
Koong CK Phase I [21]	Definitive in Unresectable	25 Gy CK	6	0%
Koong CK Phase II [23]	Definitive in Unresectable	EBRT 45Gy + 5-FU→25 Gy CK	16	6%
Koong CK Phase II [24]	Definitive in Unresectable	Gemcitabine→25 Gy CK→Gemcitabine	16	0%

EBRT – External Beam Radiotherapy; IORT – Intra-Operative Radiotherapy; 5-FU – 5-FluoroUracil chemotherapy; CK – Cyberknife radiosurgery; SMF – Streptozocin, Mitomycin, and 5-FU
* In ESPAC-1, local control not reported based on treatment arm

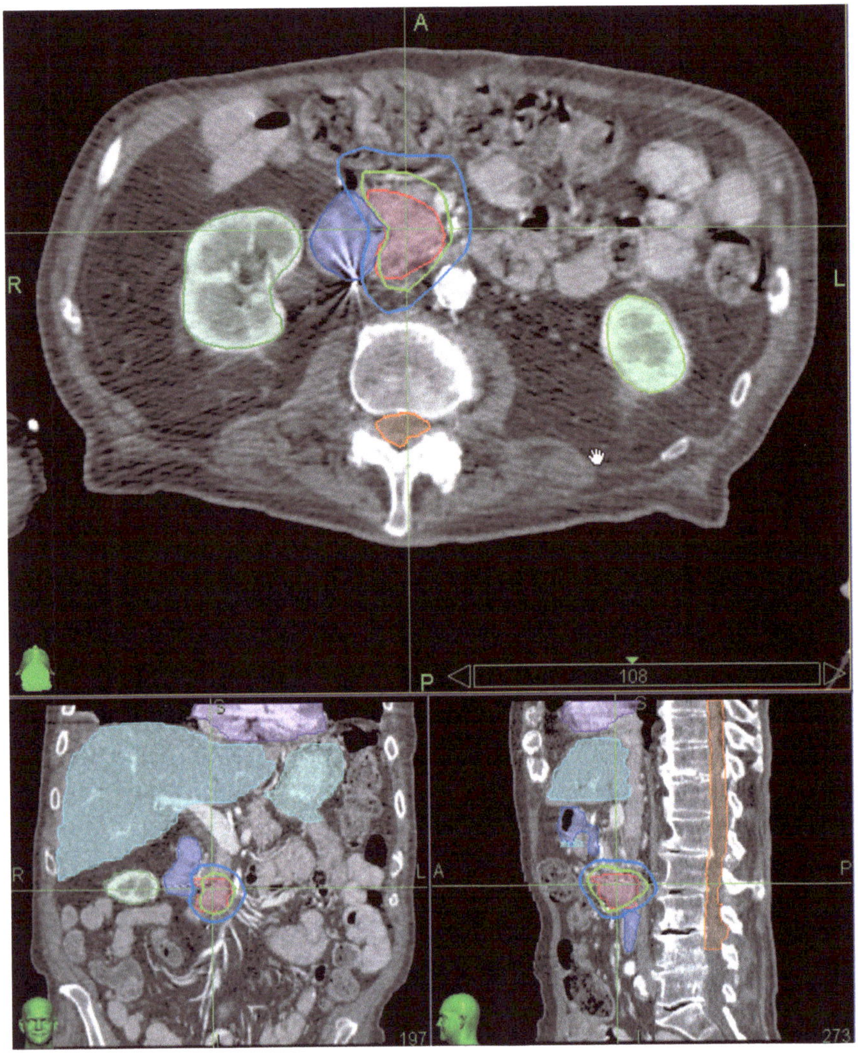

Fig. 23.3 Axial (top panel), coronal (bottom left) and sagittal (bottom right) views of a representative treatment plan on protocol. The pancreatic head GTV (*red contour*) was treated with 25 Gy, prescribed to the 75% isodose line (*green contour*). The 50% isodose line (*light blue contour*) is shown. Isodose lines are constructed to minimize dose to the duodenum (*dark blue contour*).

follow-up blood tests, including blood counts and liver function tests. No dose-limiting toxicity was observed even at the highest dose level of 25 Gy. One hundred percent local control of the primary pancreatic tumor was seen in all evaluable patients treated with 25 Gy at their last follow-up of 7 months or at the time of their death. The study, however, was stopped at 25 Gy because all 15 patients developed distant metastases as the site of first progression. Of importance, 12/15 patients experienced some clinical benefit with decreased pain and increased weight. For all patients, the median overall survival was 11 months with a median follow-up time of five months.

Given these encouraging results, the integration of radiosurgery with conventional treatment of radiotherapy and 5-FU based chemotherapy was investigated. In this Phase II trial of stereotactic radiosurgery boost after conventional chemoradiation for locally advanced pancreatic cancer, 19 patients were enrolled [23]. They received 5-FU based chemotherapy and concurrent conventionally fractionated radiotherapy to 45 Gy using an intensity modulated radiotherapy (IMRT) plan targeting the primary tumor and regional lymph nodes. Three patients progressed systemically after the 5-week course of IMRT radiotherapy. Sixteen patients underwent a single 25 Gy stereotactic boost to the primary tumor. Patients were then reassessed for surgery or they received additional gemcitabine-based chemotherapy. Of the 16/19 patients who completed the protocol, only two Grade 3 acute gastrointestinal toxicities were reported. Fifteen of sixteen patients were free from local progression until death. All patients, however, failed in distant sites as the site of first progression. The median overall survival time for all patients was 33 weeks. Due to the apparent increased GI toxicity seen using the combination of external beam radiotherapy with a radiosurgical boost, and the rapid progression of metastatic disease, the replacement of the 5 weeks of IMRT with systemic gemcitabine is our current practice.

The preliminary results of the following approach were recently reported [24]. Sixteen patients with locally advanced pancreatic cancer, following staging with pancreatic protocol CT scans and FDG PET-CT scans, received a single cycle of gemcitabine (1000 mg/m^2) followed by 25 Gy radiosurgery. Additional cycles of gemcitabine were given until disease progression. Local control was seen in all cases as determined by CT scans. FDG PET-CT scans, obtained 4–6 weeks post-radiosurgery, in 13 patients revealed decreased or stable metabolic activity in all patients following radiosurgery. One patient, however, displayed increased pancreatic tumor metabolic activity, in association with distant metastatic disease, at 35 weeks post-treatment. All patients failed with metastatic disease, with median time to progression of 36.4 weeks. Median overall survival was 41.1 weeks, with 1-year survival of 38%. Late gastrointestinal toxicity included 1 duodenal perforation (Grade 4), 3 duodenal ulcer (Grade 3), and 1 gastric ulcer (Grade 2) at a median time to development of 33 weeks post-radiosurgery. The primary advantage of this approach is the reduction of radiation treatment time to a single day, minimizing the delay in initiating systemic treatment. However, the high duodenal ulcer rate seen could be attributed to the combination of radiosurgery and gemcitabine. To minimize the ulcer rate, we currently avoid the administration of gemcitabine within 2 weeks before and after radiosurgery. Additionally, all patients are placed on prophylactic proton pump inhibitors. Additional follow-up and dose-volume histogram analysis may provide insight to diminish this complication.

Other investigator's experience with stereotactic radiosurgery for locally advanced pancreatic carcinoma, however, has not been favorable [25]. Hoyer et al. reported the use of 45 Gy in three fractions to non-resectable tumors in 22 patients. Six-month actuarial local control rate was 57%. Acute toxicity at 14 days post-treatment was significant, with deterioration of performance status and increased pain. Four patients had severe duodenal or stomach ulceration. Differences in the treatment planning technique may be a possible explanation for the increased rate of complications and decreased local control compared to the Stanford series [21, 23, 24]. Hoyer et al. noted a large median PTV volume encompassed by the 67% isodose line of 136 cm^3 for a median GTV of 32 cm^3, caused by inclusion of a relatively large CTV and PTV due to diagnostic uncertainty in delineation of the tumor [25]. We find the routine fusion of FDG-PET scans helpful for tumor delineation, as described below. Additionally, with IGRT techniques such as the generation of an internal tumor volume (ITV) based on 4D-CT scans, as well as the Synchrony tumor tracking system, the PTV expansion into normal tissue can be minimized.

Based on the three studies from our institution [21, 23, 24], there appears to be a benefit to treating patients with single fraction CyberKnife radiosurgery. As all of the evaluable patients on these trials progressed with metastatic disease, the primary benefit is in providing high rates of local control, while minimizing the delay in initiating systemic chemotherapy. Local control of pancreatic tumors provides a clinical benefit with a reduction in pain and as well as the risk of developing gastric or duodenal obstruction. Theoretically, radiosurgical ablation of the primary tumor may also prevent or reduce distant seeding from the primary tumor. Because all patients progress with metastatic disease, more effective chemotherapy is needed. Upon further study into the integration of pancreatic radiosurgery with the development of novel classes of chemotherapy or targeted agents [26], an improvement in overall survival in patients with locally advanced pancreatic cancer may be realized.

23.3 Potential Role for CyberKnife Radiosurgery in the Post-Operative Adjuvant Setting

In most medical centers in the United States, based on the GITSG trials [27, 28], and arguably, an EORTC trial [29, 30], adjuvant therapy following complete surgical resection of pancreatic cancer involves post-operative chemoradiotherapy [31]. In Europe, however, the ESPAC-1 trial [32] found a survival disadvantage with post-operative chemoradiotherapy compared to chemotherapy alone, although flaws may exist with the design and interpretation of this trial [33]. It has been postulated that the lack in benefit of adjuvant radiotherapy may be due to the delay in initiating effective chemotherapy [34]. A potential advantage of radiosurgery, as compared to conventional fractionated radiotherapy, is the minimization in this delay of effective systemic treatment.

Local tumor progression remains a significant source of treatment failure, occurring in 34% and 63% of the patients in the EORTC and ESPAC-1 trial, respectively (see Table 23.1). Given the high local control rates seen in early data with stereotactic radiosurgery in locally advanced pancreatic cancer [21, 23, 24], further research is warranted to determine the optimal method to integrate radiosurgery into the management of patients with the highest risk of local progression following resection, such as those with positive margins. The definition of an appropriate radiosurgical target volume, however, would likely be the primary obstacle in the post-operative patient. Therefore, perhaps the most promising method to integrate pancreatic radiosurgery in patients who are resectable is to administer it in a neoadjuvant setting.

23.4 Neoadjuvant CyberKnife Radiosurgery for Resectable Pancreatic Cancer

Pre-operative therapy for potentially or marginally resectable pancreatic cancer has been evaluated to improve local control rates. The advantages of a preoperative approach are given in Table 23.2.

Table 23.2 Potential Advantages of Neoadjuvant (Pre-Operative) Therapy)

- The tumor is well oxygenated, thus improving the effectiveness of radiation therapy
- There is an opportunity to sterilize cancer cells prior to surgery and to decrease intra-operative seeding
- Delaying the surgery allows for the exclusion of patients with rapidly progressive disease who would not benefit from the resection
- Pre-operative therapy increases the likelihood of negative surgical margin
- Multi-modality therapy is given regardless of post-operative recovery time
- Giving systemic chemotherapy prior to the clinical appearance of distant metastases has the potential to treat distant metastases at a time when the tumor burden is the lowest
- Treatment before surgery may be better tolerated

Coia et al. performed a pilot study of pre-operative chemoradiotherapy which showed enhanced resectability and downstaging of nodal metastases at the time of surgery [35]. Staley et al. at M.D. Anderson Cancer Center (MDACC) also found that pre-operative chemoradiation (50.4 Gy + 5-FU) with 10 Gy electron beam IORT at the time of surgery for adenocarcinoma of the pancreatic head resulted in improved local-regional tumor control [36]. The Eastern Cooperative Oncology Group (ECOG) performed a Phase II study of 53 patients with localized pancreatic cancer treated with neoadjuvant chemoradiation (50.4 Gy with 5-FU and mitomycin) [37]. Of the 41 patients taken to surgery, 24 were resectable. The median survival was 9.7 months for all patients and 15.7 months for resected patients. Pre-operative treatment was felt to be feasible, but the toxicity associated with the mitomycin was significant.

Researchers at MDACC have published their data on pre-operative chemoradiation for patients with potentially resectable pancreatic cancer. In their initial study, the use of conventional fractionation was associated with gastrointestinal toxicity severe enough to require hospitalization in 32% of patients [38]. Subsequent studies have used a rapid fractionation approach delivering 30 Gy in 3 Gy per fraction over 2 weeks [39, 40]. In these reports, the preoperative administration of chemoradiotherapy was not associated with toxicities that delayed surgical treatment and did not increase surgical morbidity or mortality. In addition, approximately 25% of patients were found to have disease progression at the time of restaging prior to pancreaticoduodenectomy and were therefore spared the morbidity of surgery which would not have been of benefit.

Breslin et al. reviewed the MDACC experience of 132 patients with potentially resectable pancreatic cancer who underwent chemoradiation followed by pancreaticoduodenectomy [41]. The majority of patients, 88/132, received rapid fractionation pre-operative radiation using 30 Gy in 3 Gy per fraction. The remainder, 44/132, received conventional fractionation to 45-50.4 Gy in 1.8 Gy per fraction. IORT was delivered to 74/132 (56%) patients. Chemotherapy most often consisted of 5-FU (105/132 cases) and the remainder received either paclitaxel, or gemcitabine. Median survival was 21 months and 32% of patients were alive with no clinical or radiographic evidence of disease at 14 months. All patients were evaluated with serial post-operative CT scans. Local recurrence was diagnosed radiographically as a component of the first recurrence site in only 8/132 (6%) patients. There was no difference in survival between patients receiving the rapid-fractionation chemoradiotherapy schedule and those who received standard-fractionation chemoradiotherapy. These investigators concluded that preoperative rapid-fractionation chemoradiation maximizes survival duration and local control in accurately staged patients.

Based on these data and the experience with the local control achieved among patients treated with CyberKnife for locally advanced pancreatic cancer, Stanford University has initiated a protocol for patients with resectable pancreatic tumors to be treated with CyberKnife stereotactic radiosurgery. The radiosurgery component is sandwiched between several infusions of weekly gemcitabine chemotherapy: 1000 mg/m^2 infused over 30 minutes. Definitive surgical resection takes place 6 weeks after 25 Gy single session stereotactic radiosurgery. The schema for this study is given in Figure 23.4.

23.5 CyberKnife Radiosurgery for Metastatic Pancreatic Cancer

Patients presenting with distant metastases from pancreatic cancer have a median survival of 3–6 months. During that time, palliation of pain may be the single most important issue in the management of these patients. External Beam Radiation Therapy (EBRT) in combination with chemotherapy has been shown to provide good palliation of pain related to pancreatic cancer [5]. Nevertheless, a conventional course of radiotherapy lasting 5–6 weeks can be difficult for patients with only a few months to live. Even the most active chemotherapy agents have limited benefit on local symptoms. The use of a single fraction of radiation as palliation is appealing in this patient population. At Stanford University, over 18 patients with metastatic pancreatic cancer who were symptomatic from local disease progression have been treated. Local control was achieved in most of the evaluable patients. Moreover, there was significant palliation of pain and improvement or stabilization in quality of life for these patients.

23.6 Technical Considerations for CyberKnife Radiosurgery in Pancreatic Cancer

23.6.1 Gold Fiducial Seeds

Radiosurgery for pancreatic cancer relies upon implantation of fiducial seeds into pancreatic tumors for targeting purposes. These 5 mm by 1 mm seeds

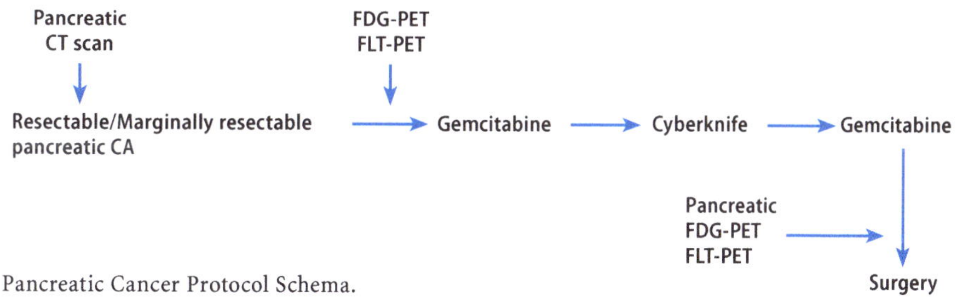

Fig. 23.4. Neoadjuvant Pancreatic Cancer Protocol Schema.

are made of gold and are radiographically visualized by kilovoltage X-rays. Although they can be implanted during laparoscopy or during an open procedure, we typically rely upon interventional radiology to implant these seeds under CT guidance. These procedures are associated with minimal complications and are all performed on an outpatient basis. We typically wait at least 5 days before performing treatment planning scans in order to allow for seed migration.

23.6.2 Respiratory Gated 4D-PET-CT Simulation

For each patient, a custom made immobilization device is constructed to hold the body in a reproducible position. With patients in the treatment position, a pancreatic protocol CT scan is performed for high resolution delineation of the tumor and surrounding structures. In addition, an FDG-PET scan is performed at the same time (GE PET-CT scanner) so that there is accurate co-registration of the pancreatic tumor with the areas of hypermetabolic activity. These scans are complimentary studies that we use to give the most accurate identification of the GTV. Furthermore, respiratory-gated scans to correct for respiratory related pancreatic tumor movement are utilized.

23.6.3 Tumor Movement and Synchrony

Respiratory induced movement of the upper abdominal organs can lead to movement in the pancreatic GTV of up to 3.9 cm [42]. Previous fluoroscopy studies demonstrated that the major movement of pancreatic tumors is in the inferior and superior directions. These studies have demonstrated that pancreas positioning was reproducible within 2.5 mm, on average, with a breath holding technique [22]. Tumor motion is compensated by tracking the position of the chest wall of patients during radiosurgery. A series of light-emitting diodes (LEDs) are placed upon the chest wall and movement can be detected by wall mounted cameras in the treatment room. Synchrony allows for real-time targeting of tumors during radiosurgery. However, only the translational respiratory-related movement of pancreatic tumors is accounted for by this system. It cannot account for tumor rotations and deformations induced by respiration. To account for this additional source of target volume uncertainty, an ITV is generated [43], using IGRT techniques such as respiratory gated 4D-CT and 4D-PET scans, as described for the treatment of lung tumors [44]. The technical issues involved with these IGRT techniques were recently reviewed [45]. The procedure used for delineation of the ITV in pancreatic tumors is identical to the approach we use for hepatic tumors. For details of our protocol, see Chapter 22 by Goodman et al. in this volume.

23.6.4 Normal Tissue Tolerance

The duodenum and stomach are considered the most radiosensitive structures near the pancreas in the upper abdomen. On our previous IRB-approved trials, when treating to 25 Gy, we generally limit no more than 5% of the duodenum or stomach to receive more than 22.5 Gy and no more than 50% of the duodenum or stomach to receive 14.5 Gy. Data for dose guidelines of other structures such as large and small bowel are evolving. Analysis of the dose volume histograms of the patients who developed treatment-associated duodenal ulcers following stereotactic radiosurgery and gemcitabine chemotherapy [24] is pending. We have not treated any tumor larger than 100 cc in volume, with the majority less than 75 cc. Although there is considerable dose heterogeneity within the tumor volume, we have not observed any adverse effects that can be attributed to dose heterogeneity.

23.6.5 Response Evaluation

Evaluating response after high-dose radiation in pancreatic tumors is especially difficult because of post-therapy inflammatory changes. Interpretation of tumor size by conventional CT scans may be obscured by these changes. We have found that metabolic imaging with FDG-PET before and after treatment is a complementary method for determining response to radiation therapy. In the group of patients treated to date at Stanford University, most pancreatic tumors have not changed significantly in size following radiosurgery. However, the majority of them have had a dramatic reduction in metabolic activity as assessed by FDG-PET scanning. Figure 23.5 is a comparison of pre-radiosurgery and post-radiosurgery FDG-PET scans in a patient with locally advanced unresected pancreatic cancer. The interval between the two scans was 6 weeks.

23.7 Conclusions and Future Directions

IGRT techniques such as Synchrony for tracking tumor translations, and respiratory-gated 4D-CT and 4D-PET for the delineation of the true ITV have been developed to account for the movement of pancreatic tumors due to respiration. With these methods, in our experience, CyberKnife radiosurgery using a single fraction of 25 Gy is a safe and effective strategy for the management of patients with locally advanced and metastatic pancreatic cancer. It has been shown to provide palliation of local symptoms and appears to provide improved local control compared to conventional treatment. The role of pre-operative radiosurgery, in combination with gemcitabine chemotherapy, has many theoretical advantages over conventionally fractionated neoadjuvant chemoradiation and is being investigated. Nevertheless, the greatest challenge in treating patients with pancreatic cancer remains their development of distant metastases. Further investigation is needed to determine how best to integrate CyberKnife radiosurgery with other innovative therapeutic approaches [15, 46]. To define fully the impact of radiosurgical treatment of patients with pancreatic tumors, the inclusion of quality of life endpoints, such as clinical benefit response and symptom palliation, is recommended in the development of these future trials.

References

1. Jemal A, Siegel R, Ward E, *et al.* Cancer statistics, 2006. *CA Cancer J Clin* 2006; 56:106–130.
2. Kelsen DP, Portenoy R, Thaler H, *et al.* Pain as a predictor of outcome in patients with operable pancreatic carcinoma. *Surgery* 1997; 122:53–59.

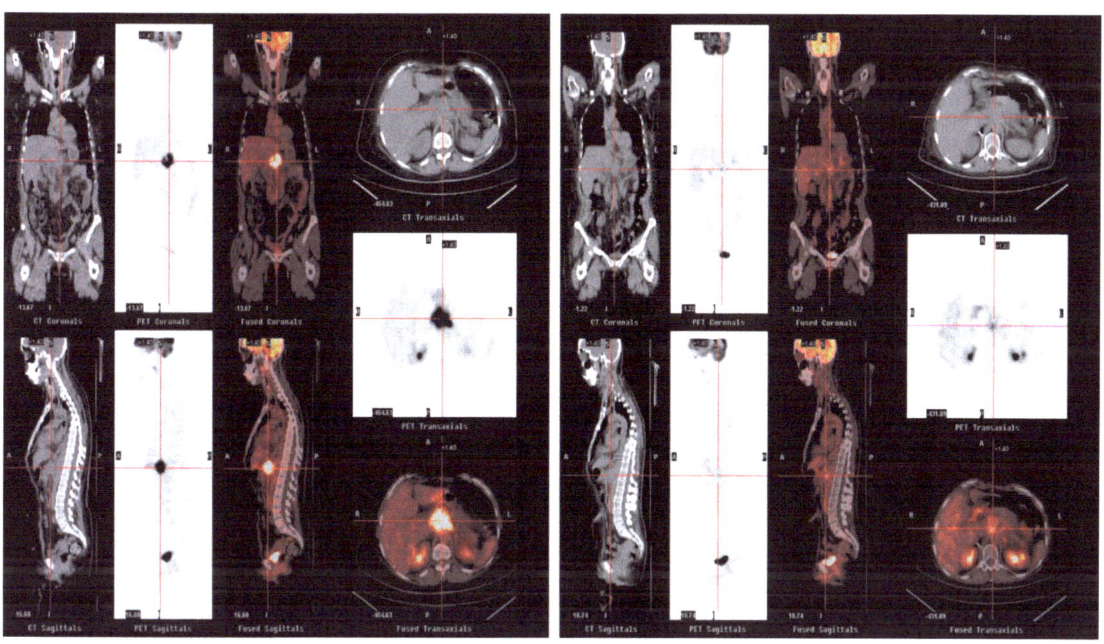

Fig. 23.5 Pre-treatment (left panel) and 6-week post-radiosurgery (right panel) FDG-PET scan in a patient with locally advanced unresected pancreatic cancer.

3. Evans DBA, J.L., Willett, C.G. Cancer of the Pancreas. In: DeVita VT, Hellman S., Rosenberg S.A.,, editor. Cancer: Principles and Practice of Oncology. 6th ed. Philadelphia, PA: Lippincott, Williams & Wilkins; 2001. pp. 1126–1161.
4. Burris HA, 3rd, Moore MJ, Andersen J, *et al.* Improvements in survival and clinical benefit with gemcitabine as first-line therapy for patients with advanced pancreas cancer: a randomized trial. *J Clin Oncol* 1997; 15:2403–2413.
5. Moertel CG, Frytak S, Hahn RG, *et al.* Therapy of locally unresectable pancreatic carcinoma: a randomized comparison of high dose (6000 rads) radiation alone, moderate dose radiation (4000 rads + 5-fluorouracil), and high dose radiation + 5-fluorouracil: The Gastrointestinal Tumor Study Group. *Cancer* 1981; 48:1705–1710.
6. Cancer Facts and Figures 2003, American Cancer Society. Atlanta: American Cancer Society; 2003.
7. Treatment of locally unresectable carcinoma of the pancreas: comparison of combined-modality therapy (chemotherapy plus radiotherapy) to chemotherapy alone. Gastrointestinal Tumor Study Group. *J Natl Cancer Inst* 1988; 80:751–755.
8. Cubilla AL, Fortner J, Fitzgerald PJ. Lymph node involvement in carcinoma of the head of the pancreas area. *Cancer* 1978; 41:880–887.
9. Leach SD, Lee JE, Charnsangavej C, *et al.* Survival following pancreaticoduodenectomy with resection of the superior mesenteric-portal vein confluence for adenocarcinoma of the pancreatic head. *Br J Surg* 1998; 85:611–617.
10. Bramhall SR, Rosemurgy A, Brown PD, *et al.* Marimastat as first-line therapy for patients with unresectable pancreatic cancer: a randomized trial. *J Clin Oncol* 2001; 19:3447–3455.
11. Bramhall SR, Schulz J, Nemunaitis J, *et al.* A double-blind placebo-controlled, randomised study comparing gemcitabine and marimastat with gemcitabine and placebo as first line therapy in patients with advanced pancreatic cancer. *Br J Cancer* 2002; 87:161–167.
12. Colucci G, Giuliani F, Gebbia V, *et al.* Gemcitabine alone or with cisplatin for the treatment of patients with locally advanced and/or metastatic pancreatic carcinoma: a prospective, randomized phase III study of the Gruppo Oncologia dell'Italia Meridionale. *Cancer* 2002; 94:902–910.
13. Moore M, Hamm, J., Eisenberg, P., Dagenais, M., Hagan, K., Fields, A., Greenberg, B., Schwartz, B., Ottaway, J., Zee, B, Seymour, L. A Comparison Between Gemcitabine (GEM) and the Matrix Metalloproteinase (MMP) Inhibitor BAY12-9566 (9566) in Patients (PTS) with Advanced Pancreatic Cancer. *Proc Am Soc Clin Oncol* 2000;19:(abstr 930).
14. Van Cutsem E, Karasek, P., Oettle, H., Vervenne, W.L., Szawlowski, A., Schoffski, P., Post, S., Neumann, H., Safran, H., Humblet, Y., van de Velde, H., Ma, Y., Von Hoff, D. Phase III trial comparing gemcitabine + R115777 (Zarnestra) versus gemcitabine + placebo in advanced pancreatic cancer (PC). *Proc Am Soc Clin Oncol* 2002;21:(abstr 517).
15. Haller DG. New perspectives in the management of pancreas cancer. *Semin Oncol* 2003; 30:3–10.
16. Tepper JE, Noyes D, Krall JM, *et al.* Intraoperative radiation therapy of pancreatic carcinoma: a report of RTOG-8505. Radiation Therapy Oncology Group. *Int J Radiat Oncol Biol Phys* 1991; 21:1145–1149.
17. Mohiuddin M, Regine WF, Stevens J, *et al.* Combined intraoperative radiation and perioperative chemotherapy for unresectable cancers of the pancreas. *J Clin Oncol* 1995; 13:2764–2768.
18. Nishimura Y, Hosotani R, Shibamoto Y, *et al.* External and intraoperative radiotherapy for resectable and unresectable pancreatic cancer: analysis of survival rates and complications. *Int J Radiat Oncol Biol Phys* 1997; 39:39–49.
19. McGinn CJ, Zalupski mm, Shureiqi I, *et al.* Phase I trial of radiation dose escalation with concurrent weekly full-dose gemcitabine in patients with advanced pancreatic cancer. *J Clin Oncol* 2001; 19:4202–4208.
20. McGinn CJ, Zalupski mm. Radiation therapy with once-weekly gemcitabine in pancreatic cancer: current status of clinical trials. *Int J Radiat Oncol Biol Phys* 2003; 56:10–15.
21. Koong AC, Le QT, Ho A, *et al.* Phase I study of stereotactic radiosurgery in patients with locally advanced pancreatic cancer. *Int J Radiat Oncol Biol Phys* 2004; 58:1017–1021.
22. Murphy MJ, Martin D, Whyte R, *et al.* The effectiveness of breath-holding to stabilize lung and pancreas tumors during radiosurgery. *Int J Radiat Oncol Biol Phys* 2002; 53: 475–482.
23. Koong AC, Christofferson E, Le QT, *et al.* Phase II study to assess the efficacy of conventionally fractionated radiotherapy followed by a stereotactic radiosurgery boost in patients with locally advanced pancreatic cancer. *Int J Radiat Oncol Biol Phys* 2005; 63:320–323.
24. Chang ST, Goodman, K.A., Quon, A., Le, Q.T., Fisher, G.A., Kuo, T., Ford, J.M., Norton, J., Yang, G.P., Koong, A.C. Phase II study to assess the efficacy of gemcitabine chemotherapy followed by stereotactic body radiotherapy (SBRT) in patients with locally advanced pancreatic cancer. *Int J Radiat Oncol Biol Phys* 2006; 66:S295–S296.
25. Hoyer M, Roed H, Sengelov L, *et al.* Phase-II study on stereotactic radiotherapy of locally advanced pancreatic carcinoma. *Radiother Oncol* 2005; 76:48–53.
26. Jafari M, Abbruzzese JL. Pancreatic cancer: future outlook, promising trials, newer systemic agents, and strategies from the Gastrointestinal Intergroup Pancreatic Cancer Task Force. *Surg Oncol Clin N Am* 2004; 13:751–760, xi.
27. Kalser MH, Ellenberg SS. Pancreatic cancer. Adjuvant combined radiation and chemotherapy following curative resection. *Arch Surg* 1985; 120:899–903.
28. Further evidence of effective adjuvant combined radiation and chemotherapy following curative resection of pancreatic cancer. Gastrointestinal Tumor Study Group. *Cancer* 1987; 59:2006–2010.
29. Klinkenbijl JH, Jeekel J, Sahmoud T, *et al.* Adjuvant radiotherapy and 5-fluorouracil after curative resection of cancer of the pancreas and periampullary region: phase III trial of the EORTC gastrointestinal tract cancer cooperative group. *Ann Surg* 1999; 230:776–782; discussion 782–774.
30. Garofalo MC, Regine WF, Tan MT. On statistical reanalysis, the EORTC trial is a positive trial for adjuvant chemoradiation in pancreatic cancer. *Ann Surg* 2006; 244:332–333; author reply 333.
31. Garofalo M, Flannery T, Regine W. The case for adjuvant chemoradiation for pancreatic cancer. *Best Pract Res Clin Gastroenterol* 2006; 20:403–416.

32. Ghaneh P, Neoptolemos JP. Conclusions from the European Study Group for Pancreatic Cancer adjuvant trial of chemoradiotherapy and chemotherapy for pancreatic cancer. *Surg Oncol Clin N Am* 2004; 13:567–587, vii-viii.
33. Koshy MC, Landry JC, Cavanaugh SX, *et al.* A challenge to the therapeutic nihilism of ESPAC-1. *Int J Radiat Oncol Biol Phys* 2005; 61:965–966.
34. Neoptolemos JP, Stocken DD, Friess H, *et al.* A randomized trial of chemoradiotherapy and chemotherapy after resection of pancreatic cancer. *N Engl J Med* 2004; 350:1200–1210.
35. Coia L, Hoffman J, Scher R, *et al.* Preoperative chemoradiation for adenocarcinoma of the pancreas and duodenum. *Int J Radiat Oncol Biol Phys* 1994; 30:161–167.
36. Staley CA, Lee JE, Cleary KR, *et al.* Preoperative chemoradiation, pancreaticoduodenectomy, and intraoperative radiation therapy for adenocarcinoma of the pancreatic head. *Am J Surg* 1996; 171:118–124; discussion 124–115.
37. Hoffman JP, Lipsitz S, Pisansky T, *et al.* Phase II trial of preoperative radiation therapy and chemotherapy for patients with localized, resectable adenocarcinoma of the pancreas: an Eastern Cooperative Oncology Group Study. *J Clin Oncol* 1998; 16:317–323.
38. Evans DB, Rich TA, Byrd DR, *et al.* Preoperative chemoradiation and pancreaticoduodenectomy for adenocarcinoma of the pancreas. *Arch Surg* 1992; 127:1335–1339.
39. Pisters PW, Abbruzzese JL, Janjan NA, *et al.* Rapid-fractionation preoperative chemoradiation, pancreaticoduodenectomy, and intraoperative radiation therapy for resectable pancreatic adenocarcinoma. *J Clin Oncol* 1998; 16:3843–3850.
40. Spitz FR, Abbruzzese JL, Lee JE, *et al.* Preoperative and postoperative chemoradiation strategies in patients treated with pancreaticoduodenectomy for adenocarcinoma of the pancreas. *J Clin Oncol* 1997; 15:928–937.
41. Breslin TM, Hess KR, Harbison DB, *et al.* Neoadjuvant chemoradiotherapy for adenocarcinoma of the pancreas: treatment variables and survival duration. *Ann Surg Oncol* 2001; 8:123–132.
42. Bussels B, Goethals L, Feron M, *et al.* Respiration-induced movement of the upper abdominal organs: a pitfall for the three-dimensional conformal radiation treatment of pancreatic cancer. *Radiother Oncol* 2003; 68:69–74.
43. Purdy JA. Current ICRU definitions of volumes: limitations and future directions. *Semin Radiat Oncol* 2004; 14:27–40.
44. Underberg RW, Lagerwaard FJ, Cuijpers JP, *et al.* Four-dimensional CT scans for treatment planning in stereotactic radiotherapy for stage I lung cancer. *Int J Radiat Oncol Biol Phys* 2004; 60:1283–1290.
45. Xing L, Thorndyke B, Schreibmann E, *et al.* Overview of image-guided radiation therapy. *Med Dosim* 2006; 31:91–112.
46. Wagman R, Grann A. Adjuvant therapy for pancreatic cancer: current treatment approaches and future challenges. *Surg Clin North Am* 2001; 81:667–681.

Stereotactic Radiosurgery for Unresectable Adenocarcinoma of the Pancreas: Initial Experience at Sinai Hospital of Baltimore

24

Christopher Hoffelt and Mukund Didolkar

CONTENTS

24.1 Introduction

Adenocarcinoma of the pancreas remains among the most lethal cancers in the United States. For most of the past four decades, little has changed in the treatment and poor survival associated with this tumor [1]. Surgery is still the only treatment associated with cure, though most patients have unresectable disease at diagnosis. Even in favorable, resectable disease, median survival is only 15–24 months [2, 3], with 15–20% long-term survival. Aggressive chemo-radiotherapy trials have demonstrated only marginal improvements in overall survival, rarely with long-term survival.

Radiation therapy doses of 30 Gy to 60 Gy in daily fractions of 180 cGy to 200 cGy are commonly given as definitive treatment, but also to alleviate symptoms such as pain or bleeding. In standard radiation therapy, the optimal target has not yet been defined [4], but usually consists of the primary tumor and regional lymph nodes. The stomach, small bowel, liver, and kidney are dose-limiting organs, especially when administered with concurrent chemotherapy. Despite dose escalation and improved radiotherapy techniques, such as intensity-modulated radiation therapy (IMRT) and 3D conformal therapy, local disease progression and/or recurrence remain significant problems.

There are several potential advantages to stereotactic radiosurgery in controlling the primary tumor. Hypofractionated radiation therapy is biologically very potent compared to standard fractionated radiation therapy. If safely delivered, stereotactic radiosurgery (SRS) likely increases the therapeutic ratio as initial treatment; the benefit may be even greater for recurrent disease. Second, the overall treatment time is shorter, which may minimize delays in systemic chemotherapy. In addition, institutions using intraoperative radiotherapy to treat unresectable disease have demonstrated promising local control and survival [5–8]. Stereotactic radiosurgery in theory represents a similar hypofractionated treatment, although it is much less invasive. Two important studies regarding stereotactic radiosurgery for adenocarcinoma of the pancreas were conducted at Stanford University [9, 10]; the results from Sinai Hospital of Baltimore are discussed in this chapter.

24.2 Methods

From February 2004 to March 2005, 25 patients were treated with stereotactic radiosurgery for unresectable adenocarcinoma of the pancreas. All patients had biopsy-proven disease in the head (17) or body (8) of the pancreas. Tumors were considered unresectable due to portal vein/superior mesenteric vein encasement, superior mesenteric artery involvement, or the presence of distant metastases. Patient characteristics are heterogeneous in this series, but can be summarized as follows:

- 11 patients: T4NM0 disease with no prior therapy
- 4 patients: Local recurrence after Whipple resection, adjuvant chemotherapy, and radiation therapy, median dose = 5040 cGy
- 5 patients: Local recurrence after prior chemotherapy and definitive radiation therapy, median dose = 5040 cGy
- 5 patients: Palliative intent for local tumor progression in patients with metastatic disease, 1 received prior palliative radiation therapy.

All patients were treated using the CyberKnife® (Accuray Incorporated, Sunnyvale, CA), with four to six gold fiducials implanted in or near the tumor for image-guidance. Most fiducials were placed by the surgeon or interventional radiologist, with two to four fiducials in the tumor and four in the paraspinal soft tissue to be used for initial rotational positioning.

At least 7 days after fiducial placement, a Vac-Loc immobilization device (Bionix Radiation Therapy, Toledo, Ohio) was then custom made for each patient. A CT scan, 1.5 mm slice thickness with oral and IV contrast, was then acquired and used for treatment planning for nearly all patients, heterogeneity corrections were not used in planning.

Contouring the gross tumor volume (GTV) was performed cooperatively by the radiation oncologist and surgeon; the GTV included only gross disease identifiable on the planning CT scan, with appropriate modification based on intraoperative or ultrasound findings. Positron emission tomography fused images were not routinely used for planning patients in this series.

Treatment planning occurred within two weeks of fiducial placement. Liver, stomach, spinal cord, and both kidneys were routinely contoured as critical structures in each patient. Dose distributions for each structure were subject to the approval of the treating radiation oncologist, and in general, doses to the majority of kidney, liver, and spinal cord were well within tolerance for greater than 90% of the organ.

Dose was prescribed to the highest percent isodose line encompassing the GTV, usually the 80% line (range 70–87%), see Figure 24.1a and 24.1b. Total doses of 15–26 Gy delivered in 1–4 fractions were used for all patients. Tumor volume (median 81 cm^3, range 28–305 cm^3) and the proportion of patients with prior radiation therapy were larger in our series than in prior studies [9, 10]; therefore single-fraction treatment was rarely used. The majority of patients received 8 Gy × 3 or 8.5 Gy × 3 with 1–3 days between treatments. Synchrony® (Accuray Incorporated, Sunnyvale, CA) software was used for nearly all treatments. Pain was assessed prior to treatment and at 3-month intervals (at least) on a zero-to-ten verbal report scale. All patients received gemcitabine-based chemotherapy within 2 months of completion of treatment at the discretion of the treating medical oncologist.

24.3 Results

At last update, surviving patients had been followed at least 3 months (median = 10 months). The median survival was 8 months from the time of SRS and 25 months from the time of diagnosis. Five patients (20%) survived over 12 months from the time of SRS.

Increase in pain symptoms or greater than 10% increase in tumor size were used as criteria for local progression. Based on these criteria, crude local control rate was 85%. The first site of progression and cause of death in most patients was distant metastases, either in the liver or peritoneal cavity.

Twenty patients had variable degrees of pain prior to SRS, and 85% of these patients had at least 50% reduction in pain symptoms following treatment. Pain relief was durable in patients surviving beyond four months. Pain eventually recurred in 3 patients with local progression as determined by CT.

Nineteen patients had an elevated CA19-9 marker prior to SRS. After treatment, the value decreased in 8 patients (42%), and was durable in 3 of these patients who survived beyond one year from the time of treatment. The CA19-9 increased in 6 patients (32%) and 4 of these patients died from rapid progression of metastatic disease within 4 months of treatment.

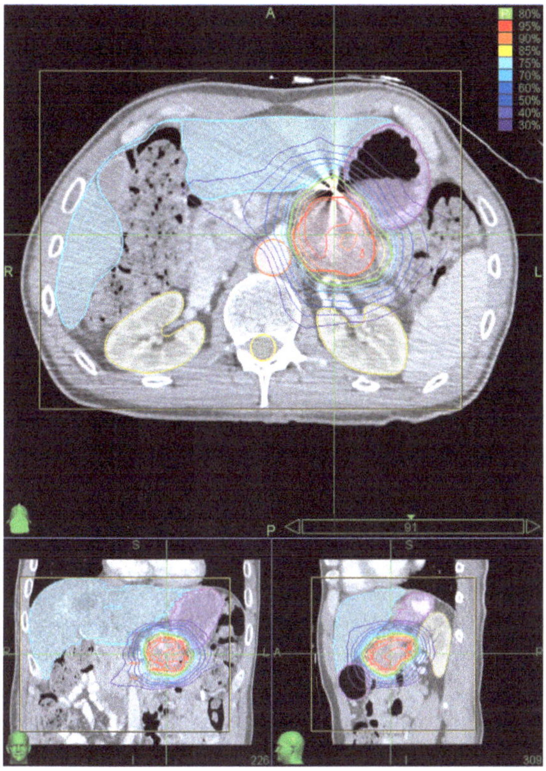

a

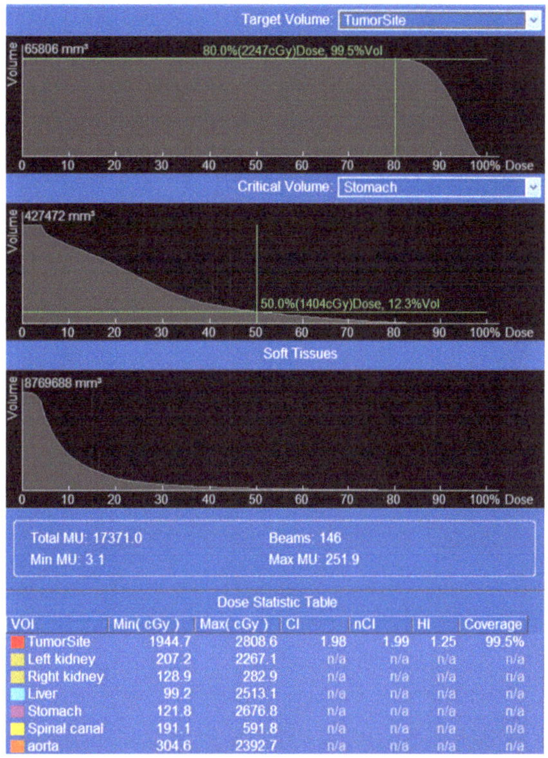

b

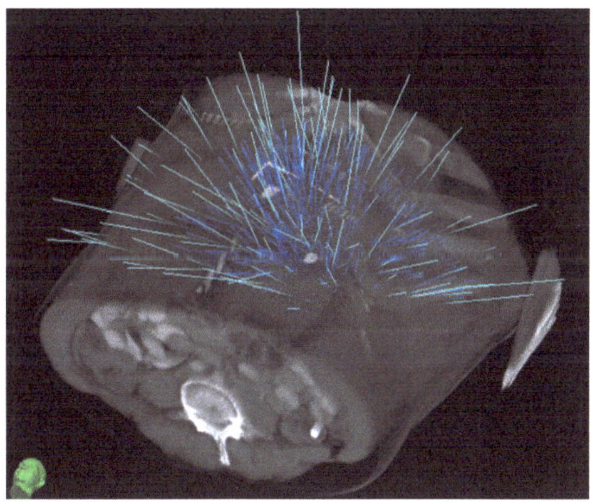

c

Fig. 24.1. a Typical treatment plan for unresectable pancreas adenocarcinoma. 24 Gy was delivered in 3 fractions of 8 Gy each prescribed to the 85% isodose line. **b** Dose volume histogram with dose statistics and **c** 3D display of pencil beam orientations. Partial response was seen on CT at follow-up. The tumor and pain symptoms were still controlled at 18 months follow-up.

Acute toxicity within one month of treatment was mild. Grade 1 nausea, fatigue, or anorexia occurred in 4 patients. One patient developed Grade 2 diarrhea which resolved within 2 weeks, and one patient developed gastric bleeding requiring hospital admission (Fig. 24.2). On endoscopy, biopsy confirmed the bleeding was at the site where tumor had ulcerated the gastric mucosa, and the bleeding resolved without further treatment.

Late toxicity was more significant. Seven of 25 patients (28%) required hospital admission for anorexia, fatigue, and epigastric pain 2–9 months after SRS. Only one of these patients had prior radiation therapy. Endoscopy in all of these patients identified either duodenitis, or gastric and duodenal ulcers. Biopsies were performed in all cases, and none had evidence of tumor. All patients recovered with parenteral support within 2 months. All Grade 3

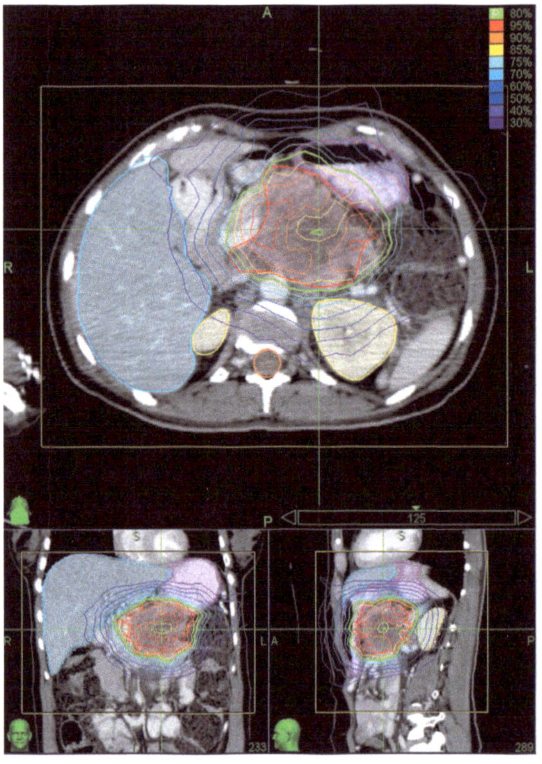

Fig. 24.2 Treatment plan of a patient who developed a Grade III ulcer in the stomach. The dose was 25.5 Gy delivered in 3 fractions of 8.5 Gy each prescribed to the 80% isodose line. Within 2 days a gastric bleed occurred as a result of tumor invasion through the gastric wall, and resolved without intervention. Also note the length of contact between the prescription isodose line and gastric wall. Pain symptoms improved, but gastritis developed within 2 months requiring feeding tube placement. Symptoms resolved over the subsequent two months.

toxicity was seen in patients with tumor (GTV) sizes greater than 70 cm^3 (range 70–305 cm^3); no other factors, including prior radiation therapy, were associated with this toxicity. No late renal, hepatic, or neurological toxicity was seen.

24.4 Discussion

There have been few meaningful improvements in survival and local control for carcinoma of the pancreas over the past 40 years. Standard treatment has largely consisted of 5-FU given concurrently with radiation therapy, resulting in a median survival of 8–12 months [11, 12]. The high rate of distant progression has led to interest in other drugs, and modest trends toward survival benefit have been demonstrated [13–15].

The relationship between local control and survival is unclear in this disease. In our relatively small, heterogeneous series, little can be inferred regarding the impact of SRS on survival. The Stanford University Phase II trial aggressively employed SRS in combination with external beam radiotherapy, and while most tumors were locally controlled, survival was still poor [9]. For Sinai patients, median survival from the time of diagnosis in 25 patients, only four of whom had resection, was 25 months, far greater than expected for patients with unresectable or metastatic disease. This is most likely a consequence of patient selection; most patients had recurrent disease after 6–16 month disease-free intervals after initial therapy, when treatment options are limited. Such a patient group clearly had either a favorable response to initial treatment or indolent disease. It is noteworthy that all patients survived a median of 8 months after radiosurgery, arguably long enough to justify treatment with an aggressive modality. In patients with no prior treatment, the median survival was greater than 10 months, very favorable in comparison to the 6–10 month median associated with major studies [11, 16, 17].

The incidence of local disease progression, when reported, is still unacceptably high, and progression is associated with significant morbidity. Improvements in endoscopic techniques have made stenting an effective and tolerable option for palliation of biliary or duodenal obstruction [18, 19]. Bleeding can often be controlled with embolization or endoscopy. Options for controlling pain, however, are few and marginally effective.

The Sinai experience suggests that the major benefit of radiosurgery may be in controlling pain related to the primary tumor. Pain was durably controlled in most patients in our series (85%), including nine patients with prior radiation therapy. Pain improvement did not correlate with tumor response by CT scan; most patients with symptom improvement had stable disease or a partial response.

Detailed information about pain control and quality of life are not often reported in major studies or in our series, but SRS can be generally compared to other palliative options. Standard radiotherapy techniques with or without concurrent chemotherapy are also 35–65% effective with acceptable toxicity [13, 20]. Less common radiotherapy techniques, including intraoperative radiotherapy and brachytherapy, are 50–80% effective [21, 22] but are not practical at most institutions.

Neurolytic celiac plexus block (NCPB), which involves the injection of alcohol into the celiac plexus, has clearly demonstrated efficacy approaching 90% in two randomized, placebo-controlled trials [23, 24]. NCPB requires an invasive procedure and is performed intraoperatively, percutaneously, or by endoscopy. Though relatively safe and effective, significant pain can recur in up to 2/3 of patients within 4 months [24]. Narcotic analgesics alone can often suffice for pain control, though drowsiness and constipation can be undesirable significant side effects.

PET was used in follow-up of only a few patients, and similar to the Stanford series, long-term survivors in general demonstrated excellent response by PET scan, regardless of CT findings. The value of PET in planning treatment has yet to be defined, but it was useful in identifying tumor that was poorly defined on CT studies. Caution must be used as PET also clearly underestimated disease extent in two patients, and falsely identified a loop of small bowel as tumor in one patient. CA19-9 was evaluable in 14 of the 25 patients. When elevated, a rapid rise after SRS was associated with progression of metastatic disease in all patients, and decreased in all patients with stable disease or a partial response.

There was no significant acute toxicity in the Sinai series. Fatigue, anorexia, and diarrhea were mild in four patients, similar to the Stanford Phase I study. Late toxicity, however, was unacceptable. Twenty-eight percent of patients developed Grade 3 gastritis, duodenitis, or duodenal ulcers 2–6 months after radiosurgery. All patients recovered with supportive care, either jejunostomy or other parenteral nutrition. This was also observed in the Stanford Phase II study, though incidence was not reported [9]. Dose, fraction size, and prior radiotherapy were not associated with toxicity. The only common factor among these patients was a tumor size greater than 70 cm^3 (Fig. 24.2). It is clear that these tumors contact greater lengths of duodenum and stomach, though this was not objectively measured here. Caution should be used in treating larger volume disease. Further study regarding optimal dose or fractionation would be useful, though standard radiotherapy techniques may suffice in such cases.

Significant respiratory motion of the pancreas occurs in most patients and can reach 35 mm [25, 26], primarily in the craniocaudal direction. Synchrony was considered essential for treating all pancreas patients at Sinai hospital, and may be the greatest advantage for the CyberKnife in comparison to other body radiosurgery systems, which must rely on respiratory gating or expanded tumor margins.

24.5 Conclusion

Adenocarcinoma of the pancreas remains a challenge, largely due to rapid progression of metastatic disease. Improvements in survival will depend immensely on advancement in systemic therapies, but considerable emphasis must also be placed on local disease control as a component of cure and for palliation. Stereotactic radiosurgery can be useful as initial local therapy, though it does not clearly benefit all patients. SRS is useful in treating residual or recurrent disease in selected patients, with appropriate tumor volume and stage of disease. Further research into optimal dose, fractionation, and timing of this local therapy may maximize the benefit.

References

1. Jemal A, Siegel R, Ward E, *et al.* Cancer statistics, 2006. *CA Cancer J Clin* 2006; 56:106–130.
2. Yeo CJ, Cameron JL, Lillemoe KD, *et al.* Pancreaticoduodenectomy for cancer of the head of the pancreas. 201 patients. *Ann Surg* 1995; 221:721–731; discussion 731–723.
3. Klinkenbijl JH, Jeekel J, Sahmoud T, *et al.* Adjuvant radiotherapy and 5-fluorouracil after curative resection of cancer of the pancreas and periampullary region: phase III trial of the EORTC gastrointestinal tract cancer cooperative group. *Ann Surg* 1999; 230:776–782; discussion 782–774.
4. McGinn CJ, Zalupski mm. Radiation therapy with once-weekly gemcitabine in pancreatic cancer: current status of clinical trials. *Int J Radiat Oncol Biol Phys* 2003; 56:10–15.

5. O'Connor JK, Sause WT, Hazard LJ, *et al.* Survival after attempted surgical resection and intraoperative radiation therapy for pancreatic and periampullary adenocarcinoma. *Int J Radiat Oncol Biol Phys* 2005; 63:1060–1066.
6. Shipley WU, Wood WC, Tepper JE, *et al.* Intraoperative electron beam irradiation for patients with unresectable pancreatic carcinoma. *Ann Surg* 1984; 200:289–296.
7. Tepper JE, Shipley WU, Warshaw AL, *et al.* The role of misonidazole combined with intraoperative radiation therapy in the treatment of pancreatic carcinoma. *J Clin Oncol* 1987; 5:579–584.
8. Garton GR, Gunderson LL, Nagorney DM, *et al.* High-dose preoperative external beam and intraoperative irradiation for locally advanced pancreatic cancer. *Int J Radiat Oncol Biol Phys* 1993; 27:1153–1157.
9. Koong AC, Christofferson E, Le QT, *et al.* Phase II study to assess the efficacy of conventionally fractionated radiotherapy followed by a stereotactic radiosurgery boost in patients with locally advanced pancreatic cancer. *Int J Radiat Oncol Biol Phys* 2005; 63:320–323.
10. Koong AC, Le QT, Ho A, *et al.* Phase I study of stereotactic radiosurgery in patients with locally advanced pancreatic cancer. *Int J Radiat Oncol Biol Phys* 2004; 58:1017–1021.
11. Moertel CG, Frytak S, Hahn RG, *et al.* Therapy of locally unresectable pancreatic carcinoma: a randomized comparison of high dose (6000 rads) radiation alone, moderate dose radiation (4000 rads + 5-fluorouracil), and high dose radiation + 5-fluorouracil: The Gastrointestinal Tumor Study Group. *Cancer* 1981; 48:1705–1710.
12. Mawdsley S, Hall M, Glynne-Jones R. Locally advanced pancreatic cancer treated with radiation and 5-fluorouracil. *Clin Oncol (R Coll Radiol)* 2002; 14:308–312.
13. Mishra G, Butler J, Ho C, *et al.* Phase II trial of induction gemcitabine/CPT-11 followed by a twice-weekly infusion of gemcitabine and concurrent external beam radiation for the treatment of locally advanced pancreatic cancer. *Am J Clin Oncol* 2005; 28:345–350.
14. Wolff RA, Evans DB, Gravel DM, *et al.* Phase I trial of gemcitabine combined with radiation for the treatment of locally advanced pancreatic adenocarcinoma. *Clin Cancer Res* 2001; 7:2246–2253.
15. Safran H, Dipetrillo T, Iannitti D, *et al.* Gemcitabine, paclitaxel, and radiation for locally advanced pancreatic cancer: a Phase I trial. *Int J Radiat Oncol Biol Phys* 2002; 54:137–141.
16. Treatment of locally unresectable carcinoma of the pancreas: comparison of combined-modality therapy (chemotherapy plus radiotherapy) to chemotherapy alone. Gastrointestinal Tumor Study Group. *J Natl Cancer Inst* 1988; 80:751–755.
17. Klaassen DJ, MacIntyre JM, Catton GE, *et al.* Treatment of locally unresectable cancer of the stomach and pancreas: a randomized comparison of 5-fluorouracil alone with radiation plus concurrent and maintenance 5-fluorouracil–an Eastern Cooperative Oncology Group study. *J Clin Oncol* 1985; 3:373–378.
18. Smith AC, Dowsett JF, Russell RC, *et al.* Randomised trial of endoscopic stenting versus surgical bypass in malignant low bileduct obstruction. *Lancet* 1994; 344:1655–1660.
19. Adler DG, Baron TH. Endoscopic palliation of malignant gastric outlet obstruction using self-expanding metal stents: experience in 36 patients. *Am J Gastroenterol* 2002; 97:72–78.
20. Haslam JB, Cavanaugh PJ, Stroup SL. Radiation therapy in the treatment of irresectable adenocarcinoma of the pancreas. *Cancer* 1973; 32:1341–1345.
21. Termuhlen P, Evans D, Willett G. IORT in Pancreatic Carcinoma. In: Gunderson LL, Willett, Christopher G., Harrison, Louis B., and Calvo, Felipe A., editor. Intraoperative Irradiation: Techniques and Results. Totowa, NJ: Humana Press; 1999.
22. Nori D, Merimsky O, Osian AD, *et al.* Palladium-103: a new radioactive source in the treatment of unresectable carcinoma of the pancreas: a phase I-II study. *J Surg Oncol* 1996; 61:300–305.
23. Wong Gy, Schroeder DR, Carns PE, *et al.* Effect of neurolytic celiac plexus block on pain relief, quality of life, and survival in patients with unresectable pancreatic cancer: a randomized controlled trial. *Jama* 2004; 291:1092–1099.
24. Lillemoe KD, Cameron JL, Kaufman HS, *et al.* Chemical splanchnicectomy in patients with unresectable pancreatic cancer. A prospective randomized trial. *Ann Surg* 1993; 217:447–455; discussion 456–447.
25. Bryan PJ, Custar S, Haaga JR, *et al.* Respiratory movement of the pancreas: an ultrasonic study. *J Ultrasound Med* 1984; 3:317–320.
26. Davies SC, Hill AL, Holmes RB, *et al.* Ultrasound quantitation of respiratory organ motion in the upper abdomen. *Br J Radiol* 1994; 67:1096–1102.

Section 6

Emerging Thoraco-Abdominal Radiosurgery Concepts

4D Treatment Optimization and Planning for Radiosurgery with Respiratory Motion Tracking

25

Jay B. West, Jong Park, John Robinson Dooley, and Calvin R. Maurer, Jr.

CONTENTS

25.1 Abstract

The CyberKnife® Robotic Radiosurgery System (Accuray Incorporated, Sunnyvale, CA) can treat targets that move with respiration using the Synchrony® Respiratory Motion Tracking System or the Xsight™ Lung Tracking System (Accuray Incorporated, Sunnyvale, CA). Alignment of each treatment beam with the moving target is maintained in real time by moving the beam dynamically with the target. The challenges of treatment planning for mobile targets are different for dynamic respiratory motion tracking than for conventional approaches such as motion-encompassing and respiratory gating methods that are common on gantry-based delivery devices. Internal motion during respiration is not rigid, and thus positions of critical structures relative to the target and hence to the beam can change during respiration. The 4D Treatment Optimization and Planning feature, which recently became available in the MultiPlan® (Accuray Incorporated, Sunnyvale, CA) Treatment Planning System, is a new approach to four-dimensional (4D) treatment planning for motion tracking. It uses a 4D-CT image study to measure respiratory tissue motion and deformation and to account for the effect of motion and deformation on dose. The individual 3D-CT images are aligned so that the target coincides in each image. A tissue motion model is computed by performing nonrigid registration of the individual 3D-CT images. Using the target-centric alignment and the deformation model, it is possible to calculate a dose distribution that takes into account both beam movement and soft tissue deformation. This dose distribution may be calculated before plan optimization and hence used to determine the desired beam geometry and weighting, or it may be calculated after plan optimization in order to review the effects of respiration on the dose isocontours and statistics for a given plan.

25.2 Introduction

Planning for radiation therapy is typically performed using a static 3D-CT image of the patient. However, during treatment delivery, the anatomy can undergo substantial tissue motion and deformation. The inclusion of organ deformation and motion in treatment planning is growing in importance, especially in thoracic and abdominal applications where breathing has a large effect on target position and shape and the positions of critical structures relative to the target [1].

Numerous approaches have been developed to address the effect of respiratory motion [2]. One straightforward approach is to enlarge the target by a margin within which the target should move during the breathing cycle [3]. A related motion-encompassing approach is the slow scanning method, in which the CT scanner is operated very slowly, or multiple CT scans are averaged such that multiple respiration phases are recorded per slice [4]. The image of the tumor should show the full extent of the respiratory motion that occurs during the scanning process, provided the acquisition time at each couch position is longer than the breathing cycle. The disadvantage of slow scan methods is the loss of resolution due to motion blurring, which potentially leads to larger observer errors in tumor and normal organ delineation, as well as estimated dose delivered to the patient. Another related approach is to acquire CT images during the end-exhale and end-inhale phases of a breathing cycle. The pair of images provides the range of target motion, which is used to provide an anisotropic enlargement of the target. All of these motion-encompassing approaches ensure that tumor motion during breathing will not affect the dose delivered to the target, but they also lead to increased dose delivered to normal tissues, which can be a particular problem when the lesion is located close to organs at risk.

Another natural approach is to add a temporal component to the process using a gated strategy, which can be based on either breath holding or free breathing. Breath-hold techniques can use simple voluntary breath holding or more sophisticated approaches such as active breathing control [5]. Treatment is delivered during the breath hold period, which can be at end-exhale or end-inhale. These methods provide higher accuracy of dose delivery than motion-encompassing methods, but breath hold repeatability and patient compliance are challenges, especially for elderly patients or patients with lung cancer or other pulmonary disease. For respiratory gating approaches, the patient continues breathing normally. The treatment delivery system tracks respiratory motion, for example, using optical tracking, and turns the radiation beam on only when the respiratory position falls within a specified range of the complete breathing cycle. The delivery of radiation during a limited portion of the breathing cycle can substantially reduce the duty cycle and prolong treatment time. Lesion motion and gating model stability are also challenges for gating methods.

Recent advances in CT imaging have led to the development of high-quality 4D-CT or respiration-correlated CT scanning methods [6, 7]. Most major CT scanner vendors offer 4D-CT capability on their currently available equipment. A 4D-CT image study consists of a set of 3D-CT images, each one of which represents a different phase of the patient's breathing cycle (Fig. 25.1). The number of phases present in a study varies, but ten is a common number. The availability of 4D imaging has led to efforts to include organ motion and deformation in the treatment planning process.

25.3 4D Planning System for Respiratory Gating

With traditional gantry-based radiation delivery systems in which the treatment beams are not moved during treatment to compensate for respiratory motion, the term "4D planning" is generally used to refer to the process of finding a compromise between the size of the Planning Target Volume (PTV) and the width of the gating window, i.e., the portion of the respiratory cycle during which the beams are turned on (Fig. 25.2).

Clearly, the wider the gating window, the greater the volume of normal tissue that must be treated in order to ensure target coverage; however, a narrow gating window significantly lengthens treatment

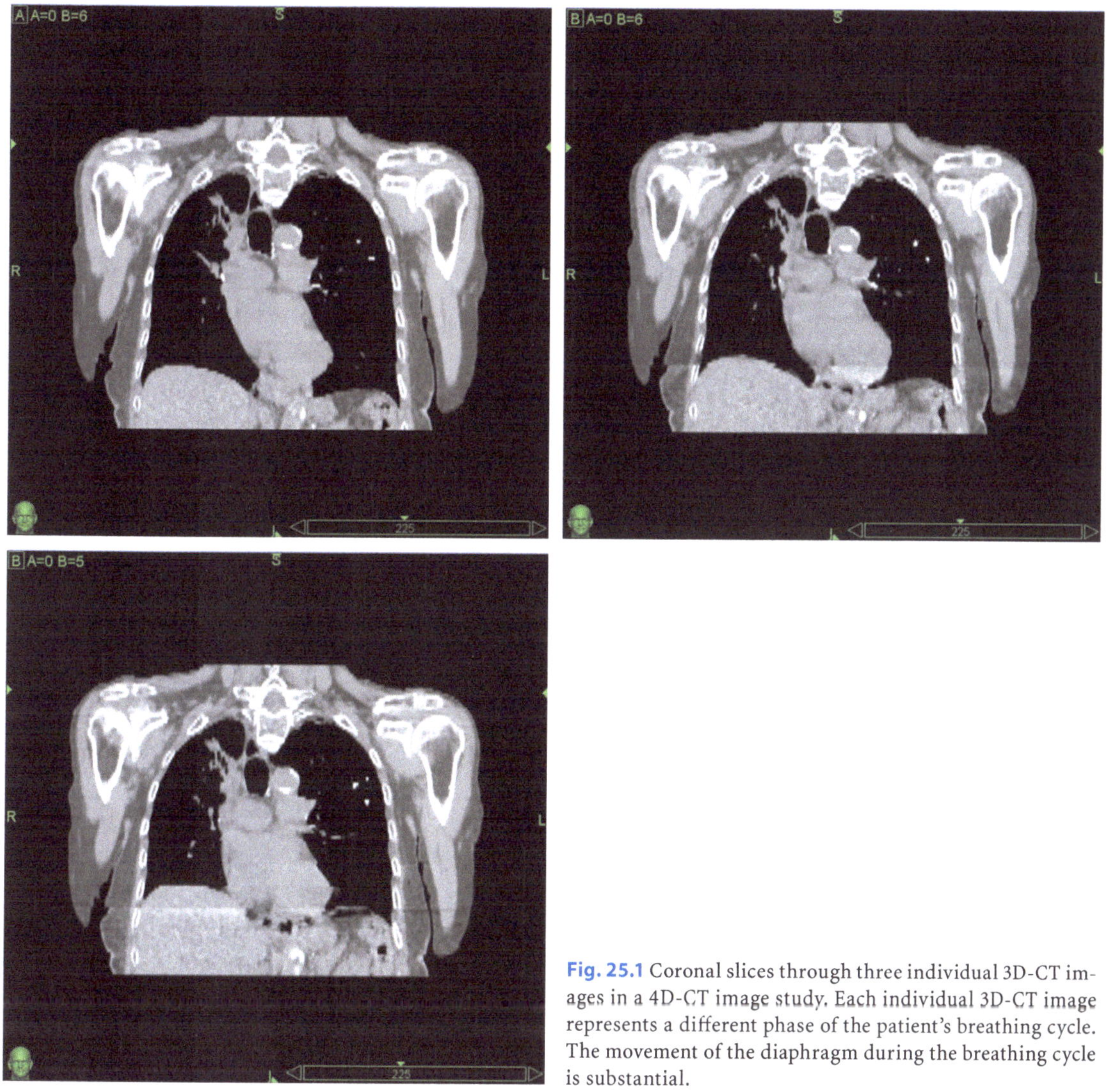

Fig. 25.1 Coronal slices through three individual 3D-CT images in a 4D-CT image study. Each individual 3D-CT image represents a different phase of the patient's breathing cycle. The movement of the diaphragm during the breathing cycle is substantial.

time. Hence it is necessary to find a compromise. The workflow for a traditional 4D planning approach would then be as follows:

- Superimpose the 4D-CT series using their native coordinate system.
- Draw the Clinical Target Volume (CTV) in at least one of the 4D-CT series.
- By examining the position and shape of the CTV in the series, expand the CTV into a PTV that covers the CTV in all series to be used. For the unused series, set the gating window so that the beam is disabled during this part of the respiratory cycle.

Some experimental work has been done [8] in an attempt to integrate automatically the effects of respiratory motion into a treatment plan without the necessity for gating, but this work essentially creates several treatment plans giving multileaf collimator positions for each of multiple phases of the breathing cycle, hence splitting the optimization and review process into several independent parts, each one tied to a portion of the cycle.

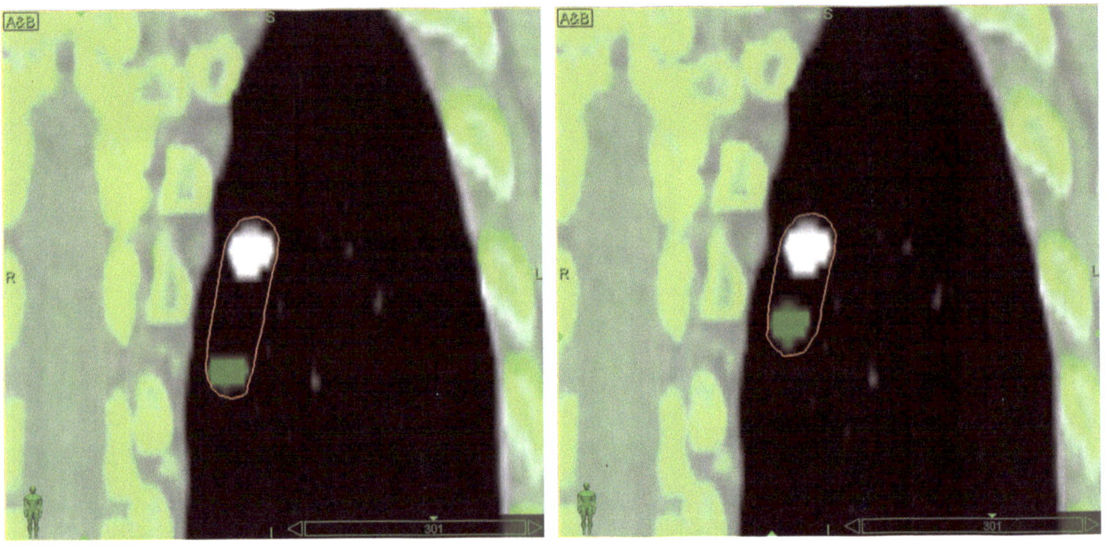

Fig. 25.2 The PTV required to ensure full coverage of the CTV for a radiation delivery system with stationary beams. In the left figure, the white and green blobs show the position of the treatment target at end-exhale (*white*) and end-inhale (*green*), and the orange contour shows the region of tissue that must be treated to fully cover the moving target. In the right figure, we show the area of tissue that must be treated if the beams are turned on only for the 20% of the respiratory cycle following end-exhale.

25.4 4D Planning System for Respiratory Motion Tracking

The CyberKnife Radiosurgery System can treat targets that move with respiration using respiratory motion tracking. The Synchrony and the Xsight Lung Tracking Systems use continuously tracked optical markers on the patient's chest to enable the treatment beams to follow respiratory movement of the target. By building a correlation model relating the position of the external optical markers to that of implanted fiducials visible in two orthogonal X-ray images (Synchrony) or to the target itself visible in the X-ray images (Xsight Lung), an estimate of the target position at any instant in time may be calculated, and the beam position updated accordingly. Hence, if the target motion is only translational, with no rotation or deformation, and the Synchrony model is perfect, the motion of the target will not affect the dose delivered to it. Because treatment targets can undergo some rotation and deformation, in addition to translational shift, during respiration, there can be a subtle change in the treated target dose relative to that computed in the treatment plan. Another potentially relevant effect is that on critical structures which do not move in the same way as the target. For example, the spinal cord can generally be considered to be stationary during respiration; because the treatment beams are moving to track the target, it is possible for the beams to "sweep" across the cord during treatment. This may affect the delivered dose to the cord in two ways – if the static plan showed the beam going directly through the cord, the dose delivered under dynamic treatment would be less than that shown in the plan, because the beam would in fact only coincide with the cord for part of the breathing cycle (Fig. 25.3). Conversely, if the static plan showed the beam missing the cord, the dose delivered under dynamic treatment would exceed that in the plan if the action of following target motion made the beam coincide with the cord for part of the cycle.

The 4D Treatment Optimization and Planning feature, which recently became available in the MultiPlan Treatment Planning System, is a new approach to 4D treatment planning that integrates the dynamic nature of respiratory motion tracking with the planning process. It uses a 4D-CT image study to measure respiratory tissue motion and deformation and to account for the effect of motion and deformation – deformation of the target, deformation of

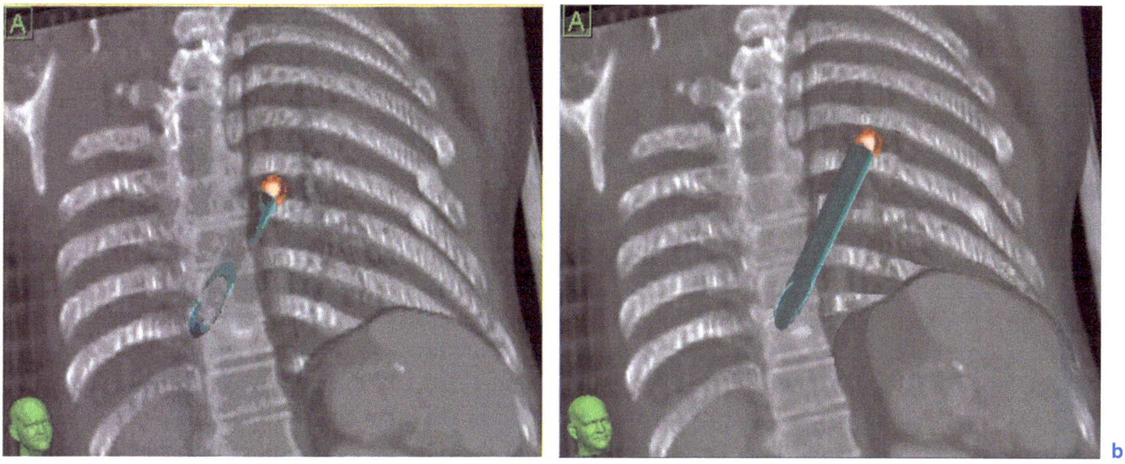

a b

Fig. 25.3a,b. The position of one of the treatment beams at two phases in the respiratory cycle. At end-inhale, (a) the beam passes through the spinal column after exiting the treatment target. At end-exhale, (b) the beam position is updated to track anterior and superior motion of the target. This results in the beam missing the spinal column during this part of the breathing cycle.

tissue outside the target, and motion of the target and critical structures relative to each other and/or to the fiducials – on dose. The individual 3D-CT images are aligned so that the target coincides in each image. A tissue motion model is computed by performing nonrigid registration of the individual 3D-CT images. One of the available CT images is chosen as the reference dataset and used to create a treatment plan. Using the target centric alignment and the deformation model, it is possible to calculate a dose distribution on the reference dataset that takes into account both beam movement and soft tissue deformation. This dose distribution is referred to as "4D dose." The deformation model maps the position of each point in the reference 3D image to the corresponding location in all other 3D images in the 4D-CT image study. The 4D dose is computed over the multiple breathing phases, weighted by the time spent in each phase, and is referred back to the reference 3D image. The 4D dose distribution may be calculated before plan optimization and hence used to determine the desired beam geometry and weighting, or it may be calculated after plan optimization in order to review the effects of respiration on the dose isocontours and statistics for a given plan. The concepts in this approach to 4D treatment planning, some of which have been recently described by others [9], are applicable to any treatment delivery system that uses respiratory motion tracking to dynamically move or shape the radiation beam.

25.5 Building a Deformation Model

25.5.1 Target Centric Alignment

In order to apply a deformation model to the dose calculation process for a treatment plan, it is necessary to define a coordinate system that takes into account the movement of the beams during treatment. Because the beam delivers the radiation dose, for the purposes of dose calculation it is most convenient to assign a constant offset to the deformation so that a value of zero is achieved by structures that are stationary with respect to the beams. To achieve this, we must translationally align the 4D-CT volumes according to the object being tracked. There are two cases that must be handled: Synchrony tracking and Xsight Lung tracking.

25.5.2 Synchrony Respiratory Tracking System

In order to perform Synchrony tracking, metal fiducials are embedded in or near the treatment target before the planning CT scan is taken. During treatment, the position of the fiducials as seen in periodically acquired orthogonal X-ray images is correlated with the locations of optical markers on the patient's chest. These markers are continuously tracked, allowing the treatment beams to be moved in real time to compensate for respiratory target motion. The beams are moved according to the motion of the centroid of the fiducials; hence the fiducial centroid is chosen as the alignment center for the 4D-CT dataset (Fig. 25.4a).

25.5.3 Xsight Lung Tracking System

For peripheral lung lesions that are 15 mm or more in diameter, real-time motion compensation may be performed without implanted fiducials, using the Xsight Lung Tracking System. Instead of finding fiducials in the orthogonal X-ray images, a sophisticated image processing algorithm is used to find directly the tumor and triangulate its spatial position. This position is then correlated with real-time measurements of tracked optical markers in a similar manner to Synchrony tracking, and the position of the treatment beam updated accordingly. In order to account for this beam motion during treatment planning, the user manually identifies the position of one or more landmarks on the target, in each of the 4D-CT series. The series are then translationally aligned so that the centroid of these landmarks is coincident on each of the images (Fig. 25.4b).

25.5.4 Deformable Registration

In order to perform a dose calculation that takes into account both beam movement and tissue deformation, it is necessary to construct a tissue deformation model that will take as input an anatomical location in the reference CT image and return as output the estimated position of the same anatomical element in each of the individual 3D-CT images in the 4D-CT study. The model is computed by nonrigid or deformable registration of all other 3D-CT images to the reference 3D-CT image. The registration process optimizes a cost function that is a combination of the distance between corresponding landmarks and the intensity difference between the reference image and deformed secondary image [10]; the deformation is represented by a combination of third-order B-splines, thus ensuring that the resulting deformation field has a continuous second derivative (i.e., that the deformation field is smooth). Similar methods for constructing a tissue

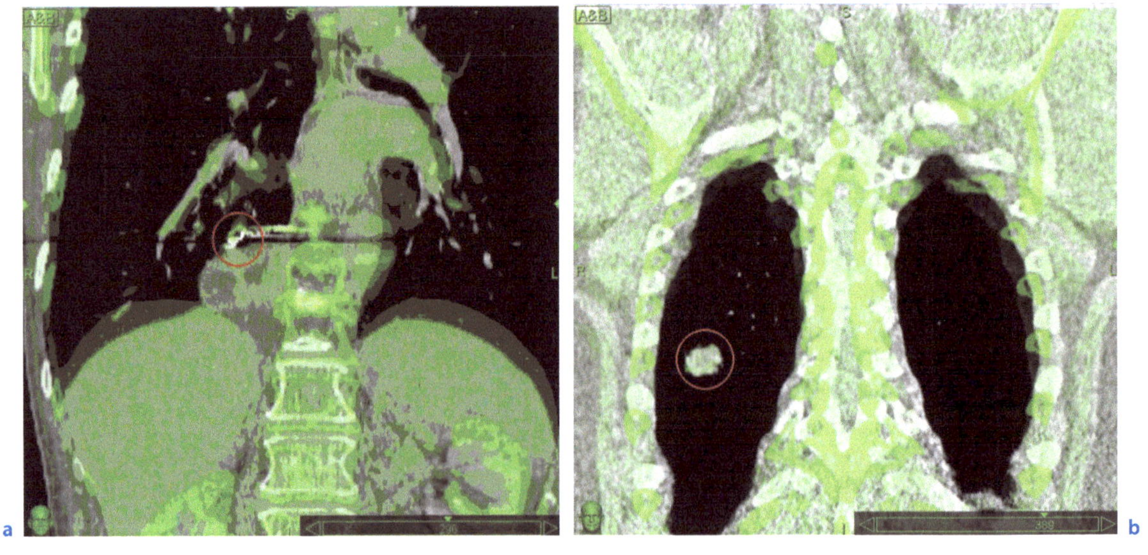

Fig. 25.4a,b. Translational target centric alignment of CT images, showing end-inhale (*grayscale*) versus end-exhale (*green overlay*). On the left, fiducials are used to align the images, and on the right, anatomical landmarks within the treatment target. In both cases, the alignment of the targeted structure (*denoted by red circle*) can be seen, whereas regions that are stationary during respiration, e.g., the spinal column, are seen to be offset.

deformation model from 4D image datasets have recently been reported [11, 12]. Because B-spline deformation modeling is typically most successful in the presence of small deformations, three steps are taken to improve the quality of the resultant tissue deformation model:

- If the reference 3D-CT image is not one of the individual 3D-CT images in the 4D-CT image study, the reference image is registered to the individual 3D-CT image in the 4D-CT dataset that is closest to the reference image in the respiratory cycle. All individual 3D-CT images in the 4D-CT image study are registered to their temporal neighbor, producing a "chain" of deformations from the reference CT to all images in the 4D-CT dataset. The deformations are added in a transitive manner to produce a mapping that goes directly from the reference CT to each image in the 4D-CT dataset. The sum of second derivative continuous deformations will have the same property (the sum of smooth deformations is still smooth).
- The deformation is calculated with respect to the native coordinate alignment of the CT images. The scanning protocol requires that there is no gross patient movement between any of the image acquisitions. The target centric alignment offset is applied after the deformable registration has been completed. In this way, structures that do not move during respiration, such as the spinal cord, are aligned in all images while the deformation is being calculated, and the deformation model needs only to represent the global deformation of structures such as the chest wall and diaphragm.
- The nonrigid registration uses as landmark points the fiducials or landmarks used in the alignment step. This ensures that the deformation is constrained to be approximately correct in the neighborhood of the treatment target. In some cases, where there is a large movement of a small target, the contribution of this movement to the overall intensity difference between images is small, and it is not correctly captured unless a landmark point is present in the cost function. Additional landmark points can be interactively added to assist the registration process as necessary.

An example of deformable registration is shown in Figure 25.5. The deformed secondary image is in much closer agreement with the reference CT than before registration, although a small artifact has been created at the intersection between the rear chest wall and the diaphragm. Shearing motion, as often occurs between the liver and the chest wall during respiration, cannot be exactly represented by a continuous representation such as third-order B-splines, hence the deformation is not captured exactly in these cases.

a

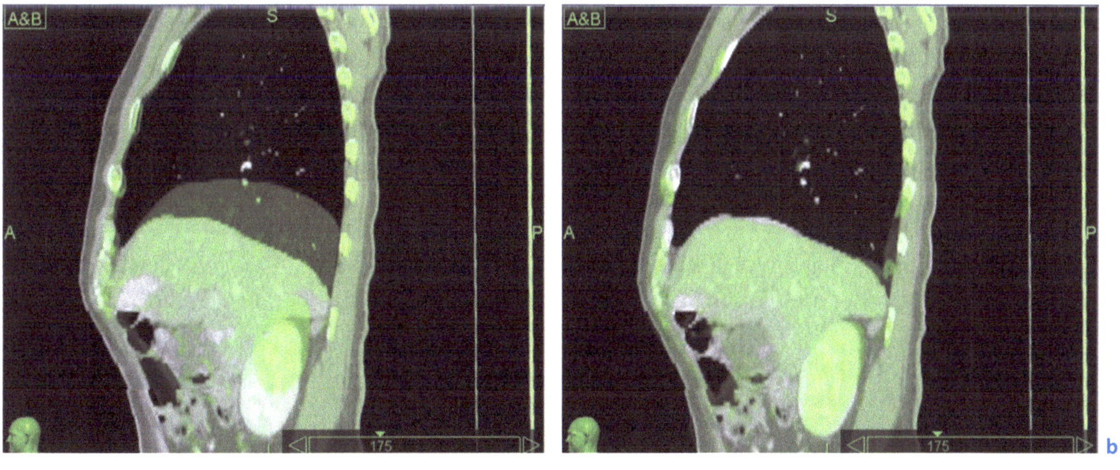

b

Fig. 25.5a,b. Overlay of an end-inhale image (*grayscale*) and end-exhale image (*green*) before (**a**) and after (**b**) deformation. Note the large discrepancy in the area of the diaphragm in the left image. The deformation process naturally resolved the diaphragm motion, but two landmark points had to be added on the superior and inferior aspects of the kidney to improve the quality of the model there. There is a small residual artifact on the patient's back near the diaphragm which results from the B-spline model being unable to perfectly represent discontinuous "shearing" motion as occurs between the liver and posterior chest wall during breathing.

It is also worth noting that two landmark points had to be added, at the superior and inferior aspects of the kidney, to achieve the deformation shown in Figure 25.5. As discussed by West et al. [10], one of the main advantages of allowing manually placed landmarks as part of the deformation optimization is that it permits us to overcome the limitations of solely intensity-based registration in regions of poor CT contrast such as the lower abdomen. The example in Figure 25.5 took approximately five minutes of computation time to calculate the deformation model relating the reference CT to the other nine 3D-CT images in the 4D-CT dataset (using a PC workstation with two dual-core processors running at 2.8 GHz).

25.6 Dose Calculation

After building a tissue deformation model that gives us the trajectory of all tissue elements throughout the breathing cycle, the next step is to apply this to calculate a dose distribution that takes into account tissue deformation and beam movement. As mentioned previously, this dose distribution is referred to as 4D dose. There are two separate effects that need to be taken into account during 4D dose calculation.

25.6.1 Effective Depth Compensation

The nature and amount of tissue traversed by a treatment beam may change as the beam moves to track the target, and as the patient anatomy moves and deforms. In the example shown in Figure 25.3, in the inhale part of the respiratory cycle, the depicted beam has a trajectory that takes it through the bone of the spinal column; in the exhale part of the cycle, the beam misses the spinal column and hence has a trajectory passing mostly through lung tissue, which is much sparser and has less attenuation than bone. In order to account for this effect, each of the individual 3D-CT images is used to create a vector describing the effective depth (equivalent depth of water) along the beam central axis at that respiratory phase, taking into account the beam tracking offset.

25.6.2 Tissue Motion

The other important effect that must be accounted for in order to perform a 4D dose calculation is the motion of tissue elements relative to the treatment beams. A dose calculation is performed for each of the 3D-CT images in the 4D-CT study. The dose information is stored using a "dose map" architecture, i.e., for each beam, a list of floating point values is stored, indexed to the dose calculation grid in the reference image, which represents the radiation dose in centiGray (cGy) delivered to each dose grid voxel by that beam, per monitor unit (MU). The tissue deformation model enables the dose to be mapped to the dose calculation grid in the reference image. The 4D dose accounting for tissue deformation is computed by summing dose over the multiple breathing phases, weighted by the time spent in each phase. In order to calculate the 4D dose map for a particular beam, the procedure used is as follows:

- Traverse the voxels in the dose calculation grid in the reference CT image.
- For a particular voxel, calculate the dose delivered per MU by the beam in the reference CT.
- For each additional 3D-CT image in the 4D-CT study, apply the deformation model to map from the reference CT space to the corresponding position in the secondary image.
- Apply the beam tracking offset to move the central axis of the beam so that it follows the target in the prescribed manner.
- Calculate the dose delivered per MU by the beam at the deformed position, using the phase-specific effective depth vector described in the previous section.
- Set the dose per MU in the reference CT voxel to be the weighted sum of the dose calculations for each of the 3D-CT images in the 4D-CT study.

25.6.3 Weighting

The weighting used to sum the independent dose calculations depends on the nature of the phase labeling for the 4D-CT volume. First, if the reference image is a breath hold image, rather than a part of the 4D-CT series, the weight for the dose contribution from the reference image is set to zero, even though it is used as the coordinate system for dose display. In some cases, the user may wish to use a breath hold image as the reference CT to take advantage of the potentially higher image quality of a breath hold image relative to the images in a 4D-CT study. However, because a breath hold image often has an unnatural anatomical configuration (e.g., the patient, on being told to inhale, takes an extremely deep breath), this image is excluded from dose calculation on the grounds that it may not accurately represent the patient's anatomy during free breathing. Second, for the 4D-CT study, the goal is to find a relative weighting that corresponds to the amount of time spent in the respiratory phase for each image in the study. When the phases correspond to a time measurement, the weighting is simple – the width of the time bin for each image is used, and normalized so that the weights sum to 1. When the phases correspond to an amplitude measurement, a correction factor is used to account for the fact that most people spend more time in the exhale part of the breathing cycle than the inhale part. Writing the respiratory phase as an angle θ, where $\theta = 0$ at full exhale and $\theta = \pi$ at full inhale, the relative weighting for amplitude-binned images is equal to the width of the amplitude bin multiplied by $1 + \cos^2(\theta/2)$.

25.6.4 Difference Between 3D and 4D Dose Calculation

The difference between 3D and 4D dose isocontours for a single beam is illustrated in Figure 25.6. As expected, the main visible change is that the low dose isocontours are spread in the superior-inferior direction with the 4D calculation, representing the fact that the beam will sweep the posterior chest wall during treatment. For example, in a typical CyberKnife treatment plan consisting of 75–150 treatment beams,

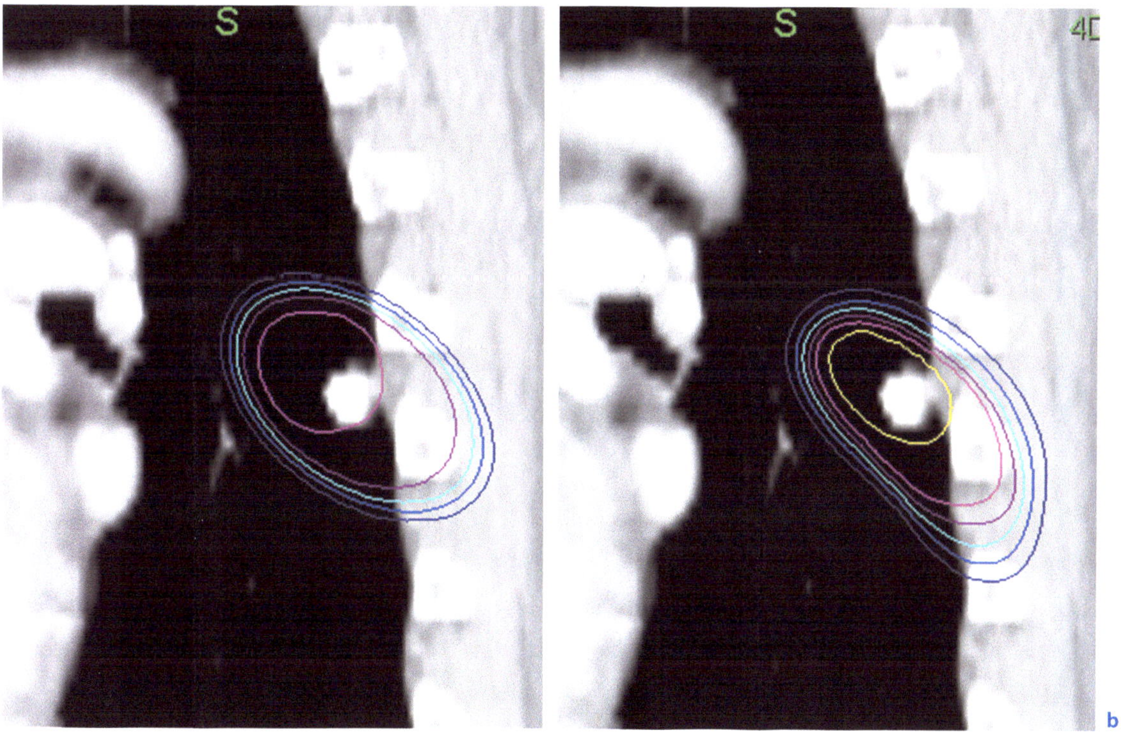

Fig. 25.6a,b. Comparison between 3D (**a**) and 4D (**b**) dose calculation for a single beam, using the end-exhale CT as the reference image. With the 4D dose calculation, in general coverage to the target has not been greatly changed, because the beam will move to track respiratory target motion. The lower dose isocontours, however, have been extended in the inferior direction for the 4D dose calculation, representing the fact that the beam will sweep the posterior chest wall as it tracks the target. Because a typical CyberKnife treatment plan contains 75–150 beams, the difference between 3D and 4D dose for a whole plan is less than that for an individual beam (Fig. 25.7).

the difference between 3D and 4D dose calculation is less than that for a single beam (Fig. 25.7).

The calculation time for 4D dose is higher than that for a standard 3D dose calculation; in fact, it is N times greater, where N is the number of individual 3D-CT images in the 4D-CT study. One advantage, however, as will be discussed in the following section, is that it allows creation of a treatment plan that is optimized for dynamic tracking and respiratory motion, with no need to manually delineate margins, contour extra "tuning" structures, or change constraints.

25.7 Optimization

Using the 4D dose calculation framework, it is possible to build an interface for 4D plan optimization that is both powerful and intuitive. The 3D optimization algorithms use the same dose map architecture described above, so generating a treatment plan using the tissue deformation model created from the 4D-CT study that accounts for the effect of respiratory motion on the dose distribution during the plan optimization process simply involves substituting the 4D dose maps for their standard 3D dose counterparts. The user specifies dose constraints in the usual way. The optimization process begins with a set of candidate beams (approximately 1200 for linear programming optimization and 500 for iterative optimization). Then the 4D dose maps are calculated for all candidate beams, and this information is passed to the optimization engine, which determines a beam geometry and weighting that best satisfies the specified dose constraints. The result of this 4D optimization is a treatment plan with dose volume histograms, dose statistics, and dose isocontours computed using 4D dose and optimized accounting for the effect of respiratory motion on the dose distribution (Fig. 25.8).

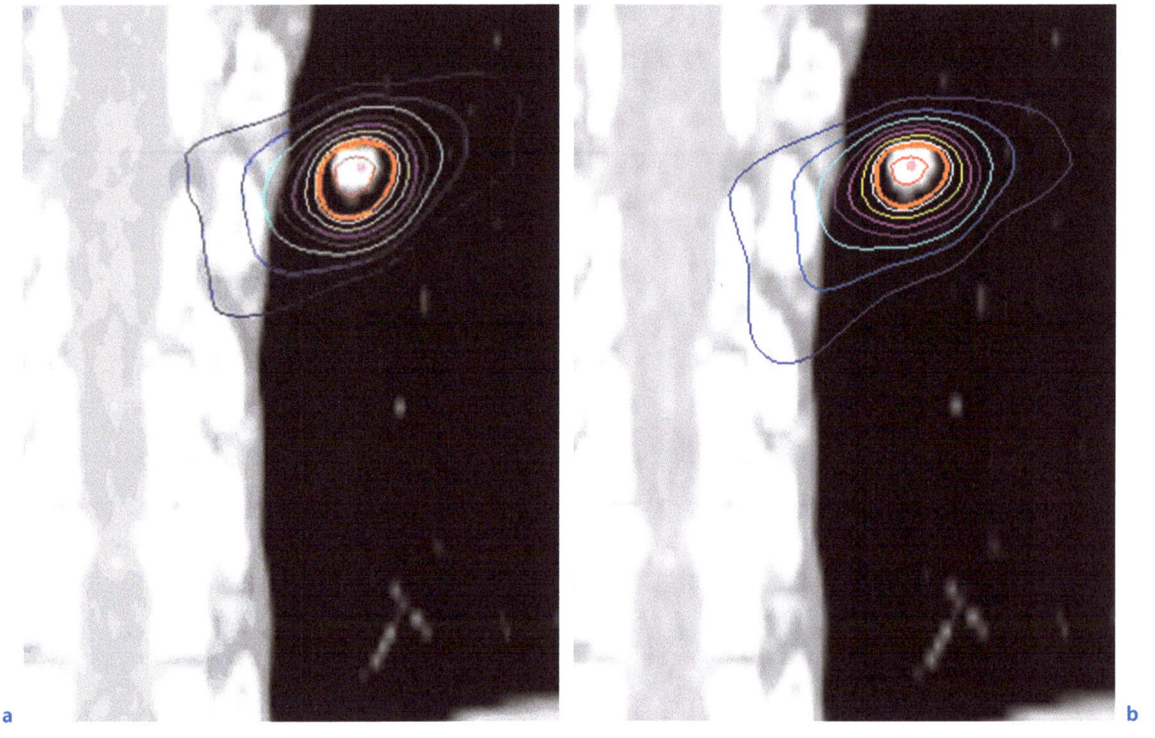

Fig. 25.7a,b. Comparison between 3D (**a**) and 4D (**b**) dose calculation for a treatment plan consisting of 84 beams. The 4D dose calculation shows an extension of the overlap of the 10% isocontour with the spinal column and a small reduction in the area covered by the higher dose isocontours, the latter effect possibly due to slight deformation and rotation of the target that is not captured by dynamic tracking. The coverage of the target by the 75% isodose was reduced from 99.6% with 3D calculation to 88.2% with 4D calculation.

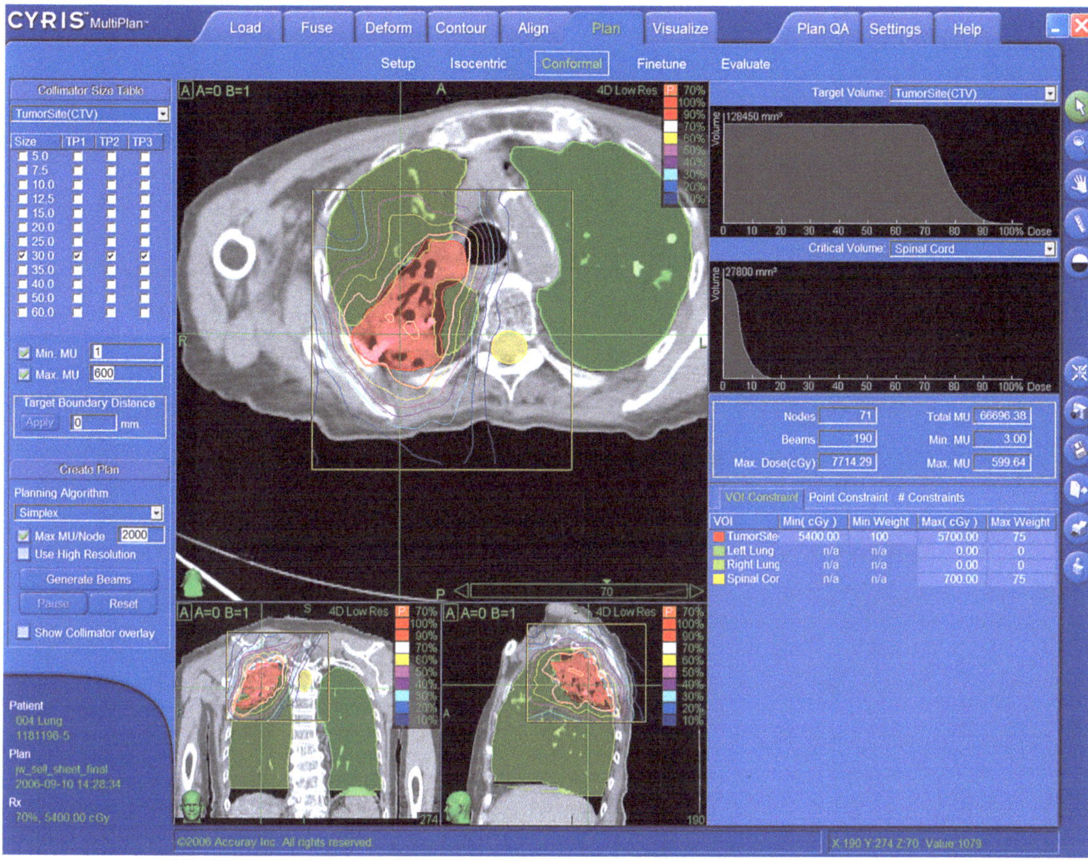

Fig. 25.8 4D treatment optimization. The user supplies dosimetric constraints in the usual manner, and the result of the optimization is a beam geometry and weighting that best obeys the constraints, taking into account tissue deformation and dynamic tracking.

25.8 Review

As described in the previous section, the 4D planning capability allows the user to produce CyberKnife treatment plans that are automatically optimized to take into account the effect of respiratory motion. However, to do this it is necessary to perform 4D dose calculation for all the candidate beams used for the optimization. This 4D dose map calculation can have a high computational cost, often more than ten minutes. In cases where the user feels that the motion effects are likely to be small – for example, lesions near the apex of the lung – it may be more time efficient to perform a normal 3D optimization and review the generated plan using a 4D dose calculation. This has the advantage that, in the review step (Fig. 25.9), only the beams with nonzero MU need to be calculated in 4D; in the example shown, there are 84 beams used in the treatment plan, in contrast to the approximately 1200 candidate beams used for optimization. If the reviewing physicist and/or physician feel that the change in dose due to respiratory motion is not clinically iportant, the plan may then be delivered as is; otherwise, a full 4D optimization may be performed.

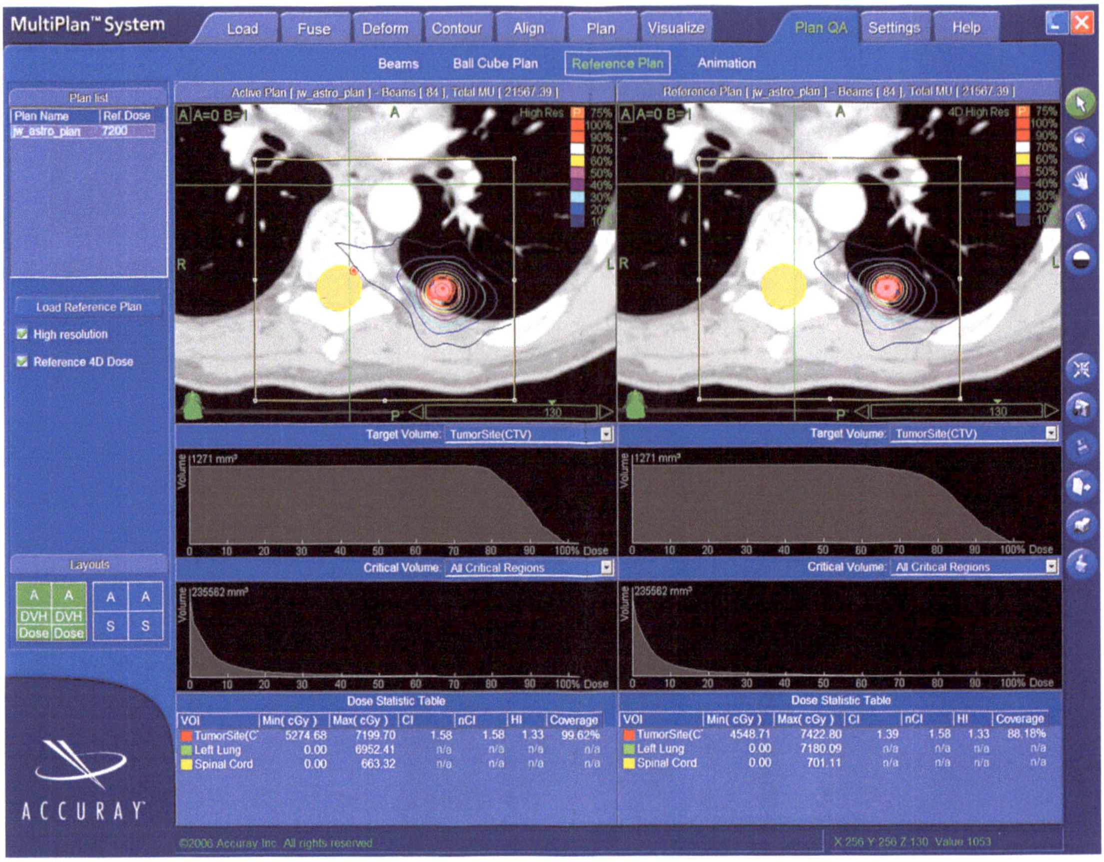

Fig. 25.9 Comparison of a treatment plan, generated using 3D optimization, with 3D dose calculation (left side) versus the same treatment plan with 4D dose calculation (right side).

25.9 Validation

25.9.1 The Phantom

Before 4D Treatment Planning and Optimization is used clinically, it is important to demonstrate that the 4D dose distribution displayed by the planning system is an accurate representation of the dose that will be received by the patient during treatment. To this end, Accuray developed a phantom for 4D planning validation. The phantom is designed to model approximately the configuration of a human torso, with an external skin, a dense cylinder representing the spinal column, and less dense regions corresponding to the lungs (Fig. 25.10). This can be seen in coronal slices through the 4D-CT study of the phantom (Fig. 25.11). The “spinal column” contains three embedded fiducials, which are used for initial global alignment of the phantom.

Also present is a ball containing two fiducials, which is moved in a regular pattern via a shaft connected to a motor (Fig. 25.10). The amplitude and period of motion may be varied; in Fig. 25.11, the amplitude was set to 30 mm between full inhale and full exhale. Dose measurements are made by means of GAFChromic® radiochromic dosimetry film (International Specialty Products, Wayne, NJ). Slices of film are embedded in coronal and axial orientations in the stationary “spinal column”, and also in the moving target. Of most interest for 4D validation is the dose distribution in the coronal stationary film, because this is where dynamic beam tracking should cause a measurable difference between the 3D and 4D dose distributions.

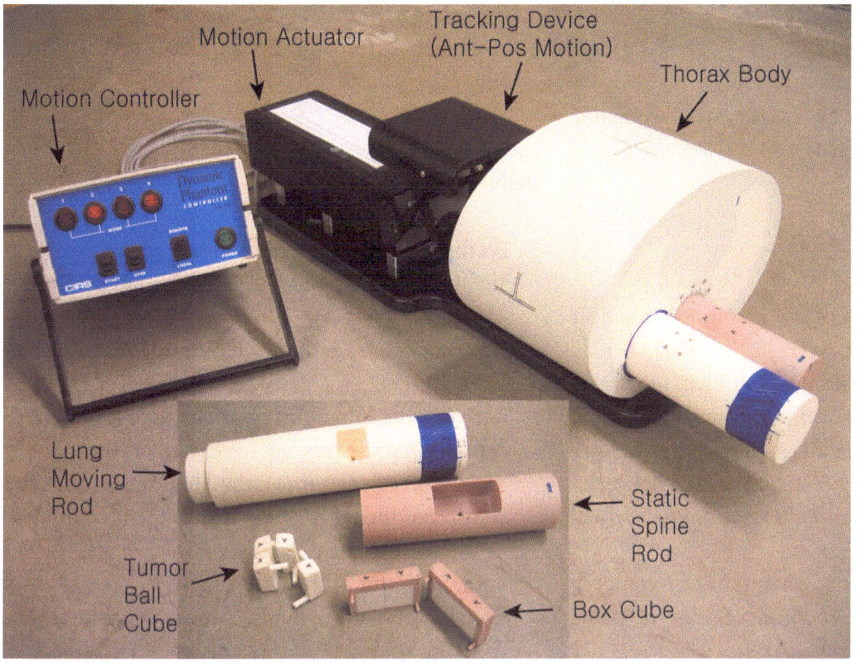

Fig. 25.10 The 4D motion phantom and its constituent parts. The motion controller sends a signal to the motion actuator, which causes it to move the lung moving rod containing the spherical target (*tumor ball cube*). The tracking device is designed to exhibit anterior-posterior motion when the lung rod is moved; this allows the 4D-CT scanner to deduce the phase, and bin the CT slices correctly, at any point in the cycle, and also facilitates Synchrony respiratory tracking when the treatment plan is delivered.

a

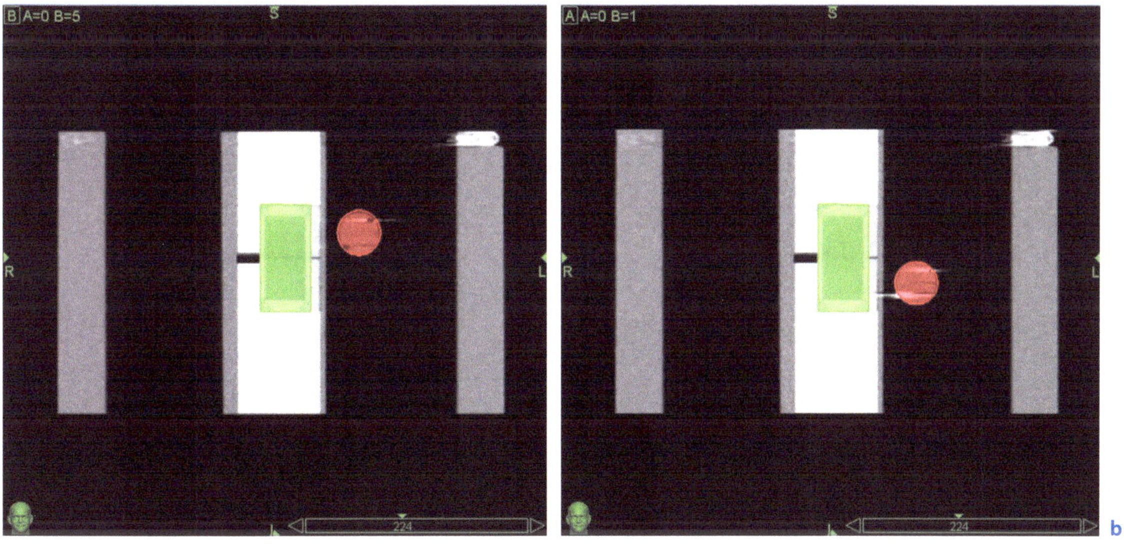

b

Fig. 25.11a,b. Coronal slices through 4D scan of the phantom in exhale (**a**) and inhale (**b**) position. The green rectangle shows the radiochromic dosimetry film embedded in the "spinal cord": the red ball has two metal fiducials implanted in it, and moves according to a regular pattern, simulating respiratory motion of the target. The treatment beams track the ball, thus sweeping across the film.

25.9.2 Validation Procedure

In order to perform the necessary measurements, the following steps are used:

25.9.2.1 Planning

1. Perform a 4D-CT scan of the phantom while it is in motion.
2. Import the 4D-CT scan into MultiPlan, preferably using a series near the central point of phantom motion as the reference image to prevent excessive tracking motion in any direction during delivery.
3. Delineate the target and the stationary coronal film.
4. Create a simple isocentric treatment plan centered on the target, using Synchrony as the indicated tracking mode.
5. Use the fine tune step to enhance or suppress individual beams to ensure that the film dose is within desired limits (we wish the maximum dose in the coronal film to be approximately 3.5 Gy for best results with the GAFChromic EBT film used for the stationary coronal measurement).
6. Calculate 3D and 4D dose distributions, exporting each dose volume as a DICOM RTDose object.

25.9.2.2 Delivery

7. Perform global alignment of the phantom using the fiducials embedded in the spinal column, either by using an "alignment plan" with only these fiducials, or by switching off the fiducials in the target.
8. Activate the phantom motion, using the same motion parameters as were set for the 4D-CT scan. If necessary, move the phantom so that the motion of the target stays within the field of view of the X-ray images. Switch off the fiducials in the spinal column, if they are enabled.
9. Deliver the plan using Synchrony System tracking.
10. Remove the film from the phantom and scan it.
11. Choose an isocontour from the RTDose object that has a significant intersection with the film. Derive a histogram of the deviation of measured dose from planned dose along the isocontour, for both the 3D and 4D dose distributions.

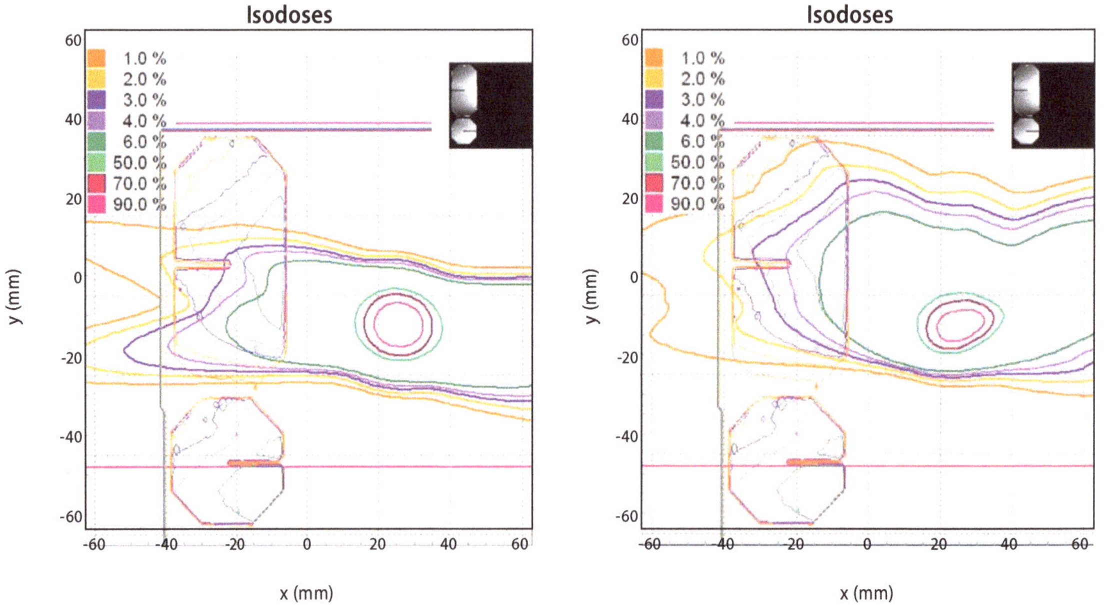

Fig. 25.12. Example overlay of dose isocontours generated by the planning system (*thick lines*) with the corresponding isocontours from the scanned film (*thin lines*). The left image shows that the 3D dose distribution does not take into account the spreading effect caused by beam movement, and hence disagrees with the measured dose. On the right, we see that the 4D distribution is in much better agreement with measured dose.

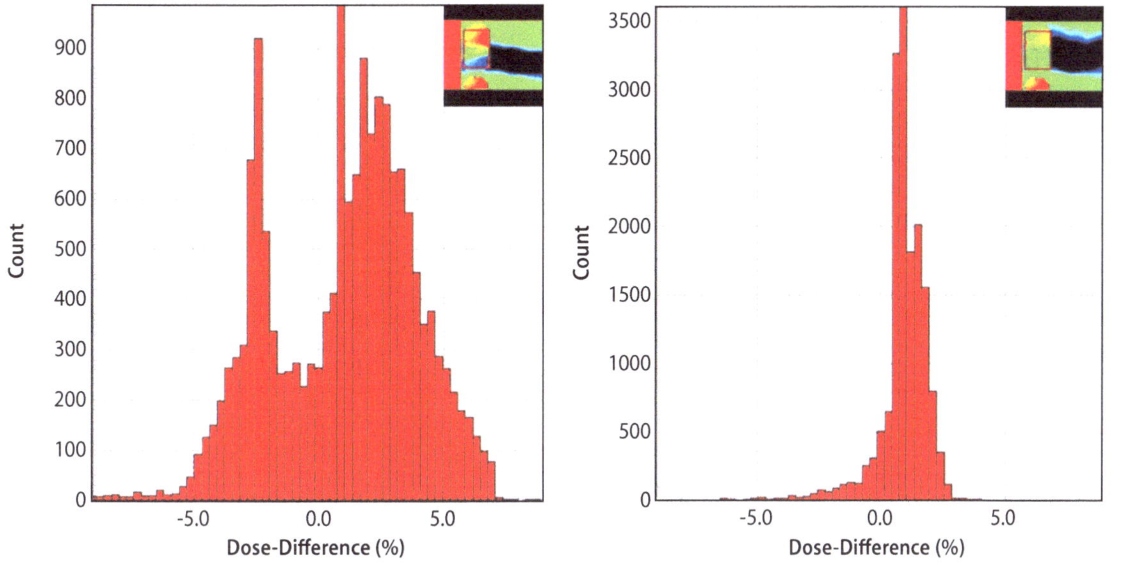

Fig. 25.13. Histogram of the deviation of measured dose from planned dose generated by the planning system. The histogram for the 3D dose distribution, shown on the left, shows a much greater averange discrepancy between measured and planned dose than the histogram for the 4D dose distribution, shown on the right.

25.9.3 Preliminary Results

In Figures 25.12 and 25.13, planned versus measured dose for the 3D and 4D cases are shown in two ways. Figure 25.12 shows the shape of the planned and measured isodose contours intersecting the stationary coronal film; Figure 25.13 shows a histogram of the deviation of measured dose from the planned dose generated by the planning system. The 3D dose distribution does not take into account the spreading effect in the superior-inferior direction caused by beam movement during treatment. The 4D dose distribution, however, does account for this effect, and for this reason; according to both methods used for comparison, the 4D distribution shows much smaller average discrepancy between planned and measured dose.

25.10 Conclusion

We have described here the framework for 4D treatment planning and optimization, which uses a 4D-CT image study to measure respiratory tissue motion and deformation and account for its effect on dose, and which has been implemented in Accuray's MultiPlan Treatment Planning System. It is a considerably different approach from 4D planning for respiratory gating that is common on traditional gantry-based radiation delivery systems in which the treatment beams are not moved during treatment to compensate for respiratory motion. The new approach integrates the dynamic nature of respiratory motion tracking, which is available for the CyberKnife Radiosurgery Treatment System, with the planning process. The CyberKnife System 4D planning capability allows treatment plans to be optimized in the presence of target motion and soft tissue deformation, with little change to the workflow when compared to standard planning on a single static CT image. Dose measurements performed using a specially designed motion phantom demonstrate that the calculated 4D dose distribution is an accurate measure of the delivered dose. Potential clinical benefits include enhanced accuracy of critical structure dosimetry, enhanced ability to treat single lesions that deform during respiration, and enhanced ability to treat multiple lesions that deform relative to each other during respiration.

References

1. Keall PJ, Mageras GS, Balter JM, *et al. Report of AAPM Task Group 76: The Management of Respiratory Motion in Radiation Oncology.* College Park, MD: American Association of Physicists in Medicine; 2006.
2. Wong J. Methods to manage respiratory motion in radiation treatment. In: Palta JR, Mackie TR, editors. *Intensity Modulated Radiation Therapy.* Madison, WI: Medical Physics; 2003. pp. 663–702.
3. George R, Keall PJ, Kini VR, *et al.* Quantifying the effect of intrafraction motion during breast IMRT planning and dose delivery. *Med Phys* 2003; 30:552–562.
4. Lagerwaard FJ, Van Sornsen de Koste JR, Nijssen-Visser MR, *et al.* Multiple "slow" CT scans for incorporating lung tumor mobility in radiotherapy planning. *Int J Radiat Oncol Biol Phys* 2001; 51:932–937.
5. Wong JW, Sharpe MB, Jaffray DA, *et al.* The use of active breathing control (ABC) to reduce margin for breathing motion. *Int J Radiat Oncol Biol Phys* 1999; 44:911–919.
6. Ford EC, Mageras GS, Yorke E, *et al.* Respiration-correlated spiral CT: a method of measuring respiratory-induced anatomic motion for radiation treatment planning. *Med Phys* 2003; 30:88–97.
7. Vedam SS, Keall PJ, Kini VR, *et al.* Acquiring a four-dimensional computed tomography dataset using an external respiratory signal. *Phys Med Biol* 2003; 48:45–62.
8. Keall PJ. 4-dimensional computed tomography imaging and treatment planning. *Semin Radiat Oncol* 2004; 14:81–90.
9. Schlaefer A, Fisseler J, Dieterich S, *et al.* Feasibility of four-dimensional conformal planning for robotic radiosurgery. *Med Phys* 2005; 32:3786–3792.
10. West JB, Maurer CRJ, Dooley JR. Hybrid point-and-intensity-based deformable registration for abdominal CT images. *Medical Imaging 2005: Image Processing 2005*; Proc. SPIE 5747: 204–211.
11. Rohlfing T, Maurer CR, Jr., O'Dell WG, *et al.* Modeling liver motion and deformation during the respiratory cycle using intensity-based nonrigid registration of gated MR images. *Med Phys* 2004; 31:427–432.
12. Stancanello J, Berna E, Cavedon C, *et al.* Preliminary study on the use of nonrigid registration for thoraco-abdominal radiosurgery. *Med Phys* 2005; 32:3777–3785.

Xsight Lung Tracking System: A Fiducial-Less Method for Respiratory Motion Tracking

26

Dongshan Fu, Robert Kahn, Bai Wang, Hongwu Wang, Zhiping Mu, Jong Park, Gopinath Kuduvalli, and Calvin R. Maurer, Jr.

CONTENTS

26.1 Abstract

The CyberKnife® Robotic Radiosurgery System (Accuray Incorporated, Sunnyvale, CA) can treat targets that move with respiration using the Synchrony® Respiratory Tracking System (Accuray Incorporated, Sunnyvale, CA). Alignment of each treatment beam with the moving target is maintained in real time by moving the beam dynamically with the target. The Synchrony system requires fiducials that are placed in or near the tumor to target the lesion and track it as it moves with respiration. The Xsight™ (Accuray Incorporated, Sunnyvale, CA) Lung Tracking System, which recently became available for the CyberKnife system, is a direct soft tissue tracking method for respiratory motion tracking of lung lesions that eliminates invasive fiducial implantation procedures, thereby decreasing the time to treatment and eliminating the risk of pneumothorax and other fiducial placement complications. This chapter presents the concepts, methods, and some experimental results of the Xsight Lung Tracking System, which is fully integrated with the Synchrony Respiratory Tracking System. Observation and analysis of clinical image data for patients previously treated with the CyberKnife indicates that many reasonably large tumors (larger than 15 mm) located in the peripheral and apex lung regions are visible in orthogonal X-ray images acquired by the CyberKnife system. Direct tumor tracking can be performed for such visible tumors by registration of the tumor region in digitally reconstructed radiographs (DRRs), generated from the planning CT image, to the corresponding region in the treatment X-ray images. Image processing is used to enhance the visibility of the lung tumor in the DRRs and X-ray images. Experiments with an anthropomorphic motion phantom and retrospective analysis of clinical image data obtained from patients who underwent CyberKnife treatment for lung lesions using implanted fiducial markers show that the accuracy of Xsight Lung tracking is better than 1.5 mm.

26.2 Introduction

The CyberKnife Robotic Radiosurgery System provides minimally invasive radiosurgery for lung tumors [1, 2]. In this process, radioopaque fiducial markers are implanted in or near the tumor region several days to a week before CT scanning for treatment planning. The fiducials, which are detectable in X-ray images, are used as reference markers to locate and track tumor location during patient alignment and treatment delivery. The Synchrony Respiratory Tracking System builds a correlation model between the positions of periodically detected fiducials and the real-time locations of optically tracked markers placed on the chest to track tumor location [3]. The robotic manipulator then continuously moves the linear accelerator to point the radiation beams accurately at the moving target. Investigators have shown this to be an effective way to direct continuous beams of focused radiation accurately to moving lung tumors [4–6].

Despite their utility, there are several drawbacks to using fiducials. Fiducial implantation is an invasive surgical operation that carries a relatively high risk of pneumothorax. Pneumothorax is a collapse of the lung due to an abrupt change in pressure within the chest cavity. Small pneumothoraces require no treatment other than repeated observation on chest X-rays. Larger pneumothoraces may require chest tube placement. A tube is inserted into the chest wall outside the lung and air is extracted using a simple one-way valve or vacuum and a water valve device, depending on severity. This allows the lung to re-expand within the chest cavity. The pneumothorax is followed up with repeated X-rays. If the air pocket has become small enough, the vacuum drain can be clamped temporarily or removed. Because of the risk of pneumothorax [7, 8], some clinicians may be reluctant to implant fiducials. In fact, it is likely that some lung tumor patients have been excluded from CyberKnife treatment for this reason. In addition to this serious complication, fiducial implantation increases overall treatment time and reduces the patient's comfort level.

Even when fiducial implantation proceeds without complication, a favorable clinical result is not assured. Fiducials must be placed to meet specific geometric criteria in order to provide correct information for tumor location. When this does not occur targeting accuracy may be compromised. In addition, fiducials may migrate from the time of CT scanning to the time of treatment, causing inconsistent references for fiducial registration.

In order to overcome these difficulties, we have developed a direct lung tumor registration method that does not require the use of fiducials. Observation and analysis of clinical image data for patients previously treated with the CyberKnife shows that some lung tumors are visible in the orthogonal X-ray images acquired by the CyberKnife system. Such tumors are typically larger than 15 mm in each dimension and are located in the peripheral or apex regions. For these visible tumors, direct tumor tracking can be accomplished by registering or matching the image intensity pattern of the tumor region in digitally reconstructed radiographs (DRRs), which are synthetic X-ray images generated from the planning CT image, to the corresponding region in the treatment X-ray images. Image processing can be used to enhance the visibility of the lung tumor in the DRRs and X-ray images. Direct tumor tracking eliminates the need for fiducial implantation, thereby decreasing the time to treatment and eliminating the risk of pneumothorax and other complications of fiducial placement. Thus, direct tumor tracking enables fully non-invasive lung tumor radiosurgery. This approach, now available as the Xsight Lung Tracking System in the CyberKnife System, is fully integrated with the Synchrony Respiratory Tracking System to provide real-time tumor tracking of lung tumors without fiducial placement.

26.3 CyberKnife Robotic Radiosurgery System

The CyberKnife Radiosurgery System has been previously described in detail [1]. Briefly, an image-guided targeting system is combined with a mechanism for precisely delivering high-energy radiation. A miniature lightweight X-band linear accelerator provides a source of 6 MV therapeutic radiation. The linear accelerator is coupled to a multijointed robotic manipulator that is capable of accurately aiming the radiation beam with six degrees of freedom. The CyberKnife targeting system uses two diagnostic X-ray generators, which are mounted to the ceiling of the CyberKnife treatment room, to illuminate two approximately orthogonal amorphous silicon X-ray detectors that generate high-resolution digital images. The X-ray generators and detectors are rigidly fixed so that their projection camera geometry is calibrated and known with very high accuracy in the treatment room coordinate system. After an initial coarse patient alignment, performed by the operator, computer algorithms automatically compare the projection images of the target region, taken from the perspective of the two X-ray cameras, with the patient's treatment planning CT scan. Using image registration methods, the spatial coordinates of the radiosurgical target are computed and relayed to the robotic manipulator, which compensates automatically for small movements of the target by realigning the radiation beam generated by the linear accelerator.

26.4 Xsight Lung Tracking System – Overview

Thoracic anatomy is composed mainly of the spinal region, rib cage, and soft tissue including the lung and heart. The spinal region of a patient in the supine position can be considered approximately static, i.e., its position changes very little during respiration. In contrast, the soft tissue and the rib cage usually move synchronously with breathing. A lung tumor is embedded in soft tissue and therefore moves continuously during respiration with an amplitude that depends partially on the tumor location. Tumor motion may be complex, involving not only translation, but sometimes also involving some rotation and non-rigid deformation. An example of breathing anatomy dynamics is illustrated in Figure 26.1 where two slices from a 4D-CT image study at different phases in the breathing cycle are superimposed.

The general approach for the Xsight Lung Tracking System differs from previous standard tracking methods used in the CyberKnife system in that patient alignment and tumor tracking are performed in two stages rather than one. This two-stage approach is based on the observation that the spinal region is approximately static for a patient in the supine position. It is similar to an approach for fiducial-based Synchrony treatment of lung lesions reported by Wu et al. at the CyberKnife Center of Miami (see chapter by Wu et al., this volume). In their approach, the patient is globally aligned (including both position and orientation) at the beginning of the treatment using the Xsight Spine Tracking System, which directly tracks skeletal structures of the spine. Align-

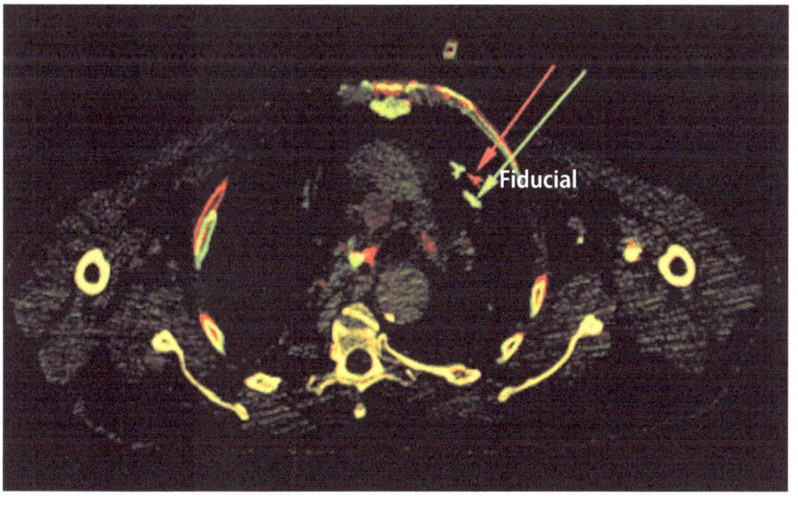

Fig. 26.1 Two superimposed image slices from a 4D-CT image study. Both images were obtained at the same scanner position but correspond to different phases in the breathing cycle. The two images, one red and one green, are window-leveled to show only bone and a metal fiducial marker that is implanted in the lung tumor. The red image corresponds to end-inhalation and the green image corresponds to end-exhalation; yellow indicates that the object is present in the same location in both images, i.e., that the object did not move between the respiratory phases. The arrows point to the fiducial marker in the two phases of the breathing cycle. The anatomy dynamics are clearly demonstrated: the spinal vertebra does not move, whereas the rib cage and tumor (fiducial) do move.

ment is performed using the spine region nearest to the lung tumor. Then the patient is moved using the treatment couch from the spine alignment center to the tumor treatment center specified in the treatment plan. Treatment delivery is performed after loading the treatment plan to do standard patient target tracking based on implanted fiducials (typically the Miami group implants a single fiducial in or near the center of the tumor) and using the Synchrony Respiratory Tracking System.

A similar approach is used for Xsight Lung Tracking, but with two fundamental differences. First, and most important, the tumor position is determined by direct tracking of the tumor rather than by tracking of fiducials implanted in the tumor. Second, the treatment planning software combines the two steps into a single plan; two CT offsets, one for global patient alignment using Xsight Spine Tracking and one for treatment, are specified and saved in the patient database. Figure 26.2 describes this workflow.

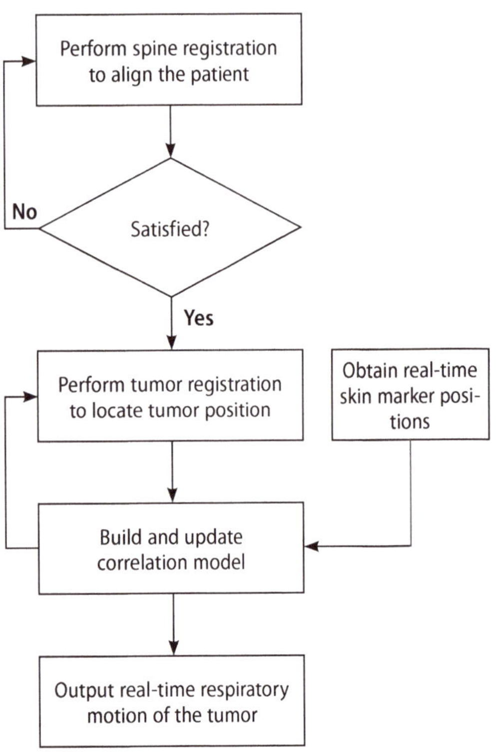

Fig. 26.2 The workflow of the Xsight Lung Tracking System.

26.5 Global Patient Alignment

The first step in Xsight Lung tracking is global patient alignment, including both position (translation) and orientation (rotation), using the region of the spine nearest the lung tumor. The patient alignment step is performed using the Xsight Spine Tracking System, a fiducial-less spine tracking method, that uses an intensity-based 2D-3D image registration algorithm to localize spinal targets by directly tracking adjacent skeletal structures, thereby eliminating the need for implanted fiducials [9, 10].

Experience has shown Xsight Spine tracking to be accurate and robust under a wide range of clinical circumstances [11, 12]. Nevertheless, two issues may substantially impair the effectiveness of this method. First, image artifacts in DRRs can be introduced by breathing during CT scanning or by structures such as ribs that have a different position relative to the spine at the time of imaging and the time of treatment. Especially in the thoracic and lumbar regions, breathing artifacts can increase registration error or, in the worst cases, make satisfactory registration difficult to achieve. Second, roll estimation can be less accurate than pitch and yaw. This is partially because in the CyberKnife imaging geometry roll corresponds to out-of-plane rotation for both X-ray images, which makes roll inherently more challenging to measure accurately than pitch and yaw. Another reason for less accurate roll estimation is that only one DRR is computed for each projection and used for image registration. The roll angle is computed from gradients in the displacement field; the gradients have small values and are sensitive to errors in the dataset. Therefore, simultaneous with the release of Xsight Lung Tracking System, we introduced an enhanced version of Xsight Spine Tracking. The two major improvements in the spine tracking method eliminate the image artifacts in the DRRs used for 2D-3D spine image registration and increase the roll estimation accuracy. We first review the techniques of the previous Xsight Spine Tracking System method, and then present the improvements in the new version.

In the CyberKnife imaging system, two orthogonal X-ray projection images are registered with two set of DRRs, respectively. A 2D-2D registration between an X-ray image and the corresponding DRR image is performed independently for each projection. The process begins by using image processing techniques to enhance skeletal features in the X-rays and DRRs and thereby improve the registration of local skeletal structure. The registration is performed using a defined region of interest (ROI). The size of the ROI is manually defined and typically includes three vertebrae. The local displacement vector that aligns a point in the DRR image with the corresponding point in the X-ray image is estimated at each node in a 9 × 9 grid or mesh laid over the ROI. The local mesh node displacements are estimated individually but constrained by displacement smoothness. Nodal displacements in the mesh form two 2D displacement fields in the two projections. Three types of parameters can be derived from the 2D displacement fields. First, the 3D positions of targets within the ROI can be calculated by interpolation of the displacement field and 2D-to-3D back projection. Second, the in-plane rotation for each projection is calculated by a least squares method in a global sense and the two in-plane rotations are converted to pitch and yaw. Finally, roll is computed from gradients in the displacement fields.

One improvement we have made to the Xsight Spine Tracking System is the removal of DRR image artifacts by segmenting the spine region. During treatment planning, the spine region is contoured in the planning CT image using model-based segmentation and subsequently saved as a spine region volume of interest (VOI) in the patient database (Fig. 26.3). The DRRs are then generated by casting rays through the CT image and restricting attenuation to voxels within the spine region VOI for both projections. Thus, the DRRs represent only spine anatomy and do not have image artifacts caused by moving soft tissue (heart and lung) and non-spinal bony anatomy such as the rib cage. An example of how DRR image quality is improved and how image artifacts are removed by using only the segmented spine region is illustrated in Figure 26.4.

Another improvement we have made to the Xsight Spine Tracking System is in the use of multiple DRRs for calculating roll. The concept of using multiple DRRs has been previously demonstrated for 6D skull tracking [10]. In the CyberKnife imaging geometry, patient pitch and yaw rotations yield in-plane rotations in the 2D X-ray projections. Changes of pitch and yaw are approximately proportional to the changes of in-plane rotations in the 2D X-ray images (the approximation is very accurate for small angles). Therefore, pitch and yaw can be directly computed from the 2D-2D registration of DRR and X-ray images. Roll, however, corresponds to an out-of-plane rotation that causes subtle changes in the X-ray images. Multiple DRRs can reflect such subtle changes and produce more

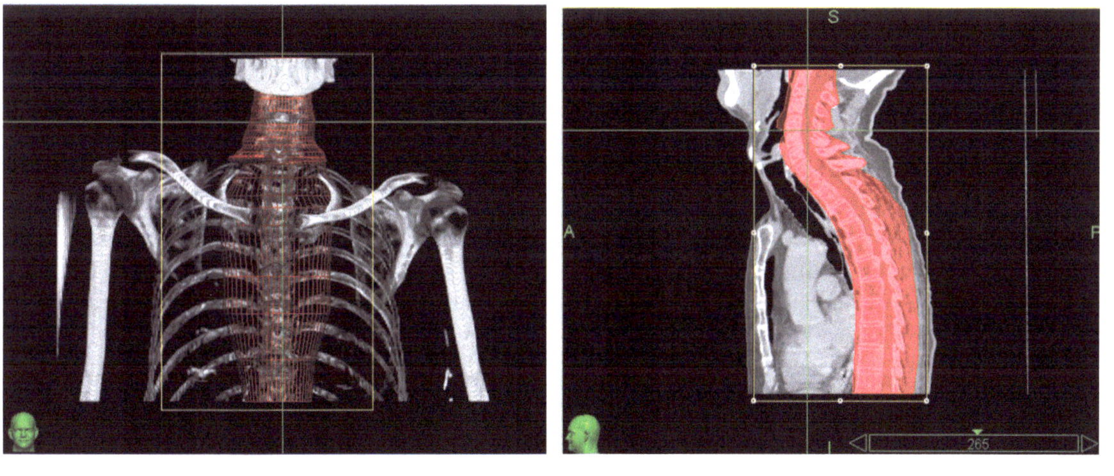

Fig. 26.3. Illustration of spine region segmentation performed in the MultiPlan treatment planning system. The spine region is contoured in the planning CT image using model-based segmentation and saved as a VOI. The spine region VOI is used during DRR generation to produce alignment DRRs that facilitate spine registration (restricting attenuation to voxels within the spine region) as well as tracking DRRs that facilitate direct tumor registration (excluding attenuation within the spine region).

accurate roll estimation than in-plane image gradients, which are sensitive to noise and artifacts in the images. Given a set of reference DRR images corresponding to various out-of-plane rotations (roll), image registration compares all of the DRRs with the X-ray image to find the DRR that best matches the X-ray image. In doing this, roll is identified as the rotation angle corresponding to the angle for the best-matched DRR. Because the rotations are expected to be small after the patient is well aligned, the roll angles are more sparsely sampled in the range of larger angles and are more densely sampled in the range of smaller angles.

26.6 Direct Tumor Tracking

After the patient is globally aligned using the Xsight Spine Tracking System, the patient is moved by the treatment couch from the spine alignment center to the tumor treatment center. The two centers are defined during treatment planning, providing the reference positions in the CT image for initial patient alignment and tumor tracking, respectively. After the patient is moved to the tumor treatment center, the tumor will be close to its reference position around which it will move as the patient breathes. For Xsight Lung tracking, we are interested in tumors that are "visible" in the X-ray images, where visible means that the tumor has an intensity pattern in the X-ray image that is distinguishable from other objects in the image. As previously described, observation and analysis of image data from a large number of patients previously treated with the CyberKnife for lung tumors shows that visible tumors are typically larger than 15 mm in each dimension and located in the peripheral or apex regions. For these visible tumors, direct tumor tracking can be performed by registration of the tumor region in the DRR to the corresponding region in the treatment X-ray images. The entire process consists of three steps: off-line operations during treatment planning, off-line DRR generation, and tumor region image registration during treatment delivery. In the following, we describe the concepts and approaches in more detail for each step.

26.6.1 Off-line Operations During Treatment Planning

Several features were added to the MultiPlan® (Accuray Incorporated, Sunnyvale, CA) treatment planning software to support Xsight Lung tracking. The new features include model-based segmentation of the spine region, the ability to define two CT centers (one for global patient alignment and one for treatment), generation of a 2D tumor silhouette in the DRRs from the 3D tumor contour in the planning CT image, and quality review of the DRR.

Several steps should be taken to achieve better global patient alignment using Xsight Spine Tracking. First, as was discussed in the previous section, the spine region is contoured in the planning CT image using model-based segmentation. Enhanced alignment DRRs for spine tracking are generated using only the spine region (Fig. 26.4). The segmented spine region is also used to improve the tracking DRRs for the Xsight Lung Tracking System, as will be discussed in the next section. Second, the CT offset for global alignment should be selected in the region of the spine nearest the lung tumor.

Contouring of the tumor is a standard operation in CyberKnife treatment planning. Several tumor contours can be defined, including the gross tumor volume (GTV) and planning tumor volume (PTV). Direct tumor tracking is performed by image registration of the tumor region in the DRRs to the corresponding region in the treatment X-ray images. Whereas the treatment plan is usually generated using the PTV, for the purpose of tumor tracking by image registration, the GTV, which represents the radiographically visible tumor volume, is preferred over the PTV in order to exclude surrounding soft tissue and skeletal structures from the registration process (the PTV typically consists of the GTV plus a margin that accounts for microscopic extension and patient setup and treatment delivery uncertainties). Additionally, the treatment CT center is automatically computed as the centroid of the tumor contour and saved in the patient database. The treatment CT center is used for two purposes: it provides the reference point to which the treatment couch is moved after global patient alignment and it defines the center of the matching windows in the DRRs for tumor registration.

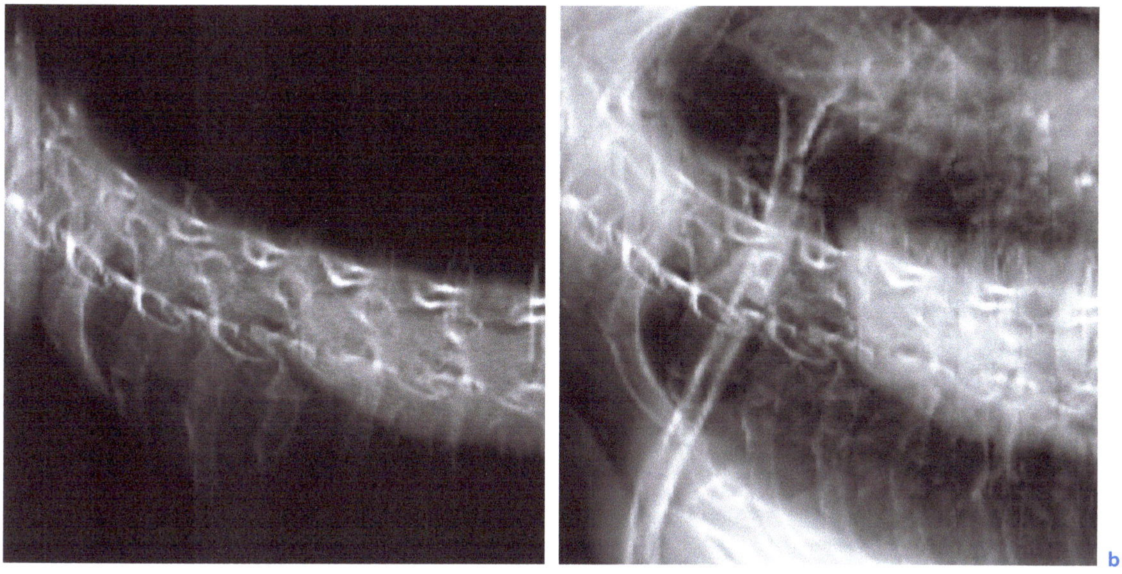

Fig. 26.4a,b. Illustration of how the alignment DRR image quality is improved by using only the segmented spine region to generate DRRs for the Xsight Spine Tracking System. The DRR generated using only the spine region (**a**) does not include attenuation of the clavicle, rib cage, and soft tissue (heart and lung), which are clearly visible in the conventionally generated DRR (**b**).

Direct tumor tracking is performed by image registration of the tumor region in the DRRs to the corresponding region in the treatment X-ray images. In order for this to be possible, the tumor must be visible in the X-ray images. The image intensity pattern of the tumor must be distinguishable from other objects in the image, which also requires the tumor to have sufficient contrast relative to the surrounding region. Two main factors that influence visibility of the tumors are their size (which influences contrast) and location (which can influence contrast if the tumor is superimposed in the X-ray image on radiodense structures such as the spine). Other factors include the size of the patient (larger patients have higher X-ray attenuation, which reduces tumor contrast) and presence of other pulmonary disease (e.g., denser lung parenchyma will reduce tumor contrast). Thus, it is important to include a patient qualification process to help determine whether a patient can be treated using the Xsight Lung Tracking System.

The MultiPlan Treatment Planning System has an added feature that provides a quality review of the tracking DRRs to help confirm patient eligibility for Xsight Lung tracking. A pair of tracking DRRs is generated from the planning CT image to simulate the X-ray images. Unlike other DRRs in the treatment planning system, these DRRs have the same high quality as the actual tracking DRRs. Figure 26.5 shows an example of the quality review step. In this example, the tumor is clearly in the middle part of both DRRs. The 2D tumor silhouettes can optionally be overlaid on the images to assist the visual assessment, as is shown in Figure 26.5 (the 2D tumor silhouette in a DRR is generated from the 3D tumor contour in the planning CT image by rendering the tumor VOI as a solid binary object and determining the silhouette edge of the projected 2D object). If the image intensity pattern of the tumor is not clearly distinguishable, the patient is not a good candidate for treatment with the Xsight Lung Tracking System. In this case, fiducials could be considered for tracking.

26.6.2 Off-line DRR Generation

A DRR is a synthetic X-ray image that is generated by computationally casting rays through the CT image according to the X-ray imaging system geometry and modeling of the X-ray imaging process as a linear attenuation of X-rays as they pass through the object. The CyberKnife system em-

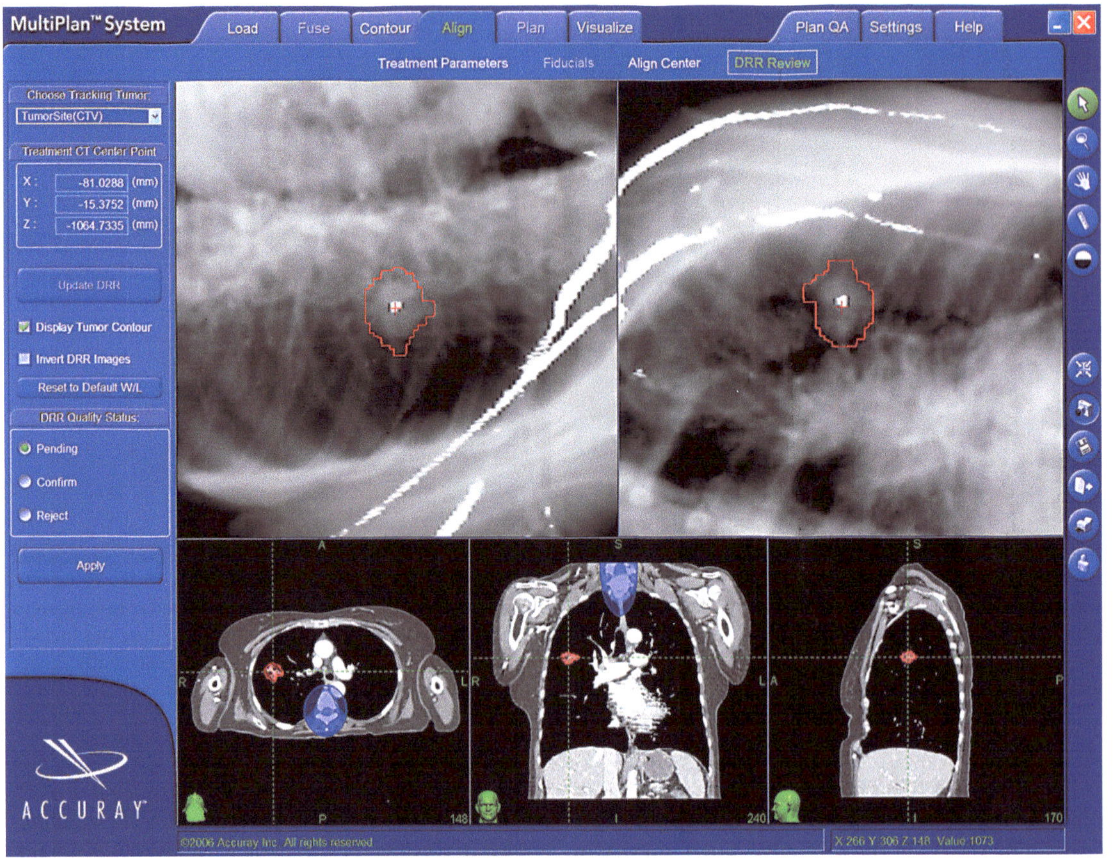

Fig. 26.5 Review of tracking DRRs in the MultiPlan Treatment Planning System. This step helps determine whether a patient can be treated using the Xsight Lung Tracking System. The top pair of images are tracking DRRs generated from the planning CT image to simulate the treatment X-ray images. The 2D tumor silhouettes (red contours) can optionally be overlaid on the DRRs to assist the visual assessment. The silhouettes are generated from the 3D tumor contour in the planning CT image (red VOI in bottom row of images).

ploys a variety of image enhancement techniques to optimize the DRR images for different clinical applications. In 6D skull tracking, the original CT image values are utilized without modification. In Xsight Spine tracking, a pre-processing operation is first employed to modify the X-ray attenuation map in CT before generating the DRR projections, and then two post-processing image filters are applied to the resulting DRR image. The purpose is to enhance spinal structures and suppress soft tissue for better spine image registration. In Xsight Lung tracking, the target to be registered is the area of tumor. In the lung, tumor intensity in CT and X-ray images is substantially greater than in surrounding tissue, but less than the intensity of bony structures such as spine and ribs. To optimize tumor visibility in DRRs, two operations are performed. First, the spine VOI is excluded in the ray-casting process during DRR generation. Second, a gamma correction is applied to emphasize the tumor contrast relative to the background of lung parenchyma. A gamma correction is also performed on the X-ray images just before registration. An example of an enhanced tracking DRR for lung tumor tracking is shown in Figure 26.6 and compared with a conventionally generated DRR. Since attenuation within the spine region VOI is excluded during generation of the tracking DRRs, it is important that the user-defined spine region VOI not include any portion of the tumor volume.

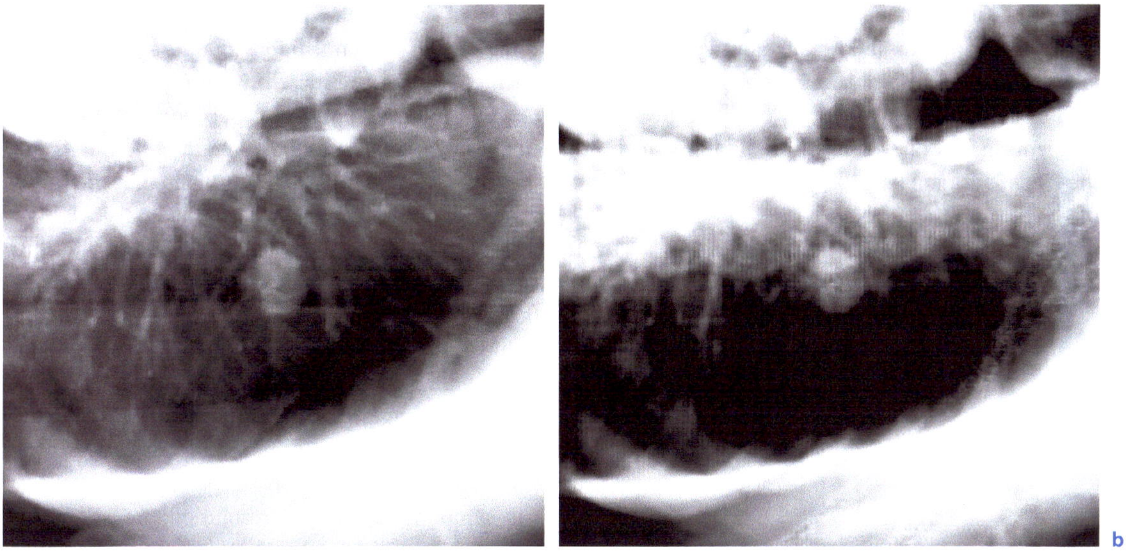

a b

Fig. 26.6a,b. Illustration of how tracking DRR image quality is improved by using the segmented spine region during DRR generation. The lung tumor is clearly visible and distinguishable from surrounding anatomy in the DRR on the left (**a**), which was generated by excluding attenuation within the spine region and by applying a gamma correction to emphasize the tumor contrast relative to the background of lung parenchyma. The right image (**b**) is a conventionally generated DRR, which includes attenuation within the spine region.

26.6.3 Tumor Tracking During Treatment Delivery

Tumor tracking is performed by image registration of the DRRs to the treatment X-ray images. Specifically, the image intensity pattern of the tumor region in the DRR is matched to the most similar region in the X-ray image. A "matching window" for the tumor is defined based on the 2D tumor silhouette in each projection. Specifically, the matching window is the smallest rectangle that encompasses the entire tumor. Since the tumor is not generally rectangular, the matching window will include some of the surrounding region. As mentioned earlier, it is preferred to use the tumor silhouette based on the GTV, which represents the radiographically visible tumor volume, in order to exclude as much surrounding soft tissue and bony anatomy as possible.

A translation-only model is assumed for registration of the matching window. Although some rotation or non-rigid deformation of the tumor may occur, our experience is that this approach still provides an accurate estimation of translation, especially for the central part of the tumor region. Also, the CyberKnife treatment beam only accounts for translation during respiratory motion tracking. The registration process is conducted separately for each projection, resulting in 2D translations for each projection. The 3D tumor translation is determined by 2D-to-3D backprojection of the 2D translations.

The image registration process involves a block matching search in which the matching window containing the tumor region from the DRR is moved throughout a search window in the treatment X-ray image until the region in the X-ray image most similar to the matching window from the DRR is found. An image similarity measure called pattern intensity [13] is used to determine how similar a region in the X-ray image is to the matching window.

The pattern intensity similarity measure examines the contents of a difference image. Thus, the difference image between X-ray and DRR images is first formed:

$$I_{diff}(i,j) = I_{Xray}(i,j) - I_{DRR}(i,j)$$

A pixel is considered to belong to a pattern or structure if it has a substantially different image intensity value from its neighboring pixels. Pattern intensity is based on the assumption that the number of patterns or structures in the difference image will be minimal when the optimal alignment has been found. The mathematical formulation of pattern intensity is described as follows. The similarity measure operates on the difference image and is expressed as the summation of functions of difference image gradients in four directions (horizontal, vertical, and the two 45° diagonals):

$$S = \sum_{i,j} \frac{\sigma^2}{\sigma^2 + \left(I_{diff}(i,j) - I_{diff}(i,j-1)\right)^2} + \sum_{i,j} \frac{\sigma^2}{\sigma^2 + \left(I_{diff}(i,j) - I_{diff}(i-1,j)\right)^2} + \sum_{i,j} \frac{\sigma^2}{\sigma^2 + \left(I_{diff}(i,j) - I_{diff}(i-1,j-1)\right)^2} + \sum_{i,j} \frac{\sigma^2}{\sigma^2 + \left(I_{diff}(i,j) - I_{diff}(i-1,j+1)\right)^2}$$

Each of these terms has a maximum value of 1 if a pixel and its neighbor have the same intensity value. The value of a term decreases asymptotically to 0 as the difference between the intensity values of a pixel and its neighbor increases. In this formulation, σ is a weighting constant. As the value of σ increases, the registration process becomes more robust, but when the value of σ is too large, small details in the images will not be reflected in the similarity measure. We evaluated the effect of σ on the matching and found that any value within the range 4 to 32 gives nearly identical results; a value of 20 is currently used.

Pattern intensity has several advantages as a similarity measure. First, the measure is insensitive to low frequency information. This is because the measure is computed on the difference image, which basically removes the uniform image content while retaining the high frequency information that consists of image features such as lines and corners. This characteristic makes the algorithm robust against image intensity differences between live and DRR images. Second, the terms in the similarity measure asymptotically approach zero as the gradient in the difference image increases. Thus, large intensity differences such as those caused by image artifacts have the same effects on the similarity measure regardless of their magnitude. This feature makes pattern intensity relatively insensitive to image artifacts.

The image registration process is illustrated in Figure 26.7 for one of the two projections. The bottom image in Figure 26.7 shows the value of the pattern intensity similarity measure for all searched blocks in the search window. The block in the X-ray image most similar (highest value of the similarity measure) to the matching window corresponds to the tumor location. All other blocks in the X-ray image have smaller similarity values. Visual assessment of the tracking result confirms that the tumor was correctly identified in the X-ray image.

Multi-resolution (also known as hierarchical or coarse-to-fine) image matching is employed as an efficient way to achieve robust and fast computation. Four image resolutions are defined. The original DRR and X-ray images are used as the second highest resolution images. The highest resolution is obtained by up-sampling the original image in each dimension. The lower resolution images are generated by sub-sampling their higher resolution images (averaging 2 × 2 blocks). The basic idea of multi-resolution matching is to match the images at each level successively, starting with the lowest resolution. The results at each resolution are used as initial values for the next higher resolution. The results are refined using the higher resolution images. In the final matching results, the accuracy of the translation depends on the spatial resolution of the highest resolution images.

In Figure 26.7, the correct match corresponds to the global maximum of the similarity measure. Some local maxima, however, also exist. Because of the possibility of an incorrect tracking result, it is very important for the user to perform visual assessment of the result (Fig. 26.8). It is also helpful to compute and display a quantitative quality measure that provides an indication of the tracking result correctness. The quality measure concept used in Xsight Lung tracking is based on the tumor region matching method. The matching window for the image registration process contains the tumor and some surrounding background. If the matching window is shifted slightly in any direction, the new window still includes most of the tumor plus some additional background. We call this new matching window a "shifted matching window." Registration performed on the original matching window and on the shifted matching window should give the same or similar result. If the results are different, this indicates an increased likelihood of an incorrect match. This can happen when there are local maxima similar in value to the global maximum, which reflects that the intensity pattern of the tumor is similar to other regions in the image. Based on

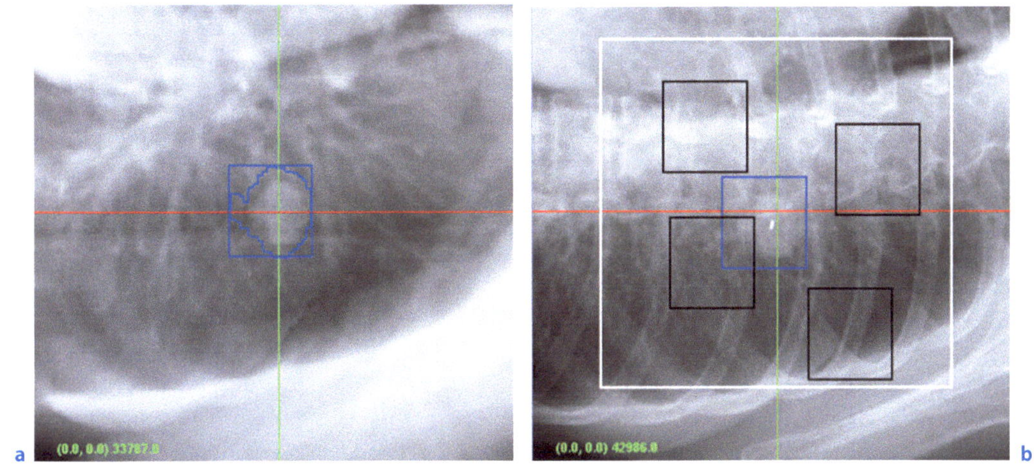

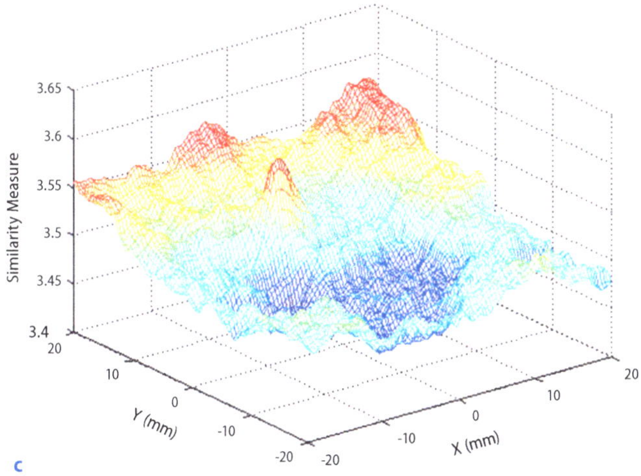

Fig. 26.7a–c. Illustration of direct tumor tracking. The image registration process involves a block matching search in which the matching window (*blue square* in **a**) containing the tumor region (*blue contour* inside *blue square*) from the DRR (**a**) is moved throughout a user-defined search window (*white square*) in the treatment X-ray image (**b**) to find the region in the X-ray image most similar to the matching window from the DRR (*blue square* in **b**). The matching window is the minimal encompassing rectangle that contains the entire tumor. The bottom image (**c**) shows the value of the image similarity measure, which is pattern intensity, for all searched blocks in the search window. The block in the X-ray image most similar (highest value of the similarity measure) to the matching window corresponds to the tumor location (*blue square* in **b**). All other blocks in the X-ray image (*black squares*) have smaller similarity values.

this concept, the quality measure is computed in the following manner. 1) Define multiple shifted matching windows, each offset from the nominal tumor location (original matching window) by a different amount. 2) Define a small search window centered at the position of the tumor detected in the X-ray image by the original matching window. 3) Repeat registration for each shifted matching window, and calculate the differences from the original translation. 4) Determine detection confidence, defined as the percentage of shifted matching windows with small differences from the original translation. If the detection confidence is larger than a threshold (i.e., if differences between the translations produced using the shifted matching windows and the original translations are small), it is more likely that the registration algorithm succeeded in computing the correct tracking result. Otherwise, the registration is judged to fail and no result is provided to the user. Figures 26.9–26.11 illustrate the process of computing and interpreting the quality measure.

26.6.4 Integration with the Synchrony Respiratory Tracking System

The Synchrony Respiratory Tracking System is a subsystem of the CyberKnife system that allows the treatment of targets that move due to respiration. The alignment of each treatment beam with the moving target is maintained in real time by moving the beam dynamically with the target. An advantage of the Synchrony system is that patients can breathe normally during treatment while the robotic manipulator moves the linear accelerator

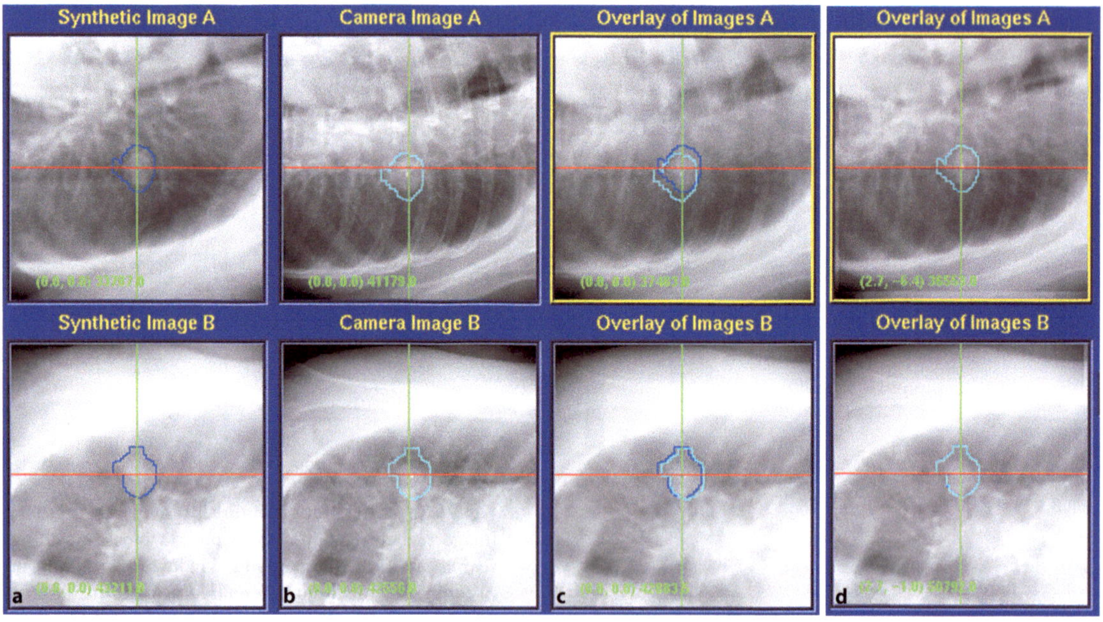

Fig. 26.8a–d. Illustration of the Xsight Lung Tracking System display created by retrospective evaluation of a patient who underwent CyberKnife treatment for a lung tumor using implanted fiducial markers. The two columns on the left (**a, b**) show the tracking DRRs (labeled Synthetic Image) and treatment X-ray images (labeled Camera Image), respectively. The 2D tumor silhouette is shown in dark blue on the DRRs and light blue on the X-ray images. The third column (**c**) shows a combination of the DRRs and X-ray images before tumor alignment. The images are combined by adding the corresponding pixel values. The right column (**d**) shows the combination of the DRRs and X-ray images after tumor alignment. It is important for the user to perform visual assessment of the tumor tracking result. The tumor silhouettes, which can be removed, can help with visual assessment. The two rows represent each of the two X-ray image projections.

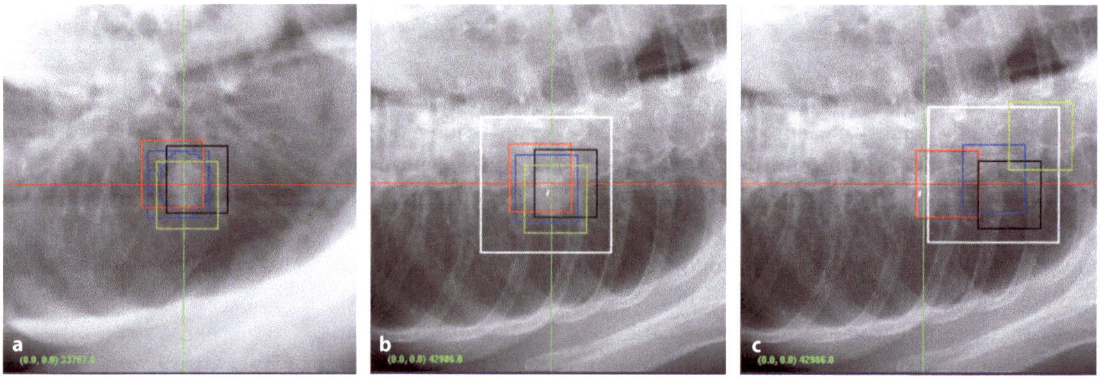

Fig. 26.9a–c. A quality measure for correctness of direct tumor tracking. The matching window for the image registration process contains the tumor and some surrounding background. Many shifted matching windows are defined by slightly translating the original matching window in the DRR (**a**). The shifted matching windows contain most of the tumor plus some additional background. It is more likely that the matching result is correct if registrations performed on the original and shifted matching windows give the same or similar results. The X-ray image (**b**) represents a correct tracking result. The shifted matching windows in the DRR registered to the X-ray image have a location pattern similar to that in the DRR. The X-ray image (**c**) represents an incorrect tracking result. The shifted matching windows in the DRR registered to the X-ray image have an inconsistent location pattern relative to the DRR.

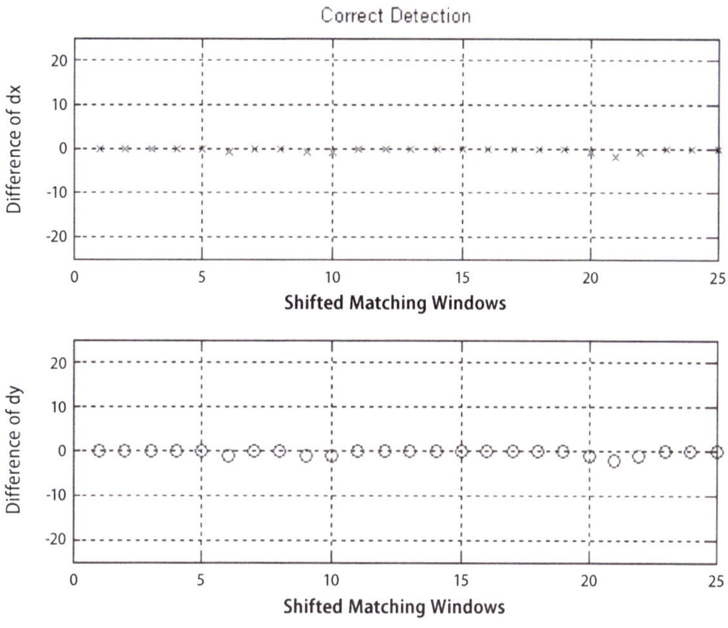

Fig. 26.10 Example of a correct detection. The differences between the translations produced using 25 shifted matching windows and the original translation are small, which indicates a high confidence in the correctness of the tracking result.

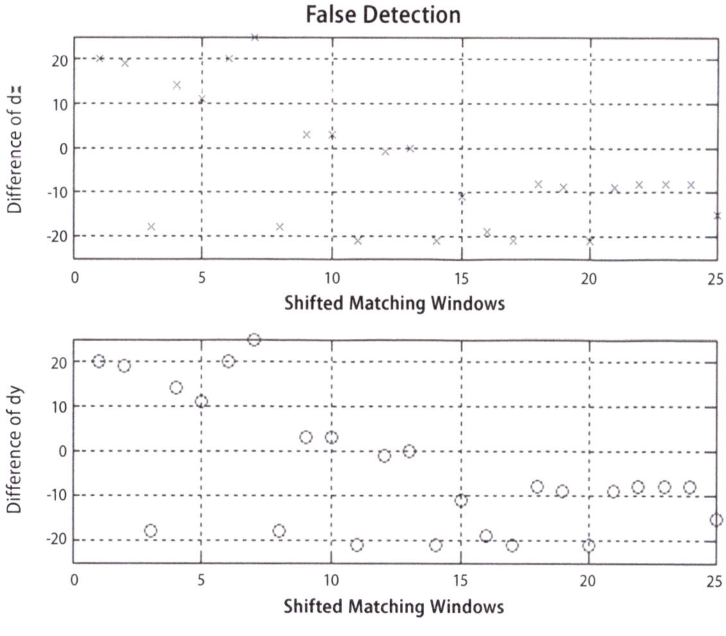

Fig. 26.11 Example of an incorrect detection. The translations produced using 25 shifted matching windows are inconsistent and different from the translation obtained using the original matching window, which indicates a low confidence in the correctness of the tracking result.

dynamically. The primary concept in the Synchrony System is a correlation model between internal tumor position and external marker position. The position of external optical markers, which are attached with Velcro to a snugly fitting vest that the patient wears during treatment, are measured continuously with a stereo camera system. At the start of treatment, the internal tumor position is measured at discrete time points by acquiring orthogonal X-ray images. A linear or quadratic correlation model is generated by fitting the 3D internal tumor positions at different phases of the breathing cycle to the simultaneous external marker positions. During treatment, the internal tumor position is estimated from the external marker positions using the correlation model, and this information is used to move the linear accelerator dynamically with the target. The model is checked and updated regularly during treatment by acquiring additional X-ray images. The internal tumor position has previously been determined as the centroid of fiducial markers implanted in and around the tumor. With the Xsight Lung Tracking System, the Synchrony System works identically as with fiducial markers, except that the internal tumor position is provided by the direct tumor tracking result from the tracking software rather than from fiducial localization software.

26.7 Experimental Results

26.7.1 Phantom Experiments

The Xsight Lung Tracking quality assurance (QA) motion phantom simulates simple respiratory motion of a lung tumor and provides radiochromic dosimetry film-based test capability at locations inside the phantom corresponding to a typical lung tumor (Fig. 26.12). The motion actuator can translate a rod in and out of the thorax, simulating superior-inferior tumor motion, while moving the tracking platform up and down, representing anterior-posterior chest motion. The thorax contains radiographically equivalent lung, spine, and ribs surrounded by soft tissue. The lung material has a Hounsfield CT number of approximately –800, representing the average lung density throughout respiration. The ribs and spine have an anthropomorphic shape and are made of inhomogeneous material to represent trabecular and cortical bones. A sphere (1 inch diameter) in the film cube represents a lung tumor. Two orthogonal films oriented in the axial and sagittal planes are inserted in the film cube. Using this film cube, one can measure beam targeting accuracy.

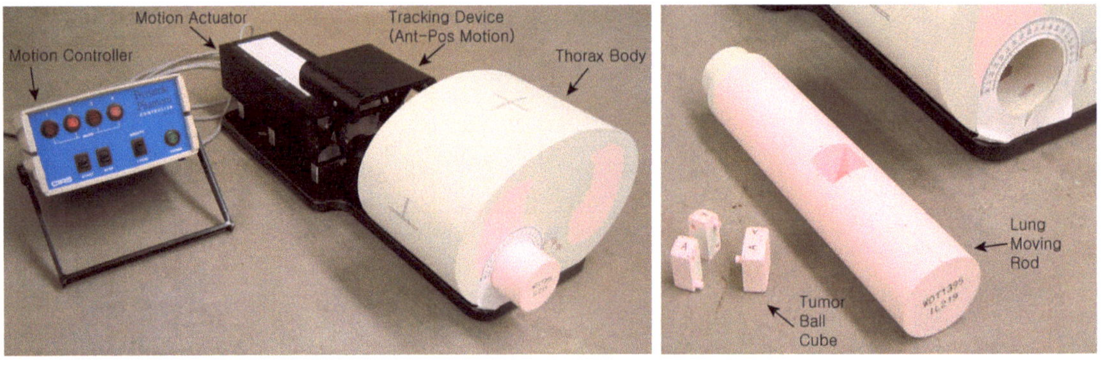

Fig. 26.12. The Xsight Lung Tracking quality assurance (QA) motion phantom and its constituent parts. The motion controller sends a signal to the motion actuator, which translates a rod containing a spherical target (tumor ball cube) in and out of the thorax, simulating superior-inferior tumor motion, while moving the tracking platform up and down, representing anterior-posterior chest motion. The thorax contains radiographically equivalent lung, spine, and ribs surrounded by soft tissue. Two orthogonal radiochromic dosimetry films oriented in the axial and sagittal planes are inserted in the film cube. The total system error is calculated as the distance between the centroid of the sphere and the centroid of the isocontour in the exposed film dose distribution corresponding to the surface of the sphere (i.e., the distance between the centroids of the planned and delivered dose distributions).

Film is loaded in the low-density ball cube and inserted into the cavity in the moving rod. Next, the moving rod is inserted into the thorax body and attached to the motor. A CT scan of the phantom is acquired while the moving rod is stationary and in its starting position. A treatment plan is generated in MultiPlan for Xsight lung tracking delivery using the 15-mm collimator. At the time of treatment delivery, the phantom is aligned on the treatment couch using Xsight Spine Tracking. Then the Xsight Lung Tracking System is used together with the Synchrony Respiratory Motion Tracking System to model and track the tumor movement during treatment (phantom irradiation). The tumor is moved in a periodic breathing cycle 30 mm in the superior-inferior direction within the thorax body. After delivering the radiation, the films are removed from the cube, the orientation of the films is labeled, and the films are scanned in a calibrated flat-bed optical scanner. Finally, the accuracy of Xsight lung tracking is analyzed using Accuray's "end-to-end test" software. The total system error is calculated as the distance between the centroid of the sphere and the centroid of the isocontour in the exposed film dose distribution corresponding to the surface of the sphere (i.e., the distance between the centroids of the planned and delivered dose distributions). Importantly, this value measures all possible errors intrinsic to the relatively complex process of image-guided robotic radiosurgery, including those originating with the CT scanner, the treatment planning system, the robot, the linear accelerator, and the tracking system.

Three tests were performed: one using fiducial tracking without tumor motion, one using lung tracking without tumor motion, and one using lung tracking with tumor motion (30 mm superior-inferior motion). The stationary fiducial test serves as a reference baseline result. The two lung tracking test results represent the total system targeting error using the Xsight Lung Tracking System. The results are listed in Table 26.1. The total system error for the Xsight Lung Tracking System with a moving target was 1.07 mm.

Table 26.1 Total system error of the CyberKnife Radiosurgery System using the Xsight Lung Tracking System measured by using an anthropomorphic lung phantom and a ball-cube film holder. Included are the components of error along the three coordinate axes (SI, superior-inferior; LR, left-right; and AP, anterior-posterior).

Tracking method	Motion	Error components (mm)			Total system error (mm)
		SI	LR	AP	
Fiducial	Static	0.34	0.31	0.71	0.84
Xsight Lung	Static	0.33	0.40	0.68	0.86
Xsight Lung	30 mm SI	0.62	0.29	0.82	1.07

26.7.2 Clinical Image Data Evaluation

The fundamental process in the Xsight Lung Tracking System is image registration. The tracking system error component of total system error is equivalent to the tumor target registration error. The Xsight Lung Tracking System error was computed by retrospectively analyzing clinical image data obtained from ten patients who underwent CyberKnife treatment for lung lesions using implanted fiducial markers. All tumors were larger than 15 mm in each direction and were located in the peripheral or apex lung regions. All tumors were visible and clearly distinguishable from other objects in the image, which was assessed by visual inspection of the tracking DRRs in the quality review step of the treatment planning system. When the patients were treated, the fiducials were used as a geometric frame of reference for targeting the lung tumor. In the retrospective analysis, the fiducial-based tracking result was used as a reference gold standard transformation to which the intensity-based (fiducial-less) registration performed by the lung tracking software was compared. The fiducials were removed from the CT scans, and thus the DRR images, using image processing techniques ("air brushing") before performing the registration to more realistically simulate the fiducial-less situation and avoid bias in the evaluation of lung tracking system error. The fiducials were not removed from the treatment X-ray images. The example shown in Figure 26.7 is from one of these ten patients. The fiducial is visible in the treatment X-ray image but not the DRR.

For each patient, the following information was obtained and used to perform the analysis: 1) the pre-treatment CT image used in treatment planning; 2) the camera calibration model and parameters for the two orthogonal X-ray imaging systems; 3) ten randomly selected pairs of X-ray images acquired during treatment; and 4) the 3D positions of the fiducial markers in the CT image. The fiducial positions were obtained from the log file generated during treatment. The lung tracking system error (tumor target registration error) was computed as the difference between the fiducial-based tracking result (translational displacement) and the lung tracking system result. The results are listed in Table 26.2. The error components and total tracking system error for each patient are the root-mean-square (RMS) average of the ten individual results (one for each of the ten randomly selected X-ray image pairs). All error components were less than 1 mm. The total lung tracking system error ranged from 0.69 mm to 1.31 mm with a mean of 1.06 mm. It is important to note that these error measurements are differences relative to fiducial-based tracking results. Fiducial tracking may not be a perfect reference gold standard for determining the tumor target registration error of the lung tracking system.

Table 26.2 Tumor target registration error, which is the lung tracking system error component of total system error, computed from clinical image data obtained from ten patients who underwent CyberKnife treatment for lung lesions using implanted fiducial markers. The error components and total tracking system error for each patient are the root-mean-square (RMS) average of ten individual results (one for each of ten randomly selected X-ray image pairs). SI, superior-inferior; LR, left-right; AP, anterior-posterior; SD, standard deviation.

Patient	Error components (mm)			Total tracking system error (mm)
	SI	LR	AP	
1	0.50	0.77	0.87	1.27
2	0.57	0.88	0.43	1.14
3	0.61	0.25	0.21	0.69
4	0.46	0.36	0.72	0.93
5	0.15	0.65	0.67	0.95
6	0.39	0.87	0.91	1.31
7	0.43	0.67	0.58	0.98
8	0.76	0.61	0.88	1.31
9	0.66	0.63	0.29	0.96
10	0.46	0.53	0.75	1.09
Mean				1.06
SD				0.19

26.8 Conclusion

The Xsight Lung Tracking System performs direct lung tumor tracking by registration of the tumor region in DRRs generated from the planning CT image to the corresponding region in the treatment X-ray images. By eliminating the need for implanted fiducials, lung tracking can greatly simplify the process of performing lung radiosurgery. Despite its technical simplicity and generally good performance, fiducial-based tracking for lung radiosurgery has numerous drawbacks. The primary shortcoming is that an initial procedure is required. Surgical implantation is both modestly painful and has the potential for complications, including a risk of pneumothorax. The additional fiducial implantation step makes the process more time consuming and complex and can complicate patient scheduling. The time, resources, and costs associated with surgical insertion of fiducials are not inconsequential. Fiducials can migrate, a risk that is often minimized by waiting several days to a week after implantation before acquiring the planning CT image. By obviating the need for implanted markers, the Xsight Lung tracking system eliminates these problems.

The accuracy of the Xsight Lung Tracking System was evaluated using an anthropomorphic lung motion phantom. The total system error (the distance between the centroids of the planned and delivered dose distributions) represents all possible errors in the treatment planning and delivery process, including error in the tracking system, the CT scanner, the treatment planning software, the robot, and the linear accelerator. The total system error was measured to be 1.07 mm. However, it is well known that the measurements in phantoms generally underestimate the actual application accuracy in the clinical environment. With the CyberKnife System concept, the component of total system error that is likely to differ most between the experimental and clinical situations is the tracking system error. To assess the extent of this inaccuracy, retrospective analysis was performed using clinical image data from ten patients who underwent CyberKnife treatment for lung lesions using implanted fiducial markers. The tracking system error was found to be 1.06 ± 0.19 mm (mean ± standard deviation). Although this measure does not include the other possible errors in the treatment delivery process, it is reasonable to assume that the other errors in a treatment delivery are similar in the experimental and clinical situations. In the worse case, in which all errors measured in the phantom studies are due to non-tracking system errors, the experimental (non-tracking) and clinical (tracking) errors add in quadrature and the clinical total system error would be 1.5 mm (square root of the sum of the squares of experimental error, 1.07 mm, and clinical error, 1.06 mm).

The Xsight Lung Tracking System is fully integrated with the Synchrony Respiratory Tracking System. Thus the alignment of each treatment beam with the moving target is maintained in real time by moving the beam dynamically with the target. This feature allows patients to breathe normally without the use of complex gating or difficult breath-holding techniques.

The primary limitation of the Xsight Lung Tracking System is that it is not, at present, appropriate for every lung tumor patient. Direct tumor tracking is performed by image registration of the tumor region in the DRRs to the corresponding region in the treatment X-ray images. The image registration process requires that the tumor be visible in the X-ray images. The image intensity pattern of the tumor must be distinguishable from other objects in the image, which also requires the tumor to have sufficient contrast relative to the surrounding region. Two main factors that influence visibility of a tumor are its size (which influences contrast) and location (which can influence contrast if the tumor is superimposed in the X-ray image on radiodense structures such as the spine). Other factors include the size of the patient (larger patients have high X-ray attenuation, which reduces tumor contrast) and the presence of other pulmonary disease (e.g., denser lung parenchyma will reduce tumor contrast). Observation and analysis of clinical image data for more than 50 patients previously treated with the CyberKnife System indicates that many reasonably large tumors (larger than 15 mm) located in the peripheral and apex lung regions are visible in the orthogonal X-ray images acquired by the CyberKnife system. But these are just guidelines. This analysis suggests that approximately 20–40% of CyberKnife lung patients are appropriate for treatment with the Xsight Lung Tracking System. The MultiPlan Treatment Planning System has an added feature that provides a quality review of the tracking DRRs to help confirm patient eligibility for lung tracking. If the image intensity pattern of the tumor is not clearly distinguishable, the patient is not a good candidate for treatment with lung tracking. Also, during treatment delivery, it is important for the user to perform visual assessment of the tracking result. Tumor silhouettes are provided, which can help with visual assessment. The system also provides a quality measure that is a quantitative indication of the likelihood that the registration algorithm succeeded in computing the correct tracking result.

References

1. Adler JR, Jr., Murphy MJ, Chang SD, *et al.* Image-guided robotic radiosurgery. *Neurosurgery* 1999; 44:1299–1306; discussion 1306-1297.
2. Bhatnagar AK, Gerszten PC, Ozhasaglu C, *et al.* CyberKnife frameless radiosurgery for the treatment of extracranial benign tumors. *Technol Cancer Res Treat* 2005; 4:571–576.
3. Schweikard A, Glosser G, Bodduluri M, *et al.* Robotic motion compensation for respiratory movement during radiosurgery. *Comput Aided Surg* 2000; 5:263–277.
4. Le QT, Loo BW, Ho A, *et al.* Results of a phase I dose-escalation study using single-fraction stereotactic radiotherapy for lung tumors. *J Thorac Oncol* 2006; 1:802–809.
5. Nuyttens JJ, Prevost JB, Praag J, *et al.* Lung tumor tracking during stereotactic radiotherapy treatment with the CyberKnife: Marker placement and early results. *Acta Oncol* 2006; 45:961–965.
6. Whyte RI, Crownover R, Murphy MJ, *et al.* Stereotactic radiosurgery for lung tumors: preliminary report of a phase I trial. *Ann Thorac Surg* 2003; 75:1097–1101.
7. Pishvaian AC, Collins B, Gagnon G, *et al.* EUS-guided fiducial placement for CyberKnife radiotherapy of mediastinal and abdominal malignancies. *Gastrointest Endosc* 2006; 64:412–417.
8. Reichner CA, Collins BT, Gagnon GJ, *et al.* The placement of gold fiducials for CyberKnife stereotactic radiosurgery using a modified transbronchial needle aspiration technique. *J Bronchol* 2005; 12:193–195.
9. Fu D, Kuduvalli G, Maurer CR, *et al.* 3D target localization using 2D local displacements of skeletal structures

in orthogonal X-ray images for image-guided spinal radiosurgery. *International Journal of Computer Assisted Radiology and Surgery (Suppl 1)* 2006:198–200.
10. Fu D, Kuduvalli G, Mitrovic V, *et al.* Automated skull tracking for the CyberKnife image-guided radiosurgery system. In: Reinhardt JM, Pluim JP, editors. SPIE Medical Imaging: Image Processing. Vol 5744. San Diego, CA: SPIE; 2005. pp. 366–377.
11. Ho AK, Fu D, Cotrutz C, *et al.* A study of the accuracy of Cyberknife spinal radiosurgery using skeletal structure tracking. *Neurosurgery* 2007; 60:147–156.
12. Muacevic A, Staehler M, Drexler C, *et al.* Technical description, phantom accuracy, and clinical feasibility for fiducial-free frameless real-time image-guided spinal radiosurgery. *J Neurosurg Spine* 2006; 5:303–312.
13. Penney GP, Weese J, Little JA, *et al.* A comparison of similarity measures for use in 2-D-3-D medical image registration. *IEEE Trans Med Imaging* 1998; 17:586–595.

CyberKnife Radiosurgery: Quality of Life and Cost Considerations

27

Donald B. Fuller

CONTENTS

27.1 Introduction

Minimally invasive treatment alternatives, such as the CyberKnife® (Accuray Incorporated, Sunnyvale, CA), are becoming increasingly important. The benefits of minimally invasive treatments are not limited to improvements in efficacy and safety. Thorough evaluation of these treatment modalities also requires in-depth examinations of financial costs and effects on the patients' quality of life. Appreciation of the CyberKnife's viability as a treatment option for thoracic malignancies such as early-stage non-small cell lung cancer (NSCLC) will depend heavily on such analyses. This chapter will focus on the steps involved in undertaking the difficult challenge of assessing costs and quality of life outcomes for CyberKnife treatment of early-stage NSCLC.

Well-designed and valid cost and quality of life analyses require the identification of the current treatment modalities for a pathology followed by positioning of the treatment under consideration within established treatment algorithms. Within this structure, the costs of pre-operative evaluation, treatment and follow-up, and treatment-related morbidity are all assessed. Quality of life, compared across the different treatments, is assessed concurrently using validated tools. Efficacy of treatment is a key factor in these analyses because a less effective treatment will invariably lead to a higher percentage of recurrences and re-treatments, which all result in a higher risk of additional morbidity, increased costs, and detrimental effects on quality of life. Throughout this book authors have described technology and methods that have yielded promising short-term results for a variety of extracranial lesions that had not been treated radiosurgically until very recently. In this chapter, we present a plan for taking the next necessary step to determine if a safe and efficacious treatment will also become the preferred one. We should stress that the analyses are, of necessity at this stage, speculative and preliminary. But they point clearly to studies that are required and may be incorporated into prospective clinical studies of CyberKnife radiosurgery for lung cancer.

27.2 General Considerations

Stage Ia and Ib NSCLC is primarily treated by surgical resection because of the documented survival benefit; 5-year survival for surgical resection approaches 60–80% in this population [1–4]. Although this treatment path is clear cut, many patients are deemed medically inoperable because of associated co-morbidities. For these patients, generally referred to as "surgically curable but medically inoperable", alternative treatments are primary conventional radiation therapy and chemotherapy, which yield 5-year survival rates of only 20–40% [5–7]. Clinical research over the last decade has shown that these patients are an appropriate population for locally ablative non-surgical therapy such as CyberKnife radiosurgery or other forms of radiotherapy, including focal conventional conformal radiotherapy (3D-CRT), intensity modulated radiotherapy (IMRT) with or without daily image guidance (IGRT), or hypofractionated stereotactic body radiotherapy (SBRT). Studies of SBRT suggest a potential efficacy similar to resection [8–14], though compared with surgical literature SBRT patient numbers are limited and median follow-up is shorter. Several of these studies have suggested local control rates in excess of 90% with 1–5 years of follow-up [8, 9, 14].

CyberKnife radiosurgery may be conceived as a more precise method of SBRT delivery. It is the only radiation delivery system currently capable of achieving radiosurgical precision in the thorax, made possible by the unique combination of inverse-planned noncoplanar dose painting, delivered with sub-millimeter accuracy, and Synchrony® (Accuray Incorporated, Sunnyvale, CA) respiratory tracking that allows the targeting beams to move in tandem with respiration [15, 16]. This enables more focal planning target volume (PTV) margins, allowing sufficient restriction of the high-dose region so that a biologically ablative dose falls only within extreme close proximity to the PTV, even for lesions that are near critical structures. The limited, very early clinical data with this method indicate a high local control rate and relatively few serious complications [15, 17, 18].

Compared with resection, there are no specific medical contraindications to CyberKnife radiosurgery, such that even "medically frail" or elderly patients may receive this treatment. Medically inoperable patients are less likely to withstand many of the morbidities associated with conventional radiation as well, such as radiation pneumonitis secondary to irradiating large volumes of normal lung tissue. CyberKnife offers an option for these patients by avoiding the risks associated with resection and conventional radiation of the involved pulmonary lobe or segment in medically frail patients. At the same time, CyberKnife treatment may offer quality of life advantages, including the preservation of more normal lung tissue (which translates to better respiratory function), a shortened recovery period, and absence of pain. In spite of its technical complexity and expense to operate, CyberKnife radiosurgery may also be more cost-effective because it is an outpatient procedure, which avoids expensive line items for resection such as hospitalization, operating room use, and post-operative recovery that normally includes a stay in the intensive care unit (ICU).

27.3 Morbidity and Mortality Risk Assessment

27.3.1 Surgical Resection

Surgical resection, whether a lobectomy or a limited wedge resection, carries risks of post-operative morbidity and mortality. Handy et al. published data on a group of 139 NSCLC patients treated with surgical resection and found an 8% occurrence of operative mortality and significant post-operative morbidity [19]. In fact, post-operatively, 18% of patients required a chest tube for 5 or more days, 11% developed atrial fibrillation, and 8% required reintubation for respiratory collapse. In addition, Handy et al. examined quality of life differences in patients before and after surgical resection and found a significant decrease in both mental and physical quality of life, likely secondary to post-treatment morbidity and surgical recovery.

27.3.2 Conventional Radiation Therapy

For decades conventional radiotherapy has been a treatment option for patients with clinical stage I NSCLC who are not surgical candidates, i.e., those with severe co-morbidities or poor pulmonary function. Conventional radiotherapy has been reported to yield 5-year survival rates ranging from 5–30% and local recurrence rates averaging about 40% [6, 20–22]. In a landmark study by researchers from Duke University, 156 patients with stage I medically inoperable NSCLC were treated solely with radiotherapy to a median dose of 64 Gy delivered in 1.2-Gy fractions twice a day to 3-Gy daily fractions. The 5-year cause-specific survival rate in the 141 evaluable patients was 32%; 55 patients failed, 42% of these at the treated site. A dose response for radiotherapy has been obtained, with better outcomes for early-stage NSCLC using conventionally fractionated doses of 65 Gy or higher [21–23].

Although most published studies show low levels of toxicity with conventional radiotherapy [21, 22, 24], the poor outcomes have lead to several dose-escalation trials using more conformal 3D delivery methods. The outcomes of RTOG trial 9311 confirmed the beneficial effects of high doses of conformal radiotherapy, but also showed that doses greater than 77 Gy (for larger treatment volumes) and 84 Gy (for smaller volumes) caused unacceptable toxicity in the form of Grade 3–4 esophagitis, pneumonitis, and pericarditis [25]. Thus, while studies repeatedly point to dose escalation as a means of improving outcomes of conventionally fractionated radiotherapy, effective doses may not be well tolerated, and this may be associated with an increase in costs and a decrease in quality of life.

27.3.3 CyberKnife Radiosurgery

Although CyberKnife radiosurgery avoids traditional surgical risk factors, it is not without risks of its own, some of which may not manifest until many months of post-treatment. Prior to treatment, for example, multiple radio-opaque fiducial markers implanted near the tumor are typically required to localize the planning target volume (PTV) for real-time target tracking. These fiducials are usually placed percutaneously with a needle under CT guidance; similar percutaneous procedures (e.g., needle biopsy) carry a risk of pneumothorax that has run as high as 35% in some series [26, 27]. Although pneumothorax with this approach appears common, the magnitude of this complication is typically small, requiring simple observation in the majority, with overnight hospital admission for pulmonary re-expansion via chest tube placement in approximately 4–5% of patients [27, 28]. Rates of chest tube placement can vary widely, from less than 1% [29] to as high as 17% [30]. Very rarely, through evolution to tension pneumothorax, hemopneumothorax, massive hemoptysis, or air embolism, this becomes a potentially lethal complication [26]. To date, however, a fatal pneumothorax has not been reported with CyberKnife fiducial placement. In a very large Japanese study of nearly 10,000 patients, the incidence of fatal CT-guided lung biopsy complications measured 0.07% [26].

For patients in whom a pneumothorax would pose grave risk, such as those with severe emphysema and limited pulmonary reserve, there are methods of fiducial placement that can reduce the pneumothorax risk. These include limiting the procedure to a single needle pass (which can decrease target tracking accuracy due to the smaller number of fiducials), use of a blood patch technique to seal the pleural lining, or placing the fiducial markers transbronchially or transesophageally [31–33], an application that may be useful for more central lesions but is of limited utility for small peripheral lesions. If no safe fiducial placement method presents itself, the patient may be better served by a non-fiducial-based radiation approach, such as 3D-CRT, IMRT, IGRT or other forms of non-fiducial-based SBRT. There is also a new CyberKnife application known as XSight Lung™ (Accuray Incorporated, Sunnyvale, CA) that represents a non-invasive method for targeting and tracking lung tumors without implanted fiducial markers in patients who satisfy the imaging parameter requirements.

Potentially fatal hemoptysis or hemothorax warrants greater attention in patients who take anticoagulant medications. In these patients, the risk of temporary anticoagulant withdrawal, lesion location (central versus peripheral) and number of needle passes will need to be individually assessed.

If a patient has an absolute contraindication to anticoagulant withdrawal, particularly if their lesion also resides in a difficult location, it may be wise to consider a treatment method that does not require fiducials. To date, fatal hemoptysis or intrathoracic bleeding following fiducial placement has not been reported, though these complications were indicated as potentially fatal in the Tomiyama et al. study [26].

27.3.4 Post-Treatment Morbidity and Mortality with the CyberKnife

Possible serious complications of thoracic SBRT, a subset of which represents CyberKnife radiosurgery, include radiation pneumonitis, bronchial necrosis, stenosis or fistula, delayed hemoptysis, cardiotoxicity, esophageal ulceration or stenosis, spinal cord injury, and chest wall injury [8–14]. For small peripheral lesions, a specific SBRT dose-limiting toxicity has not been observed in a sequential dose escalation trial, while dose-limiting bronchopulmonary toxicity has been observed at 72 Gy in 3 fractions for stage T2 pulmonary lesions [8]. This study suggests both a dose response curve and a volume response for complications, with doses in excess of 66 Gy in 3 fractions and stage T2 lesions being at higher risk [8].

In addition to target volume, central target lesion location also appears to be a predictor of normal tissue toxicity. In 2003 Timmerman et al. reported on a series of 37 patients in a dose escalation study where 3 patients developed areas of atelectasis distal to central tumors irradiated at 18 Gy×3 fractions and 20 Gy×3 fractions [13]. In a more recent study Timmerman et al. found 11 of 70 patients developed Grade 3–5 pulmonary toxicity when treated with doses of 20-22 Gy×3 fractions [12]. All 11 of these patients had centrally located tumors. Onishi et al. [11], in a series of 245 patients, reported excellent local control with no severe morbidity when treating centrally located lesions to a dose of 48 Gy delivered in 4 fractions, which suggests that safe treatment of central tumors can be achieved with lower doses delivered in more fractions.

Overall, when patient selection has excluded patients large lesions or central lesions, the incidence of severe radiation-induced complications of SBRT [11] or CyberKnife radiosurgery [15, 18] has ranged from 0% to 2.3%, consisting primarily of rare cases of bronchial injury, radiation pneumonitis, or rib fracture. In a recent Phase I dose escalation study of CyberKnife radiosurgery by Le et al. [17], 32 patients with inoperable T1-2N0 NSCLC or solitary lung metastases were treated in a single fraction with doses of 15–30 Gy. At one year, freedom from local progression was 91% in patients treated with doses over 20 Gy and 54% at doses less than or equal to 20 Gy. Higher doses (25–30 Gy) were also associated with serious late complications, contributing to death in three patients with prior radiotherapy, larger midline tumors, or who were treated with adjuvant chemotherapy. Thus, the morbidity profile for CyberKnife radiosurgery has generally mirrored that for SBRT; for the vast majority of patients the approach is safe and effective, though caution is clearly advised for some patients. Although CyberKnife-specific data remain preliminary, it appears that severe complications may be more likely in patients treated with a large, single dose [17].

Direct comparison of the morbidity and mortality of resection versus CyberKnife radiosurgery or radiation therapy is complicated by the fact that, by definition, resection patients will usually be healthier and have superior preexisting pulmonary function. Patients deemed medically inoperable are often selected to receive alternative treatment based on the very nature of their impaired health and pulmonary function.

As Table 27.1 demonstrates, while both resection and CyberKnife radiosurgery have a potential for complications, the incidence of life-threatening and fatal sequelae are clearly lower in the CyberKnife radiosurgery patients, even taking into consideration their preexisting adverse health.

27.4 Quality of Life Considerations

For the treatment of a potentially lethal condition such as lung cancer, consideration of the effects of treatment on quality of life follow assessments of the treatment's efficacy and risks in the decision-

Table 27.1 Inventory of complications by modality; incidence of fatal pneumothorax derived from CT-guided biopsy literature.

Complication	Resection	SBRT (non-fiducial-based)	CyberKnife Radiosurgery
Operative Mortality	5.8%	0	0
Cardiac[1]	15-20%	0	0
Pulmonary[2]	30%	0	0
Infection[3]	15%	0	0
Prolonged ICU stay	12%	0	0
Prolonged CT requirement > 5 days	26%	0	0
Grade ≥ 3 Radiation Sequellae	0%	0–15.7%	0–12.5%[4]

[1] Cardiac: arrhythmias (ventricular and atrial), MI
[2] Pulmonary: PE, lobar atelectasis, tracheostomy, reintubation, prolonged ventilatory time
[3] Infection: pneumonia, wound infection
[4] The upper bound of this range was observed in a single study [17] that employed a single-fraction regimen with a relatively high dose (30 Gy).

making schema. They nevertheless warrant serious consideration in their own right. If factors such as efficacy and safety are comparable, then the treatment with the smaller adverse impact on quality of life will naturally emerge as the preferred method. The following discussion describes some of the quality of life issues that may be considered when comparing resection to CyberKnife radiosurgery.

27.4.1 Loss of Time and Activity

The initial quality of life consideration is that of "down time" for the patient, meaning the length of time actually devoted to preparing for, receiving, and recovering from the treatment, until patients are able to resume their normal daily activities without significant limitation. Both resection and CyberKnife radiosurgery have minimal adverse effects on quality of life in the pre-treatment stage. A brief comparison of the pretreatment period for these two modalities is listed in Table 27.2.

Table 27.2 shows that treatment preparation extracts a similar time demand from patients for each modality of therapy, though the preparative steps are not identical. The resection option requires a larger commitment of time for the treatment itself, due to the need for hospitalization. Recovery time will be variable for both methods of treatment, and these line items in Table 27.2 are more speculative than the other line items, as there are no specific published references to draw upon. The average hospital stay for lobectomy can range widely, but is typically 3–7 days, with an estimated 30-day recovery period. The situation is drastically different for CyberKnife patients, the majority of whom require no hospital stay and return to pre-treatment activity levels shortly after the procedure. There may be an occasional exception if a patient has unusually severe acute side effects of radiosurgery, such as a COPD exacerbation or significant esophagitis incident to the treatment of a central lesion.

Table 27.2. Time invested by patient according to treatment.

	Resection	CyberKnife Radiosurgery
Preparation for treatment		
Specialist Consultation	2 hours[1]	4 hours[2]
Radiologic Work-up[3]	4 hours	4 hours
Medical Work-up	6 hours[4]	1 hour[5]
Fiducial Placement	N/A	2 hours
Total	12 hours	11 hours
Treatment		
Hospitalization	7 days[6]	N/A
Radiosurgery Appointments	N/A	6 hours[7]
Recovery from treatment		
Post-op activity limitation	30 days[8]	N/A
Post-treatment fatigue	N/A	14 days[9]

[1] Thoracic Surgeon
[2] Thoracic Surgeon, Radiation Oncologist
[3] Assumes PET-CT and Cranial Imaging for both treatment types
[4] Pulmonary Function Tests, EKG, Comprehensive Lab, Type and cross match, Medical Clearance by Primary MD
[5] Lab (PT, PTT – R/o bleeding risk prior to fiducial placement)
[6] Average hospital stay for an uncomplicated thoracotomy
[7] Assumes three fractions lasting two hours per fraction
[8] Limitation primarily due to need for narcotic analgesia to control pain
[9] Highly variable from patient to patient – Also an estimate based on experience

In summary, CyberKnife radiosurgery will typically prove less disruptive to a patient's life and activities and be followed by a more rapid recovery. A formal quality of life instrument, such as the SF-36 or UCSD Dyspnea Questionnaire, administered at specific time intervals following treatment would be required to quantify these post-treatment results.

27.4.2 Post-Procedural Pain

The only invasive aspect of CyberKnife radiosurgery is the placement of fiducial markers within the tumor target. With CyberKnife radiosurgery, there are no rib spreading or other maneuvers which pose a high risk of post-treatment neuropathic pain, as seen with classic lobectomy. Video-assisted thoracoscopic surgery (VATS) represents a minimally invasive surgical means to reduce postoperative pain compared with open lobectomy, though this form of resection has a lower efficacy than lobectomy [34]. The use of validated instruments for assessing pain is essential to detect the actual level of post-treatment pain.

27.4.3 Post-Procedural Fatigue

A majority of patients undergoing CyberKnife treatment have complained of acute fatigue immediately following the procedure. This onset of fatigue has been difficult to predict, and its severity and duration vary from patient to patient. Fatigue reduces a patient's capacity to function in the workplace and at home, and may secondarily lead to other negative outcomes, such as exacerbating any tendency the patient may have towards depression. At this time there is no specific published comparison of this quality of life outcome between these modalities; it would be an appropriate domain to evaluate in a formal quality of life study.

27.4.4 Respiratory Function

Treatment of primary lung cancer or pulmonary metastases by any local method will create a loss of functioning respiratory tissue. For a typical small pulmonary lesion, such as a stage I NSCLC, CyberKnife treatment confines radiation to the target tumor and a planned margin, minimizing dose to normal lung parenchyma and outlined critical structures. Obviously, the dose to normal lung parenchyma will increase with tumor size and as tumor shape deviates from spherical. The tissue irradiated by CyberKnife typically does not follow anatomic boundaries, and as such, the amount of tissue affected may be comparable to that removed by surgical wedge resections. In lobectomy, the preferred surgical resection method, tissue is excised based on anatomic boundaries, and as such more tissue is removed than by limited wedge resections. The majority of patients undergoing surgical resection are noted to have some decrease in post-operative pulmonary function tests (PFTs), although patients with upper lobe lesions and concurrent COPD will often show improvements in post-treatment PFTs [35, 36]. The explanation for this improvement is based on the theory that, because COPD is predominately an upper lobe disease, the surgical resection or irradiation of the upper lobe lesion may function in the same capacity as lung reduction procedures, thus improving V/Q matching in the remainder of the lung parenchyma [35].

Pulmonary function quality of life domains have been defined for patients undergoing resection [37–39]. For example, the Functional Assessment of Cancer Therapy-Lung (FACT-L) has been established as a reliable and valid instrument for quality of life assessments in lung cancer trials [39]. It assesses general categories of physical, functional, social, and emotional well-being, and includes a lung-cancer-specific subscale that assesses cough, breathlessness, fatigue, loss of appetite, weight loss, chest pain or discomfort, constipation, and poor sleep. Summing physical and functional well-being scores with the subscale scores produces the Trial Outcome Index, which is highly correlated with performance status post-treatment [39]. It would be reasonable to use these same outcome domains, as well as sequential formal pulmonary function and exercise tolerance testing, to measure and compare more fully the respiratory quality of life outcome using CyberKnife radiosurgery relative to resection. Unfortunately, post-treatment respiratory comparisons will be complicated by the fact that radiosurgery and radiotherapy patients usually have more impaired pulmonary function at baseline than resection patients. A much more accurate comparison would include pulmonary function testing, exercise tolerance measurement, and self-administered pulmonary quality of life assessment as part of the pre-treatment and post-treatment analysis in a randomized trial between healthy resectable patients that are prospectively assigned to surgery or radiosurgery. One can anticipate that such a trial will be conducted if 5-year tumor control and survival rates for radiosurgery patients are shown to approximate those obtained with surgery.

27.4.5 Overall Satisfaction with the Treatment Choice

The final quality of life outcome comparison between different therapy modalities involves the patient's overall satisfaction with their treatment choice. Do they think they are cured or are there excessive lingering doubts? Were the side effects or complications worse than they expected? Did the recovery take longer? Would they choose this same treatment again having now gone through it? A global assessment of "satisfaction with treatment" domain may also be considered as the final component in the design of a formal quality of life assessment between the modalities.

27.5 Economic Comparison Between Resection and CyberKnife Radiosurgery

Despite their differences, resection and CyberKnife radiosurgery are both capital-intensive treatments. Each requires a series of steps to plan and accomplish, and a cost may be assigned to each of these steps, allowing a comparison of the total cost required to accomplish either method of treatment.

27.5.1 Method

The respective cost of modalities may be compared using a number of resource-based versus charge-based methodologies. For the purpose of this manuscript, our cost assessment comparison exercise is based on the 2006 U.S. Medicare allowable charge rates for each procedure, where these data are available, and estimated prevailing charge data where specific Medicare data do not exist.

Although a specific correlation between U.S. Medicare allowable charges and the actual cost of delivering the service may be debated, this method does assure a consistent method applied to both service lines, and is adopted by many private insurers in the United States as a "benchmark" against which their own reimbursement policy may be calculated. This exercise does not constitute billing advice, nor is every line item necessarily accurate for every local practice situation. Instead, it represents a reasonable attempt to quantify typical costs associated with resection versus CyberKnife radiosurgery.

This comparison does not address differences in cost created by differences in long-term efficacy or delayed complications, which will only be obtained with longer term follow-up for radiosurgery patients. Perhaps the best way to deal with the delayed outcome uncertainty is to evaluate only the respective planning and treatment costs of each modality, initially assuming efficacy parity, and subsequently to evaluate the long-term cost of relapse, continuously updating the analysis as efficacy data mature.

27.5.2 Result

As outlined in Table 27.3, CyberKnife radiosurgery costs an estimated $3,087.80 more at the preparation phase, due to the requirement for treatment planning imaging studies and interventional radiologic-guided fiducial placement, for which there is no corresponding surgical requirement. For treatment delivery itself, due to the elimination of expensive surgical and hospital line items, definitive resection appears $20,855.98 more expensive than CyberKnife radiosurgery (though the latter is, admittedly, also an expensive procedure). In summary, combining the cost of treatment preparation and delivery, CyberKnife radiosurgery creates an apparent net savings of $17,771.59 per case relative to resection.

Table 27.3. Respective Cost Components – Resection versus CyberKnife Radiosurgery

	Resection	CyberKnife
Pre-Treatment Direct Costs		
Specialist Consult[1]	\$149.84[(2)]	\$289.07[(3)]
Radiological Work-Up[4]	\$752.68	\$1752.68
Medical Work-up [5]	\$390.56	\$291.91[(6)]
Treatment Planning[7]	N/A	\$2,047.22
Sub-Total	\$1,293.08	\$4,380.88
Treatment Direct Costs		
1st hour OR level III; Major[8]	\$4,175	N/A
1/4 incremental hour OR level III[8]	\$1,043	N/A
3 Hour Procedure OR cost	\$12,519	N/A
OR Supplies[9]	\$3,095.41	N/A
Blood Transfusion[10]	\$1,273.00	N/A
ICU Critical Room Charge per day[11]	\$13,237.00	N/A
Post operative labs[12]	\$1,034.58	N/A
Room and Bed Med Surg (per day)[13]	\$1,519.00	N/A
Post Operative Radiology[13]	\$326.70	N/A
Robotic Linac SRS 1st session[14]	N/A	\$5,491.23
Robotic Linac SRS 2-5 session[15] (2 units)	N/A	\$6,190.00
Treatment device, design and Construction[16] (2 units)	N/A	\$423.18
Therapeutic Radiology Port Film [17]	N/A	\$44.30
Sub-Total	\$33,004.69	\$12,148.71
Physician Fees		
Thoracic Surgeon[18]	\$1,537.50	\$1,205.00
Radiation Oncologist[19]	N/A	\$1,542.54
Anesthesiologist[20]	\$1,213.45	N/A
Sub-Total	\$2,750.95	\$2,747.54
Total	\$37,048.72	\$19,277.13

[1] A consultation is a type of service provided by a physician whose opinion or advice regarding evaluation and/or management of a specific problem is requested by another physician or other appropriate source. CPT® 2007 Professional Edition. Copyright American Medical Association

[2] 2005 Medicare Claims data; Mean Cost. CPT 99245: Office consultation for a new or established patient, which requires these key components: A comprehensive history; a comprehensive examination; and medical decision making of high complexity

[3] 2005 Medicare Claims data; Mean Cost. CPT 99245: Office consultation for a new or established patient, which requires these key components: A comprehensive history; a comprehensive examination; and medical decision making of high complexity. CPT 99215:Office or other outpatient visit for the evaluation and management of an established patient, which requires at least two of these three key components: a comprehensive history ; a comprehensive examination; medical decision making of high complexity

[4] Diagnostic Radiology; 2005 Medicare Claims Data: CPT 71030: Radiologic examination, chest, complete, minimum of 4 views; CPT 71275: Computed tomographic angiography, chest (noncoronary), without contrast material(s), followed by contrast material(s) and further sections, including image post processing; CPT 71555: Magnetic resonance angiography, chest (excluding myocardium) with or without contrast materials. Also includes fiducial placement at \$1,000.00 per case (unlisted procedure on fee schedule).

[5] 2007 Medicare Clinical laboratory fee schedule: CPT 80053: Comprehensive Metabolic Panel, CPT 80051: electrolyte panel, CPT 80061 Lipid Panel, CPT Renal Function Panel, CPT 8100 Urinalysis, CPT 82040-84630 (8 specific tests selected), CPT 85025: Complete (CBC), automated (Hgb, Hct, RBC, WBC and platelet count), CPT 85345 Coagulation time, CPT 85610 Prothrombin Time , CPT 85670 Thrombin Time, CPT 93000 Electrocardiogram, routine ECG, CPT 94620 Pulmonary stress testing ; complex

[6] 2007 Medicare Clinical laboratory fee schedule: CPT 85025: Complete (CBC), automated (Hgb, Hct, RBC, WBC and platelet count), CPT 85345 Coagulation time, CPT 85610 Prothrombin Time , CPT 85670 Thrombin Time, CPT 93000 Electrocardiogram, routine ECG, CPT 94620 Pulmonary stress testing complex

[7] 2005 Medicare Claims Data; Mean Cost; CPT 77295 Therapeutic radiology simulation-aided field setting; 3D; CPT 77300 (10 units) Basic radiation dosimetry calculation; CPT 77370Special Medical Radiation Physics consult

[8] 2005 CDM- Chinese Hospital, Rev code 360-Office of Statewide Health Planning and Development Healthcare Quality & Analysis Division; http://www.oshpd.ca.gov/hqad/hospital/chargemaster/2005/chrgmstrCa_Ch.htm

[9] 2005 CDM- Alhambra Hospital, Office of Statewide Health Planning and Development Healthcare Quality & Analysis Division; http://www.oshpd.ca.gov/hqad/hospital/chargemaster/2005/chrgmstrCa_Ch.htm

[10] 2005 CDM- Alameda Hospital, Office of Statewide Health Planning and Development Healthcare Quality & Analysis Division; http://www.oshpd.ca.gov/hqad/hospital/chargemaster/2005/chrgmstrCa_Ch.htm

[11] 2005 CDM- Cedar Sinai Medical Center; Office of Statewide Health Planning and Development Healthcare Quality & Analysis Division; http://www.oshpd.ca.gov/hqad/hospital/chargemaster/2005/chrgmstrCa_Ch.htm (2 day charge)

[12] 2005 CDM- Stanford Hospital and Clinics; Office of Statewide Health Planning and Development Healthcare Quality & Analysis Division; http://www.oshpd.ca.gov/hqad/hospital/chargemaster/2005/chrgmstrCa_Ch.htm

[13] 2005 CDM- Citrus Valley Hospital; Office of Statewide Health Planning and Development Healthcare Quality & Analysis Division; http://www.oshpd.ca.gov/hqad/hospital/chargemaster/2005/chrgmstrCa_Ch.htm (2 day Charge)

[14] 2005 Medicare Claims Data; Mean Cost; G0339

[15] 2005 Medicare Claims Data; Mean Cost; G0340

[16] 2005 Medicare Claims Data; Mean Cost; CPT 77334

[17] 2005 Medicare Claims Data; Mean Cost; CPT 77417

[18] Physician Fee Schedule 2006: CPT 32440 Removal of Lung, total pneumonectomy; CPT 61793 Stereotactic Radiosurgery, one or more sessions

[19] Physician Fee Schedule 2006: CPT 77263, 77295 (10 units), 77300, 77334, 77435

[20] Physician Fee Schedule 2006: CPT 00546 Anesthesia for pulmonary resection

27.5.3 Further Discussion

This analysis is oversimplified for a number of reasons. For example, should a delayed local failure or complication pattern emerge with CyberKnife radiosurgery that is not seen with resection, any early cost advantage of CyberKnife radiosurgery would be eroded or completely negated by the high cost of secondary treatment evaluation and management. Conversely, it may be that patients who are destined to fail radiosurgical treatment would be secondarily treated far earlier if they had taken the resection option, because adverse pathologic findings are immediately apparent and can lead to adjuvant chemotherapy or radiotherapy being pursued without delay. Corresponding patients treated with CyberKnife radiosurgery would have no such adjuvant therapy, as their adverse pathologic presentation would not become apparent until they relapsed.

The potentially high cost of adjuvant post-operative treatment based on findings of surgical pathology is not addressed in this analysis. Nevertheless, this issue needs to be noted, because it could partially or completely offset any negative cost difference that accrues against CyberKnife radiosurgery should radiosurgery prove to have a higher late failure rate.

Finally, there is a definite incidence of serious surgical complications the costs of which are not addressed in this analysis. Any incidence of serious post-operative infection, prolonged ventilator dependence, thromboembolic sequellae, etc., would also accrue a far greater cost to the resection side of the ledger, but is beyond the scope of this analysis to itemize in detail.

27.5.4 Conclusion Regarding Cost Analysis

In this era of ever-increasing medical cost awareness, assessment of the respective cost of treatment delivery and the added cost of treatment failure would be reasonably included as part of a prospective comparison trial between resection and CyberKnife radiosurgery or between radiosurgery and radiotherapy. Such a comparison would require very careful consideration in its design to capture properly all of the cost components of each treatment option.

27.6 Overall Conclusions Regarding Resection Versus CyberKnife Radiosurgery

Compared with resection by lobectomy, CyberKnife radiosurgery appears to have a lower risk of serious acute morbidity and mortality. CyberKnife radiosurgery also has a potentially superior quality of life outcome, manifested by a lack of post-operative pain, faster recovery, and the possibility of more well-preserved respiratory capacity. Of course, the quality of life impact of these respective modalities awaits formal comparison by appropriate, validated quality of life instruments, sequential pulmonary function testing, and exercise tolerance testing, all of which would be reasonably included as part of a larger prospective trial. Pending long-term confirmation of clinical effectiveness comparable to resection, CyberKnife radiosurgery also appears more cost effective.

Although CyberKnife radiosurgery appears to have safety, quality of life, and cost advantages, as well as encouraging short-term results [15, 17, 18], its use will likely typically remain restricted to patients who are medically unable to withstand the definitive lung cancer operation until its long-term efficacy is established as being comparable to resection. If efficacy parity is ultimately confirmed, the safety, quality of life, and cost advantages of CyberKnife radiosurgery could then cause it to emerge as the treatment of choice, even for healthy patients who are fully capable of withstanding a surgical procedure. This would represent a paradigm shift from current practice.

References

1. Gharagozloo F, Tempesta B, Margolis M, *et al.* Video-assisted thoracic surgery lobectomy for stage I lung cancer. *Ann Thorac Surg* 2003; 76:1009–1014; discussion 1014–1005.

2. Flehinger BJ, Kimmel M, Melamed MR. The effect of surgical treatment on survival from early lung cancer. Implications for screening. *Chest* 1992; 101:1013–1018.
3. Dominioni L, Imperatori A, Rovera F, *et al.* Stage I nonsmall cell lung carcinoma: analysis of survival and implications for screening. *Cancer* 2000; 89:2334–2344.
4. Mountain CF. Revisions in the International System for Staging Lung Cancer. *Chest* 1997; 111:1710–1717.
5. Kaskowitz L, Graham MV, Emami B, *et al.* Radiation therapy alone for stage I non-small cell lung cancer. *Int J Radiat Oncol Biol Phys* 1993; 27:517–523.
6. Qiao X, Tullgren O, Lax I, *et al.* The role of radiotherapy in treatment of stage I non-small cell lung cancer. *Lung Cancer* 2003; 41:1–11.
7. Chen M, Hayman JA, Ten Haken RK, *et al.* Long-term results of high-dose conformal radiotherapy for patients with medically inoperable T1-3N0 non-small-cell lung cancer: is low incidence of regional failure due to incidental nodal irradiation? *Int J Radiat Oncol Biol Phys* 2006; 64:120–126.
8. McGarry RC, Papiez L, Williams M, *et al.* Stereotactic body radiation therapy of early-stage non-small-cell lung carcinoma: phase I study. *Int J Radiat Oncol Biol Phys* 2005; 63:1010–1015.
9. Nagata Y, Takayama K, Matsuo Y, *et al.* Clinical outcomes of a phase I/II study of 48 Gy of stereotactic body radiotherapy in 4 fractions for primary lung cancer using a stereotactic body frame. *Int J Radiat Oncol Biol Phys* 2005; 63:1427–1431.
10. Nyman J, Johansson KA, Hulten U. Stereotactic hypofractionated radiotherapy for stage I non-small cell lung cancer-mature results for medically inoperable patients. *Lung Cancer* 2006; 51:97–103.
11. Onishi H, Araki T, Shirato H, *et al.* Stereotactic hypofractionated high-dose irradiation for stage I nonsmall cell lung carcinoma: clinical outcomes in 245 subjects in a Japanese multiinstitutional study. *Cancer* 2004; 101:1623–1631.
12. Timmerman R, McGarry R, Yiannoutsos C, *et al.* Excessive toxicity when treating central tumors in a phase II study of stereotactic body radiation therapy for medically inoperable early-stage lung cancer. *J Clin Oncol* 2006; 24:4833–4839.
13. Timmerman R, Papiez L, McGarry R, *et al.* Extracranial stereotactic radioablation: results of a phase I study in medically inoperable stage I non-small cell lung cancer. *Chest* 2003; 124:1946–1955.
14. Xia T, Li H, Sun Q, *et al.* Promising clinical outcome of stereotactic body radiation therapy for patients with inoperable Stage I/II non-small-cell lung cancer. *Int J Radiat Oncol Biol Phys* 2006; 66:117–125.
15. Nuyttens JJ, Prevost JB, Praag J, *et al.* Lung tumor tracking during stereotactic radiotherapy treatment with the CyberKnife: Marker placement and early results. *Acta Oncol* 2006; 45:961–965.
16. Yu C, Main W, Taylor D, *et al.* An anthropomorphic phantom study of the accuracy of Cyberknife spinal radiosurgery. *Neurosurgery* 2004; 55:1138–1149.
17. Le QT, Loo BW, Ho A, *et al.* Results of a phase I dose-escalation study using single-fraction stereotactic radiotherapy for lung tumors. *Journal of Thoracic Oncology* 2006; 1.
18. Whyte RI, Crownover R, Murphy MJ, *et al.* Stereotactic radiosurgery for lung tumors: preliminary report of a phase I trial. *Ann Thorac Surg* 2003; 75:1097–1101.
19. Handy JR, Jr., Asaph JW, Skokan L, *et al.* What happens to patients undergoing lung cancer surgery? Outcomes and quality of life before and after surgery. *Chest* 2002; 122:21–30.
20. Dosoretz DE, Galmarini D, Rubenstein JH, *et al.* Local control in medically inoperable lung cancer: an analysis of its importance in outcome and factors determining the probability of tumor eradication. *Int J Radiat Oncol Biol Phys* 1993; 27:507–516.
21. Dosoretz DE, Katin MJ, Blitzer PH, *et al.* Radiation therapy in the management of medically inoperable carcinoma of the lung: results and implications for future treatment strategies. *Int J Radiat Oncol Biol Phys* 1992; 24:3–9.
22. Krol AD, Aussems P, Noordijk EM, *et al.* Local irradiation alone for peripheral stage I lung cancer: could we omit the elective regional nodal irradiation? *Int J Radiat Oncol Biol Phys* 1996; 34:297–302.
23. Martel MK, Ten Haken RK, Hazuka MB, *et al.* Estimation of tumor control probability model parameters from 3-D dose distributions of non-small cell lung cancer patients. *Lung Cancer* 1999; 24:31–37.
24. Sibley GS, Jamieson TA, Marks LB, *et al.* Radiotherapy alone for medically inoperable stage I non-small-cell lung cancer: the Duke experience. *Int J Radiat Oncol Biol Phys* 1998; 40:149–154.
25. Bradley J, Graham MV, Winter K, *et al.* Toxicity and outcome results of RTOG 9311: a phase I-II dose-escalation study using three-dimensional conformal radiotherapy in patients with inoperable non-small-cell lung carcinoma. *Int J Radiat Oncol Biol Phys* 2005; 61:318–328.
26. Tomiyama N, Yasuhara Y, Nakajima Y, *et al.* CT-guided needle biopsy of lung lesions: a survey of severe complication based on 9783 biopsies in Japan. *Eur J Radiol* 2006; 59:60–64.
27. Yamagami T, Kato T, Hirota T, *et al.* Pneumothorax as a complication of percutaneous radiofrequency ablation for lung neoplasms. *J Vasc Interv Radiol* 2006; 17:1625–1629.
28. Chojniak R, Isberner RK, Viana LM, *et al.* Computed tomography guided needle biopsy: experience from 1,300 procedures. *Sao Paulo Med J* 2006; 124:10–14.
29. Montaudon M, Latrabe V, Pariente A, *et al.* Factors influencing accuracy of CT-guided percutaneous biopsies of pulmonary lesions. *Eur Radiol* 2004; 14:1234–1240.
30. Anderson CL, Crespo JC, Lie TH. Risk of pneumothorax not increased by obstructive lung disease in percutaneous needle biopsy. *Chest* 1994; 105:1705–1708.
31. Pishvaian AC, Collins B, Gagnon G, *et al.* EUS-guided fiducial placement for CyberKnife radiotherapy of mediastinal and abdominal malignancies. *Gastrointest Endosc* 2006; 64:412–417.
32. Reichner CA, Collins BT, Gagnon GJ, *et al.* The placement of gold fiducials for CyberKnife stereotactic radiosurgery using a modified transbronchial needle aspiration technique. *Journal of Bronchology* 2005; 12:193–195.
33. Savides TJ. EUS-guided fine-needle insertion of radiopaque fiducials:×marks the spot. *Gastrointest Endosc* 2006; 64:418–419.
34. Martini N, Bains MS, Burt ME, *et al.* Incidence of local recurrence and second primary tumors in resected stage I lung cancer. *J Thorac Cardiovasc Surg* 1995; 109:120–129.

35. Korst RJ, Ginsberg RJ, Ailawadi M, *et al.* Lobectomy improves ventilatory function in selected patients with severe COPD. *Ann Thorac Surg* 1998; 66:898–902.
36. Carretta A, Zannini P, Puglisi A, *et al.* Improvement of pulmonary function after lobectomy for non-small cell lung cancer in emphysematous patients. *Eur J Cardiothorac Surg* 1999; 15:602–607.
37. Aaronson NK, Ahmedzai S, Bergman B, *et al.* The European Organization for Research and Treatment of Cancer QLQ-C30: a quality-of-life instrument for use in international clinical trials in oncology. *J Natl Cancer Inst* 1993; 85:365–376.
38. Hollen PJ, Gralla RJ, Kris MG, *et al.* Measurement of quality of life in patients with lung cancer in multicenter trials of new therapies. Psychometric assessment of the Lung Cancer Symptom Scale. *Cancer* 1994; 73:2087–2098.
39. Cella DF, Bonomi AE, Lloyd SR, *et al.* Reliability and validity of the Functional Assessment of Cancer Therapy-Lung (FACT-L) quality of life instrument. *Lung Cancer* 1995; 12:199–220.

Abbreviations

A

AAPM American Association of Physicists in Medicine
ABC Active breathing control
ABG Arterial blood gas
AC Attenuation correction
AFB Acid-fast bacilli culture
AP Anterior-posterior
aPTT Activated partial thromboplastin time
AWD Alive with disease

B

BAL Bronchoalveolar lavage
BED Biologically effective dose
bFGF Basic fibroblast growth factor
BIDMC Beth Israel Deaconess Medical Center, Boston MA
BTV Biologic tumor volume
BudR Bromodeoxyuridine

C

CA 19-9 Cancer antigen 19-9 tumor marker
CAD Coronary artery disease
CBC Complete blood count
CBR Clinical benefit response
CCC Cholangiocellular cancer
CEA Carcinoembryonic antigen
cGy Centigray (1 cGy = 10^{-2} Gy)
CHF Congestive heart failure
CI Conformity index
CK CyberKnife®
CNS Central nervous system
CMS The Centers for Medicare and Medicaid Services
COPD Chronic obstructive pulmonary disease
CpI Shape complexity
CR Complete response
3D-CRT 3D conformal radiotherapy
CT Computerized tomography (e.g., 4D-CT)
CTCAE Common toxicity criteria adverse event
CTV Clinical target volume

D

DAH Dose allocation histograms
DIBH Deep inspiration breath hold
DLCO Diffusion capacity
DLT Dose-limiting toxicity
DMH Dose-mass histogram
DNA Deoxyribonucleic acid
DOD Dead of disease
DRR Digitally reconstructed radiograph
DVH Dose volume histogram

E

EBRT External beam radiotherapy
ECOG Eastern Cooperative Oncology Group, Boston MA
EORTC European Organization for the Research and Treatment of Cancer
ESTOP Emergency stop
EUS Endoscopic ultrasound

F

^{18}F Radioactive isotope ^{18}Fluorine
FACT-L Functional assessment of cancer therapy-lung
FDA United States Food and Drug Administration
FDG Fluordeoxyglucose (2-fluoro-2-deoxy-D-glucose)
FEV_1 Forced expiratory volume in 1 second
FNA Fine-needle aspiration
FOV Field of view
FSU Functional sub-units
5-FU 5-Fluorouracil
fx Fraction

G

G3 Generation 3
G4 Generation 4
GI Gastrointestinal
GITSG Gastrointestinal Tumor Study Group
GTV Gross tumor volume
GUH Georgetown University Hospital, Washington DC
Gy Gray (unit of absorbed dose)
Gy_n Corrected dose in Gy based on an α/β ratio of n

H

HCC Hepatocellular carcinoma
HDR High dose rate
HTV Healthy tissue volume
HTVRI Healthy tissue volume covered by the prescription isodose

I

ICU Intensive care unit
ICRU International Commission on Radiation Units
IGRT Image-guided radiation therapy
IHCC Intrahepatic cholangiocarcinoma

IMRT	Intensity-modulated radiotherapy
INR	International normalized ratio
IORT	Intra-operative radiation therapy
ITV	Internal tumor volume
IV	Intravenous

K

KGF	Keratinocyte growth factor
kHz	Kilohertz (1 kHz = 10^3 Hz)
KPS	Karnofsky performance status (or score)
kV	Kilovolts (e.g., 130 kV X-rays)

L

LEDs	Light emitting diodes
LINAC	Linear accelerator
LITT	Laser induced thermotherapy
LLL	Left lower lung (or lobe)
LN	Lymph node
LQ	Linear quadratic
LR	Left/Right
LV	Lesion volume
LVRI	Lesion volume covered by the prescription isodose
LUL	Left upper lung (or lobe)

M

MAC	Mycobacterium avium complex
MDACC	M.D. Anderson Cancer Center, Houston, TX
ME	Microscopic extension
MICU	Medical intensive care
MLC	Multileaf collimator
MLD	Mean lung dose
MNTLD	Mean normalized total lung dose
MRI	Magnetic resonance imaging
MTD	Maximum tolerated dose
MU	Monitor unit
MV	Million volts (linear accelerator)

N

N	Nodal stage (e.g. NO, N1, N2)
NCI	New conformity index
NCPB	Neurolytic celiac plexus block
NPO	Nothing by mouth
NSCLC	Non-small cell lung cancer
NTCP	Normal tissue complication probability
NTD	Normalized total dose

O

OAR	Organ at risk

P

PD	Progressive disease
PEI	Percutaneous ethanol injection
PET	Positron emission tomography
PFT	Pulmonary function tests
PIV	Planning isodose volume
PR	Partial response
PT	Prothrombin time
PTV	Planning target volume

Q

QA	Quality assurance

R

RFA	Radiofrequency ablation
RILD	Radiation-induced liver disease
RLL	Right lower lung (or lobe)
RMS	Root-mean-square
RML	Right middle lobe
ROI	Region of interest
RP	Radiation pneumonitis
RT	Radiation therapy
RTOG	Radiation Therapy Oncology Group
RTP	Radiation therapy planning
RUL	Right upper lung (or lobe)

S

SBRT	Stereotactic body radiotherapy
SD	Stable disease
SDD	Source-to-detector distance
SEERS	Surveillance, epidemiology, and end results
SF	Surviving fraction
SI	Superior-inferior
SMA	Superior mesenteric artery
SMF	Streptozocin, Mitomycin, and 5-FU
SMV	Superior mesenteric vein
SPECT	Single-photon-emission computed tomography
SRS	Stereotactic radiosurgery
SSD	Source-to-surface distance
SST	Skeletal structure tracking
StRal	Stereotactic radiation therapy of liver metastases
SUV	Standardized uptake value

T

T	Tumor stage (e.g. T1, T2, T3)
TACE	Transarterial chemoembolization
TBNA	Transbronchial aspiration needle
TCP	Tumor control probability
TIV	Target isodose volume
TNM	Tumor nodal metastasis staging (e.g. T2N1M0)
TOMO	Tomographic approach
TR	Therapeutic ratio

V

V/Q	Ventilation perfusion
V_n	Percentage of the total lung volume that received a dose of n Gy or less
VATS	Video-assisted thoracoscopic surgery
VEGF	Vascular endothelial growth factor
V_{eff}	Lung effective volume
VOD	Veno-occlusive disease
VOI	Volumes of interest

W

WB	Whole-body

Subject Index

C

List of Contributors

Alan Alfieri, MSc
Clinical Associate Professor
Department of Radiation Oncology at
Montefiore Medical Center and the
Albert Einstein College of Medicine
111 East 210 Street
Bronx, NY 10467
USA

Beatriz E. Amendola, MD, FACR
Director
Brachytherapy Institute of South Florida
9300 SW 87th Ave Suite 3
Miami, Fl 33176
USA

Marco Amendola, MD, FACR
Professor of Radiology
Department of Radiology
Leonard M. Miller School of Medicine
University of Miami
P.O. Box 016960 (R-109)
Miami, FL 33101
USA

Eric D. Anderson, MD, FCCP
Assistant Professor of Medicine
Pulmonary Critical Care & Sleep Medicine
Georgetown University Hospital
3800 Reservoir Road, NW
Washington, DC 20007
USA

Simon Ashiku, MD
Division of Cardiothoracic Surgery
Beth Israel Deaconess Medical Center
Instructor in Surgery
Harvard Medical School
110 Francis Street (LMOB) 2A
Boston, MA 02215
USA

Filip Banovac, MD
Clinical Director of Imaging Sciences &
Information Systems Center
Assistant Professor of Radiology
Division of Interventional Radiology
Georgetown University Hospital
3800 Reservoir Road, NW
Washington, DC 20007
USA

Brian D. Berger, MD
Associate Medical Director
Baylor Radiosurgery Center
Baylor University Medical Center
3500 Gaston Avenue
Dallas, TX 75246
USA

Stuart M. Berman, MD
Department of Radiation Oncology
Beth Israel Deaconess Medical Center
Instructor in Radiation Oncology
Harvard Medical School
330 Brookline Avenue
Boston, MA 02215
USA

James W. Blalock, MD
Department of Pulmonary Medicine
Frank C. Love Cancer Institute
St. Anthony Hospital
1011 North Dewey
Oklahoma City, OK 73102
USA

Phillip Boiselle, MD
Director, Thoracic Imaging Section
Beth Israel Deaconess Medical Center
Associate Professor of Radiology
Harvard Medical School
330 Brookline Avenue
Boston, MA 02215
USA

Elizabeth Bossart, PhD, DABR
Assistant Professor, Division of Medical Physics
Department of Radiation Oncology
University of Miami
Leonard M. Miller School of Medicine
1475 NW 12 Ave
Miami, FL 33136
USA

Joseph Both, PhD
Assistant Professor, Division of Medical Physics
Department of Radiation Oncology
University of Miami
Leonard M. Miller School of Medicine
1475 NW 12 Ave
Miami, FL 33136
USA

Darren Brennan, MD
Department of Radiology
Beth Israel Deaconess Medical Center
Instructor in Radiology
Harvard Medical School
330 Brookline Avenue
Boston, MA 02215
USA

William T. Brown, MD
CyberKnife Center of Miami
Voluntary Professor of Surgery
Leonard M. Miller School of Medicine
University of Miami
6300 SW 104 ST
Miami, FL 33156
USA

Richard D. Bucholz, MD, FACS
Chairman and Professor
Division of Neurosurgery
Department of Surgery
St. Louis University Medical Center
1320 S. Grand Boulevard
St. Louis, MO 63104
USA

Michael S. Buff, MD
Clinical Fellow in Hematology and Medical Oncology
Beth Israel Deaconess Medical Center
Harvard Medical School
330 Brookline Ave
Boston MA 02215
USA

Steven A. Burton, MD
Assistant Professor
Department of Radiation Oncology
University of Pittsburgh Medical Center
5200 Centre Avenue #715
Pittsburgh, PA 15232
USA

Thomas C. Chang, MD
Assistant Professor of Radiology
Division of Interventional Radiology
Department of Radiology
Georgetown University Hospital
3800 Reservoir Road, NW
Washington, DC 20007
USA

Aline Charabaty-Pishvaian, MD
Assistant Professor
Georgetown University Hospital
Division of Gastroenterology
3800 Reservoir Road, NW
LL Bles Building
Washington, DC 20007
USA

Neil A. Christie, MD
Assistant Professor
Heart Lung Esophageal Surgery Institute
University of Pittsburgh Medical Center
5200 Centre Avenue #715
Pittsburgh, PA 15232
USA

Brian T. Collins, MD
Assistant Professor
Department of Radiation Medicine
Georgetown University Hospital
3800 Reservoir Road, NW
Washington, DC 20007
USA

Sean P. Collins, MD, PhD
Assistant Professor
Department of Radiation Medicine
Georgetown University Hospital
3800 Reservoir Road, NW
Washington, DC 20007
USA

Lori Copsey, CMD, BS
Medical Physicist
CyberKnife Center
St. Joseph's Hospital
Saint Paul, MN 55102
USA

Malcolm DeCamp, MD
Chief, CardioThoracic Surgery
Beth Israel Deaconess Medical Center
Associate Professor of Surgery
Harvard Medical School
110 Francis Street (LMOB) 2A
Boston, MA 02215
USA

ALBERTO DE LA ZERDA, PhD, DABR
Assistant Professor, Division of Medical Physics
Department of Radiation Oncology
University of Miami
Leonard M. Miller School of Medicine
1475 NW 12 Ave
Miami, FL 33136
USA

RICHARD DESI, MD
Fellow, Department of Medicine
Division of Gastroenterology
Georgetown University Hospital
Washington, DC 20007
USA

LISA DIAS, MD
Howard University School of Medicine
530 W. Street, N.W.
Washington, DC 20059
USA

MUKUND DIDOLKAR, MD
Director, Surgical Oncology
Sinai Hospital of Baltimore
2401 West Belvedere Ave
Baltimore, MD 21215-5271
USA

SONJA DIETERICH, PhD
Assistant Professor
Director of CyberKnife Physics
Georgetown University Hospital
Department of Radiation Medicine
3800 Reservoir Road NW
Washington, DC 20007
USA

JOHN ROBINSON DOOLEY, PhD
Manager, Treatment Planing Software
Accuray Incorporated
1310 Chesapeake Terrace
Sunnyvale, CA 94089
USA

KELLY ERICKSON, BA
Medical Student
University of Southern California
Keck School of Medicine
Los Angeles, CA 90089
USA

ARMIN ERNST, MD
Chief, Interventional Pulmonology
Beth Israel Deaconess Medical Center
Associate Professor of Medicine
Harvard Medical School
One Deaconess Road
Deaconess Building-201
Boston, MA 02215
USA

FAHED FAYAD, MD
Attending Physician
Cedars Medical Center
1400 NW 12th Street
Miami, FL 33136
USA

DAVID FELLER-KOPMAN, MD
Director, Medical Procedure Service
Interventional Pulmonology
Division of Pulmonary and
Critical Care Medicine
Beth Israel Deaconess Medical Center
Assistant Professor of Medicine
Harvard Medical School
One Deaconess Rd.
Boston, MA 02215
USA

ANDREW S. FINK, MD
Medical Director
Department of Surgery
St. Joseph's Hospital
Saint Paul, MN 55102
USA

DONGSHAN FU, PhD
Senior Imaging Scientist
Accuray Incorporated
1310 Chesapeake Terrace
Sunnyvale, CA 94089
USA

DONALD B. FULLER, MD
Radiation Oncologist
Radiosurgery Medical Group (RSMG)
San Diego CyberKnife Center
5395 Ruffin Road Suite 103
San Diego, CA 92123
USA
and
Radiation Oncologist
Radiation Medical Group, Inc. (RMG)
2466 First Ave.
San Diego, CA 92101
USA

GREGORY J. GAGNON, MD
Associate Professor and
CyberKnife Program Director
Interim Chairman, Department of
Radiation Medicine
Georgetown University Medical Center
3800 Reservoir Road, NW
LL Bles Building
Washington, DC 20007
USA

Sidhu P. Gangadharan, MD
Staff Surgeon, Division of Cardiothoracic Surgery
Beth Israel Deaconess Medical Center
Instructor in Surgery, Harvard Medical School
110 Francis Street (LMOB) 2A
Boston, MA 02215
USA

Silvio García, MD
CyberKnife Centers of Miami
Clinical Professor of Radiation Oncology
Leonard M. Miller School of Medicine
University of Miami
7867 N Kendall Drive, Suite 105
Miami, FL 33156
USA

Madhur Garg, MD
Clinical Director
Assistant Professor, Department of Radiation Oncology
at Montefiore Medical Center and the
Albert Einstein College of Medicine
111 East 210 Street
Bronx, NY 10467
USA

Michael Goldstein, MD
Beth Israel Deaconess Medical Center
Assistant Clinical Professor of Medicine
Harvard Medical School
330 Brookline Ave.
Boston, MA 02215
USA

Robert M. Goldstein, MD
Director, Abdominal Radiosurgery Services,
Hepatobiliary-Transplant Surgeon
Baylor University Medical Center
3500 Gaston Ave
Dallas, TX 75246
USA

Karyn A. Goodman, MD
Assistant Attending Radiation Oncologist
Department of Radiation Oncology
Memorial Sloan-Kettering Cancer Center
1275 York Avenue
New York, NY 10021
USA

Chandan Guha, MD, PhD
Associate Professor
Vice Chairman, Department of Radiation Oncology
at Montefiore Medical Center and the
Albert Einstein College of Medicine
111 East 210 Street
Bronx, NY 10467
USA

Nadim G. Haddad, MD
Associate Professor of Medicine
Director of Fellowship
Department of Medicine
Division of Gastroenterology
Georgetown University Hospital
Washington, DC 20007
USA

Hoke T. Han, MD
Hollywood Radiation Oncology
6300 Hollywood Blvd.
Hollywood, FL 33024
USA

Vikki Harriet, RN, BSN
Oncology Research Coordinator
Department of Radiation Oncology
Frank C. Love Cancer Institute
St. Anthony Hospital
1011 North Dewey
Oklahoma City, OK 73102
USA

Julie Hensley, RN
Nurse Navigator
Frank C. Love Cancer Institute
St. Anthony Hospital
1011 North Dewey
Oklahoma City, OK 73102
USA

Klaus K. Herfarth, MD
Associate Professor of Radiation Oncology
Vice Chairman of Radiation Oncology
Department of Radiation Oncology
University of Heidelberg Medical School
Im Neuenheimer Feld 400
69120 Heidelberg
Germany

Michael J. Hervey II, MD
Department of Family Medicine
University of Minnesota
Minneapolis, MN 55455
USA

Christopher Hoffelt, MD
Assistant Professor
Radiation Oncology
Oregon Health & Science University
Portland, Oregon
Medical Director, Radiation Oncology
Southwest Washington Medical Center
400 NE Mother Joseph Pl
Vancouver, WA 98664
USA

Misha S. Hoogeman, PhD, MSc
Medical Physicist
Department of Radiation Oncology
Erasmus MC Daniel den Hoed Cancer Center
Postbus 5201
3008 EA Rottterdam
The Netherlands

Zhicong Huang, MD, MSc
Department of Biomedical Engineering
Florida International University
11200 SW 8 St
Miami, FL 33199
USA

Mark Huberman, MD
Beth Israel Deaconess Medical Center
Assistant Professor of Medicine
Harvard Medical School
330 Brookline Avenue
Boston, MA 02215
USA

Sanjay R. Jain, MD, PhD
Division of Hematology-Oncology
Beth Israel Deaconess Medical Center
Instructor in Medicine
Harvard Medical School
330 Brookline Avenue
Boston, MA 02215
USA

Carlos Jamis-Dow, MD
Assistant Professor
Chief of Cardiothoracic Imaging
Department of Radiology
3800 Reservoir Road, NW
Washington, DC 20007
USA

Walter Jean, MD
Assistant Professor
Director, Surgical Neuro-Oncology
Department of Neurosurgery
3800 Reservoir Road NW
PHC Building, Room 1
Washington, DC 20007
USA

Huaying Ji
Department of Radiation Medicine
Georgetown University Hospital
3800 Reservoir Road, NW
Washington, DC 20007
USA

Robert Kahn, MS
Senior Software Engineer
Accuray Incorporated
1310 Chesapeake Terrace
Sunnyvale, CA 94089
USA

Shalom Kalnicki, MD, FACR
Professor and Chairman
Department of Radiation Oncology
Montefiore Medical Center
Albert Einstein College of Medicine
111 East 210 Street
Bronx, NY 10467
USA

Warren Kilby, MSc
Senior Clinical Research Scientist
Accuray Europe
Tour Atlantique 25e
1 Place de la Pyramide
92911 Paris La Défense Cedex
France

June A. Kim, MD
Medical Director
Radiation Medicine
St. Joseph's Hospital
Saint Paul, MN 55102
USA

Albert C. Koong, MD, PhD
Assistant Professor
Department of Radiation Oncology
Stanford Cancer Center
875 Blake Wilbur Drive
Stanford, CA 94305-5847
USA

John J. Kresl, MD, PhD
Arizona Oncology Services at
St. Joseph's Hospital & Medical Center
Department of Radiation Oncology
CyberKnife Center
Barrow Neurological Institute
Gamma Knife Center
350 West Thomas Road
Phoenix, Arizona 85013
USA

Gopinath Kuduvalli, PhD
Director of Engineering
Accuray Inorporated
1310 Chesapeake Terrace
Sunnyvale, CA 94089
USA

Charles L. Lee, PhD
Medical Physicist
San Diego CyberKnife Center
5395 Ruffin Road; Suite 103
San Diego, CA 92123
USA

Peter C. Levendag, MD, PhD
Professor and Chairman
Radiation Oncology
Erasmus MC Daniel den Hoed Cancer Center
Postbus 5201
3008 EA Rotterdam
The Netherlands

James D. Luketich, MD
Sampson Family Endowed Professor of Surgery
Chief, The Heart, Lung & Esophageal Surgery Institute
University of Pittsburgh Medical Center
PUH, C-800
200 Lothrop Street
Pittsburgh, PA 15213
USA

Timothy Mahoney, RRT
Special Procedures Therapist
Department of Respiratory Care
3800 Reservoir Road, NW
Washington, DC 20007
USA

William T. Main, PhD
Manager of Development Physics
Accuray Incorporated
1310 Chesapeake Terrace
Sunnyvale, CA 94089
USA

Shakun Malik, MD
Associate Professor of Medicine
Division of Hematology/Oncology
Georgetown University Hospital P Lombardi
3800 Reservoir Road, NW
Washington, DC 20007-2197
USA

Jodi L. Mammenga, BA
CyberKnife Center
St. Joseph's Hospital
Saint Paul, MN 55102
USA

Arnold M. Markoe, MD, ScD
Professor and Chairman
Department of Radiation Oncology
University of Miami
Leonard M. Miller School of Medicine
1475 NW 12 Ave
Miami, FL 33136
USA

Calvin R. Maurer Jr., PhD
Director, Research
Accuray Incorporated
1310 Chesapeake Terrace
Sunnyvale, CA 94089
USA

D. Ross McBride, MD
Medical Director
CyberKnife Center
St. Joseph's Hospital
Saint Paul, MN 55102
USA

Danielle McDonald, NP
Thoracic Oncology Program Coordinator
Beth Israel Deaconess Medical Center
330 Brookline Avenue
Boston, MA 02215
USA

Ronald C. McGarry, MD, PhD
Clinical Associate Professor
Department of Radiation Medicine
Room C114C
University of Kentucky
Lexington, KY 40356
USA

Donald A. McRae, PhD
Associate Professor
Director of Physics
Department of Radiation Medicine
Georgetown University Hospital
3800 Reservoir Road, NW
Washington, DC 20007
USA

Clinton A. Medbery, III, MD
Director, Department of Radiation Oncology
Frank C. Love Cancer Institute
St. Anthony Hospital
1011 North Dewey
Oklahoma City, OK 73102
USA

Matthew D. Miller, DO
Hematology Oncology Fellow
Division of Hematology Oncology
Department of Internal Medicine
St. Louis University Medical Center
St. Louis University
3635 Vista At Grand- FDT 2
St. Louis, MO 63110
USA

Astrid E. Morrison, MD
Radiation Oncologist
Department of Radiation Oncology
Frank C. Love Cancer Institute
St. Anthony Hospital
1011 North Dewey
Oklahoma City, OK 73102
USA

Zhiping Mu, PhD
Imaging Software Engineer
Accuray Incorporated
1310 Chesapeake Terrace
Sunnyvale, CA 94089
USA

Marc W. Münter, MD
Department of Radiation Oncology
German Cancer Research Center
Im Neuenheimer Feld 280
69120 Heidelberg
Germany

Elena A. Nedea, MD
Department of Radiation Oncology
Beth Israel Deaconess Medical Center
Instructor in Radiation Oncology
Harvard Medical School
330 Brookline Avenue
Boston, MA 02215
USA

Nghi Nguyen, MD, PhD
PET Fellow
Division of Nuclear Medicine
Department of Radiology
St. Louis University Medical Center
St. Louis University
3635 Vista At Grand- FDT 2
St. Louis, MO 63110
USA

Walter Nikesch, PhD, DABR
Chief Physicist
CyberKnife Center of Palm Beach
10335 N. Military Trail Suite C
Palm Beach Gardens, FL 33410
USA

Joost J. Nuyttens, MD, PhD
Radiation Oncologist
CyberKnife Medical Coordinator
Department of Radiation Oncology
Erasmus MC Daniel den Hoed Cancer Center
Postbus 5201
3008 EA Rottterdam
The Netherlands

John K. O'Connor, MD
Baylor Radiosurgery Center
Baylor University Medical Center
3500 Gaston Ave
Dallas, TX 75246
USA

Dana A. Oliver, MT(ASCP), MPH
Biostatistician
Co-Director, Tissue Procurement Lab
St. Louis University Cancer Center
St. Louis University
3635 Vista At Grand- FDT 2
St. Louis, MO 63110
USA

Medhat M. Osman, MD, SCM, PhD
St. Louis University Medical Center
Assistant Professor, Director of PET
Department of Radiology
Division of Nuclear Medicine
St. Louis University
3635 Vista At Grand- FDT 2
St. Louis, MO 63110
USA

Lech Papiez, PhD
Associate Professor
Department of Radiation Oncology
University of Texas Southwestern Medical Center
5801 Forest Park Road, NF3.302B
Dallas, TX 75390
USA

Jong Park, PhD
R&D Engineer
Accuray Incorporated
1310 Chesapeake Terrace
Sunnyvale, CA 94089
USA

J. Anthony Parker, MD, PhD
Department of Radiology
Beth Israel Deaconess Medical Center
Associate Professor of Radiology
Harvard Medical School
330 Brookline Avenue
Boston, MA 02215
USA

Arjun Pennathur, MD
Assistant Professor
Heart Lung Esophageal Surg Inst
University of Pittsburg Medical Center
PUH C-800
200 Lothrop Street
Pittsburgh, PA 15213
USA

Mark Perman, MD
North Florida Radiation Oncology Center
1021 NW 64th Terrace
Gainesville, FL 32605
USA

Jean-Briac Prévost, MD
Fellow
Department of Radiation Oncology
Erasmus MC Daniel den Hoed Cancer Center
Postbus 5201
3008 EA Rottterdam
The Netherlands

Cristina A. Reichner, MD, FCCP
Assistant Professor of Medicine
Pulmonary Critical Care & Sleep Medicine
Georgetown University Hospital
3800 Reservoir Road, NW
Washington, DC 20007
USA

David H. Roberts, MD
Assistant Professor
Harvard Medical School
Clinical Director
Pulmonary & Critical Care Division
Beth Israel Deaconess Medical Center
KSB 23, 330 Brookline Avenue
Boston, MA 02215
USA

Jill Rossinow, MD
Resident
Department of Radiation Oncology at
Montefiore Medical Center and the
Albert Einstein College of Medicine
111 East 210 Street
Bronx, NY 10467
USA

Nancy Rumplik, RN, MSN
Clinical Nurse IV, Hematology Oncology
Beth Israel Deaconess Medical Center
330 Brookline Avenue
Boston, MA 02215
USA

Sohail Sayeh, MS
Technical Manager – Robotics & Motion Technologies
Accuray Incorporated
1310 Chesapeake Terrace
Sunnyvale, CA 94089
USA

Raymond A. Schulz, MSc
Principal
RAenterpriseS
(Radiology And Image-Guided Surgery Consultants)
65 Mission Drive
San Mateo CA 94402
USA

Susan Schumer, MD
Beth Israel Deaconess Medical Center
Instructor in Medicine
Harvard Medical School
330 Brookline Avenue
Boston, MA 02215
USA

James G. Schwade, MD, FACO
Executive Director
CyberKnife Centers of Miami & Palm Beach
Clinical Professor of Radiation Oncology
Leonard M. Miller School of Medicine
University of Miami
7867 N Kendall Drive, Suite 105
Miami, FL 33156
USA

Hua Shao, PhD, DABR
Assistant Professor
Division of Medical Physics
Department of Radiation Oncology
University of Miami
Leonard M. Miller School of Medicine
1475 NW 12 Ave
Miami, FL 33136
USA

Guy R. Sherwood, DABR
Senior Medical Physicist
CyberKnife Center
St. Joseph's Hospital
Saint Paul, MN 55102
USA

Penny J. Sinner, MPH
Senior Statistician
HealthEast Research and Education
Saint Paul, MN 55102
USA

Scott G. Soltys, MD
Clinical Instructor
Department of Radiation Oncology
Stanford Cancer Center
875 Blake Wilbur Drive
Stanford, CA 94305-5847
USA

Yelin Suh, MS
Virginia Commonwealth University
Department of Radiation Oncology
401 College St.
Richmond, VA 23219
USA

Euan Thomson, PhD
President and Chief Executive Officer
Accuray Incorporated
1310 Chesapeake Terrace
Sunnyvale, CA 94089
USA

Robert L. Thurer, MD
Harvard Medical International
Associate Professor of Surgery
Harvard Medical School
1135 Tremont Street
Boston, MA 02120
USA

Harold C. Urschel Jr., MD
Chair of Cardiovascular and Thoracic Surgical Research, Education and Clinical Excellence
Baylor University Medical Center
1201 Barnett Tower
3600 Gaston Avenue
Dallas, TX 75246
USA

Allison Wall, MD
Hematology Oncology Fellow
Division of Hematology Oncology
Department of Internal Medicine
St. Louis University Medical Center
St. Louis University
3635 Vista At Grand- FDT 2
St. Louis, MO 63110
USA

Bruce J. Walz, MD, FACR
Professor and Director
Department of Radiation Oncology
St. Louis University Medical Center
St. Louis University
3635 Vista At Grand- FDT 2
St. Louis, MO 63110
USA

Bai Wang, PhD
Sr. Software Engineer
Accuray Incorporated
1310 Chesapeake Terrace
Sunnyvale, CA 94089
USA

Hongwu Wang, PhD
Sr. Software Engineer
Accuray Incorporated
1310 Chesapeake Terrace
Sunnyvale, CA 94089
USA

James Wang, PhD
Manager, Technical Business Development
Accuray Incorporated
1310 Chesapeake Terrace
Sunnyvale, CA 94089
USA

Jay B. West, PhD
Senior Research and Development Engineer
Accuray Incorporated
1310 Chesapeake Terrace
Sunnyvale, CA 94089
USA

Kenneth H. Wong, PhD
Assistant Professor
Georgetown University Medical Center
Department of Radiology
ISIS Center
2115 Wisconsin Avenue NW, #603
Washington, DC 20007
USA

Xiaodong Wu, PhD, DABR
Associate Professor and Chief
Division of Medical Physics
Department of Radiation Oncology
University of Miami
Leonard M. Miller School of Medicine
1475 NW 12 Ave
Miami, FL 33136
USA

Marianne M. Young, MD
Department of Radiation Oncology
Frank C. Love Cancer Institute
St. Anthony Hospital
1011 North Dewey
Oklahoma City, OK 73102
USA

MIX
Papier aus verantwortungsvollen Quellen
Paper from responsible sources
FSC® C105338

If you have any concerns about our products,
you can contact us on
ProductSafety@springernature.com

In case Publisher is established outside the EU,
the EU authorized representative is:
Springer Nature Customer Service Center GmbH
Europaplatz 3, 69115 Heidelberg, Germany

Printed by Libri Plureos GmbH
in Hamburg, Germany